Selected Readings on Telecommunications and Networking

Jairo Gutierrez
University of Auckland, New Zealand

Information Science REFERENCE
INFORMATION SCIENCE REFERENCE
Hershey · New York

Director of Editorial Content: Kristin Klinger
Managing Development Editor: Kristin M. Roth
Senior Managing Editor: Jennifer Neidig
Managing Editor: Jamie Snavely
Assistant Managing Editor: Carole Coulson
Typesetter: Lindsay Bergman
Cover Design: Lisa Tosheff
Printed at: Yurchak Printing Inc.

Published in the United States of America by
Information Science Reference (an imprint of IGI Global)
701 E. Chocolate Avenue, Suite 200
Hershey PA 17033
Tel: 717-533-8845
Fax: 717-533-8661
E-mail: cust@igi-global.com
Web site: http://www.igi-global.com

and in the United Kingdom by
Information Science Reference (an imprint of IGI Global)
3 Henrietta Street
Covent Garden
London WC2E 8LU
Tel: 44 20 7240 0856
Fax: 44 20 7379 0609
Web site: http://www.eurospanbookstore.com

Library of Congress Cataloging-in-Publication Data

Selected readings on telecommunications and networking / Jairo Gutierrez, editor.

p. cm.

Includes bibliographical references and index.

Summary: "This book presents quality articles focused on key issues concerning the planning, design, maintenance, and management of telecommunications and networking technologies"--Provided by publisher.

ISBN 978-1-60566-094-3 (hard cover) -- ISBN 978-1-60566-095-0 (ebook)

1. Wireless communication systems. 2. Information networks. I. Gutiérrez, Jairo, 1960-

TK5103.2.S45 2009

621.382--dc22

2008019468

British Cataloguing in Publication Data
A Cataloguing in Publication record for this book is available from the British Library.

Table of Contents

Section V
Critical Issues

Section VI
Emerging Trends

Detailed Table of Contents

This chapter provides a brief overview of wireless systems and standards. The evolution of wireless systems from voice-centric circuit-switched systems to data-centric packet-switched systems is discussed. The first- and second-generation wireless systems were designed primarily for voice service. The data rate supported by these systems is very limited. The 2.5G systems were developed to retrofit second-generation systems to support higher data rate applications. The third-generation systems are designed to meet the demands of the Internet era. A wide range of IP-based services is provided using these systems. IEEE 802.16 standard-based systems, commonly referred to as WiMAX, are being proposed as an alternative to third-generation systems for carrying data traffic. Popular wireless LAN and wireless PAN standards are also discussed.

In this chapter, visions of mobile communication are explored by focusing on idealised concepts surrounding wireless technology. By examining sources on the development, marketing, and use of wireless technology, the author contextualises these visions within earlier accounts of ideal communication found in media history and isolates the regularities that are part of these accounts. On close examination, a paradox reveals itself in these regularities, one that can be described as resulting from an uneasiness in the human communication psyche: an unfulfilled desire for divine togetherness that clashes with individual communication needs.

Organizations are beginning to apply data mining and knowledge discovery techniques to their corporate data sets, thereby enabling the identification of trends and the discovery of inductive knowledge. Since traditional transaction databases are not optimized for analytical processing, they must be transformed. This chapter proposes the use of modular components to decrease the overall amount of human processing and intervention necessary for the transformation process. Our approach configures components to extract data-sets using a set of "extraction hints." Our framework incorporates decentralized, generic components that are reusable across domains and databases. Finally, we detail an implementation of our component-based framework for an aviation data set.

This chapter tells the development story of Waleli, a high-tech company utilizing the latest proven developments in wireless communications to bring innovations to the market. It presents the journey of the firm through the entrepreneurial process, from initial idea right through to value creation. Furthermore, it looks at the roles and origins of network contacts and how these both impact the entrepreneur's effort to accumulate the necessary resources to establish and subsequently develop the firm.

This chapter describes the detailed configuration and LAN infrastructure design at the University of Bahrain (UOB). The authors describe the configuration based on the new setup and migration requirements and indicate how the design satisfies those requirements. The chapter explains the detailed configuration of the design process of the distribution layer switches and shows how these switches can be configured in the final implementation. The chapter also discusses the modifications that occurred during the implementation/migration phase. The design of the network at UOB campuses incorporates resiliency into the network core in order to manage problems effectively.

This chapter reviews the literature on networks and, more specifically, on the development of community telecommunication networks. It strives to understand the collaboration needed for innovative projects such as intelligent networks. Guided by a change management framework, collaboration within a community network is explored in terms of the formation and performance phases of its development. The context, content, and process of each phase is analyzed, as well as the interaction of the two phases. User involvement and technology appropriation are discussed and collaboration challenges are identified and linked to the sustainability of the community network.

In recent years, many technologies have converged to integrated solutions and one of the most frequently discussed topics has been the deployment of wireless personal area networks (WPANs). In this chapter, the authors present a generic architecture scheme that allows voice and other real-time traffic to be carried over longer distances. The proposed scheme, Bluetooth Promoted Multimedia on Demand (BlueProMoD) is a novel framework that combines a wired backbone network including Bluetooth access points (APs) with the mobile Bluetooth-enabled devices of the end users.

This chapter investigates how to create ad hoc audience response systems using nonspecialist devices. The chapter revolves around two case studies: one involving the use of mobile phones, and the other based on PDAs. Both case studies are carried out in tertiary education institutions, showing how these devices can be used to facilitate audience participation using devices that students might, themselves, bring to lectures. Both are evaluated from the perspective of the student and the educator, using a mixture of observational and interview-based techniques.

The purpose of this chapter is to provide guidelines for city governments considering implementing large-scale wireless networks to provide Internet access for their citizens and businesses. Case studies of cities in the United States that have implemented wireless networks will be evaluated in the context of opportunities and potential challenges. Some key considerations discussed in this chapter involve free versus fee-based models, security considerations, conflicts with local telecommunications companies, and network support.

Discrete sequence modeling and prediction is an important goal and challenge for Web environments, both wired and wireless. Web clients' data-request forecasting and mobile location tracking in wireless cellular networks are characteristic application areas of sequence prediction in such environments. Accurate data-request prediction results in effective data prefetching, which combined with a caching mechanism can reduce user-perceived latencies as well as server and network loads. This chapter presents information-theoretic techniques for discrete sequence prediction. It surveys, classifies, and compares the state-of-the-art solutions, suggesting routes for further research by discussing the critical issues and challenges of prediction in wired and wireless networks.

This chapter presents a computer-aided integration tool, iCAD, that can predict a network evolution. iCAD is connected to four device libraries, each of which contains a distinct set of network-technology devices, such as Ethernet hubs, ATM switches, IP routers, and gateways. As a network technology changes, each device library is updated. This chapter presents empirical results from 1999 until 2005 recording the network evolution progress, where the lower and upper bounds of network evolution came out to be 10% to 25% and 57% to 74% respectively in terms of network-design cost reduction.

In optical networks, survivability becomes very important because of the huge amount of traffic carried by fibers. Therefore, the prevention of service interruption, or the reduction of the service loss, must now be an integral part of the network design and operations strategy. In this chapter, optical network survivability is discussed and special emphasis has been given to WDM lightpath protection.

Fractal geometry can help us to describe the shapes in nature (e.g., ferns, trees, seashells, rivers, mountains) exceeding the limits imposed by Euclidean geometry. Fractal geometry is quite young: The first studies are the works by the French mathematicians Pierre Fatou (1878-1929) and Gaston Julia (1893-1978) at the beginning of the 20th century. However, only with the mathematical power of computers has it become possible to realize connections between fractal geometry and other disciplines. Recent studies apply this geometry to controlling traffic in computer networks (LANs, MANs, WANs, and the Internet). The aim of this chapter is to present fractal geometry, its properties (e.g., self-similarity), and their applications in computer science.

This chapter briefly reviews two important goals in online education: interaction and presence. These are important goals in online education because they are linked to learning and motivation to learn. The chapter provides guidelines and an extended example of how to design an online course in information security in a manner that will enhance interaction and presence. This chapter's contribution is to provide guidelines with a corresponding extended and concrete example for those who are tasked with designing and delivering online courses. Although the guidelines and example were targeted to the field of information security, they can be readily adopted by other disciplines.

This chapter proposes a mobile intelligent agent-based e-business architecture that allows buyers and sellers to perform business at remote locations. This proposed architecture promises a number of benefits. First, it provides great convenience for traders as business can be conducted anytime and anywhere. Second, since the task of finding and negotiating with appropriate traders is handled by a mobile, intelligent agent, the user is freed from this time-consuming task. Third, this architecture addresses the problem of limited and expensive connection time for mobile devices: A trader can disconnect a mobile device from its server after generating and launching a mobile intelligent agent. Later on, the trader can reconnect and call back the agent for results, therefore minimizing the connection time. Finally, by complying with the standardization body FIPA, this flexible architecture increases the interoperability between agent systems and provides high scalability design for swiftly moving across the network.

Prologue

INTRODUCTION

The field of telecommunications and networking continues to produce a rich set of contributions to industry and the research community as this selection of readings demonstrates. The book has compiled a series of recent submissions that cover a range of aspects in this dynamic field. The chapters are divided in six themes: (1) Fundamental Concepts and Theories, (2) Development and Design Methodologies, (3) Tools and Technologies, (4) Application and Utilization, (5) Critical Issues, and 6) Emerging Trends. The resulting volume is a valuable and multi-faceted resource that gives the reader a good insight about where the technology is going and a sample of some of the most interesting applications and developments exploiting that emerging technology base. In the next few paragraphs, I'd like to highlight some of the most interesting issues being raised in telecommunications and networking research and perhaps speculate somewhat about directions for further enquiry that may have a long-term effect of how we use and deploy telecommunications and networking solutions that contribute to an increasingly connected world.

UBIQUITOUS NETWORKING

The exponential growth of the Internet has diminished the difficulties associated with communication between distant places, allowing people to participate in the digital economy regardless of their geographical limitations. Additionally, developments in wireless technologies are freeing people from using wires for communicating. For example, the conveniences of wireless connections have converted mobile phones in a commodity, rather than a luxury item (Weatherall and Jones, 2002). Juniper Research's global m-commerce revenue projections for 2009 include global revenues of US$88 billion, ticket purchases of US$39 billion, and phone-based retail POS sales US$299 million (Juniper Research, 2008). The idea of "anywhere, anytime, by anything and anyone" (or 4As) networking is at the core of a new emerging networking technology, referred to as a "Ubiquitous Networking". The concept of Ubiquitous Networking originated from the concept of Ubiquitous Computing, which was aimed to "make many computers available throughout the physical environment, while making them effectively invisible to the user" (Weiser, 1993; Wikipedia, 2008). Additionally, Weiser highlighted 4 important "networking" issues when focusing on ubiquitous computing, namely: wireless media access, wide-bandwidth range, real-time capabilities for multimedia over standard networks, and packet routing. All of these research areas are well-covered in this selected readings collection.

The world of Ubiquitous Networking creates new business opportunities for the development of services and applications that maximise the freedom offered by the technology. Evolving networking technologies will change our daily lives, both in social and economical terms. The potential demand includes, but is not limited to services such as ubiquitous health / concierge systems, ubiquitous automobile systems and ubiquitous education / learning systems.

NETWORK INFRASTRUCTURE ISSUES

The core of a ubiquitous networking environment is the underlying networking infrastructure that is capable of meeting the requirements of users. In general, networking technologies can be divided into two broad categories, (1) wired networks and (2) wireless networks. The key functional distinction between the two types of networks is the speed, where wired networking technologies are capable of providing much higher speed communications compared to wireless networking technologies. The technologies supporting wireless networks will always be "resource-poor" when compared with those supporting wired networks (Satyanarayanan, 1996). Below is the list of key wired and wireless networking technologies.

- Fibre Technologies (e.g., FTTH, FTTC)
- Wireless LANs (802.11g/n, 802.16)
- Wireless MANs (Fixed wireless, LMDS, 802.20)
- Wireless WANs (3G/4G, GPRS, UMTS, GPS, mesh networks)
- Short-distance wireless communication technologies (Wireless PAN) (e.g., Bluetooth, 802.15.3)
- Satellite Systems (e.g., Motorola's Iridium)
- Sensor networks

Drew (2002) argues that wired networks are necessary while considering wireless networks as a supplement. However, in a ubiquitous network environment, the role of wireless networks matches the importance of wired networks. In a ubiquitous networking environment, wired networks act as the backbone providing high to very high speed connection to end-users. However, wired networks are very restricted in terms of mobility. Wireless networks address this deficiency, and provide high mobility to users although with lower speeds than wired networks. Thus, in a ubiquitous networking environment, wired and wireless networking technologies form a symbiotic relationship, by compensating for the deficiencies of each other.

In a ubiquitous networking environment, the networking technologies are not limited to providing "person to person" communications; they also need to address "person to object" and "object to object" communications (MIAC, 2005). The convergence of wired and wireless technologies will provide an appropriate use of networking technologies for those three different types of communications, maximising the potential of ubiquitous networking. That convergence is a key characteristic of environments which support the spontaneous appearance of entities, in other words "persons" or "objects" that move into range and initiate participation in the ubiquitous network (Kindberg & Fox, 2002). A vital area of research is associated with the viable business models along with the potential applications and services that could be offered as part of a ubiquitous networking environment.

EVOLUTION OF TELECOMMUNICATIONS AND NETWORKS

The promising benefits and attractive proposals of applications and services for ubiquitous networking environments have sparked the interest of many countries around the world and have also attracted the attention of the International Telecommunications Union (ITU), the organisation that is well known for its standardisation efforts in the telecommunications field. Recently, the ITU have addressed the idea of the "Ubiquitous Network Society" as part of their "New Initiatives Programme" which aims to identify the emerging trends in the telecommunications environment (ITU, 2005a). In that ITU programme South Korea and Japan were selected to illustrate early implementations of ubiquitous networking.

Why these two countries? As discussed above the networking infrastructure is the critical factor for ubiquitous networking. According to the ITU, South Korea is the world's broadband leader by a significant margin; along with a high number of mobile subscribers that even outnumbers fixed line subscribers (ITU, 2005b). Similar statistics apply to Japan (ITU, 2005c). The high penetration rate of broadband and the widely use of wireless technologies around the country allows facilitates the implementation of "anywhere, anytime by anything, anyone" ubiquitous networking. The success factors for these two countries are explained in detail in the ITU's case studies (ITU, 2005a; ITU, 2005b). In South Korea, the Ministry of Information and Communication has the intention of realising their "digital home plan", in which digital home appliances with communications capabilities are installed in apartment houses as a total, integrated system (Murakami, 2004). In the private sector, plans for ubiquitous networking are also emerging: the Dongtan Ubiquitous Networking city plan, supported by the Samsung group, involves 40,000 households (Murakami, 2004). At the university level a number of institutions have successfully implemented ubiquitous networking environments, usually labelled "u-campuses". For example, at Sukmyung Woman's University, students can download "credit-card" functionality to their PDA or mobile phone and use the device as a medium for payment (Jung, 2004).

The strong focus and national level of support towards ubiquitous networking in South Korea and Japan are driving other countries to adopt similar strategies. In Europe, the project Amigo is addressing the idea of ubiquitous networking, stating that it is an *"integrated project that will realize the full potential of home networking to improve people's lives"* (Amigo Project, 2004). Active participants in this project include companies from various countries, among them France, Germany, Spain and the Netherlands. However, these efforts are limited to the commercial sector without the stronger government and country-wide support found in South Korea and Japan. According to the ITU, Italy and Singapore are the two other countries that are actively participating in achieving a ubiquitous networking environment (ITU, 2005d; ITU 2005e), with relatively well- established infrastructure throughout their territories.

The ubiquitous networking technology is still in its very early stages and there are numerous issues that need to be addressed before achieving a perfect operating environment. One of the major issues is to maintain interoperability between different networking technologies. For example, an office employee may have a Bluetooth device that connects with her laptop, use a Wireless LAN based on 802.11g, a Wireless WAN based on 3G, and a wired connection using ADSL. To maximise the benefits from a ubiquitous networking environment, these various technologies should be able to communicate without any disruptions. Additionally, processing power of mobile devices and issue of security is one of other concerns for true ubiquitous networking environment. Currently, significant research emphasis is given to the security and middleware side of ubiquitous networking to address this, and it is highly related with improvements in the processing power of mobile devices.

The selection of networks in a ubiquitous networking environment is one of the main operating issues with this technology. For example, in a ubiquitous networking environment, a cordless phone may substitute your mobile phone when you are outside the house. Choosing the best network based solely on the user requirements complicates the selection of the "ideal" network for a particular connection time and location. The user-initiated selection of a provider also generates the issue of billing. Currently customers "subscribe" to the desired services, and get billed based on the usage. However, in a ubiquitous networking environment, there is no need to "subscribe" for a desired service, but rather users have the capability to employ ad-hoc type services when needed. This adds complexity to existing billing systems however these requirements need to be addressed to achieve a truly ubiquitous networking environment.

Another key issue for the success of ubiquitous network services is the issue of assigning prices to those services. Furthermore, ubiquitous services based on a network of complementary technologies, both fixed and wireless, have created the expectation of services that can be obtained dynamically and automatically with the minimum possible of interaction between the users and potentially complex network systems. Intelligent agents would negotiate the best conditions to make sure the user obtains always the best possible connection (Voinov and Valladares, 2003). This best possible connection would be selected by comparing the different services, quality of the services offered, prices and reaching a decision based on the policies the user has configured in her intelligent agent and in conjunction with the policies being presented by the different service providers.

It is clear that, from the technical point of view, the scenario depicted above is feasible. There has been continued progress on the integration of technologies such as WiFi, "Mesh" and "Ad-Hoc" networks with the traditional phone networks and fixed sub-networks based on the TCP/IP family of protocols. Telecommunication companies have exploited the popularity of WiFi "hot spots" as access ramps to their 3G services (Legard, 2003). However, there is work to be done in the area of agreeing how to price network services, especially when that "network" is supplied by different organizations and potential users may not have contractual agreements with all the players involved.

The current telecommunications environment, in which virtual operators re-sell network services, in which some firms are customers of a traditional "Telco" while at the same time offering services to many other smaller organizations, forces us to redefine many of the business models that had been used so far. Long term contracts are being challenged by many other arrangements that give more flexibility to the users. These changes, in most cases promoted by users' requirements and further "pushed" by competitive, and innovative, new entrants into the telecommunications arena have resulted on a profound transformation in the way services are acquired and billed. This fact will always clash with the tendency of traditional "Telcos" to keep billing as simple as possible (Meyer, 2003).

It is entirely possible that the much vaunted convergence of the Internet and Telco worlds will push companies competing in that field to adjust the way they do business (Panagiotakis et al., 2005). An optimistic view of these changes argues that network operators will be able to obtain additional revenues by pricing quality services (with guaranteed levels of performance or guaranteed security) at a premium and that selected customers will be more that willing to foot the bill for a better service.

The ubiquitous networking environment creates new challenges in security and requires development of new approaches to address both existing and new security problems (Van Dyke and Koc, 2003). Heterogeneous networking environments add several levels of complexity to existing security mechanisms, and different techniques needs to be developed to ensure optimum levels of security in the ubiquitous networking environment (Privat, 2005).

The advancements in handheld devices are one of the key drivers of ubiquitous networking, and these devices are improving its capabilities at exponential rates. However, due mainly to their size restrictions, these devices suffer from a number of limitations. These limitations include but are not limited to: inadequate processing capability, restricted battery life, limited memory space, slow expensive connections and confined host bandwidth (Sharmin, Ahmed & Ahmed, 2006). To address these limitations, middleware can be play an essential role. For example, rather than delegating processing responsibility to the light-weight handheld devices, core processing can be performed by the middleware applications. Currently developed middleware applications are capable of providing services such as security, data backup services, resource discovery services and ad-hoc communication services, to list a few (Sharmin, Ahmed and Ahmed, 2006). Given that middleware is the most viable solution to minimise limitations of handheld devices, a large number of middleware applications are under research by both academics and practitioners (Yau, Wang and Karim, 2002; Murphy, Picco and Roman, 2001; Sharmin, Ahmed and Ahmed, 2006).

Finally, security has always been a critical issue within the area of networking, and this is not an exception for the evolving telecommunications scenarios. In fact, security in this type of environments requires more emphasis than what has been perceived in traditional networks. The convenience of handheld devices, such as PDAs, means that people are storing personal data on these devices, which means that more stringent security mechanisms to protect these data are required. The typical characteristics of handheld devices also create security concerns (Raghunathan, Ravi, Hattangady and Quisquarter, 2003):

- Mobile communications uses a public transmission medium, which creates opportunity for hackers to eavesdrop communications more easily than with secured private connections
- Mobile devices are vulnerable to theft, loss and corruptibility
- Processing power limitations on mobile devices can imply restrictions on security features (e.g. Algorithm selection)

To address these issues various methods have been proposed and refined during the last few years but numerous challenges, associated with the proliferation of different networks and the secure and seamless integration of those technologies, are still being actively investigated in research and development facilities throughout the world.

CONCLUSION

It is clear that "anywhere, anytime by anything and anyone" (4As) networking will not arrive overnight. It will be an incremental process that requires commitment, careful planning and preparation. Viable implementations will initially take place at a relatively small scale, in apartment buildings or university campuses before extending to a wider environment. Eventually, a world-wide level of ubiquitous networking will be reached, allowing people to use the same device everywhere.

Developments in networking have changed our everyday lives, especially with the introduction and popularity of the Internet. High demand from individual users and organisations has brought the concept of ubiquitous networking to the front as seen in the cases of South Korea and Japan. The importance of the underlying networking infrastructure is one of the key success factors for establishing ubiquitous networking environments along, of course, with successful applications and services. Ubiquitous networking is not without problems and there are a number of critical issues that need to be resolved. However, as illustrated by several of the readings included in this selection, the technology offers great promise and it will eventually reach us, converting our "4As" networking dreams into reality.

REFERENCES

Amigo Project (2004). Ambient intelligence for the networked home environment: Summary. *Amigo Project*. Dec. 2004. Available from the World-Wide Web: http://www.hitech-projects.com/euprojects/amigo/

Drew Jr, W. (2002). Wireless Networks: New meaning to ubiquitous computing. *University of New York*. Jun 2002, pp. 1 - 11

Juniper Research (2008). Global mCommerce Revenue Projections For 2009. Available from the World-Wide Web: http://epaynews.com/statistics/mcommstats.html#49

International Telecommunication Union (2005a). New Initiatives Programmes, Feb 2005. *International Telecommunication Union*. Available from the World-Wide Web: http://www.itu.int/osg/spu/ni/

International Telecommunication Union (2005b). ITU Workshop on Ubiquitous Network Societies: The Case of the Republic of Korea. *International Telecommunication Union*. Apr 2005, pp. 10 – 45

International Telecommunication Union (2005c). ITU Workshop on Ubiquitous Network Societies: The Case of the Japan. *International Telecommunication Union.* Apr 2005, pp. 13 – 45

International Telecommunication Union (2005d). ITU Workshop on Ubiquitous Network Societies: The Case of the Italian Republic. *International Telecommunication Union*. Apr 2005, pp. 25 – 50

International Telecommunication Union (2005e). ITU Workshop on Ubiquitous Network Societies: The Case of the Republic of Singapore. *International Telecommunication Union.* Apr 2005, pp. 21 – 39

Jung, K. U. (2004). Community, S&A Research Group, Research Group Notice Board: Ubiquitous Campus in Korea. *Ubiquitous IT Korea Forum.* Dec. 2004, Available from the World-Wide Web: http://www.ukoreaforum.or.kr/

Kindberg, T. and Fox, A. (2002). System Software for Ubiquitous Computing. *IEEE Pervasive Computing*, January, pp. 70-81

Legard, D. (2003). SingTel, NTT DoCoMo tie up to push 3G, IDG News Service.

Meyer, D. (2003). AT&T wireless tries simple rate plans, fewer freebies. *RCR Wireless News, 22(7).*

Ministry of Internal Affairs and Communications, Japan (2005). Tokyo Ubiquitous Networking conference: *Program (Session 5) Ubiquitous Network Society*. May 2005. Available from the World-Wide Web: http://www.wsis-japan.jp/session5_e.html

Murakami, T. (2004), NRI Papers: Ubiquitous Networking: Business Opportunities and Strategic Issues, *Nomura Research Institute*. Aug 2004, pp. 1 – 24

Murphy, A L.; Picco, G P.; Roman G C. (2001). Time: A Middleware for Physical and Logical Mobility. *Proceedings of the 21st International Conference on Distributed Computing Systems,* May.

Panagiotakis, S., Koutsopoulou, M. and Alonistioti, A. (2005). Business Models and Revenue Streams in 3G Market. IST Mobile Communication Summit 2005, Dresden, Germany, June.

Privat, G. (2005). Ubiquitous Network & Smart devices: New Telecom Services & Evolution of Human Interfaces. *ITU Ubiquitous Network Societies Workshop*. Apr 2005. Available from the World-Wide Web:

http://www.itu.int/osg/spu/ni/ubiquitous/Presentations/8_privat_applications.pdf

Raghunathan, A.; Ravi, S.; Hattangady, S.; Quisquater J. (2003). Securing Mobile Appliances: New Challenges for the System Designer. *Proceedings of the Design, Automation and Test in Europe Conference and Exhibition.*

Sharmin, M. Ahmed, S. Ahamed, S I. (2006). MARKS (Middleware Adaptability for Resource Discovery, Knowledge Usability and Self-healing) for Mobile Devices of Computing environments. *Proceedings of Third International Conference on Information Technology: New Generation (ITNG 2006),* April, Las Vegas, NE, USA, 6 pages

Van Dyke, C., Koc, C. K. (2003). On Ubiquitous Network Security and Anomaly Detection. *Electrical & Computer Engineering, Oregon State University.* Feb 2003, pp. 1 – 5

Voinov, I. A. and Valladares, T. R. (2003). An enhanced socket interface for next generation systems. In the 1st International Workshop on Wireless, Mobile & Always Best Connected pp. 73–85. University of Strathclyde, Glasgow, UK.

Weatherall, J.; Jones, A. (2002). Ubiquitous networks and its applications. *IEEE Wireless Communications.* 9(1). Feb 2002, pp.18 – 29

Weiser, M. (1993).Some Computer Science Problems in Ubiquitous Computing. *Communications of the ACM,* Vol. 36(7), July, pp. 75-84

Wikipedia (2008), Article: Ubiquitous Computing. *Wikipedia: The Free Encyclopaedia.* April 2008. Available from the World-Wide Web: http://mobileman.projects.supsi.ch/glossary.html

Yau, S.; Wang, Y.; Karim, F. (2002). Development of Situation Aware Application Software for Ubiquitous Computing environments. *Computer Software and Application Conference.* pp. 233-238

About the Editor

Jairo Gutierrez is a senior lecturer in information systems at the University of Auckland and coordinator of its Cisco Networking Academy Program. He teaches data communications and computer networking. He is the editor-in-chief of the *International Journal of Business Data Communications and Networking* (IGI Global). His current research topics are in network management systems, viable business models for mobile commerce, programmable networks, and QoS issues associated with Internet protocols. He received a Systems and Computer Engineering degree from The University of The Andes (Colombia, 1983), a master's degree in computer science from Texas A&M University (1985), and a PhD (1997) in information systems from The University of Auckland (New Zealand).

Section I
Fundamental Concepts and Theories

Chapter I
A Brief Overview of Wireless Systems and Standards

Sundar G. Sankaran
Atheros Communications, USA

ABSTRACT

This chapter provides a brief overview of wireless systems and standards. The evolution of wireless systems from voice-centric circuit-switched systems to data-centric packet-switched systems is discussed. The first- and second-generation wireless systems were designed primarily for voice service. The data rate supported by these systems is very limited. The 2.5G systems were developed to retrofit second-generation systems to support higher data rate applications. The third-generation systems are designed to meet the demands of the Internet era. A wide range of IP-based services is provided using these systems. IEEE 802.16 standard-based systems, commonly referred to as WiMAX, are being proposed as an alternative to third-generation systems for carrying data traffic. Popular wireless LAN and wireless PAN standards are also discussed.

INTRODUCTION

Wireless systems have been around for over a century. Guglielmo Marconi successfully transmitted Morse code from Cornwall, England to St-John's, Canada in 1901. The wireless technology has come a long way since then. The proliferation of Internet in every aspect of life resulted in rapid convergence of computing and communication industries fueling an explosive growth of wireless communication in the mid-1990s. Now, mobile computing—the use of a portable computing device capable of wireless networking—is a reality. For example, today's PDAs and cell phones have the capability for Internet surfing. Consequently, one can use the PDAs and cell phones to do everything from stock trading to finding driving directions. WiFi enabled laptops allow the users to connect to Internet from WiFi hotspots, which are becoming ubiquitous. Emerging standards such as WiMAX aim to provide high-speed wireless data access from anywhere at anytime. This chapter describes various wireless standards that have made mobile computing a reality.

FIRST- AND SECOND-GENERATION CELLULAR SYSTEMS

The first-generation cellular systems, developed in the late 1970s, use analog modulation techniques. These systems are designed primarily to carry analog speech. Very low-rate data transmission is possible in these systems. The advance mobile phone service (AMPS) system, developed by AT&T Bell Labs, is an example of first-generation wireless systems. A good fraction of cellular systems currently deployed around the world are based on AMPS. For example, AMPS is still being used in some rural parts of the U.S.

Starting in the early 1990s, wireless operators started deploying second-generation cellular systems that use digital modulation (Yacoub, 2001). The second-generation systems use advanced digital-signal-processing algorithms to process signals. The transition to digital from analog allowed the second-generation cellular systems to offer higher capacity[1] than the first-generation analog systems. The second-generation systems offer services such as text messaging, also known as short message service (SMS), and circuit switched data (CSD), in addition to legacy voice service.

Some of the popular second-generation cellular systems include global system mobile (GSM), interim standard 136 (IS-136), and interim standard 95 (IS-95). The GSM system (Mouly, 1992) was originally designed and deployed in Europe to solve the fragmentation problems[2] of the first cellular systems in Europe. Now, GSM is the most widely deployed wireless system in the world, with deployments in Europe, Asia, Australia, South America, and some parts of the U.S. IS-136, the American counterpart of GSM, is a digital evolution of the first-generation AMPS system. It is often, albeit imprecisely, referred to as the TDMA standard since it uses time division multiple access (TDMA) air interface. However, it should be noted that many other standards, including GSM, use TDMA. The IS-136 systems are widely deployed in North America. IS-95, pioneered by Qualcomm, is the popular second-generation system based on code division multiple access (CDMA). It is also known as *cdmaOne*. These systems are in wide use in North America, South Korea, India, and China.

The second-generation cellular systems were rolled out before the dawn of the Internet era. Consequently, these systems are not efficient in carrying data: these systems transfer data with circuit switching, which is not as efficient as packet switching, used by systems of later generation. Furthermore, the data rate provided by these systems is very limited. For example, GSM systems provide a maximum data rate of 14.4 kbps.

EVOLUTION FROM SECOND GENERATION TO 2.5G

New 2.5G technologies were developed in an effort to retrofit the second-generation systems to be able to support the higher data rates that are required by modern Internet applications. These technologies enable cellular service providers to support features such as Web browsing, e-mail, mobile commerce (m-commerce), and location-based mobile services using existing second-generation infrastructure, with minimal hardware and software upgrades to base stations and handsets.

The popular 2.5G systems include high speed circuit switched data (HSCSD), general packet radio service (GPRS), and enhanced data rates for GSM evolution (EDGE) (Halonen, Romero, & Melero, 2003). HSCSD is a 2.5G upgrade to GSM. It needs just a software upgrade at the base stations. With this upgrade, the maximum data rate per user can be increased to 57.6 kbps. However, the data transfer is still done with circuit switching. GPRS is a 2.5G upgrade to both GSM and IS-136. These systems use packet switching and provide access to the Internet at a maximum data rate of 171.2 kbps. EDGE is a more advanced upgrade, requiring addition of new hardware and software at the existing base stations, to both GSM and IS-136. The EDGE systems can provide a maximum per user data rate of 384 kbps.[3]

The cellular systems belonging to 2.5G and above support a new Web browsing format language called wireless application protocol (WAP). This language was designed to meet the challenges of Web browsing from a wireless handset, which usually has small displays and limited memory. The wireless devices usually run WAP-based microbrowsers. These browsers allow the users to access the Web using one hand without requiring a keyboard.

THIRD-GENERATION CELLULAR SYSTEMS

The third-generation cellular systems (Mandyam & Lai, 2002) aim to provide a wide range of Internet protocol (IP) based services, along with voice and data services, using a single handset, whether driving, walking, or standing still in an office setting. The key attributes of third-generation systems include 144 kbps or higher data rate in high mobility (vehicular) traffic, 384 kbps for pedestrian traffic, 2 Mbps or higher for indoor traffic, and *capability to determine geographic position of mobiles and report it to both the network and the mobile terminal.* Some of the third-generation wireless standards include CDMA2000 and universal mobile telecommunication system (UMTS, also known as W-CDMA) (3GPP2 Web site, n.d.; UMTS Forum Web site, n.d.). CDMA2000 is the 3G evolution of IS-95, while W-CDMA is the 3G evolution of GSM and IS-136.

The CDMA2000 family of standards includes CDMA2000 1xRTT (also known as 1xRTT), CDMA2000 1xEV-DO (often referred to as EV-DO), and CDMA2000 1xEV-DV (also known as EV-DV). The 1xRTT (short for single-carrier radio transmission technology) systems provide data rates up to 144 kbps, while the latest generation of EV-DO (short for evolution data optimized) systems are capable of providing data rates up to 3.1 Mbps. 1xRTT and EV-DO have been rolled out in the U.S. by Verizon and Sprint. The EV-

Figure 1. Evolution of wireless standards

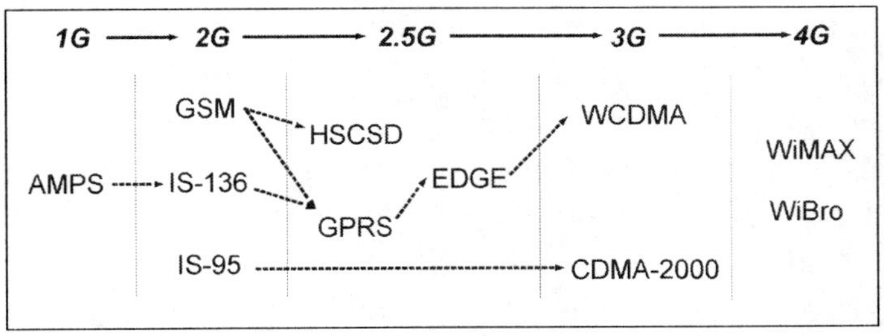

DV (short for evolution data voice) supports both voice and data users simultaneously. Due to lack of interest by carriers, the EV-DV development is currently on hold.

UMTS is the European and Japanese counterpart of CDMA2000. The UMTS system has been deployed in Japan by NTT DoCoMo. An evolution of UMTS known as high speed downlink packet access (HSDPA) has been rolled out in the U.S. by Cingular.

WIRELESS LOCAL AREA NETWORKS

Wireless local area networks (WLAN) are rapidly replacing wires within homes and offices. The most common wireless LANs are based on the IEEE 802.11 standard (Gast, 2005; Wi-Fi Web site, n.d.). They are commonly referred to as WiFi networks. The phenomenal growth of Internet combined with increased use of portable, laptop computers also contributed to the rapid acceptance of WLANs. Some of the other factors that can be attributed to their widespread adaptation are their low cost and the ease of installation. They can be easily deployed by individuals within buildings without a license (since these devices usually operate in license-free band).

The original 802.11 standard, adopted in 1997, provided data rates up to 2 Mbps. Higher data rate enhancements to 802.11, commonly referred to as 802.11a, 802.11b, and 802.11g, emerged later. The 802.11b system operates in the 2.4 GHz band and provides data rates up to 11 Mbps. One of the drawbacks of 802.11b is its limited capacity due to the limited bandwidth availability in the 2.4 GHz band. This capacity limitation restricted the use of WLANs in office environments, where higher capacity is usually needed. Furthermore, the 2.4 GHz is already "crowded" with other devices such as cordless phones, Bluetooth, and microwaves. These drawbacks were mitigated with the emergence of 802.11a devices, which operate in the 5 GHz band and provide data rates up to 54 Mbps. The wider bandwidth available in the 5 GHz band allows using 802.11a based WLANs in office environments. The 802.11g standard, adopted in 2003, provides data rates up to 54 Mbps in the 2.4 GHz band. The emerging 802.11n standard promises data rates up to 600 Mbps. The 802.11n draft compliant products are expected to reach the market in the second half of 2006.

WiFi has proven to be a way of providing reliable in-building wireless access. The cellular phone manufacturers are launching dual mode GSM/WiFi devices that can make and receive calls on both cellular network and enterprise WLAN.

The European counterpart of WiFi is HiperLAN (short for high performance radio local area network), standardized by European Telecommunications Standards Institute (ETSI). HiperLANs have similar physical layer as Wi-Fi and are used chiefly in Europe.

WIRELESS METROPOLITAN AREA NETWORK

Wireless metropolitan area networks (WMAN) aim to provide always on broadband access to Internet at anytime from anywhere. The most successful wireless MAN technology is based on the IEEE 802.16 standard, which is often referred to as WiMAX networks (Ohrtman, 2005; WiMAX Forum Web site, n.d.). WiMAX is an alternative to third-generation cellular systems to provide broadband connections over long distances, and it is considered to be a fourth-generation technology. While the third-generation systems support both circuit and packet switching, the fourth-generation systems support packet switching only. The first-generation WiMAX products are expected to support data rates up to 40 Mbps per channel at vehicular speeds.

High performance radio metropolitan area network (HiperMAN) is the European alternative to WiMAX. Korea adopted a flavor of WiMAX known as wireless broadband (WiBro) as its wireless MAN standard.

WIRELESS PERSONAL AREA NETWORKS

Wireless personal area networks (WPANs) are designed to replace the wires that connect devices to one another (such as printer cables, headphone cables, cables that run from personal computers to mouse, cables that run from set-top boxes). Bluetooth (Bluetooth Web site, n.d.) is the most widely embraced WPAN standard. It operates in the 2.4 GHz unlicensed band and provides short-range communication within a nominal range of 10 meters. Many laptops and cell phones currently support Bluetooth connections. Some car manufacturers, including Acura and BMW, are installing Bluetooth in cars. The latest Bluetooth devices consume extremely low power and provide reliable communication at data rates up to 2.1 Mbps.

ZigBee (ZigBee Web site, n.d.) is another competing technology for wireless PAN. ZigBee devices are expected to be cheaper and simpler than Bluetooth devices. The target applications for ZigBee include general-purpose, inexpensive, self-organizing mesh network that can be shared by medical devices, smoke and intruder alarms, and building and home automation.

CONCLUSION

The wireless revolution is just beginning. The confluence of communication and computing has created many new applications for wireless systems. The throughput, capacity, and range of wireless systems are constantly being improved. This chapter summarized the evolution that has happened over the last 2 decades.

The first- and second-generation systems were primarily designed for voice applications. The 2.5G systems were designed to be retrofit to 2G systems to support data. The third-generation systems were designed to support both voice and data, while the evolving fourth-generation systems are being designed to support primarily data. The WLAN system based on WiFi has seen tremendous growth and success over the last 5 years.

REFERENCES

Bluetooth Web site. (n.d.). Retrieved February, 2006 from http://www.bluetooth.com

Gast, M. (2005). *802.11 Wireless networks: The definitive guide* (2nd ed.). Sebastapol: O'Reilly Media, Inc.

Halonen, T., Romero J., & Melero J. (2003). *GSM, GPRS, and EDGE performance: Evolution towards 3G/UMTS*. West Sussex: John Wiley & Sons.

Mandyam, G., & Lai J. (2002). *Third generation CDMA systems for enhanced data services.* San Diego: Academic Press.

Mouly, M., & Pautet, M. (1992). The *GSM system for mobile communications.* Palaiseau: Telecom Publishing.

Ohrtman, F. (2005). *WiMAX handbook*. New York: McGraw Hill Professional.

3GPP2 Web site. (n.d.). Retrieved February 2006 from http://www.3gpp2.org

UMTS-Forum Web site. (n.d.). Retrieved February 2006 from http://www.umts-forum.org

Wi-Fi Web site. (n.d.). Retrieved February 2006 from http://wi-fi.org

WiMAX FORUM Web site. (n.d.). Retrieved February 2006 from http://www.wimaxforum.org

Yacoub, M. (2001). *Wireless technology: Protocols, standards, and techniques*. Boca Raton: CRC Press.

ZigBee Web site. (n.d.). Retrieved February 2006 from http://www.zigbee.org

ENDNOTES

[1] Capacity is the number of active users that can be supported at the same time.

[2] Before GSM deployment, different parts of Europe used different cellular standards and it was not possible to use the same handset everywhere.

[3] This is the maximum data rate per GSM channel. By combining many GSM channels together, it is possible to achieve data rates up to several megabits per second.

APPENDIX: GLOSSARY

1xRTT	Single Carrier Radio Transmission Technology
AMPS	Advance Mobile Phone Service
CDMA	Code Division Multiple Access
CSD	Circuit Switched Data
EDGE	Enhanced Data Rates for GSM Evolution
EV-DO	Evolution Data Optimized
EV-DV	Evolution Data Voice
GPRS	General Packet Data Service
GSM	Global System Mobile
HiperLAN	High Performance Radio Local Area Network
HSCSD	High Speed Circuit Switched Data
HSDPA	High Speed Downlink Packet Access
IP	Internet Protocol
IS-136	Interim Standard 136
IS-95	Interim Standard 95
PSD	Packet Switched Data
SMS	Short Message Service
TDMA	Time Division Multiple Access
UMTS	Universal Mobile Telecommunication System
WAP	Wireless Application Protocol
WAP	Wireless Application Protocol
WiFi	Wireless Fidelity
WiMAX	Worldwide Interoperability for Microwave Access
WiBRO	Wireless Broadband
WLAN	Wireless Local Area Network
WPAN	Wireless Personal Area Network
WMAN	Wireless Metropolitan Area Network

This work was previously published in E-Business Process Management: Technologies and Solutions, edited by J. Sounderpandan and T. Sinha. 148-154, copyright 2007 by IGI Publishing, formerly known as Idea Group Publishing (an imprint of IGI Global).

Chapter II
Propagating the Ideal: The Mobile Communication Paradox

Imar de Vries
Utrecht University, The Netherlands

ABSTRACT

In this chapter, visions of mobile communication are explored by focussing on idealised concepts surrounding wireless technology. By examining sources on the development, marketing, and use of wireless technology, I contextualise these visions within earlier accounts of ideal communication found in media history and isolate the regularities that are part of these accounts. On close examination, a paradox reveals itself in these regularities, one that can be described as resulting from an uneasiness in the human communication psyche: an unfulfilled desire for divine togetherness that clashes with individual communication needs. While the exact nature of this paradox—innate and hardwired into our brains, or culturally fostered—remains unknown, however, I assert that the paradox will continue to fuel idealised ideas about future communication technology. I conclude with the observation that not all use of mobile technology can immediately be interpreted as transcendental, and that built-in locational awareness balances the mobile communication act.

INTRODUCTION

In October 2003, two British climbers were caught in a blizzard on a Swiss mountain. Rachel Kelsey and her partner Jeremy Colenso, both experienced climbers, were forced to stop behind a large rock at 3000 meters up and wait for the weather to clear. They soon realised that their chances of finding the abseil points in heavy snow were very slim, which meant they were stuck. They texted five friends, one of whom received the message in London at 5 a.m. and immediately notified the rescue services in Geneva. After having to wait another 36 hours because the conditions were too severe for the rescue team to pick them up, the two climbers were finally rescued (Allison, 2003).

The idea that Earth is becoming entirely networked is not new,[1] but the characteristics of mobile communication media have—just as with the first wireless revolution in the beginning of the 20th century[2]—fiercely fuelled the Western notion that through better communication technology all problems of communication will—finally—be solved (Peters, 1999). The "anywhere, anytime, anyhow, anyone" slogan, subliminally attached to every mobile apparatus, opens up a vision of a universally accessible communication space, in which the exchange of information comes to stand for the single most important condition of human progress. More than at any other time in history, this human progress is thought to depend on technological progress.

Rescue stories as those described in the opening paragraph play their part in keeping the idea alive that improvement through technological progress can be measured. The conventional wisdom is that human lives are the single most valuable things we can think of, and if new technology can help save them, it must be treasured. Moreover, if new technology such as mobile telephony makes possible a way of life that is never forsaken of human contact—which therefore is taken as *safe* because there will always be someone who can help—this technology is surely poised to be seamlessly adapted to and integrated in our being (Katz, 2003). Through the remediation of older dominant forms of communication and entertainment technology, the mobile device (or personal digital assistant (PDA) or smart phone, as it is increasingly being called by mobile phone operators and providers) does seem to try to provide an ultimate extension of the natural balance of our sense organs (Levinson, 1997, 2004). Future visions of mobile communication strive for setting up globally accessible meeting points that cater bodiless but perfect interaction, and ultimately for opening up a communication space in which everyone is represented.

This is the inherently human dream of reaching an ideal state, which is cunningly exploited by advertisements, telecom operators, service providers, and the like. We know it is a dream, and we know that we are confronted by it day after day. It will probably haunt us for centuries to come. However, just as "our desire for each other [is] a poor substitute for the primary Eros—and therefore doomed to fail" (Campe, 2000), so are our telecommunication media substitutes for the primary *closeness*—and bound to fail (Vries, 2005). The end result of this is a tragic search for ideal communication through a continuous so-called improvement of communication technologies, a search that will never end.

This chapter will investigate the paradox of this eternal futile quest that we seem to keep embarking on, and will do so by looking at how mobile discourse is framed within quest-ending narratives. By analysing texts from influential scholars such as Pierre Lévy, Howard Rheingold, and Paul Levinson, we will get a grasp of how idealised ideas of the power of new communication technology have pervaded the mobile realm. From there, an attempt is made to single out the recurrent elements in those ideas, whose pervasiveness in our culture will then be examined. Finally, we will look at a few current trends in mobile cooperation techniques that potentially realise certain ideals of communication, albeit in a more pragmatic sense than a sublime one.

UNWIRING THE KNOWLEDGE SPACE

So far, it has mainly been cyberspace and its accompanying access points in the form of personal computers and laptops that are associated with potentially establishing the universally accessible communication realm. However, with the amount of mobile phones growing at an enormous pace,[3] the mobile device has with stunning speed become an essential tool to establish and maintain social networks, as well as managing all kinds of data flows. In this capacity, the device seems perfectly poised to morph itself into the logical choice of medium when accessing the ever-expanding Übernetwork, the Internet.[4] Wherever, whenever, whatever: downloading or uploading information on the move, sharing news events as they happen with your carefully filtered online friends, checking in on your favourite weblog while lying on the beach; it is already possible and will be even more so when the devices grow into always-on mode. It is at this point where Pierre Lévy's (1997) imaginative *collective intelligence*, located in what he calls the knowledge space, starts to come into its own on an immense scale.

Lévy describes the evolution of earthbound living as being immersed in a succession of four types of space, in which man's identity is determined by the tools and symbols predominantly available in that space (see Table 1). The knowledge space is the fourth—and final—space in which we have come to live, and can best be seen as an informational cloud, a "space of living-in-knowledge and collective thought" (Lévy, 1997, p. 140). An important premise for its existence, growth, and preservation is that people interact with the informational cloud by adding, changing, and retrieving data in whatever way possible.[5] It is to "unfold and grow to cover an increasingly vast and diverse world" (Lévy, 1997, pp. 111-112), ultimately creating a universally accessible information realm. Already, we can recognise this vision in descriptions of the multiple thrusts behind both the Internet and the mobile revolutions, such as those found in marketing publicity and open source movements' manifests alike.

Lévy's hierarchical description of the four levels of space invoke Borgmann's (1999) distinction between information *about* ("my shed can be found next to the willow tree"), *for* ("this is how you build a cathedral"), and *as* reality ("hi, I am Imar's avatar, shall we start exchanging data?"). Both Lévy and Borgmann show us historical shifts that expose a dematerialising transition of the dominant form of information. Although—as is conspicuously evident from the title of his book *Holding on to Reality*—Borgmann warns us for a Baudrillard-like potentially dangerous split between information about/for reality and information as reality, Lévy is not so much concerned about the danger of leaving reality behind, as he frames the knowledge space firmly within the other three spaces: "[It is n]ot exactly an earthly paradise, since the other spaces, with their limitations, will continue to exist. The intention of collective intellect is not to destroy the earth, or the territory, or the market economy" (Lévy, 1997, p. 141).

Paradise or not, Lévy cannot help but describe the knowledge space in terms of "a u-topia ... waiting to be born," "a cosmopolitan and borderless space," "an electronic storm," and "a sphere of artifice shot through with streaks of light and mutating signs" (Lévy, 1997, pp. 138-141), thereby mimicking the eccentric cyberpunk style of William Gibson's *Neuromancer*. There is undeniably a religious element visible in the way Lévy writes about the knowledge space, in which information is to be uncoupled from its static base. This dematerialising movement fits perfectly with the transcendental nature of going wireless: liberating things by releasing them from their carriers (be it wires, paper, or the brain) promises more

Table 1. Succession of spaces according to Lévy (1997)

Space	Identity
Nomadic Space of Earth	totems, lineage
Territorial Space	territorial inscription
Commodity Space	position within the domains of production and consumption
Knowledge Space	skill, nomadic cooperation, continuous hybridization

opportunities to interconnect those liberated entities, as they form free-floating nodes in a dynamic network. In the end, in its most radical form, the idea is that every node can be connected to all others, providing instant and perfect transferral of whatever form of data.

As asserted previously, although the knowledge space is self-regulated and its transcendental nature gives rise to the supposition that it might leave the other spaces behind, Lévy holds that it can not be entirely separated from the three preceding spaces. Moreover, in a circular movement—"a return of the earth to itself," as Lévy (1997, p. 141) calls it—the knowledge space connects back to the first space through the recurrence of the nomadic identity. Again, this is a characteristic that is typically found in the mobile device, as has been shown by scholars in recent literature (Gergen, 2003; Kopomaa, 2000; Meyrowitz, 2003). The multiple social roles we possess are called upon in increasingly diverse geographical and social environments when a mobile device is carried along: we can perform parental tasks while at work, we can keep in touch with friends while on vacation, and we can consume entertainment while sitting in classrooms. Slowly, urban design is responding to the diminishing need to build strict and fixed divisions between sites for work, leisure, and family, creating heterogeneous zones in which the individual's social status is defined by the type of communication he or she engages with. The use of mobile technology therefore does not entail a full-circle return to the nomadic in the sense that it forces *us* to change location in order to find more fertile ground, as was the case in Lévy's first earthly space, but it forces our *locations* to adapt to our dynamic modes of being.

The transcendental and nomadic nature of the knowledge space calls for an intricate investigation of the points where it meets other spaces, and of the materiality of these meeting points. Considering the ease with which the mobile device has found its place as the essential data tool, such meeting points, which according to Rheingold (2002) seem to call for a "marriage of bits and atoms" (p. 100) or for us to be able to "click on reality," (p. 95) are set to be facilitated by the smart phones of the future. Or, as we will see in the next section, this is how it is envisioned in idealised ideas of communication.

THE LURE OF THE IDEAL

Although he admits to being utopian, and has subsequently tried to capture the dynamics of the collective intelligence in a formal language in order to make it more visible and tangible, Lévy has been criticised for painting an exaggeratedly pretty picture, ignoring the tough reality of political, economic, social, and other factors that influence the way communication technology is developed, produced, distributed, and used. In the fourth chapter of their book *Times of the Technoculture: From the Information Society to the Virtual Life*, Robins and Webster (1999) accuse Lévy of "promot[ing] and legitim[ising] the prevailing corporate ideology of globalization," and hold that "there is a desperate need for a richer debate of knowledges in contemporary societies — in place of the shallow, progressivist marketing that attaches itself to the cyberculture slogan (and reflects the hegemony of corporate interests)" (Robins & Webster, 1999, pp. 225, 227). In the same chapter, the aforementioned Rheingold receives similar flak for his—supposedly uncritical—belief in the Internet as a means of restoring communities.

However, Lévy and Rheingold are influential writers and are certainly not alone in taking an optimistic and idealised view on the possible contributions new communication technology can make to finally bring people together in an intelligent collective—nor will they be the last. If the years between the launch of the world's first graphic Internet browser in March 1993 and the crash of the dotcom boom in early 2000 marked the building up of the cyberspace hype, then the subsequent years can be characterised as having been labelled the new and improved mobile or wireless era: countless press releases, research papers, news articles, advertisements, books, radio shows, and television programmes have heralded mobile technology as the ideal solution to many communication problems. Two books I would like to bring to the fore in this respect are *Smart Mobs: The Next Social Revolution* by Howard Rheingold (2002) and *Cellphone* by Paul Levinson (2004), as their structures show interesting similarities with Lévy's (1997) approach—and with it, the same dangerous tendency to overestimate communication technology's power to fulfill longtime ideals of communication.

Comprised of a large series of anecdotal, interview, and travel journal material, *Smart Mobs* intends to uncover the characteristics of the "next social revolution," which is to be cranked up by the new mobile devices that "put the power of instant and ubiquitous communication — literally—within everyone's grasp" (Rheingold, 2002, back cover). Describing an impressive amount of trends, experiments, news reports, and commercial projects within the global realm of mobile telephony and computing, Rheingold shows how "technologies of cooperation" have an inherent tendency to group people together—and where there is a group, there are opportunities to learn, create, or topple over. The well-known (albeit somewhat overused) example of the protest demonstration in the Philippines in 2001, in which more than 1 million people were rallied by text messages to oppose Joseph Estrada's regime, is used by Rheingold as a key argument in describing a pivotal cultural and political moment: the power of mobile, ad hoc social networks is not to be underestimated; it can even influence politics on a momentous scale! To be fair, Rheingold's argument does not hinge upon this example alone; next to three other activist movements, he also mentions the squads of demonstrators that, thanks to mobile coordination, *won* the "Battle of Seattle" during a World Trade Organization meeting in 1999. These *movements*, however, have been minor in impact and longevity, and do not appeal to the imagination as much as the Philippine regime change does. It is therefore that *Smart Mobs* focuses mainly on events and projects that contain a clearly visible potential to change things; after all, what better way is there to show that the social impact of mobile technology is not only measurable, but can also be described in terms of setting in motion an unstoppable voyage towards a better future?

Other examples of what the consequences of ubiquitous mobile communication might be are equally carefully chosen for their provocative nature. Among the phenomena that await us, Rheingold (2002) names WiFi neighbourhoods; wearable computing that makes our environment aware of our presence and can react accordingly; RFID tags that provide contextual information on any object; and swarm intelligence that makes possible useful emergent behaviour. He does his best to convince us of the inherent potential of these things to fundamentally change the way we are living—and does so with an obligatory nod to the possibility that some of those changes might not be as pleasurable as we would like—but fails to go much further beyond stating the mantra *together is good*. The majority of Rheingold's examples, however tangible and useful they may be within their own context, are used to construct a vision of a futuristic world in which the possibility to connect things (people and machines) is most highly rated. To connect is to solve, to evolve, to come closer to the ideal of sublime togetherness.

Levinson's *Cellphone*[7] is another very good example of how opportunistic ideas found in much cyberculture literature have been transferred to the mobile realm. Not wasting any time, the book's subtitle, which is as subtle as it is provocative, already promises to tell us "[t]he story of the world's most mobile medium **and how it has transformed everything**" (bold in original). Working from within his Darwinian approach to media evolution—only the fittest media persist in the human environment—Levinson holds that "the cellphone has survived a human test," and that the human need it satisfies is "as old as the human species — the need to talk and walk, to communicate and move, at the same time" (Levinson, 2004, p. 13). This need, which "even defines the human species" (Levinson, 2004, p. 13), is satisfied by the mobile device to such an extent that Levinson foresees the end of the digital divide; the rise of new and more honest forms of news gathering and dispersal; and the birth of a smart world.

The most important (and obvious) characteristic Levinson stresses is that the mobile device blurs the boundary between inside and outside, rendering it unnecessary to confine ourselves to brick and mortar rooms when we want to call someone or find information. The consequence of this blurring is that it will enable us to "do more of what we want to do, be it business or pleasure, pursuit of knowledge, details, companionship, love," and that it will make "every place in the world in which a human may choose to thread ... well-read, or 'intelligent'" (Levinson, 2004, pp. 60-61). Dubbing this intelligent world a "telepathic society"—accompanied by the obligatory but hollow disclaimers that "our progress ... will be tough going at times" (Levinson, 2004, pp. 60-61) and that the mobile device not only solves things but generates new problems of privacy as well—Levinson sides with previous visions of emerging all-encompassing intelligence that have proved to be vulnerable to easy critique, including the Noosphere of Teilhard de Chardin (1959), the morphic fields of Sheldrake (1989) and the global brain of Bloom (2000). As we will see in the next section, the recurrence of these ideas is not coincidental.

RESEARCHING REGULARITIES

Clearly, optimistic visions of new futures are often met with scepticism, but this does not stop them from reoccurring through time; especially when new information and communication media find the limelight. To understand why this "almost willful, historical amnesia," as Mosco (2004, p. 118) calls it, occurs, it is necessary to investigate the underlying regularities of such idealised claims, and to map the basic elements that make up those regular elements. By focussing not on a new medium itself—nor on what it is that makes it unique—but on the path that lies before that medium, we can get a detailed view of the moments in time that mark significant contributions to the medium's earlier discourse. This can best be achieved using the so-called media archaeology approach, which aims to prevent historical amnesia by "(re)placing [the histories of media technologies] into their cultural and discursive contexts" (Huhtamo, 1994). Doing so, the emphasis is shifted "[f]rom a predominantly chronological and positivistic ordering of things, centered on the artefact, ... into treating history as a multi-layered construct, a dynamic system of relationships" (Huhtamo, 1994). It is these relationships that can clarify the intricate ways in which idealised regularities in the dynamic communication media discourse may have changed face, but not their core.

Huhtamo proposes to call the regularities *topoi*, or topics, which he defines as "formulas, ranging from stylistic to allegorical, that make up the 'building blocks' of cultural traditions." He stresses that these topoi are dynamic themselves: "they are activated and de-activated in turn; new topoi are created along the way and old ones (at least seemingly) vanish" (Huhtamo, 1994). In other words, topoi are highly political and ideologically motivated. As an example of a topos found in media history, Huhtamo considers the recurrent "panicky reactions" of public being exposed to visual spectacles, and finds these in illustrations of the Fantasmagorie shows at the end of the 18th century, in reports of the showing of the arriving train in the Lumière brother's *L'Arrivee d'un train a La Ciotat* (1895) and in the stereoscopic movie spectacle *Captain EO* in Disneyland. There is, of course, a danger of over-interpreting historical sources that may well have served another function than to give an accurate account of what actually happened, but this is exactly Huhtamo's point: "unrealized 'dream machines,' or discursive inventions (inventions that exist only as discourses), can be just as revealing as realized artefacts" (Huhtamo, 1994). The Lumière showing may well not have created any panic at all, but it still remains a poignant reference, a media myth that is repeatedly used in numerous books, articles, and essays in which the reception and impact of new media is discussed. Media archaeology tries to expose these dubious but persistent stories, to collect and dust off forgotten elements of a medium's history by looking at discursive connections, however weak those connections may be. By looking at the many levels on which the discursive construction of a communication technology presents itself, media archaeology bridges the revolutionary gaps that are often found in teleological historiographies of that technology.

This archaeological approach has been put to practice by several scholars in recent years,[8] and has so far been successful in revealing and critically analysing media topoi such as the desires for immediacy, presence, liveness, and simultaneity. The most powerful (or overarching) topos, however, is the gnostic longing to transcend earthly life by improving technology, and to create a Universal Brotherhood of Universal Man. This ultimate topos unites every imaginable description of fulfillment, perfection, pureness, and harmony, and can be found in accounts of every communication medium, in every stage of its development, production, distribution, and use. The dream to finally fulfill the ultimate topos through improvement of communication technology can be comprehensively traced through media history, as many scholars (Mattelart, 2000; Mosco, 2004; Peters, 1999) have already shown. As I have written elsewhere, "[w]ireless telegraphy was seen as 'the means to instantaneous free communication'; telephony seemed to promise banishment of distance, isolation and prejudice; radio would pave the way for contact with the dead and television would transform its viewers into eyewitnesses of everything that went on in the world" (Vries, 2005, p. 11). With every development, be it technological, political, economical, or social, the regularities in discursive accounts of older media have been passed on to newer versions, thereby changing form but not essence.

The argument here is that mobile technology fits into a long line of media in which a limited set of regularly used *modes of reflection* determines the discursive domain of media reception. By analysing the discursive construction of mobile technology and comparing it to that of previous communication media, we can get a grasp of the topoi that have flourished or been revived—be it essentially unchanged or in disguised form—and of those that have floundered or been abandoned. Some of the most interesting indicators of these topoi are to be found in rationalisation techniques people use when explaining why they buy mobile phones, or what they are mainly going to use them for. On the surface, these explanations mostly point to very pragmatic reasons. Field study has shown that common justifications for acquiring a mobile phone are business, safety, and security (Palen, Salzman, & Youngs, 2000). On a deeper psychological level, however, these pragmatic reasons can be tied to fears of solipsism, a desire to increase the amount and strength of communication channels in the social network, and a wish for greater control over one's overall connectivity and availability. Just as we have seen in Rheingold's *Smart Mobs*, a need for the potential to increase *togetherness* is expressed in the mobile discourse, reflecting the ultimate topos of ideal communication.

The hints of religious elements present in these uncovered communication ideals is not surprising; just as Ludwig Andreas von Feuerbach stated in the middle of the 19th century that

God is the projection of the human essence onto an ideal, so is an ultimate communicative Being One a projection of a human essence onto communication ideals. The religious motifs continue to exist today: authors such as Erik Davis (1998) and David Noble (1997) have written elaborate accounts of how contemporary technological discourses are still undeniably intertwined with religious beliefs, despite the widely held notion that since the Enlightenment these categories have slowly but surely separated. Such is the case with the topos of ultimate togetherness: the fears and desires disseminated by that topos are exponents of a mixture of the autonomous behaviour of the liberated Cartesian subject on the one hand, and a dream of a bodiless sharing of minds, described by Peters (1999) as angelic communication, on the other. This is a deeply paradoxical mixture, however. Angelic communication shows all the hallmarks of a divine togetherness: with no physical borders and direct one-on-one mappings of minds, every entity will ultimately know and be the same. This loss of individuality collides with the search for more control over ones individual connectivity found in the modern subject's autonomous behaviour. Both angelic communication and complete autonomy are idealised opposite poles on the same scale, and will therefore remain forever out of reach.

THINKING THROUGH PARADOX

The crux of the communication paradox can be described as an uneasiness in the human communication psyche, born out of the tension between the desire for ideal communication and the knowledge of never being able to reach that goal. This is not to say that every individual always wants to strive for perfection. Moreover, reaching perfection may not be what would actually be beneficial for human kind, as many dystopian answers to utopian projects, proposals, and literature have shown; there is no room for individuals or deviations in a society that can only function perfectly if every citizen is synchronised in the grand scheme.[9] Still, the paradox holds, as even in dystopian visions the utopian looms; in the end, Armageddon, the ultimate dystopian event, does nothing more than to destroy old structures in order to lay the foundation for a new, perfect one. A similar argument can be made for a dominant part of the communication media discourse: New media strive for the abolishment of old media in order to provide improved togetherness (Bolter & Grusin, 1999).

As we have seen in the previous section, the successive observations that the development phase and subsequent promotion of communication media are almost always framed within idealised expectations, that these are always accompanied by dystopian rebuttals, and that this process of touting and dismissing keeps reoccurring through time, give rise to the assumption that there is a steady undercurrent present, a topos that can be described as an idea of ideal communication that drives humankind to keep searching despite guaranteed failure. The objection to this assumption might be that this process is merely a marketing mechanism, but such a mechanism can only work if it addresses a human longing, one that is sensitive to promises of solving the communication tension.[10] The question, then, is whether the paradoxical attitude towards communication technology is innate, or if it is just a temporary, culturally sustained concept of progress left over from the Enlightenment, which, at some time in the future, is to be replaced by another concept. If it is innate, we will not be able to escape it; if it is not, we might be able to understand how to change or manipulate the structures in which the paradox resides.

To ask the question of innateness is to enter the realm of epistemology, the study of how we can know the world around us. Until the middle of the 18th century, this field had known two fairly opposed visions: the rationalist and the empiricist view. The rationalist Innate Concept thesis holds that there are some concepts that are already in our minds when we are born, as part of our rational nature. The notion that we can have a priori knowledge, that we have some innate awareness of things we know to be true that is not provided by experience, rests on the premise that the concepts used to construct that knowledge are also innate. Empiricists, however, argue that there are no innate concepts, and that experience alone accounts for the raw material we use to gain knowledge. The most well-known proponent of empiricism, John Locke, wrote that humans are born with a blank mind, a tabula rasa, which is *written onto* by experience. Knowledge, therefore, is not brought to consciousness by experience, but is provided by that experience itself.

This distinction largely disappeared toward the end of the 18th century when the two views were brought together by Emmanuel Kant, who divided reality into the phenomenal world (in which things are what they appear to us to be, and can empirically be known) and the noumenal world (in which things are what they are *in themselves*, and where rationalism rules). According to Kant's transcendental idealism, innate concepts do exist, but only in the noumenal world, where they remain empirically unknowable. Arguably, these innate concepts are philosophical in nature and therefore proof of their existence remains hard to formulate, but this does not mean *innateness* is always metaphysical. For instance, genetic theory, a late 20th century science, claims to provide empirical evidence for the existence of innate mechanisms in cognitive evolution: Human brains are not tabula rasa, but prestructured in specific ways so that they can learn things other organisms can not. While some elements of evolutionary psychology (EP) are highly controversial,[11] it is increasingly accepted that we all come wired with what Chomsky (1957) has called a Language Acquisition Device (LAD): Not only do we possess an innate capacity to learn, but also an innate set of universal language structures. This means that, independent of our social, cultural, or ethic environment, we already *know* how language works before we even speak it. It is on this level that we have to look for the communication paradox if we believe it to be innate: Are we in some way hard-wired to have a tendency to long for goals that are impossible to reach, to be fascinated by things that are and yet are not? Is there some sense of divine togetherness that we come programmed with, that is at some point in time to be fulfilled but keeps slipping away when we think we come close? The long history of trying to overcome distance and time through the use of media makes a strong argument for such a claim, especially when looking at the positivist discourse this search is usually framed in.

Seen this way, the topos of increased togetherness through idealised communication is but one manifestation of a central paradoxical tendency generated by our brains, albeit one of the most dominant. An imaginative account of how this paradoxical core pervades all aspects of life is found in Hofstadter's (1979/1999) *Gödel, Escher, Bach: An Eternal Golden Braid*. In the new preface in the 20th anniversary edition Hofstadter stresses the paradoxical motive for writing the book by stating that he had set out to "say how it is that animate beings can come out of inanimate matter" (Hofstadter, 1979/1999, p. xx). Introducing so-called strange loops, instances of self-reference that can often lead to paradoxical situations, Hofstadter shows that these loops can not only be found in math, perspective drawings, and music, but also—and this is his main argument—in the very essence of conscious existence itself. Without paradoxes, it seems, life as we know it could not exist. A similar argument is made by Seife (2000), who explores our uneasy relationship with zero and infinity in *Zero: The Biography of a Dangerous Idea*. Innocent as they might seem, in many situations in many times the notions of zero and infinity have been difficult to grasp, use, and explain; to such an extent even that people have equated them with the work of God and ignored them as not allowed by God at the same time. It was through the use of zero and the infinite that Zeno could create his paradoxical race, in which Achilles never overtakes the tortoise, and it is zero and the infinite that plague contemporary physicists' current understanding of our universe. Opposite poles that invoke as well as fight the paradoxical will always be with us, because we are born out of a paradox, Seife concludes.

EP is a relatively young field, and as such has not yet found very stable ground. The argument that there is a universally active module in our brain that triggers—or is even responsible for—a life with paradoxes is therefore to be very cautiously approached. As asserted previously, it may well be that our paradoxical attitude towards communication is not the manifestation of an innate concept, but of a culturally constructed one. A helpful nongenetic argument for the paradoxical inclination is found in existentialist theories, especially in Heidegger's treatment of *Gelassenheit* (releasement) and Sartre's description of *mauvaise foi* (bad faith). Whereas the former concept deals with fully accepting one's Being-in-the-world as something that has no intrinsic goal or pregiven content, as something that can only receive its significance through the meaning one chooses to give to it, the latter is the result of *not* accepting the open-ended nature of our existence, of continuously asking "why"? and trying to find the answer outside of one's own will. Such a denial of things-as-they-are and things-as-they-happen actively feeds and sustains a two-pole system, in which paradoxes reside: There is no coincidence when everything happens for a reason, and there is no sense when everything is contingent. People with bad faith—and there are a lot, according to Sartre—often face and cannot accept the most fundamental paradox: Sometimes things are just what they are, even when they are not.

Now all these observations may seem a far cry from our day-to-day experience of using mobile phones, but whenever we transfer any information in any way we are positioned as a node in a communication network, one that exists foremost because we as humans seek contact. We hope and strive for this contact to be instantaneous, clear, under control, and ideal, even when we want to mislead or deceive the other person; if we manage to use the medium and channel in such a way that it serves our intent, the contact has been ideal for its purpose. The desire is for a technologically induced complete fulfillment, which is omnipresent in mobile discourse. There is never any certainty about having reached this ideal state, however, as we have seen. The communication paradox makes sure that something always gets in the way of pure experience.

THE RETURN OF LOCATION

In light of this knowledge, the best way we can act, as Peters (1999) also argues, is to embrace the impossibility of ideal communication and make do with what forms of communication we *can* realise. The transcendental nature of wireless technology may at times lure us into thinking we have come close and need just a little push in the right direction, but this would be like chasing a mirage. What then are the elements of more appropriate pragmatic approaches to using new communication technology, ones that defy the urge to hand out idealised promises? Some interesting trends in recent innovative wireless concepts show that the independency of locality, the characteristic that seemingly constitutes the *essence* of mobile telephony, can be turned on its head. Where the most pure form of communication is equated with a bodiless presence and is therefore situated in a nondescriptive *anywhere,* part of the current crop of wireless projects inject exactly this sense of locality into the mobile communicative act. The resulting location based services (LBS) are put to use in a variety of ways: backseat games that merge road context with virtual content (Brunnberg & Juhlin, 2003), portable devices that support the tourist experience by supplying on the spot information (Brown & Chalmers,

2003), systems that provide virtual annotation of physical objects (Persson, Espinoza, Fagerberg, Sandin, & Cöster, 2002), and mobile phone applications that can *sense* the proximity of people on your buddy list (Smith, Consolvo, Lamarca, Hightower, Scott, Sohn, et al., 2005). Of course, all these projects in some way reflect a drive towards making things easier, quicker, better, or simply more enjoyable, and therefore do not completely escape paradoxical idealised thinking, but they do not ostentatiously try to transcend our present experience of communication by denying its inherent grounding in lived space and time.

Another area where mobile phones are undeniably making a difference without having to resort to metaphysical musings is in developing countries. By leapfrogging older communication technology—in most cases this concerns landlines that had been too expensive to be installed nationwide—mobile technology is used to quickly set up cheap networks, thereby facilitating measurable boosts to local economies and communities. The mobile networks do not instantly connect all parts of a country, but remain localised in existing urban or rural environments. This localisation is further strengthened by the fact that, less tempted to use the mobile device to mix different social locales into one heterogeneous zone, as is more the case in Western metropolitan areas, people in these developing countries tend to see the mobile more as a landline that happens to be wireless. If there would have been a landline the impact would have largely been the same, something communication theorist Jonathan Donner (2003) concurs with. He conducted several field studies in Rwanda, and found that the use of mobile phones by Rwandan entrepreneurs enhanced their ability to do business, but also to satisfy their emotional and intrinsic needs. This is mostly due to the mere presence of a communication channel, and not to the mobile's intrinsic essence. Again, the underlying idealised implication is that appointments, deals, and transactions can occur faster and more streamlined when people are increasingly brought together in whatever way, but in cases such as those in Rwanda the results of introducing wireless technology are clearly visible and do not remain mostly theoretical.

CONCLUSION

With the global proliferation of mobile communication devices, a reinvigorated sense of ubiquitous connection possibilities has emerged. Covering large parts of the Earth, a networked informational skin seems set to revolutionise our way of living. The key new paradigm that is stressed in this "mobilisation" of the world is the ability to tap into an all-encompassing knowledge space, thereby making information addition, retrieval, and communication virtually instantaneous. The fundamental driving force behind this endeavour can be ascribed to a desire for establishing connections to everyone or everything in whatever way possible, a bodiless omnipresence. The radical consequences of this—almost angelic—desire are affecting traditional modes of interaction such as dialogue and dissemination.

This dream of idealised communication is subconsciously stressed by the dominant image of wireless communication that is found in advertisements, press releases, books on social change, government policies, and the like. Promises that things will get better, fuel our impatience when contemporary technology fails to deliver. In other words, the desire for ideal communication itself is part of a paradoxical system found in all layers of our existence. The dream can never be realised, and will therefore continue to recur through time. Whether we will be able to change our attitude towards this strange loop depends on its nature: If it is hard-wired into our brains, we will have to live with the paradox forever. If it is not, who knows, we might come to see mobile communication for exactly what it is, a specific but not definitive "Being" of communication.

REFERENCES

Allison, R. (2003, October 7). Climbers on Alpine ridge rescued by text message. *The Guardian*. Retrieved May 16, 2005, from http://www.guardian.co.uk/uk_news/story/0,3604,1057271,00.html

Analysys Press Office (2005, May 5). Mobile penetration in Western Europe is forecast to reach 100% by 2007, says Analysys. *Analysys*. Retrieved May 16, 2005, from http://www.analysys.com/Articles/StandardArticle. asp?iLeftArticle=1897

Bloom, H. K. (2000). *The global brain: The evolution of mass mind from the big bang to the 21st century*. New York: Wiley.

Bolter, J. D., & Grusin, R. A. (1999). *Remediation: Understanding new media*. Cambridge, MA: MIT Press.

Borgmann, A. (1999). *Holding on to reality: The nature of information at the turn of the millennium*. University of Chicago Press.

Brown, B., & Chalmers, M. (2003). Tourism and mobile technology. In K. Kuutti & E. H. Karsten (Eds.), *Proceedings of the 8th European Conference on Computer Supported Cooperative Work* (pp. 335-355). Dordrecht: Kluwer Academic Press.

Brunnberg, L., & Juhlin, O. (2003). *Movement and spatiality in a gaming situation: Boosting mobile computer games with the highway experience*. Interactive Institute. Retrieved May 16, 2005, from http://www.tii.se/mobility/Files/BSGFinal.pdf

Campe, C. (2000). *Spheres I: An introduction to Sloterdijk's book*. Goethe-Institut Boston. Retrieved May 16, 2005 from http://www.goethe.de/uk/bos/englisch/Programm/archiv/2000/enpcamp100.htm

Chomsky, N. (1957). *Syntactic structures*. The Hague, The Netherlands: Mouton.

Clark, T. (2004). *Mobile communications and the wireless Internet: The Japanese experience*. Receiver 11. Retrieved May 16, 2005 from http://www.receiver.vodafone.com/11/articles/pdf/11_02.pdf

Davis, E. (1998). *Techgnosis: Myth, magic, mysticism in the age of information*. New York: Harmony Books.

Day, R. E. (1999). The virtual game: Objects, groups, and games in the works of Pierre Lévy. *The Information Society, 15*(4).

Donner, J. (2003). What mobile phones mean to Rwandan entrepreneurs. In K. Nyíri (Ed.), *Mobile democracy: Essays on society, self and politics* (pp. 393-410). Vienna: Passagen.

Gergen, K. (2003). Self and community in the new floating worlds. In K. Nyíri (Ed.), *Mobile democracy: Essays on society, self and politics* (pp. 103-114). Vienna: Passagen.

Hofstadter, D. R. (1999). *Gödel, Escher, Bach: An eternal golden braid*. New York: Basic Books. (Original work published 1979)

Huhtamo, E. (1994). *From kaleidoscomaniac to cybernerd: Towards an archeology of the media*. De Balie Dossiers Media Archaeology. Retrieved May 16, 2005 from http://www.debalie.nl/dossierartikel.jsp ?dossierid=10123&articleid=10104

Katz, J. E. (2003). *Machines that become us: The social context of personal communication technology*. New Brunswick, NJ: Transaction Publishers.

Katz, J. E., & Aakhus, M. A. (2001). *Perpetual contact: Mobile communication, private talk, public performance*. Cambridge, UK; New York: Cambridge University Press.

Kopomaa, T. (2000). *The city in your pocket: Birth of the mobile information society*. Helsinki, The Netherlands: Gaudeamus.

Levinson, P. (1997). *The soft edge: A natural history and future of the information revolution*. London; New York: Routledge.

Levinson, P. (2004). *Cellphone*. New York: Palgrave Macmillan.

Lévy, P. (1997). *Collective intelligence: Mankind's emerging world in cyberspace*. New York: Plenum Publishing Corporation.

Malik, K. (1998, December) The Darwinian fallacy. *Prospect, 36*, 24-30.

Mattelart, A. (2000). *Networking the world, 1794-2000*. Minneapolis: University of Minnesota Press.

Medosch, A. (2004). Not just another wireless utopia. *RAM5*. Retrieved May 16, 2005, from http://www.rixc.lv/ram/en/public07.html

Mosco, V. (2004). *The digital sublime: Myth, power, and cyberspace*. Cambridge, MA: MIT Press.

Noble, D. F. (1997). *The religion of technology: The divinity of man and the spirit of invention*. New York: A.A. Knopf.

Palen, L., Salzman, M., & Youngs, E. (2000). Going wireless: Behavior and practice of new mobile phone users. In W. A. Kellogg & S. Whittaker (Eds.), *Proceedings of the 2000 ACM Conference on Computer Supported Cooperative Work* (pp. 201-210). New York: ACM Press.

Persson, P., Espinoza, F., Fagerberg, P., Sandin, A. & Cöster, R. (2002). GeoNotes: A location-based information system for public spaces. In K. Höök, D. Benyon, & A. Munro (Eds.), *Designing information spaces: The social navigation approach* (pp. 151-173). London; New York: Springer.

Peters, J. D. (1999). *Speaking into the air: A history of the idea of communication*. University of Chicago Press.

Rheingold, H. (2002). *Smart mobs: The next social revolution*. New York: Perseus Publishing.

Robins, K., & Webster, F. (1999). *Times of the technoculture. From the information society to the virtual life*. London; New York: Routledge.

Seife, C. (2000). *Zero: The biography of a dangerous idea*. New York: Viking.

Sheldrake, R. (1989). *The presence of the past: Morphic resonance and the habits of nature*. New York: Vintage Books.

Smith, I., Consolvo, S., Lamarca, A., Hightower, J., Scott, J. Sohn, T., et al. (2005). Social disclosure of place: From location technology to communication practices. In H. W. Gellersen, R. Want, & A. Schmidt (Eds.), *Proceedings of the 3rd International Conference on Pervasive Computing* (pp. 134-141). London; New York: Springer.

Standage, T. (1998). *The Victorian Internet: The remarkable story of the telegraph and the nineteenth century's on-line pioneers*. New York: Walker and Co.

Teilhard de Chardin, P. (1959). *The phenomenon of man*. New York: Harper.

Vries, I. de (2005). Mobile telephony: Realising the dream of ideal communication? In L. Hamill & A. Lasen (Eds.), *Mobiles: Past, present and future*. London; New York: Springer.

ENDNOTES

1. See Standage (1998) for a comparison of the telegraph age with the rise of the Internet.
2. See Medosch (2004) for an account of how both wireless eras are very similar in the way the technology was received.
3. Mobiles in Europe are predicted to exceed Europe's population in 2007 (Analysys Press Office, 2005).

4 See Clark (2004) for an account of how "educational policy, peer pressure, and most importantly, soaring use of internet-enabled mobile handsets" drive young people in Japan to use mobile phones instead of computers when sending and receiving e-mail.

5 A fitting current example of an implementation of such a cloud would be Wikipedia, which thrives on user input and moderation. Other methods of knowledge storage and retrieval such as Google and archive.org rely on algorithms and filters, which makes them more archival than dynamic modes of knowledge preservation.

6 See http://www.aec.at/en/festival2003/wvx/FE_2003_PierreLevy_E.wvx for a Webcast of his lecture at the 2003 Ars Electronica conference, in which he presented the system of this formal language.

7 Levinson prefers to call the device a *cell-phone* instead of a *mobile phone*, because "[it] is not only mobile, but generative, creative." On top of that, it "travels, like organic cells do," and it "can imprison us in a cell of omni-accessibility" (Levinson, 2004, p. 11). I tend to use *mobile device*, as this category includes not only the mobile (or cell) phone, but also smart phones and PDAs.

8 Huhtamo names Tom Gunning, Siegfried Zielinski, Carolyn Marvin, Avital Ronell, Susan J. Douglas, Lynn Spiegel, Cecelia Tichi, and William Boddy (Huhtamo, 1994).

9 Eager to show that a collective intelligence does not mean a loss of individuality, Lévy acknowledges that it is important to ask, in Day's words, "how we can pass from a group mentality characterised by a modern notion of the mass (and with that, mass broadcasting) to a collective intelligence wherein persons may remain individual and singular" (Day, 1999, p. 266).

10 Claims that support the idea of a universal disposition towards what mobile communication is supposed to be about can be found in Katz and Aakhus (2001).

11 Malik (1998) criticises EP because it can be used to explain sexual and racial discrimination as "biologically meaningful." Because our genes have not been able to keep up with cultural evolution, the EP argument goes, we are "stone age men in a space age world," and therefore cannot help but to exhibit hunter-gatherer behaviour. Malik claims that this would completely deny the fact that culture has evolved out of natural selection too, and that we consciously make choices.

This work was previously published inInformation Communication Technologies: Concepts, Methodologies, Tools, and Applications, edited by C. Van Slyke, copyright 2008 by Information Science Reference, formerly known as Idea Group Reference (an imprint of IGI Global).

Chapter III
Teaching and Learning Wi-Fi Networking Fundamentals Using Limited Resources

Wilson Siringoringo
Auckland University of Technology, New Zealand

Nurul I. Sarkar
Auckland University of Technology, New Zealand

ABSTRACT

Wi-Fi networking has been becoming increasingly popular in recent years, both in terms of applications and as the subject of academic research papers and articles in the IT press. It is important that students grasp the basic concepts of both Wi-Fi networking and wireless propagation measurements. Unfortunately, the underlying concepts of wireless networking often intimidate students with their apparently overwhelming complexity, thereby discouraging the students from learning in-depth this otherwise exciting and rewarding subject. This chapter provides a tutorial on Wi-Fi networking and radio propagation measurements using wireless laptops and access points. Various hands-on learning activities are also discussed.

LEARNING OBJECTIVES

After completing this chapter, you will be able to:

- Describe the architecture of Wi-Fi networks.
- Discuss the evolution of IEEE 802.11 standards.
- Set up Wi-Fi networks for class demonstration.
- Suggest future enhancements to the practical activities described in the chapter.

INTRODUCTION

In recent years, Wi-Fi networks (also called IEEE 802.11b) have been gaining in popularity, both in business and in home networking applications. With the growing proliferation of mobile equipment, this trend is likely to continue in the future. It is therefore important for students of information and telecommunication technologies to cover the fundamentals of wireless networking technologies as part of their curriculum.

Many people find that networking technology in general is somewhat arcane and difficult to understand. Similarly, the apparently overwhelming complexity of the underlying concepts of wireless networking often intimidates students. This perception can easily discourage the students from learning in-depth this otherwise exciting and rewarding subject.

This chapter attempts to overcome these problems by providing a hands-on introduction to Wi-Fi networking. A tutorial is also included to guide learners on how to set up Wi-Fi networks using relatively few computing resources. Although a host of problems are to be expected, given the technical limitations of commercially available hardware, students are encouraged to gain a hands-on learning experience in setting up Wi-Fi networks. The chapter also discusses the effectiveness, measured by student feedback, of Wi-Fi projects.

BACKGROUND AND MOTIVATION

Background

Nowadays business organizations rely heavily on computer networks for their operation. The trend towards mobile communication and computing drives the networking industry further towards wireless technology — particularly Wi-Fi technology. As explained later in this chapter, the term Wi-Fi refers to the IEEE 802.11 standard for wireless LAN (WLAN). Therefore the terms Wi-Fi, IEEE 802.11, and WLAN are used interchangeably in this chapter.

Kaczman (2002) reports that an estimated 1 to 1.5 million Wi-Fi communication cards and Wi-Fi-enabled laptops were sold every month during 2002. Vaxevanakis et al. (2003) offer similar sales projections in their reports.

Wireless networks, especially the ones employing Wi-Fi technology, are gaining popularity not only in the business domain but also with home users (Vaxevanakis et al., 2003). The reasons for the popularity of wireless networks over the wired ones are highlighted below (Proxim, 1998):

- **Mobility**: Wireless LANs can provide users with real-time information within their organization without the restrictions inherent with physical cable connections.
- **Installation speed and simplicity**: The installation of wireless LANs does not involve the tedious work of pulling cables through walls and ceilings.
- **Installation flexibility**: Wireless LANs allow access from places unreachable by network cables.
- **Cost of ownership**: Overall installation expenses and life-cycle costs of wireless LANs are significantly lower than wired LAN. The discrepancy is even higher in dynamic environments requiring frequent moves and changes.
- **Scalability**: Wireless LANs can be configured relatively easily since no physical arranging of network cables is required.

Although wireless networks may never completely replace wired networks, they will gain in importance as business assets in the future. Howard (2002) reports that the use of wireless networks for mobile Internet access is also becoming big business, as is indicated by the rising number of wireless Internet service providers in the United States. The increasing number of public hotspots also opens the possibility of providing continuous connection to a roaming business traveller (Vaughan-Nichols, 2003).

Motivation

Wireless technology is likely to continue evolving at rapid a pace in the near future.

Further developments will mainly involve bandwidth increase and optimization, which enable performance improvement in terms of throughput and reliability. In anticipation of the availability of increased bandwidth, various network-based business and multimedia applications are also being developed. Prasad and Prasad (2002) discuss some applications such as teleconferencing, telesurveillance, and video-on-demand operating on wireless network backbones.

Given the importance of research and implementation of Wi-Fi networking in the future, students need to prepare themselves by gaining a thorough understanding about the technology. An overview is presented to introduce the students to the significance of wireless networking technology from a technical standpoint. To facilitate a hands-on learning experience, a tutorial for setting up a Wi-Fi network is also provided.

IEEE 802.11 WIRELESS LAN

A wireless network is a data communications system which uses electromagnetic media such as radio or infrared waves instead of cables. Wireless networks are implemented either to complement or as an alternative to a wired network.

The most prominent feature of wireless networks is mobility, which provides users with freedom of movement while maintaining connectivity. Flexibility is another great advantage of wireless networks since they allow connectivity to places physically inaccessible to network cables.

The application of wireless networks is not confined only to substituting for wired networks. It also enables communication schemes not available in wired networks. With the proliferation of mobile computers and handheld devices such as PDAs and cellular phones, the role of wireless networks is becoming more important as a means of data exchange.

In more general terms, there are currently three primary applications for wireless network technology:

1. **Wireless local area network (WLAN)**: A class of wireless LAN which provides wireless connectivity within an organization. WLAN enables access to a corporate network by individual users without hindering their mobility. Access to WLAN is provided within a defined physical region, such as within an office building.
2. **Wireless personal area network (WPAN)**: The main role of WPAN is to simplify the communication process by eliminating the need for physical connection through cables at home. For example, linking all electronics devices at home using radio technology operating at 2.4 GHz.
3. **Wireless wide area network (WWAN)**: This provides wireless access to users outside the boundaries of WLAN. Currently, cellular technologies play the main role in making WWAN feasible. Typical example of WWAN use is when a user gets access to e-mail or Web sources through a mobile phone.

Several standards have emerged to implement the wireless network technology in the three areas above. The Bluetooth standard (IEEE 802.15) is now dominating the WPAN implementation, whereas the IEEE 802.11 family of specifications is now standard for WLAN implementations. In the WWAN domain, several mobile communication technologies such as GSM, GPRS, and CDMA 2000 are still competing with each other to become the ultimate standard.

Overview

The Institute of Electrical and Electronics Engineers (IEEE) 802.11 standards define the protocols at the physical (PHY) and MAC layers. The MAC layer makes the lower part of the data link layer in the OSI model as shown in Figure 1.

At the physical layer, three different transmission techniques are used by 802.11: frequency hopping spread spectrum (FHSS), direct sequence spread spectrum (DSSS), and infrared (IR). For the 802.11b specification, only the direct sequence spread spectrum is used. Other popular standards, such as Bluetooth, use frequency hopping spread spectrum instead, although it shares the 2.4 GHz band used by 802.11b.

The original 802.11 standard uses an 11-bit chipping code called the Barker sequence to send data via radio waves. Each 11-bit sequence represents one single bit of data. A transmission technique called binary phase shift keying (BPSK) is used to transmit the sequences at the rate of 1 million per second, which corresponds to the basic 1 Mbps data transfer rate of the 802.11. A more sophisticated technique referred to as quadrature phase shift keying (QPSK) is used to achieve a 2 Mbps data transfer rate.

Instead of Barker sequences, the 802.11b uses Complementary Code Keying (CCK). When CCK is used, a 16-bit sequence transmitted over the radio channel contains either 4 or 8 information bits and can achieve data transfer rates of 5.5 Mbps and 11 Mbps, respectively.

At the MAC layer, the fundamental protocol component specified by the 802.11 is the distributed coordination function (DCF). The DCF

Figure 1. IEEE 802.11 in the OSI layer

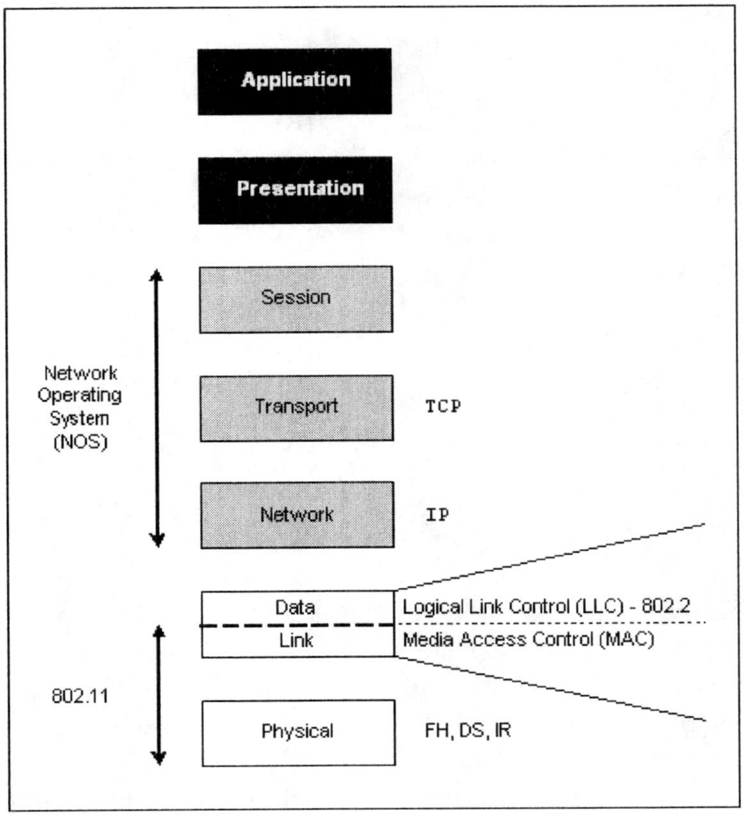

utilizes a Carrier Sense Multiple Access with Collision Avoidance (CSMA/CA) channel access method. When a station realizes that no other has transmitted within a predetermined time called the interframe space (IFS), it transmits its own frame. In unicast transmission, the receiving station is expected to reply with an acknowledgment (ACK), in a similar fashion to the standard automatic repeat request (ARQ) control mechanism. A mechanism called backoff procedure is used by DCF to prevent frame collision, which results from stations transmitting simultaneously (Golmie et al., 2003).

To complement the DCF, an optional component at the MAC layer, referred to as point coordination function (PCF), provides a centralized, polling-based access mechanism. To utilize PCF, an AP is required to perform the point coordination function. The PCF is an optional feature of the IEEE 802.11 MAC layer and is not supported by all Wi-Fi devices.

In PCF, the AP acts as the central access coordinator. It applies the round robin algorithm to poll the stations within the service set. Unlike DCF, stations wishing to transmit data must first obtain permission from the AP using the request-to-send and clear-to-send (RTS/CTS) scheme. When a polled station does not have data to transmit, it sends a null frame. Otherwise the station is allowed to transmit its data frame (Youssef et al., 2002).

Wi-Fi Transmission Issues

The use of the unlicensed 2.4 GHz radio band for Wi-Fi transmission medium leads to interference problems. In most countries the 2.4 GHz spectrum is crowded with other radio devices such as cordless phones, microwaves, and Bluetooth WPAN devices. Due to the effect of radio interference, Wi-Fi performance degrades. With the ever increasing popularity of Bluetooth WPAN, the potential problem of interference becomes sufficiently serious to compel researchers to find mechanisms that allow Wi-Fi and Bluetooth to coexist (Ophir et al., 2004).

Wi-Fi networks have limited range, typically covering up to 46m indoors and 100m outdoors from the nearest access point (Ferro & Potorti, 2005). While significantly wider than Bluetooth coverage, which is 10m, performance degradation can be an issue when a station is located near the limit of Wi-Fi range from the other station or the access point.

The use of radio broadcast also raises security concerns. Unlike wired network where access is limited by physical cable connections, any device within range of Wi-Fi network can intercept the packets and potentially intrude into the network. This potential threat necessitates the use of data encryption in order to minimize the risk of intrusion.

Wi-Fi Network Design Issues

The basic building block of a wireless network is the basic service set (BSS). A BSS consists of a set of fixed or mobile Wi-Fi stations. The simplest network configuration is the independent BSS (IBSS), which is an ad hoc network consisting of two or more stations without any structure other than peer-to-peer cooperation.

When a set of BSSs are connected by a fixed network infrastructure, the resulting structure is called an extended service set (ESS). An ESS is a set of one or more BSSs connected by a distribution system (DS), which is not specified by the IEEE 802.11 standard. The DS can take the form of wired networks or some other form of wireless network. A Wi-Fi network with all the typical components is illustrated in Figure 2.

The stations connected to the DS are called access points (APs). An AP offers distribution system services, which allow data exchange between stations of different BSSs. An AP normally performs a function called point coordination (PC), which employs a round robin policy to poll each station for data transmission.

Figure 2. A typical Wi-Fi network with ESS structure

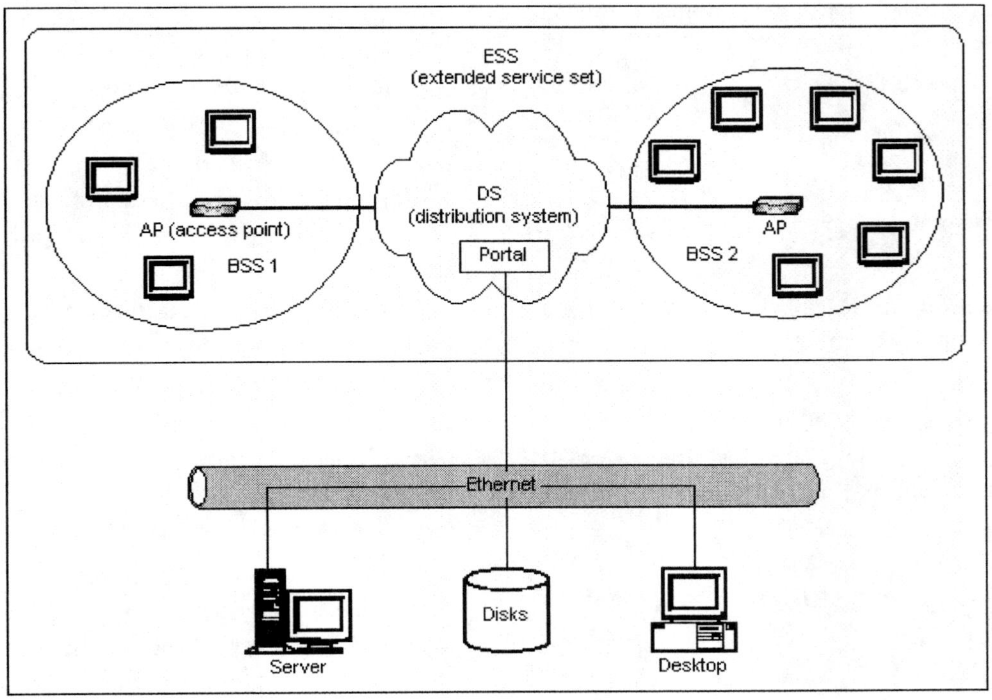

802.11 Standard Family

The IEEE released the 802.11 standard for wireless LAN (WLAN) in 1997. The specification requires a data transfer rate of from 1 Mbps up to 2 Mbps while retaining compatibility with existing LAN hardware and software infrastructure. The standard defines protocols for MAC layer and physical transmission in the unlicensed 2.4 GHz radio band. After successful implementation by commercial companies such as Lucent Technologies, amendment was made for a better performance in the same year. The resulting standard was IEEE 802.11b, which specifies higher data transfer rates of 5.5 and 11 Mbps.

The IEEE 802.11b differs from the 802.11 in the MAC layer even though it retains compatibility with its predecessor. The physical layer is left unchanged. The 802.11b standard was approved in 1999, and during that year the term wireless fidelity or Wi-Fi was introduced. The IEEE 802.11b has proven to be very successful in the commercial domain. The majority of Wi-Fi devices nowadays are made to the 802.11b standard.

In parallel with the IEEE 802.11b, another variant of the original 802.11 was also made. This variant is referred to as IEEE 802.11a, which differs from both IEEE 802.11 and IEEE 802.11b by using the 5 GHz band rather than the 2.4 GHz band. The 5 GHz radio band is unlicensed in the United States but not in many other countries, especially in Europe. The IEEE 802.11a provides up to 54 Mbps, which is much faster than both 802.11 and 802.11b. However the use of different radio frequencies denies compatibility between 802.11a and 802.11/802.11b. Nevertheless the 802.11a was found satisfactory and approved in 1999.

To resolve the incompatibility problem between the standards, an amendment to IEEE 802.11a was approved in 2003. The new standard is referred to as IEEE 802.11g, which operates at 2.4 GHz radio band while retaining the 54 Mbps data transfer rate of the 802.11a. There are other standards in the 802.11 family, as summarized in Table 1.

EXPERIMENT DETAILS

In this section we describe the implementation aspect of an IEEE 802.11 wireless LAN using available wireless equipment. We first describe Wi-Fi networks set up both in ad hoc (IBSS) and infrastructure (ESS) modes. We then focus on the experiments and Wi-Fi performance measurement.

Hardware and Software Requirements

A number of hardware and software applications are used in the experiment. The requirement for IBSS, BSS, and ESS differs mainly in the presence of the AP. The following is the more detailed description of the resources used in the experiment.

Hardware

For the basic service set, at least two Wi-Fi capable computers are required. This is a fairly loose requirement, which can be satisfied by most commercial computer hardware. However, the configuration for wireless laptops and APs that we used in the experiment is shown in Tables 2 and 3.

A different configuration is used for the extended service set (ESS). In an ESS network, an AP is used in addition to two Wi-Fi stations. A third computer is also required to simulate the wired network the AP is attached to (see Table 3).

Software

The IEEE 802.11b protocol automatically establishes data-link-level connection whenever the Wi-Fi-enabled hosts are active and within range to each other. To measure the actual throughput, however, connection at the application layer is also required. Special software applications are used to enable such connection to take place.

Colligo WE software from Colligo Networks Inc. enables users of Wi-Fi–capable devices to

Table 1. IEEE 802.11 standards family (Ferro & Potorti, 2005)

Standard	Description	Status
IEEE 802.11	WLAN; up to 2 Mb/s; 2.4 GHz	Approved 1997
IEEE 802.11a	WLAN; up to 54 Mb/s; 5 GHz	Approved 1999
IEEE 802.11b	WLAN; up to 11 Mb/s; 2.4 GHz	Approved 1999
IEEE 802.11g	WLAN; up to 54 Mb/s; 2.4 GHz	Approved 2003
IEEE 802.11e	New coordination functions for QoS	Task group development
IEEE 802.11f	IAPP (Inter-AP Protocol)	Approved 2003
IEEE 802.11h	Use of the 5 GHz band in Europe	Approved 2003
IEEE 802.11i	New encryption standards	Approved 2004
IEEE 802.11n	MIMO physical layer	Task group development

Table 2. Basic service set hardware specification

System	Specification	
Host 1 (H1)	Make:	Toshiba
	CPU:	Mobile Intel® Celeron® 2.4GHz
	RAM:	496MB
	OS:	Windows XP Professional 2002 SP 1
	Wi-Fi:	WLAN adapter D-Link 650+ 802.11b
Host 2 (H2)	Make:	Toshiba
	CPU:	Mobile Intel® Celeron® 2.4GHz
	RAM:	496MB
	OS:	Windows XP Professional 2002 SP 1
	Wi-Fi:	WLAN adapter Cisco 350 802.11b

Table 3. Extended service set hardware specification

System	Specification	
Host 1 (H1)	Make:	Toshiba
	CPU:	Mobile Intel® Celeron® 2.4GHz
	RAM:	496MB
	OS:	Windows XP Professional 2002 SP 1
	Wi-Fi:	WLAN adapter D-Link 650+ 802.11b
Host 2 (H2)	Make:	Dell
	CPU:	Intel® Pentium® M 1.4GHz
	RAM:	512MB
	OS:	Windows XP Professional 2002 SP 1
	Wi-Fi:	Intel® PRO/Wireless LAN 2100 3A Mini Adapter
AP Controller M	ake:	Toshiba
	CPU:	Mobile Intel® Celeron® 2.4GHz
	RAM:	496MB
	OS:	Windows XP Professional 2002 SP 1
	Wi-Fi:	D-Link DWL-900AP+ connected through crossover Ethernet cable

exchange information. The communication can take place either in ad hoc mode or infrastructure mode. It is necessary that every participant of a Colligo network session has a copy of the software installed on the computer. More complete information regarding the software can be obtained from the company's Web site at http://www.colligo.com/products/workgroupedition/. Evaluation copy of the software is also available to download from the Web site.

The software has many useful features, such as interactive chat, unidirectional message delivery, virtual whiteboard, and file transfer. The file transfer feature is crucial for this experiment, as it provides the means of measuring the link throughput (Colligo, 2005).

Colligo has been designed for nontechnical users. The implication is that Colligo is relatively easy to set up in a typical setting, such as in a wireless ad hoc network. On the other hand, very little control is provided for technical users to customize the settings. This restriction makes it especially difficult for setting up a wireless network in infrastructure mode.

To set up an infrastructure network, appropriate application is required to configure and control the AP. In this particular example, the AP hardware comes with AirPlus Access Point Manager configuration software. The AirPlus Access Point Manager is a Window-based program which allows setting up the access point parameters from a computer through a wired connection. The AP manager therefore is needed only in setting up the network in infrastructure mode. As an alternative, the AP can also be managed using an Internet browser such as Microsoft Internet Explorer.

IBSS Practical Setup

The following setup procedure creates a Colligo basic service set consisting of two Wi-Fi stations, as shown in Figure 3. Setting up such a Colligo ad hoc network session is relatively simple and straightforward.

To initiate a session, all participating stations must activate their Wi-Fi adapter and run the Colligo software. When Colligo is used for the first time, it is also necessary to enter a user name and other personal details for identification purposes.

Figure 4a shows the Colligo main screen for user *benny* after the program starts. Clicking the Create Instant Network button in the upper-right corner, as shown in Figure 4b, will either create a new BSS or join an existing ad hoc network. The same procedure is repeated by other users wishing to join the network.

Following a successful ad hoc network connection, the user names of other stations will be listed in the main window. The time it takes for hosts to detect each other varies from a few seconds to several minutes. Figure 5a shows that a remote user named *wilson* has been detected. The name of the remote user is written in italics to indicate that the user has not yet been authenticated and therefore is unable to communicate or gain access to the network resources.

Right-clicking on the remote user name will activate the authentication menu. The ensuing interactive authentication sequence ensures that the remote user is a legitimate member of the BSS. When the authentication is successful, the remote user name is no longer written in italics, indicating that access to network has been granted to the user, as shown in Figure 5b.

After the remote user is authenticated, all communication features of Colligo can be used. The next step is to set up a file transfer session in which the link throughput can be measured. Figure 6a shows the Colligo main menu to invoke the file transfer feature. Selecting the Send Files item brings a file transfer window similar to that shown in Figure 6b.

Colligo allows us to transfer a batch of files to a number of users. The empty lists in Figure 6b must be correctly populated before the file transfer can begin. Clicking on the Add Users button brings another window where all remote users are listed (see Figure 7a). In this example, only one remote user called *wilson* is selected. Now select the file(s) from the local drive for transmission (see Figure 7b). Clicking on the Send button will initiate a confirmation sequence, which is immediately followed by the actual file transfer.

Figure 3. Simple two-station BSS

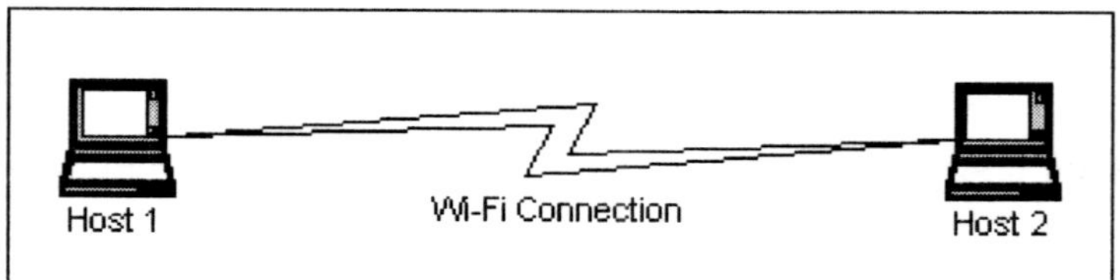

Figure 4. Colligo start-up main window

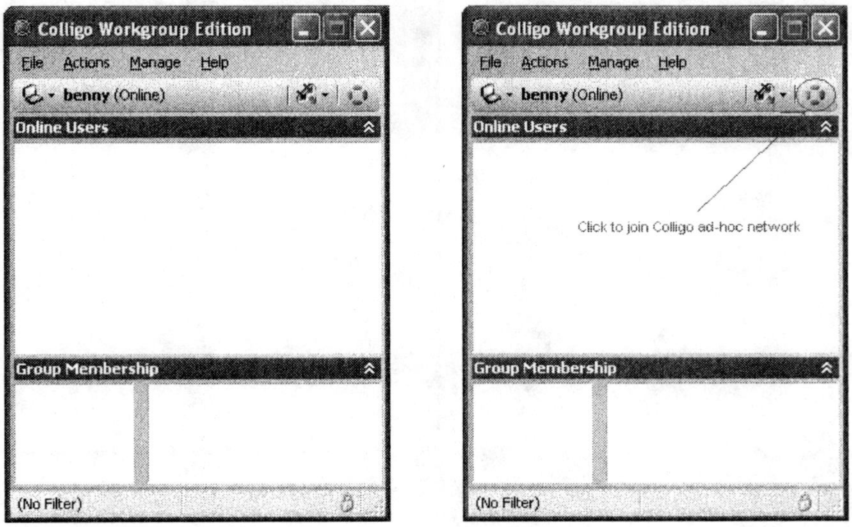

a. Initial window *b. Starting an ad hoc network*

Figure 5. Authenticating a remote user

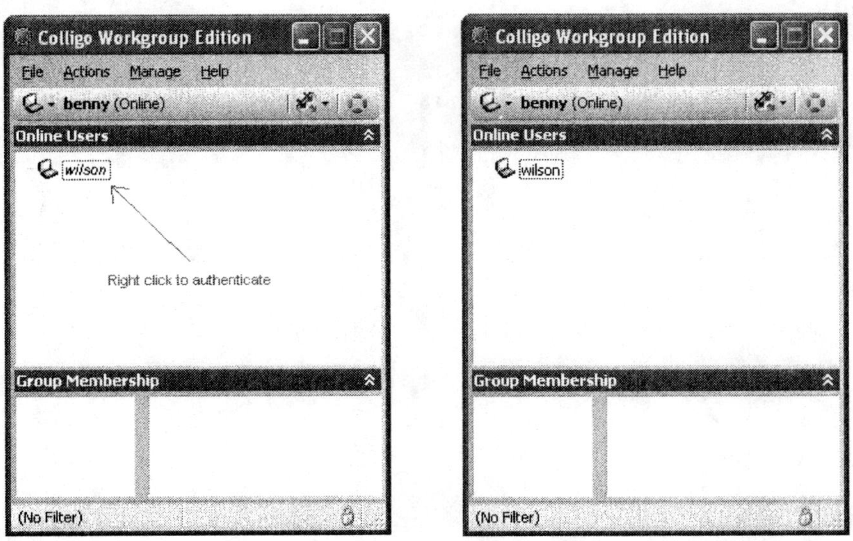

a. Remote user has no network access *b. Access to network has been granted*

Figure 6. Initiating file transfer

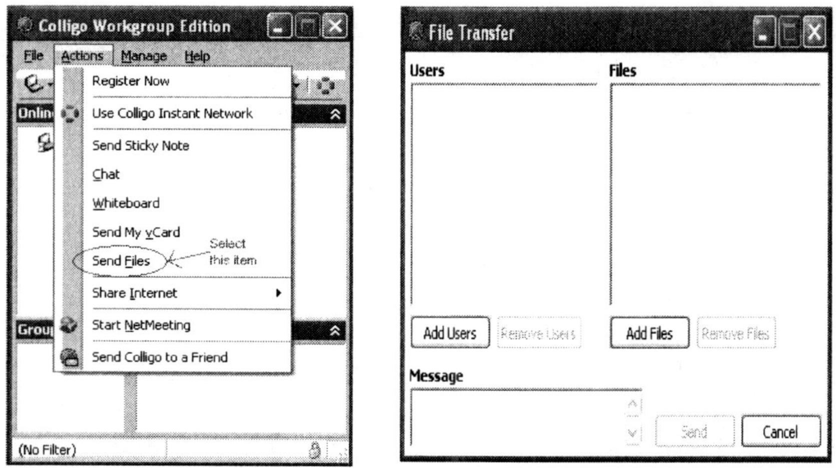

a. Action menu *b. File transfer window*

Figure 7. Completing file transfer setup

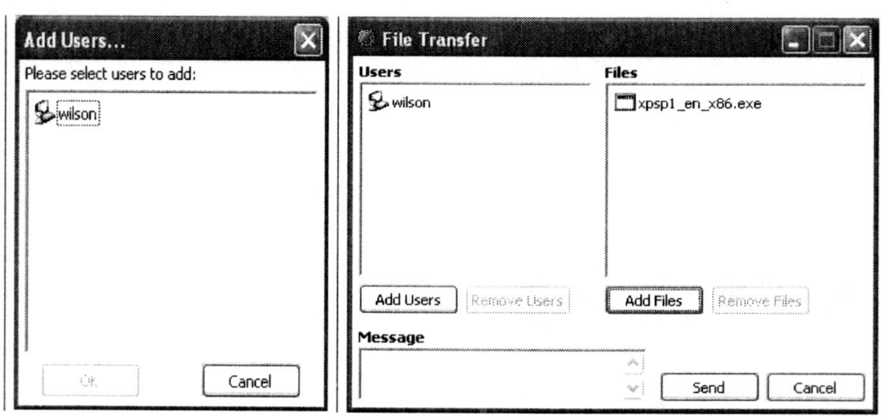

a. Selecting recipient *b. Completed file transfer setup*

BSS and ESS Setup

In the infrastructure mode, the wireless network operates under the coordination of an AP. The AP itself is connected to a wired network such as a LAN. In a simple home network, the AP can be directly connected to a router, which provides access to the Internet (see Figure 8). Two stations such as Host 1 and Host 2 communicate through an AP to access the Internet.

The AP is connected to a wired network even when access to the Internet is not required. In our

Figure 8. ESS for home use

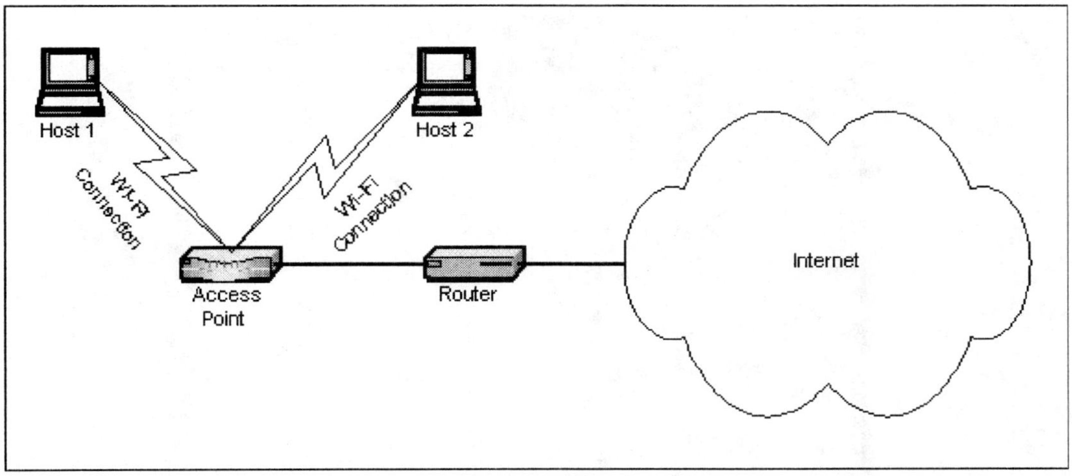

Figure 9. Simplified ESS

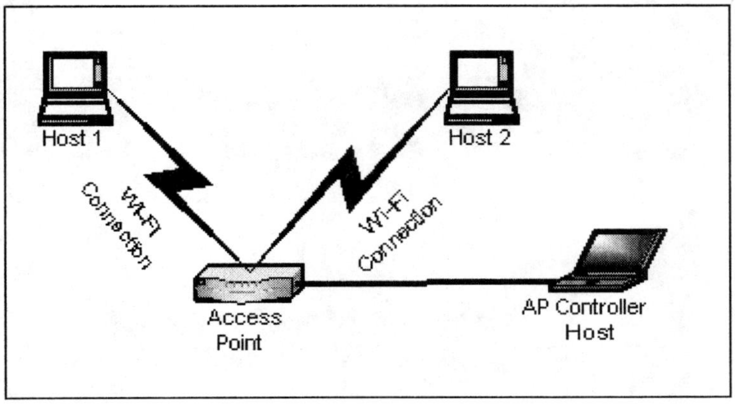

case, no access to the Internet is required. The wired network is simulated by a third host called AP controller. The AP controller is connected to the AP with a crossover Ethernet cable, as shown in Figure 9.

The AP must be configured in such a way that allows a Colligo network to be established. Configuring the AP can only be completed from the AP controller host using either the AP manager or the Web browser as discussed. When the AP is active and connected to the controller host, launching the AP manager will show the summary tab (see Figure 10). The screen capture shows the parameters set to run a Colligo wireless network in infrastructure mode.

Some of the parameters shown in the summary tab are static; whereas, others are variables that need to be set with correct values.

- **AP name** is the name assigned to the particular AP. In larger networks several APs may be used, which makes it necessary to assign unique names to each to avoid naming conflict.
- **SSID** stands for service set ID, which is a unique name for the service set. The Colligo

Figure 10. AP manager summary tab

Figure 11. AP setting tab

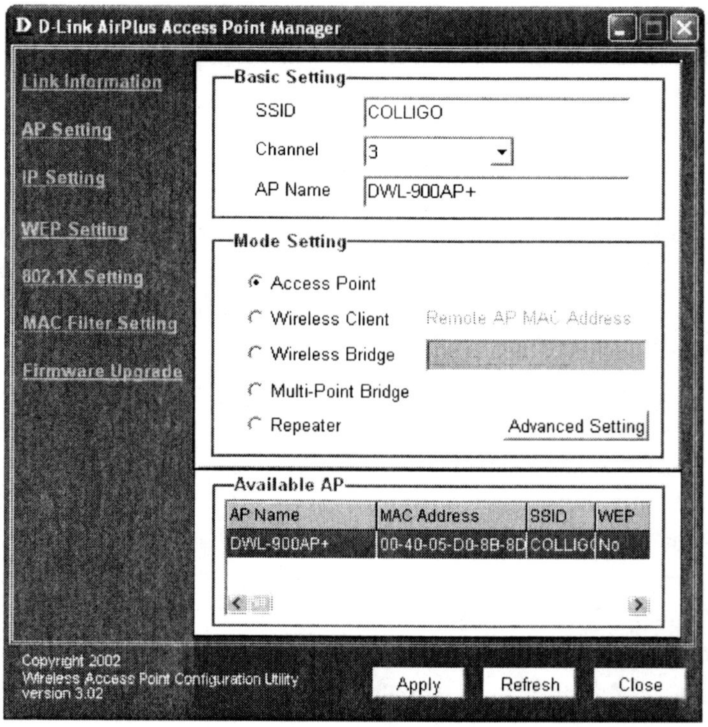

software works correctly if the SSID is set to *COLLIGO*, as shown Figure 11. Setting the AP's SSID to anything other than COLLIGO will prevent the Colligo wireless network from establishing.

- **IP address** in this context is the address of the AP for the wired connection to the controller host. The default factory value is 192.168.0.50, which indicates that the AP is designed for the use in small networks only. The controller host must use a compatible IP address to communicate with the AP through the Ethernet cable.
- **MAC address** displayed in the summary tab is the one used in the wireless connection. This value is permanently assigned to the AP and cannot be changed.
- **Channel** needs to be set to a number that does not conflict with another nearby AP. Since there is no other AP within range, an arbitrary channel number can be used.
- **WEP security** refers to the Wired Equivalent Privacy security scheme. When turned on, all devices in the network must use the same encryption key. For simplicity, this feature is turned off.

The most important settings are configured in the AP Setting tab. Figure 11 shows the configuration already set for a Colligo network in infrastructure mode.

The Mode Setting selector defines what role the AP device will assume in its operation. The device cannot perform on more that one mode at a time.

1. **Access point**: The default operation mode of the device, which creates a wireless LAN on infrastructure mode.
2. **Wireless client**: The AP acts as an adapter, which transforms an 802.3 Ethernet device into a 802.11b wireless client.
3. **Wireless bridge**: Used to utilize two APs to connect two wired LANs with wireless medium.
4. **Multi-point bridge**: An extension of the wireless bridge, where multiple APs are used to connect more LANs.
5. **Repeater**: The device is used to extend the range of the network when the wireless hosts are out of range of the actual LAN-connected AP.

In our experiment we set up the mode as Access Point. In practice, setting the mode to other than Access Point always results in the network failure. Once the AP is configured, the wireless infrastructure network is ready to operate, and the wireless hosts can join the network. The Colligo software automatically sets the SSID for each host to COLLIGO every time it is started. To join the ESS, a host only needs to launch Colligo and wait for the connection to establish automatically. In contrast to the procedure with the BSS, the Create Instant Network button should not be clicked. The user name of other hosts will appear on the main Colligo window list a few minutes after a host enters the ESS coverage. The routine for authenticating other users, sending files, and other actions is identical to that in ad hoc mode.

EXPERIMENT RESULTS

A text file of size 137 MB was transferred from Host 1 to Host 2 using the Send Files feature of the Colligo Workgroup Edition 3.2, which allowed us to obtain the file transmission time (Colligo, 2005). Different host formations were used to examine the effects of distance on the throughput performance. Tables 4 and 5 summarize the experiment results.

Table 4. File transfer time in ad hoc mode

Distance (m) O	bstacle	Transmission time	Throughput (Mbps)
1	0	3m50s	4.66
1	1	4m00s	4.46
16	0	3m57s	4.52

Table 5. File transfer time in infrastructure mode

Distance (m) H1-H2	Distance (m) H1-AP	Distance (m) H2-AP	Transmission time	Throughput (Mbps)
1	1	2	10m45s 1	.66
1	2	2	9m50s	1.82
1	1	1	10m20s 1	.73
2	1	1	15m10s 1	.18
32	12	20	9m33s	1.87
17	12	12	10m30s 1	.70

ANALYSIS AND INTERPRETATION

As seen in Table 4, a 4.5 Mbps throughput rate was achieved in ad hoc mode although the bandwidth was 11 Mbps. The poor performance was due to the retransmission of a large number of data packets that had been lost. Conducting the experiment within a room may also have contributed to the low throughput. At 2.4 GHz, the radio wavelength is only 12.5cm. At such a short length, signals echoed by various surfaces within the room will reach the receiver in various phases. The result is that at any given time, signals with varying phases and amplitudes will either reinforce or cancel each other as they reach the receiver's antenna. The original signal is therefore received in a distorted form. This problem is called multipath interference and is common in mobile communication (Stallings, 2002).

Increased distance and the presence of obstacles also degrade network performance noticeably. This effect is indeed unavoidable since the radio signals used have very low power. Overcoming obstacle and distance problems by increasing the power is not really an option since it will also increase the chance of interfering with other wireless networks nearby.

A drastic performance drop is observed when the network operates in infrastructure mode. On average the throughput in infrastructure mode is only around 40% that of ad hoc mode. Since the network uses PCF to synchronize the communications, the network is more efficient in using the available bandwidth by the use of fewer control packets. The poor throughput can therefore be explained by the mechanism by which data packets are transmitted from sender to the AP and then from the AP to the receiver. It is very likely that the AP uses a store-and-forward scheme, which results in the channel being occupied for twice as long.

On the other hand, the infrastructure mode makes it possible for wireless hosts to communicate when they are out of range from each other as long as both are within the AP's range. This implies that more flexibility and mobility are achieved at the expense of lower throughput.

EVALUATION BY STUDENT FEEDBACK

The Wi-Fi projects were offered to two graduate students as part of their summative assessment towards Net-centric Computing, a postgraduate course. The project learning outcomes include: (1) setting up and testing Wi-Fi projects; (2) demonstration of working prototype to project supervisors; (3) development of teaching resources for classroom use; (4) written report containing summary of findings and reflective statements; and (5) oral presentation to the staff and students.

The Wi-Fi projects were completed successfully, and students indicated that they had learned a great deal about Wi-Fi fundamentals in completing hands-on learning activities included in the Wi-Fi projects. This is evident from the students' reflective statements, as follows:

Student 1:

The Wi-Fi project has helped me immensely to develop a sound knowledge of Wi-Fi technology. During the practical project, I discovered that there are various issues with regard to the deployment of Wi-Fi technology. The level of exposure and knowledge that I gained from the hands-on experiment, I would have never achieved from just a theoretical mode of learning.

There are many positives that can be drawn from a hands-on project like the Wi-Fi propagation measurement. A project such as this helps you to understand and appreciate technology better, which may not (be) possible from a mere theoretical form of learning. I would strongly recommend hands-on projects to others who are interested to gain in-depth knowledge and experience of the technology.

Student 2:

By taking up the hands-on project, I realised that with a correct approach, the subject of networking in general and Wi-Fi in particular is no more difficult than other subjects in computing science. By understanding the basic issues of the technology, one can make sense on why the designers built the devices the way they did. Further, it challenges us to think about how various aspects of the technology can be improved.

With an engineering background, I never believe in any new technology unless I see it actually working. The project convinced me that the Wi-Fi technology does work, and any person willing to invest time studying and experimenting can learn to set up his or her own Wi-Fi network. Also the experience enriches us with another skill that may be important in our career later.

I would recommend similar projects to anyone interested in practical experiments and is not shy of technical challenges. They are also suitable for veterans of theoretical investigations who want to enrich their skill and experience with practical ones.

CONCLUSION AND FUTURE WORK

Through a series of experiments and measurements made at various locations of the AUT's WY office building, we gained an insight into the performance of Wi-Fi links in an office environment. This project involves both literature review and practical investigation regarding the IEEE 802.11b protocol known as Wi-Fi. A number of conclusions can be drawn from the findings of this study.

The use of electromagnetic waves as the medium instead of cables presents many technical challenges. To begin with, the available radio band is limited, and most ranges are licensed by governments across the world. This restriction forces various wireless devices to crowd into the same unlicensed bands. Also problems inherent in radio communication, such as noise, interference, and security issues, also affect wireless networks. The implication is that, despite its growing importance in recent years, the wireless network is not going to replace the wired network. It is most likely that both wired and wireless networks will coexist in the future.

The experiments reveal that data transfer rate through the wireless medium is much lower than the wired network. The wireless connection is also fragile, which necessitates that the whole transmission process be repeated whenever the connection drops during file transfer session. Because of these limitations, wireless networks at present serve mainly as a connectivity solution rather than as a performance solution. This may change in the future, however, as new wireless technologies supporting quality of service (QoS) are also being developed.

For Wi-Fi networks, it is found that the ad hoc mode provides better throughput in a low-populated network. The same network operating in infrastructure mode provides only about half the throughput of the ad hoc network. It is very likely that the AP uses a store-and-forward algorithm in delivering the data packets, which results in the drastic performance drop. The AP, however, is indispensable when the stations are out of range of each other. The technical difficulties encountered during the experiment suggest that Wi-Fi technology is not yet mature. This is indicated by the complexity of the setting-up procedure and the incompatibility that is common between Wi-Fi devices.

As it grows, wireless technology will provide research opportunities in several areas. Future research relevant to the scope of this project will mainly involve bandwidth increase and optimization, which are aimed at throughput improvement. At present the IEEE 802.11b system has many limitations since it provides services on "best effort" basis. The development of new wireless standards providing QoS is the most effective way of achieving satisfactory network performance (Prasad & Prasad, 2002).

In anticipation of the increased available bandwidth, various network-based business and multimedia applications are also being developed. Prasad and Prasad discuss applications such as teleconferencing, telesurveillance, and video-on-demand operating on wireless network backbones. The required bandwidths for delivering the data in various presentation formats are also provided in their discussion.

SUMMARY

The IEEE 802.11 wireless LAN standard has been gaining popularity worldwide in recent years. The IEEE 802.11b, commonly referred to as wireless fidelity (Wi-Fi), is by far the most successful commercially. There are other standards in the IEEE 802.11 family being developed to overcome the limitations of IEEE 802.11b.

This chapter focused on the teaching and learning aspects of Wi-Fi networking using limited hardware resources. A number of hands-on learning activities were reported, including setting up both ad hoc and infrastructure networks. We found that the throughput performance of Wi-Fi network under ad hoc mode is slightly better than the infrastructure mode. This is mainly due to the store-and-forward mechanism used by the access point.

KEY TERMS AND DEFINITIONS

Ad hoc network: A class of wireless LAN where nodes communicate without wireless access points. A wireless network operating in ad hoc mode is also called an independent basic service set (IBSS).

AP: Or access point. A device which acts as a bridge between wireless LANs and a wired backbone network. An AP coordinates communication among the mobile stations.

BSS: Or basic service set. A wireless network consisting of mobile stations connected through an access point and any number of fixed stations.

Colligo software: A commercial software package that allows users to run collaborative applications such as interactive chat, unidirectional message delivery, virtual whiteboard, and file transfer.

DCF: Or distributed coordination function. DCF is the main channel access method used in the IEEE 802.11 protocol, which uses the Carrier Sense Multiple access with Collision Avoidance (CSMA/CA) protocol.

ESS: Or extended service set. A set of BSSs connected to a wired network infrastructure.

IBSS: Or independent basic service set. A wireless LAN configuration without access points. An IBSS is also referred to as an ad hoc mode wireless network.

IEEE 802.11: A family of wireless LAN standards.

Infrastructure network: A class of wireless network in which mobile stations are connected to the wired backbone network through wireless access points.

PCF: Or point coordination function. An optional component of the MAC layer protocol which provides a centralized mechanism for packet collision avoidance during transmission via radio waves.

SSID: Or service set ID. A unique name that must be assigned to a service set before the wireless network can operate.

Wi-Fi: Or wireless fidelity. A trade or commercial name for wireless networking equipment using IEEE 802.11b standard.

REVIEW QUESTIONS

1. Discuss the main differences between wireless PAN and wireless LAN.
2. Describe the main limitations of current Wi-Fi technologies.
3. Explain why Wi-Fi equipment operates at 2.4GHz in most countries.
4. What are the main challenges in increasing bandwidth in wireless networking?
5. Describe the architecture of a typical Wi-Fi network.
6. Discuss the key characteristics of IEEE 802.11b standard.
7. Explain the significance of SSID in setting up a Wi-Fi network.
8. Can a Wi-Fi network simultaneously operate on both ad hoc and infrastructure modes? Provide sufficient argument or evidence to support your answer.

REFERENCES

Anonymous. (2005). *Introduction to wireless LAN and IEEE 802.11.*

Colligo. (2005). Colligo Networks, Inc. — Workgroup Edition, Colligo Inc. 2005 [online]. Retrieved from http://www.colligo.com

Ferro, E., & Potorti, F. (2005). Bluetooth and Wi-Fi wireless protocols: A survey and a comparison. *IEEE Wireless Communications, 12*(1), 12-26.

Golmie, N., Van Dyck, R., et al. (2003). Interference evaluation of Bluetooth and IEEE 802.11 systems. *Wireless Networks, 9*(3), 201-211.

Howard, D. (2002). It's a Wi-Fi world. *netWorker, 6*(3), 26-30.

Kaczman, J. (2002). Wi-Fi hotspot networks sprout like mushrooms. *IEEE Spectrum, 39*(9), 18-20.

Ophir, L., Bitran, Y., et al. (2004). Wi-Fi (IEEE 802.11) and Bluetooth coexistence: Issues and solutions. In *15th IEEE International Symposium on Personal, Indoor and Mobile Radio Communications, PIMRC 2004.*

Prasad, N., & Prasad, A. (2002). *WLAN systems and wireless IP for next generation communications.* Boston: Artech House.

Proxim. (1998, March). *What is a wireless LAN?* (White Paper No. 7680). Mountain View, CA.

Stallings, W. (2002). *Wireless communications and networks*. NJ: Prentice Hall.

Vaughan-Nichols, S. (2003). The challenge of Wi-Fi roaming. *Computer, 36*(7), 17-19.

Vaxevanakis, K., Zahariadis, T., et al. (2003). A review on wireless home network technologies. *ACM SIGMOBILE Mobile Computing and Communications Review*, *7*(2), 59-68.

Youssef, M. A., Vasan, A., et al. (2002). Specification and analysis of the DCF and PCF protocols in the 802.11 standard using systems of communicating machines. In *Proceedings of the 10th IEEE International Conference on Network Protocols*.

ENDNOTE

1 There is no line of sight between H1 and H2.

This work was previously published in Tools for Teaching Computer Networking and Hardware Concepts, edited by N. I. Sarker, pp. 154-178, copyright 2006 by Information Science Publishing (an imprint of IGI Global).

Chapter IV
Mobile Advertising: A European Perspective

Tawfik Jelassi
Ecole Nationale des Ponts et Chaussées, France

Albrecht Enders
Friedrich-Alexander-Universität Erlangen-Nürnberg, Germany

ABSTRACT

This chapter is based on research conducted in cooperation with 12Snap, the leading European mobile marketing company, which has implemented large-scale mobile advertising campaigns with companies such as McDonald's, Nestlé, Microsoft, Coca-Cola, Adidas, and Sony. To set the overall stage, we first discuss the advantages and disadvantages of the mobile phone in comparison to other marketing media. Then we propose a framework of different types of advertising campaigns that can be supported through the usage of mobile devices. These campaign types include (1) mobile push campaigns, (2) mobile pull campaigns, and (3) mobile dialogue campaigns. Building on this framework, we analyze different campaigns that 12Snap implemented for different consumer goods and media companies. Drawing from these experiences we then discuss a number of key management issues that need to be considered when implementing mobile marketing campaigns. They include the following themes: (1) the choice of campaign type, (2) the design of a campaign, (3) the targeting of the youth market, and (4) the combination of different media types to create integrated campaigns.

INTRODUCTION

The market for mobile phones has expanded rapidly during the past decade and continues to grow quickly. In some European countries such as Finland, Sweden, Norway, and Italy, the mobile phone has reached almost ubiquitous penetration with levels of 80% and higher (*Economist*, 2001). In Germany, mobile phones are more widely used than fixed-line connections (Brechtel, 2002). In addition to voice communications, German users send out 2.2 billion text messages through their mobile phone every month (Brinkhaus, 2002).

The fast spread of mobile phones has created immense profit expectations in the telecommunications industry. Telecommunication companies in many countries have invested large sums of money into acquiring third-generation licenses and building the necessary infrastructure. Yet, as it turns out, it is more difficult to generate revenues than initially anticipated.

In addition to call charges, there are three main revenue sources in mobile communications: (1) **transactions**, (2) **information**, and (3) **advertising**. Transactions are of high interest, yet as of now only to a limited extent, because of the small size of the screen and the clumsy usage of the keypad. With information services (such as weather forecasts or banking services) the crucial issue is the user's willingness to pay for these types of services.

Does mobile advertising have the potential to be a significant source of revenue in the future? First studies on this new advertising medium indicate that mobile advertisement campaigns can be very successful, generating response rates as high as 40%, compared with the 3% response rate generally expected for direct mail and less than 1% for Internet banner ads (Borzo, 2002).

Because of the novelty of the technology, using mobile phones for advertising campaigns presents some challenging questions for marketing departments:

- What are the strategic advantages of the mobile phone in comparison to other advertising media?
- What campaign types can leverage these characteristics?
- What critical issues need to be considered when launching a mobile advertising campaign?

In the remainder of this chapter, we discuss these questions drawing on field research conducted in cooperation with the German mobile marketing company 12Snap.

ADVERTISING THROUGH MOBILE PHONES

With the increasing number of media types, it has become more and more difficult for marketing managers to find appropriate strategies to target potential customers with their messages. First, while it was possible in the past to capture a large segment of society by placing advertisements with the main TV networks, the rise of private channels has led to a high degree of fragmentation, thereby complicating access to consumers. Similar fragmentation can be observed with other mass-media types such as print or radio. As a result, getting time and attention from their audience has turned into a major challenge for advertisers (Davenport & Beck, 2000).

Second, different media types require different approaches because of differences regarding their reach or richness. **Reach** is a function of how easily customers, or in this case, participants in advertising campaigns, can be contacted through a given medium. **Richness**, on the other hand, is defined by (1) bandwidth, that is, the amount of information that can be moved from sender to receiver in a given time, (2) the degree of individual customization of the information, and (3) interactivity, that is, the possibility to communicate bidirectionally (Evans & Wurster, 1997). The communication of rich marketing information, that is, information that ranks high on all three aspects, has traditionally required physical proximity to customers and/or channels specifically dedicated to transmitting the information (see Figure 1).

How does the mobile phone fare within the richness versus reach framework? It can serve as a powerful platform to get in touch with end consumers because it simultaneously provides expanded reach and a number of richness advantages vis-à-vis most other media types:

- **Ubiquitous Access:** Mobile phone users always have their phone with them and

Figure 1. The trade-off between richness and reach in advertising (adapted from Evans & Wurster, 1997)

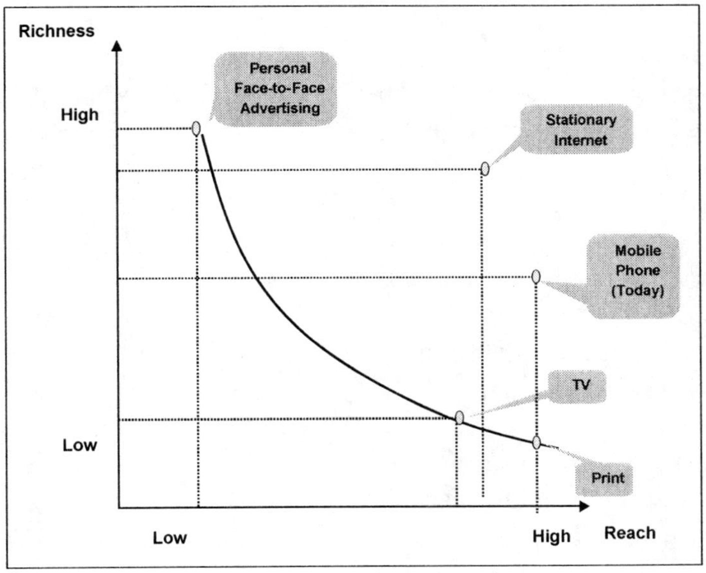

turned on at almost all times (Balasubramanian, Peterson, & Jarvenpaa, 2002; Magura, 2003). This is especially true for teenagers and young users who use the mobile phone to stay in touch with their peers—primarily through SMS (Bughin & Lind, 2001). Ubiquitous access becomes especially important in places like buses, trains and subways, airport lounges, and so forth. The time that people spend traveling is prime time for marketing since it presents a time when people are not occupied with other activities and are thus receptive to other kinds of entertainment. A study by the Boston Consulting Group (2000) found that among private users, the categories "having fun" (71%) and "killing time" (55%) belong to the main motivators for using mobile phones—ranking only behind "keeping in touch with friends" (85%).

- **Detailed user information:** While traditional marketing campaigns only have access to very limited customer information, mobile campaigns can draw on extensive and individual information about each user (such as age, sex, usage profile, etc.). This information helps to launch highly targeted campaigns for specific products and services based on individual preferences of the user.
- **Integrated response channel:** The mobile phone presents the opportunity to interact directly with the user and elicit responses through the same medium. This has two advantages. First, it provides the opportunity for rich interaction. The interactivity and ubiquity of the mobile phone opens up the possibility to turn existing traditional media formats (such as the TV, radio, print, or packaging) interactive. For instance, companies can contact consumers via TV and then subsequently, stay in touch with each one of them through the mobile phone. Second, the integrated response channel also allows mobile marketing companies to measure precisely the impact of their campaigns and then to adapt their strategies accord-

ingly—something that is much more difficult to do with traditional marketing media. For instance, a customer buys a product—with a mobile phone number on the packaging—at a retailer, and as s/he exits the store, s/he completes a quick survey of the shopping experience, which is then transmitted immediately to corporate headquarters. This not only allows the consumer the satisfaction of immediate feedback if they had a positive or negative experience, but it also allows the company to measure quality control in an extremely timely and cost-effective manner (Carat Interactive, 2002).

- **Personal channel:** Unlike other advertising media such as TV, radio, or billboards, the mobile phone belongs to only one person. Therefore, it receives much more attention and, if handled properly (see risks below), can be much more powerful than other, less personal media channels. John Farmer, a cofounder of the SMS application and service provider Carbon Partners, points out that the personal character of the mobile phone is especially important to teenagers (Haig, 2001): "The mobile phone presents the teenage market with the distinct opportunity to take control of their own communications, free from the previous limitations of the home phone or computer, which were more closely monitored by parents."

Brian Levin, CEO of Mobliss, a U.S. wireless marketing firm, sums up the advantages (Stone, 2001): "When you have a little time to spare, such as in the airport or at the bus stop—then you want to be engaged or entertained. Once you are there, the proximity of this device [the mobile phone] to your face, the intimacy there, is very powerful both in terms of direct response and in terms of branding." At the same time, however, the mobile phone also presents shortcomings and risk factors:

- **Limited media format:** Mobile phones today still have to cope with a very limited set of visual and audio capabilities. In second-generation (2G) phones, screens are typically small, have only low resolution, and are typically not in color. Sound effects are also limited due to the small speakers, and text messages cannot be longer than 160 characters. The challenge is then to ensure at this stage that consumers do not expect an identical experience to what they receive through other devices such as the TV or PC (Carat Interactive, 2002).
- **Private sphere:** The fact that mobile phones belong to only one person does not only present an opportunity but also a challenge for mobile advertisers. Unlike the TV or Internet, the mobile phone is a very personal device to which only family, friends, coworkers, and a selected few others will gain access. Thus, "spamming" is considered much more intrusive than in other media formats (Carat Interactive, 2002).

DEVELOPING EFFECTIVE MOBILE ADVERTISING CAMPAIGNS

One of the main challenges and opportunities for mobile advertising companies is the personal nature of mobile phones. Advertising campaigns over mobile phones are very sensitive and companies that engage in this type of marketing need to be careful not to offend users. Will Harris, global marketing director for Genie, British Telecom's mobile Internet service, emphasizes (Pesola, 2001): "Sending unsolicited messages is tantamount to brand suicide. Our business is entirely dependent on the goodwill of our customers."

The mobile advertising industry is trying to protect mobile phone users by establishing guidelines for responsible advertising. The main feature of these guidelines is consent, that is, consumers agree or opt-in to receive the advertisements. In addition, they must have a clear understanding of what their personal information is being used for, and if they wish, be removed from the advertiser's databases.

As a result, mobile advertisers have to find ways to entice customers to opt into their campaigns. Cyriac Roeding, 12Snap's marketing director, explains why many companies have difficulties attracting mobile phone users (Pesola, 2001): "A lot of companies make the mistake of coming to this from a technological angle, rather than thinking about what the consumer wants. If advertising is entertaining, if it engages the emotions, it will be accepted."

Although mobile advertising is a relatively recent phenomenon, a number of large corporations, including McDonald's, 20th Century Fox, and Sony, are using this medium in their marketing mix, especially to target young customers. These campaigns differ according to the degree of active involvement of advertiser and recipient (see Figure 2). Level of activity refers here to the involvement both advertiser and consumer show throughout the course of an advertising campaign. Traditional campaigns, which still present the most prominent advertisement type, display low levels of activity on both the advertiser's and the consumer's side since they consist of noninteractive, one-way advertisements in the form of TV spots, radio or print ads, or posters. Mobile campaigns, on the other hand, show high levels of activity either on the side of the advertiser, the consumer, or both. High level of activity on the side of the advertising company implies that the consumer is approached proactively, whereas a high level of activity on the side of the consumer implies that s/he reacts actively to an advertisement or a newspaper ad, for instance, by soliciting further information via the mobile phone.

Through the built-in response channel, mobile phones are suitable both for push and pull campaigns. According to the mobile advertising framework, mobile campaigns can be categorized as follows:

Figure 2. Mobile advertising framework

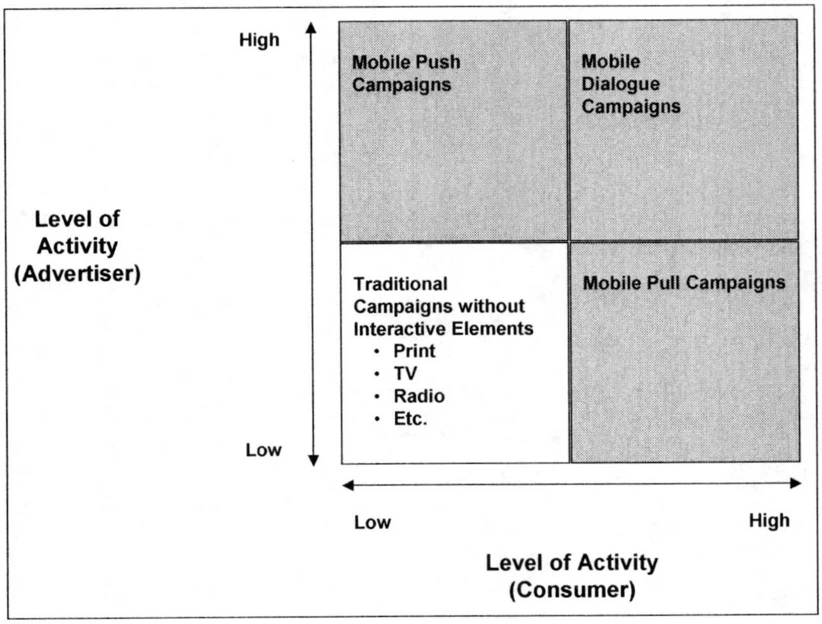

- **Mobile Push Campaigns:** Push advertising is categorized as messages that are proactively sent out to wireless users. Companies use databases with existing customer profiles, which can be proprietary or purchased externally, to address their target groups (Carat Interactive, 2002). Because of the sensitivity of the mobile phone, it is important to ensure that all members of the database have agreed beforehand (i.e., given their opt-in) to receive mobile advertising. In addition, for the success of a campaign, it is essential to ensure that the selected target group from the database is interested in the specific advertising, which requires extensive profiling of the database (Pearse, 2002). Doing so avoids the alienation of uninterested users, while at the same time maximizing the impact of the advertising budget on those targeted.
- **Mobile Pull Campaigns:** Applying a pull approach, advertisers use their traditional marketing media mix, such as TV, radio, print, or packaging, to promote an interactive mobile campaign. For instance, a telephone number on a French fries box might invite a customer to participate in a raffle for free food through his/her mobile phone. By calling, the consumer gives the "opt-in"—an explicit consent to the campaign—and can then participate.
- **Mobile Dialogue Campaigns:** Dialogue campaigns differ from the above-mentioned campaign types in their duration and the intensity of interaction between advertiser and customer. While simple push and pull campaigns tend to last only 2 to 4 weeks and center around one single theme such as a raffle or a game, dialogue campaigns last several months and include various different themes that build on one another. Their goal is to establish a long-lasting relationship with consumers so as to generate extensive insights into consumers' preferences. A mobile horoscope service, for instance, allows the advertiser to capture the

	Push Campaigns	Pull Campaigns	Dialogue Campaigns
Set-up	Targeted SMS to user in existing database • Company-owned database • Database from external provider	Advertisements for mobile campaigns through other media types, e.g. • Flyer or "on-pack" ad • TV ad	Continuous interaction between advertiser and user
Opt-In	Need to have explicit "opt-in" prior to sending out SMS	Users "opt-in" by calling the phone number	Generation of "Opt-in" either through pull or push
Scope	Single theme • Game • Raffle • Etc.	Single theme • Game • Raffle • Etc.	Multiple themes during the course of a campaign • Different games • Greetings
Length	Short duration (2-4 weeks)	Short duration (2-4 weeks)	Extended duration (several months)
Implementation	• Wella Hair Fashion • Sony • 20th Century Fox • Adidas	• Nestlé KitKat Chunky • Toyota	• McDonald's

birthday of the consumer, which can then be used for sending out personal birthday greetings later on. The in-depth consumer information serves then to distribute mobile coupons—for instance, a free candy bar as a birthday present—to introduce new products or to do market research in a very targeted fashion.

Building on this framework, it is now possible to categorize the actual implementations of mobile advertising campaigns (see Figure 3).

Push Campaigns

Wella, a German manufacturer of hair-care products, developed a push campaign, which featured a "mobile kiss." Wella sent an SMS to members of an externally acquired database offering them to send a kiss message to their friends, who received a voice file with a kiss sound. This was followed by an SMS revealing who had sent the kiss and also providing details on how to return the kiss or send it to someone else. The maximum number of kisses sent by one person was 160. Other components of this mobile campaign included an SMS quiz and a free kissing-lips logo for the mobile phone.

Sony launched a push campaign in the UK which integrated e-mail, Internet, and traditional print media to promote a new PC-compatible MiniDisc Hi-Fi system. The campaign, which was based on MiniDisc Island—an online, interactive adventure playground—had the objective of driving large numbers of users to the Web site. Through an initial SMS 100,000 participants from a database of 14 million permission-based, profiled users were selected and invited to enter a competition to win a stereo system and the PC Link product. Interested participants replied via SMS and were mailed a winning number, with which they could then go to the Web site to see if they had won. Throughout the course of this campaign, 18% of those originally contacted responded to the initial SMS. When contacted with the winning number, over 9% logged on to explore the online adventure game and see if their numbers had come up.

The movie studio **20th Century Fox** launched a push campaign in partnership with the mobile phone operator Vodafone to advertise the UK release of *Planet of the Apes*—a post- apocalyptic movie where apes rule over humans who struggle to survive. The campaign, which started 2 weeks prior to the UK release of the movie, targeted the 2 million 16- to 24-year-old Vodafone customers. They received messages, which invited them to survive a variety of challenging interactive voice response and text games—with names such Ape S-cape and Ape@tak, where callers are asked questions relating to the movie and have to shoot down apes using the keypad when they hear a roar of an ape.

The sports article manufacturer **Adidas** used interactive betting game for the Soccer World Cup 2002 to promote its products in Germany. Users bet on games and received immediate notification after the game about how they did, how they ranked overall within the betting competition, and if they had won a prize such as a shirt from the soccer idol Zinedine Zidane or a personal meeting with the German national soccer team.

Pull Campaigns

Nestlé used a mobile pull campaign to promote the KitKatChunky chocolate bar in Germany. The campaign, which lasted 2 weeks, complemented the overall marketing presence consisting of TV and radio spots and the Web site www.chunky.de. The campaign worked as follows: an SMS offered community members the opportunity to win a 1-year supply of KitKatChunky if they called a specified number. Then, callers were shown the face of a taxi driver who also appears in a KitKatChunky TV commercial, and two other new characters on their screen who were presenting riddles to them. An automated voice then explained: "Each of the three protagonists names a number which makes him shut up. Once you have discovered the number, push the appropriate button on your mobile and a KitKatChunky is stuck in his mouth and you go on to the next round." In the first round, 400,000 users were identified to receive a kick-off SMS at the beginning of the campaign. In the following rounds, only those players who had actively opted-in in the previous round received an SMS. In order to maximize the number of responses, users received alert messages the day before the ad's TV premiere and again 30 minutes before the TV show in which the questions were sent. This illustrates to what extent different media types—here the TV and the mobile phone—can be interlinked, using the respective strengths of each medium, to generate a seamless and entertaining marketing experience for the end user.

The car manufacturer **Toyota** also launched an interactive TV pull campaign during the Soccer World Cup 2002 that displayed a quiz question at the bottom of the TV screen. The question asked viewers to find the license plate number of the Toyota shown in the TV ad and to send this number in via SMS. Within seconds they received notification whether they had made it to the final drawing. In addition to having the chance to win a prize, all callers also received a Toyota ring tone for their mobile phone.

Dialogue Campaigns

Extensive dialogue campaigns are still a rarity because of the novelty of the mobile phone as an advertising medium. In Germany, **McDonald's** launched a mobile dialogue campaign with a focus on interactive mobile games and an evaluation of McDonald's products. The campaign targeted mobile phone users who were informed through in-store flyers placed in McDonald's restaurants. By activating the service (active "opting-in"), participants received automatic messages when music CDs or vouchers were raffled off. In addition, they also received SMS promotions of McDonald's products. The goal of the campaign was (1) to increase in-store traffic, (2) to build a McDonald's customer database of mobile phone numbers, and (3) to increase overall brand awareness. Following this initial pull activity, registered users continued to receive other services such as horoscopes, which in turn allowed McDonald's to capture users' birthdays and to send them personalized birthday greeting subsequently. An additional part of the campaign was a viral activity: McDonald's sent the Christmas greeting "Rockin' Rudi" to users who could then forward it to their friends. The recipient then listened to a taped version of the "Rockin' Rudi" song in combination with a short message from the sender and from McDonald's.

OUTLOOK AND MANAGEMENT ISSUES

There are plenty of opportunities in mobile advertising for companies that thoroughly understand how consumers can benefit from these types of services. The new technology will not be very useful, however, if companies simply use their existing advertising approaches and translate them to the mobile world without addressing the specific characteristics of this new medium (Nohria & Leestma, 2001).

The different innovative types of mobile advertising campaigns mentioned above offer many useful benefits—for instance, highly targeted advertising and interactivity—to those companies that want to add a mobile component to their advertising approach. They also illustrate the difficulties and challenges that are associated with this new approach. Therefore, before embarking on mobile advertising campaigns, managers need to carefully address the following questions.

Which Campaign Type Should We Employ?

For starting a mobile advertising campaign, there are two basic options: push or pull. A push campaign requires an extensive database of customers. Some companies such as telcos or retailers have built up these types of databases in the past through CRM efforts and can now tap into them. However, they always need to keep in mind the personal nature of the mobile phone when doing so. "Spamming" existing customers with unwanted SMS is a sure way to alienate them. Another option is to buy existing profiles from other companies. MTV, for instance, markets its permission-based database through an external mobile advertising company to other companies that want to target the attractive youth market. These companies benefit since they can tap into an extensively profiled, permission-based database of their target group while MTV generates additional revenues.

Setting up a pull campaign is not as sensitive regarding the opt-in, since consumers themselves decide whether they want to participate when they see the advertisement printed on a poster or watch it on TV. Here, the challenge is much more to create compelling advertisements that have the desired pull effect to entice consumers to call in and participate.

How Should We Design Attractive Mobile Advertising Campaigns?

The challenge for any mobile marketing company is to create enough interest within the target group to justify the required investment. Based on the campaigns we have analyzed, four key success factors need to be considered when launching a mobile advertising campaign (Brand & Bonjer, 2001):

- **Interactivity:** just like the Internet, the mobile phone allows advertisers to solicit immediate feedback when contacting recipients. Since the mobile phone is usually always turned on, the inherent interactivity of mobile phones should be integrated in mobile marketing campaigns where possible. The interaction can have many different facets: the number pad can be used to answer riddles or mental agility can be tested through reaction tests. A mobile marketing campaign that does not integrate interactivity would be the equivalent to the broadcasting of a slide show on TV. It would leave a main asset of the medium untapped.
- **Entertainment:** interaction is only fun for users if they find the advertisement exciting. Therefore, mobile campaigns need to combine advertising and entertainment in such a way that users are willing to lend their time to an advertisement. In this respect, the creation of mobile campaigns is similar to more traditional campaigns on TV, for instance. TV viewers watch advertisements mainly because they are entertaining. Ideally, they do not just watch them but they also talk about them to friends thereby creating a viral effect in which the message is passed on by people other than the original sender—as was the case in the Wella and McDonald's campaigns. Therefore, the inclusion of entertaining elements such as a game or a story ought to present an integral part of a mobile marketing campaign.
- **Emotion:** the inclusion of emotional elements—such as visual sequences or music clips in TV ads that aim beneath the conscious understanding of the viewer—has long presented a valuable marketing tool to subliminally reinforce the intended message with consumers. In mobile marketing, however, text, especially if shown on a small mobile phone display, can hardly carry this emotional dimension. Here, just like with TV advertising, it is necessary to leverage the admittedly limited resources

of the mobile phone to create "emotion." This can be achieved through the combination of voice and sound. For instance, music jingles such as a short sequence of the soundtrack of the movie *Titanic* can be used as the opening for a partner test or an activity aimed at single people. Again, it is not primarily the technology that drives the quality of any given campaign but instead the creative combination of different effects that ultimately determines its success.

- **Incentive:** the offering of incentives such as product samples increases the willingness of consumers to participate in interactive mobile games. The prospect of winning a prize is especially important due to the above-mentioned opt-in nature of mobile marketing campaigns, as it provides the potential participants with a direct and tangible incentive to participate in a mobile marketing campaign. However, although instant-win competitions are effective in driving volume, they are less suitable to generate a long-term relationship with consumers, since they do not offer incentive to return (Cowlett, 2002).

The overall goal of combining these four factors is to create a game, an image or a jingle that, despite the limitations of the small screen and tiny ring tone of the mobile phone, is so compelling that it is no longer seen as an ad, but takes on a value of its own.

How Should We Target the Difficult-to-Reach Youth Market?

Addressing the lucrative youth market gives marketers a perennial headache, since they do not only vary in their habits, interests, and attitudes and are swayed by rapidly changing fashion trends. They are also hard to pin down, since they do not primarily watch three or four TV stations anymore as was in the past. Instead, their media usage is fragmented between hundreds of TV stations, radio, magazines, newspapers, and the Internet. One thing is generally guaranteed, though—they almost certainly carry a mobile phone and consider SMS an intrinsic part of their lifestyle since it allows them to stay in touch with their peers in a cost-effective and entertaining way (Cowlett, 2002).

Mobile campaigns can effectively leverage the characteristics of this new youth market. Viral effects, used in the McDonald's and Wella campaigns, fulfill the desire to communicate with peers in a fun way. Teens enjoy quizzes or greeting cards they can pass on to friends, because this type of promotion focuses on using the mobile phone for what it was made to do—communicate with other people (Centaur Communications, 2002). In addition, viral elements help to expand significantly the group of recipients beyond the database of the company conducting the campaign and it increases the impact since marketing messages sent from a friend are, because of their personal nature, much more effective than those sent directly from the company itself (Haig, 2001; Kenny & Marshall, 2000). The communication from consumer to consumer helps to generate "buzz"—explosive self-generated demand—where people share their experiences with a product or a service amongst one another (Dye, 2000). At the same time, this approach helps to lower costs since users themselves target new consumers.

How Should We Combine the Mobile Phone with Other Media Types to Create Integrated Campaigns?

Because of its limitations regarding screen size, sound, and handling, the mobile phone is not suitable for stand-alone campaigns. Instead, it should be used to extend the presence of a company into an additional channel (Carat Interactive, 2002). Doing so, the mobile phone plays the role of the natural glue between other media types because of its ubiquitous nature: it is handy and turned on when watching TV, looking at a billboard on the subway, buying groceries at the supermarket, or listening to the radio. All the campaigns mentioned above make extensive usage of this cross-linking of different media types, leveraging the unique strengths of each. It is not only other media types that benefit from the integration of the mobile in multichannel advertising campaigns: tangible support mechanism from other media types that have been around for years—such as a flyer or an in-store promotion—give mobile campaigns higher legitimacy because they have a physical component (Enders & Jelassi, 2000).

From a market research perspective, the inclusion of mobile components in advertising campaigns has the added benefit that it allows to measure directly the effect of different advertising approaches. Take, for instance, a TV advertisement that is aired on different channels and broadcasting times, or a billboard advertisement placed in different locations that asks viewers to participate in an SMS contest. Based on the measurement of actual response rates in different channels or locations, it becomes possible to steer placement more effectively than via traditional indirect measurements.

REFERENCES

Balasubramanian, S., Peterson, R., & Jarvenpaa, S. (2002). Exploring the implications of m-commerce for markets and marketing. *Journal of the Academy of Marketing Sciences, 30*(4), 348–361.

Borzo, J. (2002). Advertisers begin dialing for dollars. *Asian Wall Street Journal*, February 18.

Boston Consulting Group. (2000). Mobile commerce—winning the on-air consumer. November.

Brand, A., & Bonjer, M. (2001, November). *12Snap: Mobiles Marketing im Kommunikations-Mix innovativer Kampagnenplanung* (White paper). Munich: 12Snap AG.

Brechtel, D. (2002). Bei Anruf Werbung. *Horizont, September 12*, 80–81.

Brinkhaus, G.B. (2002). Keine Massenmailings: Wie Mobile Marketing Funktioniert. *FAZ-online, September 29*. Retrieved from *www.faz.net*

Bughin, J., Lind, F. et. al. (2001). Mobile portals mobilize for scale. *McKinsey Quarterly, March*, 118–125.

Carat Interactive. (2002). *The future of wireless marketing* (White paper).

Centaur Communications. (2002). Good text guide. *In-Store Marketing, October 7*, 23–27.

Cowlett, M. (2002). Mobile marketing—"text messaging to build youth loyalty." *Marketing, October 31*, 29–34.

Davenport, T., & Beck, J. (2000). Getting the attention you need. *Harvard Business Review, September*, 118–125.

Dye, R. (2000). The buzz on buzz. *Harvard Business Review, November*, 139–144.

The Economist. (2001). The Internet untethered. *The Economist, October 13*, pp. 3–26.

Enders, A., & Jelassi, T. (2000). The converging business models of Internet and bricks-and-mortar retailers. *European Management Journal, 18*(5), 542–550.

Evans, P., & Wurster, W. (1997). Strategy and the new economics of information. *Harvard Business Review, September–October*, 71–82.

Haig, M. (2001). KIDS—talking to the teen generation. *Brand Strategy, December*.

Jelassi, T., & Enders, A. (2005). *Strategies for e-business: Creating value through electronic and mobile commerce*. Essex, UK: Financial Times/Prentice Hall.

Kenny, D., & Marshall, J. (2000). Contextual marketing: The real business of the Internet. *Harvard Business Review, November*, 119–124.

Magura, B. (2003). What hooks m-commerce customers? *MIT Sloan Management Review, Spring*, 9.

Nohria, N., & Leestma, M. (2001). A moving target: The mobile-commerce customer. *Sloan Management Review, 42*, 104.

Pearse, J. (2002). NMA wireless—mobile conversations. *New Media Age, October 31*, pp. 37–43.

Pesola, M. (2001). The novelty could quickly wear off. *Financial Times.com*, July 17.

Stone, A. (2001, January 3). Mobile marketing strategies Q & A. Retrieved from *www.mcommercetimes.com/Marketing/200*

ENDNOTE

- This chapter is based on a teaching case study that features the German wireless advertising company 12Snap (see Jelassi & Enders, 2005).

This work was previously published in Unwired Business: Cases in Mobile Business, edited by S. J. Barnes and E. Scornavacca, pp. 82-95, copyright 2006 by IRM Press (an imprint of IGI Global).

Section II
Development and Design Methodologies

Chapter V
Developing a Telecommunication Operation Support System (OSS): The Impact of a Change in Network Technology

James G. Williams
University of Pittsburgh, USA

Kai A. Olsen
Molde College and University of Bergen, Norway

EXECUTIVE SUMMARY

The Telecommunications Act of 1996 opened competition in the telecommunications market in the U.S. and forced the incumbent telecommunications companies to open both their physical and logical infrastructure for Competitive Local Exchange Carriers (CLECs). In this case study we focus on the problems that face a CLEC with regard to designing an information system and getting a back office system, called an Operations Support Systems (OSS), operational in a highly competitive, complex, fast-paced market in a compressed time frame when a change in a critical telecommunications network component, namely the central office switch, is made after 75% of the system implementation was completed. This case deals with the factors that led to this change in central office switches, its impact on the IT department, its impact on the company, and the alternatives considered by the IT department as possible solutions to the many problems created by this change.

ORGANIZATIONAL BACKGROUND

Starting in the 1970s, there have been many deregulation efforts in many sectors of the U.S. economy as well as internationally. The basic objectives have been to increase competition, improve service, and lower prices (Perez, 1994).

In the telecommunications sector, an abundance of new firms have emerged since the Telecommunications Act of 1996, both to provide new services such as data networks and wireless, but also to compete with established wire line telephone services. While deregulation opened the telecommunications sector for competition in these areas, many of the new services were made possible by the advent of new technologies: wireless services, broadband on a twisted wire pair (DSL), optical fiber, digital switchboards, the Internet and the Web standards. In many cases, the new entrants (CLECS) were the first to apply these newer technologies.

In the telecommunications sector, the Telecommunications Act of 1996 opened up competition for local voice and data services. The incumbents in the U.S., the Regional Bell Operating Companies (RBOC) called Incumbent Local Exchange Carriers (ILECs), were forced to lease infrastructure to the new entrants, namely, Competitive Local Exchange Carriers (CLECs). Many CLECs managed to get their business and associated networks installed and running in a remarkably short period of time. However, as Martin F. McDermott discusses in his book *CLEC* (McDermott, 2002), problems occurred primarily in other areas. One area that caused major problems was operations support systems (OSS) and its associated provisioning and billing related functions.

Thus, by 1999, there were political rulings, court rulings, and FCC orders that laid a foundation for competition in the local exchange (CLEC) telecommunications sector in the U.S. This was a go-ahead signal for many new companies. By 2000, there were more than 700 CLECs. Some of these were sales only companies (Total Resale) and owned no infrastructure but used the ILEC infrastructure to sell telecommunications services using different market plans and lower prices since the ILECS had to sell services to the CLECS at a discounted (wholesale) price. Other CLECs were facility based and developed a network and switching infrastructure; in many cases using new types of equipment and technologies. For example, they used DSL (digital subscriber line) to provide both data and phone services on the standard local loop (2-wire pair).

Broadstreet Communications, Inc., an entrant into the facilities based CLEC arena in February, 2000, was formed by eight individuals who had experience working for ILECS, CLECS, cable companies, or teaching and consulting for the telecommunications industry. The founders determined that there was a reasonable market for combined voice and data services for small and medium sized businesses over broadband facilities using DSL technology and formulated a business plan based on this technology as a foundation for the company. Small and medium sized businesses were defined as having between 1 and 100 employees. Based on the business plan, the founders were able to acquire 62 million dollars in venture capital from 3 different venture capitalist companies. In addition, Lucent Technologies provided $120 million dollars in financing for a total of $182 million dollars of available capital. The company was headquartered in an industrial park about 30 miles southeast of Pittsburgh, Pennsylvania and established its service area as the mid-Atlantic states of Pennsylvania, Maryland, and Virginia, as well as Washington, DC.

A major part of the business plan was the utilization of information technology to contain costs and provide a high level of service to internal users as well as customers. This was the ultimate goal of the information system design but with the need to remain within the boundaries of the business plan. The difficulties of building an information system that would integrate all aspects of the highly complex telecommunications industry are well known but the task becomes even more difficult when, after 9 months of system development on a 12 month completion schedule, a major change is made in the most critical component of the telecommunications network, namely the central office switch. The impact of this change in network components is the focus of this case study and includes the technological, organizational, managerial, industry, and economic issues that all interact in making system design decisions when a major change occurs in the environment that impacts many of the originally envisioned system requirements. This includes issues related to hardware, system software, application software, networking, scalability, reliability, buy vs. make decisions, requirements engineering, Flow through Provisioning, interfaces with the public telephone network (PSTN), reciprocal relationships with other telephone companies, and the difficulties associated with adopting new packet switched technologies for voice.

Products/Services Provided

Broadstreet Communications, Inc. was a telecommunications company offering voice and data services to small and medium sized businesses employing packet switched, broadband, digital subscriber line (DSL) technologies. This technology offered significant cost and service advantages over the traditional analog loops and leased lines traditionally used to provide voice and data services to a customer's premise.

BroadStreet provided an integrated suite of business communications services including high-speed data, local and long distance voice, voice messaging, e-mail, Internet, as well as Web and application hosting. Through the development of a next generation network, BroadStreet was among the first service providers to deliver voice and data solutions over an Internet Protocol (IP) network, and leverage digital subscriber line (DSL) technologies for last mile connectivity to customers. Local and long distance voice services included all the services provided by a commercial central office switch such as caller ID, call forwarding, conference calling, voice messaging, E911, 800 numbers, and so forth. Data services included Local Area Network (LAN) services, Internet access, e-mail, Web site hosting, and application hosting.

Management Structure

Broadstreet Communications had a fairly flat but hierarchical organizational structure as shown in Figure 1.

The board of directors was composed of representatives from the Venture Capitalist groups (3) and 2 members of Broadstreet Communications, namely, the CEO and president, plus one

Figure 1.

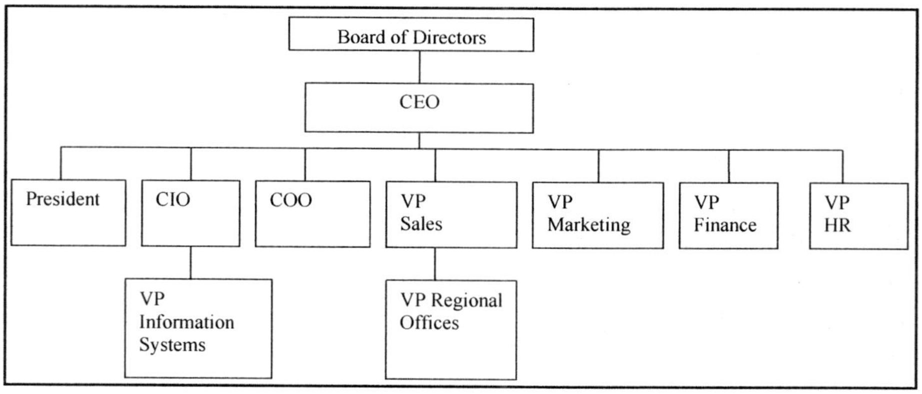

outside members agreed upon by those members. The CEO was also the Chairman of the Board of Directors. The two anomalies in the structure were that the CIO and the VP of Sales had junior VPs for information systems and regional sales offices based on the demands of the individuals who fulfilled those roles.

Financial Status

At startup, Broadstreet Communications had $62 million dollars of venture capital from three different venture capital firms and $120 million dollars of financing from Lucent Technologies. The original business plan called for Broadstreet to begin delivering services and realizing revenue after one year of development. But, it took Broadstreet approximately 18 months to become fully operational and during that time it operated on the venture capital and finance funding provided with no revenue from products or services. After 18 months of developing an infrastructure of personnel, sales offices, networks, and information systems; Broadstreet began to offer products and services to small and medium sized business customers. This began in the Pittsburgh, Pennsylvania area and then expanded to Baltimore, Maryland; Washington, DC; Richmond, Virginia; and Norfolk, Virginia. Within 7 months of offering services to customers, Broadstreet had approximately 1,400 customers with revenues of slightly over $1 million per month. Broadstreet was growing at approximately 20% per month. Based on covenants agreed to between Lucent Technologies and Broadstreet, as well as operating costs, the company needed to have revenues of approximately $2 million per month to cover costs. It was quite clear that had Broadstreet started offering services and realizing revenue 4 to 6 months earlier, the finance covenants and operating costs could have been met. Had the Back-office OSS system been operational 4 to 6 months earlier, Broadstreet would have survived the economic downturn that began in 2000, but the delay caused by the introduction of a new central office switch did not make this possible. It was late in 2001 when the "dot com" bust and the telecom sector's severe downturn caused Lucent Technologies to terminate the financing agreement based on the covenants, and one of the venture capitalist that suffered large losses in the "dot com" bust also decided to terminate their investment. This made operating impossible due the lack of resources and Broadstreet made a decision to close the business after approximately 2 ½ years. At that point in time Broadstreet had over 180 employees as well as a number of subcontractors who were dependent upon BroadStreet for their livelihood.

Strategic Planning

In early 1999, the VP of operations for Adelphia Business Solutions decided that the Telecom Act of 1996, the advent of DSL technologies, and the telecommunications needs of small and medium sized businesses made the telecommunications market an attractive investment with large revenue opportunities (New Paradigm Resources Group, Inc., 2002). He contacted individuals who had special expertise in the areas of telecommunications technologies and networking, sales, marketing, finance and accounting, human resources, and information systems and technologies (IT). He asked the IT expert to gather data related to potential customers in major cities along the east coast of the U.S. as well as data on ILECS and other CLECs serving the same region. This data was used to determine the potential revenue for offering telecommunications services. At a meeting convened by this individual, all of the recruited individuals expressed an interest in leaving their current positions and forming a startup company to offer voice and data services over DSL to small and medium sized businesses

Each individual in this group was assigned the task of documenting a plan for getting their area of responsibility operational. This included the activities that needed to be accomplished, the schedule for completing the activities, the resources required to become operational, the cost of operating, and policies and procedures that would be followed in their area of responsibility. The marketing individual was tasked with defining the products and services, market service areas, the expected number of customers, and the estimated revenue. The individual in charge of sales was tasked with determining where sales offices would be located, the staffing required for each office, the sales methods to be used, and the policies and procedures from the time a sales order was acquired, to provisioning the service, to billing, and finally to customer follow-up. The telecommunications technology expert was tasked with determining what technologies would be used to offer the DSL service and interface with the Public Switched Telephone Network, how these technologies would be networked together, how the network would be monitored and controlled, how the products and services would be provisioned after a sale, and how repairs would be made when an outage occurred. The Human Resources expert was charged with determining the policies and procedures for managing the personnel issues related to hiring, termination, benefits, payroll, expense reimbursement, and work place safety and health issues. The finance and accounting expert was tasked with exploring sources of revenue including venture capitalists, bank loans, and other financing options, as well as establishing an accounting system with appropriate policies and procedures. The information systems expert was tasked with developing a plan for what information technologies were required to support and integrate all the other plans. This, of course, meant that the IT plan could not be fully formulated until all the other plans had been developed and required working closely with all the other groups to assess needs and offer advice as to what technologies could be used to support their areas. While others were getting started on their plans, the IT expert began to examine what operations support systems other CLECs and ILECS had either developed or acquired from software vendors. A major effort was to evaluate how competitors were offering converged local voice, long distance voice, data, and Internet services (Emre, 2001).

One of the strategic decisions made by the IT expert was to minimize the number of hardware, software, and network vendors involved and attempt to make sure that the interfaces between information system components were at the database level and minimize application program interfaces (APIs) at the program module level. This would provide flexibility in acquiring the best of breed or developing applications in-house, since the data needed by an application was available at the database level. Another strategic decision was to only use technologies that adhered to standards such as SQL compliant databases, TCP/IP protocols, telecommunications industry standard formats, and so forth. Outsourcing was also considered and rejected as an approach to getting the OSS functional, (Bhandari & Mania, 2005).

The decision to buy or build applications was decided by several factors:

1. The amount of time available to build an application that was known and controlled vs. the amount of time to install, configure, and learn a purchased application.
2. The level of knowledge required to build an application such as billing that was reliable and stable vs. the amount of time to install, configure, and learn how to control a purchased application.
3. The resources required to build, operate, and maintain an application vs. the resources required to purchase, install, configure, operate, and maintain a purchased application.
4. Whether the functional capabilities as determined by the organization could be fulfilled

by a purchased application or whether the application needed to be built with desired customized features.

Organizational Culture

The organization had a culture where micromanagement was typically not done and where individuals were valued for their capabilities and the results produced. Of course, there were exceptions. The CEO was a charismatic person who liked people but also valued hard work and honesty. He was a good motivational leader and knew all aspects of the telecommunications business better than anyone else in the organization. His charisma was demonstrated by getting personnel who were well established in their careers with established companies to take a career risk by resigning their position and joining a startup company. He also had the ability to boost employee morale when situations became difficult by giving highly motivational speeches and offering sound advice and additional resources where needed. The turnover in personnel was nearly zero. The CEO had many years of experience in the telecommunications business and had started two other telecommunications companies before joining Adelphia Cable's Business Solutions division after one of the companies he started was purchased by Adelphia Cable. Broadstreet had an executive committee that met on a weekly basis and made recommendations regarding resource allocations, policies and procedures, as well as business strategies.

The organization was driven by the sales and marketing people who were longtime friends of the CEO. This caused many IT decisions to be based on look and feel and resulted, in some cases, of selecting form over functionality. For example, the president spent over $400,000 on furniture and decorating the headquarters office so that customers would be impressed when they visited the company. Of course, this almost never happens with telephone companies. Another "form over functionality" decision that cost the IT department time and money was the president's decision to have the format of the customer bill changed so that it was more aesthetically pleasing since he viewed this as an important medium of communication from the company to the customer. Since the billing system was a purchased system, the vendor had to be contracted to make the changes. This took 60 days and $200,000. One other example of "form over functionality" was that the VP for marketing was determined to present the DSL technology model in sales presentations, service/product offerings, costing, and billing so as to impress the customers with this new broadband technology utilizing the standard telephone line. Nearly all the billing systems on the market had an underlying model of one line, one device (telephone, fax, PC) based on the old technology, and to make any of the billing systems accommodate the one line, multiple devices model of the DSL technology required either having the vendors modify their systems or finding work arounds in their models. This caused several months delay and nearly a million dollars in professional service charges from the billing system vendor.

Overall, the organizational environment was, on one hand, relaxed, but on the other hand, fast paced and highly stressful for task-oriented people like engineers and software developers. For example, it took over six months for marketing and sales to get the products and services defined and prices established which made order entry, flow through provisioning, and billing applications difficult to get underway by IT in a timely manner. Except for a few individuals, people cooperated with one another and strived to make the company a success. An example of the relaxed yet stressful nature of the company is that the entire headquarters staff frequently played softball on Friday afternoons but worked seven days a week, 12-16 hours a day. All employees were granted stock in the company and therefore had a vested interest in making the company successful.

Not all technology related decisions were made based on good technical criteria. For example, the central office switching technology initially selected by the Chief Technology Officer (CTO) was a new product from Lucent Technologies called a PathStar. This switch cost approximately 50% of the older, but proven, 5ESS switch used by other telephone companies. After nine months of struggling to get the PathStar switch to function correctly and reliably, Lucent decided to remove it as a central office replacement switch for the 5ESS due to lack of functionality, scalability, and reliability. After nine months, Broadstreet replaced the Pathstar switch with the 5ESS switch. The industry standard 5ESS switches were installed and functioning in approximately two months. This left Broadstreet's IT/IS department in the difficult position of having to completely reengineer, reconfigure, redesign, and rewrite software used to perform functions such as capturing call detail records for billing, controlling flow-through provisioning, performing network monitoring and control, as well as making changes to Order Entry, Sales Force Automation, inventory management, and other smaller applications.

Economic Climate

The economic climate going into 1999 appeared to be extremely good since the technology sector stocks were continuing to increase dramatically, and new technology based products and services in the e-commerce area were being created on an almost daily basis. Also, telecommunications was a critical component of nearly all the new technology products and services and was growing in demand. Thus, in 1999, it was relatively easy to get the venture capital and financing necessary to start a company, especially with the experience and charisma of the CEO and highly experienced management team. By the time all the financing agreements had been signed in early 2000, the "dot com" crisis was starting to become a reality and the telecommunications industry was a prime victim of the overvalued companies and stocks. By mid-2001, CLECS were going bankrupt at an alarming rate, but Broadstreet was gaining momentum in terms of acquiring customers and increasing revenue.

By mid-2001, the company had grown to more than 160 employees in six markets and Broadstreet had become more proficient in its internal processes and dealing with its external partners and customers. Things were looking very positive for the company, but Lucent Technologies stock was decreasing in value at an alarming rate, and one of the venture capitalist who had invested in several of the overvalued "dot com" companies had taken large losses and was under pressure to get out of the telecommunications sector. Almost simultaneously, Lucent Technologies and this venture capitalist announced that they were terminating their agreements. The other venture capitalists could not provide additional funding and refused several others who wanted to invest because these potential investors wanted too large a share of the company.

After nine months of design, development, and testing of the telecommunications network, OSS software, and other software systems, a decision was made to change the central office switch because the one initially selected could not be made to function adequately and lacked many features needed by potential customers. This became a major factor in the survival of the company. The change in central office switches caused a nine month delay as network and software system personnel reworked all systems to accommodate the new central switch with its added capabilities and features. This delay consumed financial resources without the benefit of planned income and forced Broadstreet to close its doors in late 2001 because it could not meet the finance covenant agreements with Lucent Technologies and the declining confidence the investors had in the telecommunications industry.

Setting the Stage

While the U.S. Federal Communications Commission (FCC) orders based on the Telecommunications Act of 1996 seemed reasonable enough from a CLEC's point of view, the implementation of these orders was not simple and straightforward. Telecommunications is an extremely complex business. On the plain old telephone network, customers expect to be able to pick up any phone, at any time, and call anybody, within the country or internationally, independent of which phone company they or the recipient uses. While the technical issues of this connection are most often handled by the central office switches and network routers of the incumbent telephone companies, the OSS system of the CLEC must at least handle the provisioning, 911 access, call detail record processing, network monitoring, controls, alarms, and repairs as well as billing. This is quite a complex matter as many different companies and an abundance of procedures, data exchanges, standards, service level agreements, and price policies are involved.

Billing is a critical and extremely complex part of the functionality that OSS systems must provide and CLECs must also have functionality in place for provisioning new customers (often customers that earlier were connected to an ILEC), or for de-provisioning, when they lose a customer to a competitor as well as monitoring and controlling telecommunications network components, switches, routers, circuits, and so forth. While deregulation has opened up competition, there are other regulations in place that must be followed. For example, all telephone provider companies must provide 911 (emergency) services. This includes the ability to tell the emergency facility where the caller is located. Other services, such as "caller ID" and "800 numbers" also involve the ability to access and update national databases. This would be an easy task if all the standards were in place and followed, but the standards are compromised by the incumbents and the CLEC must accommodate many different formats and processes.

To perform all these services, a CLEC needs reliable back office systems. In principle, these can be developed in-house, or be leased or bought from vendors. In practice, only the latter two alternatives are feasible if a CLEC wants to be operational in a very short period of time. One of the keys to the success of a telecommunications company that offers a range of narrow and broadband voice and data services is how effectively and efficiently the back office operations support system functions. This system has been defined as the set of hardware, software, procedures, policies, and personnel that support the following main activities:

- Network Design and Inventory
- Network Monitoring and Control
- Provisioning and Activation of Services
- Service Assurance
- Interconnection Management
- Customer Care & Billing
- Work and Workforce Management

One of the more obvious characteristics that stand out from the list presented above is the widely diverse but highly interrelated nature of these activities. But there are many details associated with of each of these functions and their relationships. Understanding the technology of telecommunications is one thing, understanding the *business* of telecommunications is quite another. Appendix A illustrates the technology architecture for DSL technology and Appendix B illustrates the business of telecommunications.

The "natural monopoly" of telecommunications, that is, the idea that there are advantages to having only one company, has been challenged (Perez, 1994), and the business complexity of having many companies "sharing" parts of a common infrastructure has perhaps not been fully understood.

New entrants into the telecommunications market see the potential for using new technologies to take customers from the incumbents and make huge profits. The number of CLECs that have failed show that most of those who are involved with these new companies do not understand the details of the business and consistently underestimate the cost, time, skill, and knowledge that it takes to offer and maintain a wide array of telecommunications services with an adequate Quality of Service and fulfill Service Level Agreements.

CASE DESCRIPTION

In order to design, implement and operate an OSS, it is necessary to understand not only the technology of telecommunications and the technology of information systems, but the business of telecommunications as well. This includes understanding the requirements of every technology, function, service, and product involved and incorporating these requirements into every design, development, testing, and documentation decision. One of the critical components in a telecommunications network is the central office switch because it dictates the services and associated features that can be offered, how provisioning is done, how network monitoring and control is performed, the interconnection with the PSTN, and what data is collected about each call for rating and billing purposes. When a decision was made to change the type of central office switch after 75% of the OSS has been developed, tested, and documented, it caused a major disruption to the IT implementation plan. Broadstreet Communications experienced such an event 9 months into a 12 month IT implementation schedule which caused IT to develop a new plan to evaluate what information system and associated OSS components were impacted and what measures were necessary to change its acquired and in-house developed software to be compatible with the new central office switch and the new services and features provided by the new switch that sales and marketing now wanted to offer potential customers. The issue facing IT was how to recover from such a decision and still try to meet budgetary and schedule constraints imposed by management.

Technology Concerns

In attempting to recover from a change in a critical component in the telecommunications network that not only provides services to customers and interfaces with trading partners (ILECS) but also captures critical data for billing, network monitoring, network control, and provisioning, a complete halt in current system development occurred. All components in the OSS and related systems had to be evaluated to determine what, if any, impact the new switch would have. But, of course, this change in central office switches also caused marketing, sales, engineering, help desk, and so forth to all reexamine how their functions would be impacted. The following describes some of the areas of concern related to the change in central office switches.

When a CLEC acquires a customer from an incumbent (ILEC), an exchange of information between the incumbent's OSS and the competitor's OSS must take place to order facilities. The ordering process (Local Service Request [LSR] and Access Service Request [ASR]) requires knowledge of how the telephone business operates, the business rules used by the incumbent, and the special language used by the ILEC and the industry as a whole (Garifo, 2001). For instance, when ordering a local loop, you must know the CLLI (Common Language Location Identifier) code of the central office to which the customer will be connected (Telcordia Technologies, Inc., 2000). A new switch introduces new terminology as well as new port and jack labels which are critical for the ILEC to connect a local loop to the CLECs switch.

The ordering of Unbundled Network Elements (UNE's), their installation by the incumbent, the installation of equipment at the customer's premise, disconnecting the current incumbent's service, and the testing and activation of the new service must be scheduled and monitored carefully so as not to leave a customer without any service. For example, a telephone service must provide 911 capabilities. This requires a trunk from a telephone company's local central office to a 911 center (called a PSAP—Public Safety Answering Point) and this, in turn, requires that the telephone company maintain a database of addresses where telephone lines are terminated along with the telephone number associated with each line. Since telephone numbers can be "ported" (i.e., customers can take their telephone numbers with them when they move within a region), there is a national database that must be updated with this porting information. If a customer wants an 800 number (dial free), this also requires interactions with other vendors and updating a national database. Likewise, if caller ID is desired by the customer, this requires yet another national database be updated as well. A new switch changes the flow-through provisioning components of an OSS to accommodate these features.

If a calling card service is to be offered to customers, then an agreement with the Centralized Message Distribution Service (CMDS) must be established and Call Detail Records (CDR) or billing records must be exchanged on a timely basis. Since most customers want a long distance service, interconnection arrangements must be made with the long distance carriers and if convergent billing is offered, the ability to acquire and exchange CDRs with the inter-exchange carriers (IXC or long distance—LD) is a must. Likewise, the equal access regulation requires the exchange of CARE information (Customer Account Record Exchange) to notify the LD carriers when they are losing or gaining a customer. This provisioning of services is one of the most complex components for an OSS to accommodate (Jethi, 2005) and a new switch can change the procedures and data formats necessary for this provisioning.

Although there are data exchange standards for the format of these records, every vendor has its own interpretation or use of various fields within the record which causes back office systems to have many translation software packages for transforming call detail and billing records into a format that can be processed by their own OSS. A new switch can have a different format and data element interpretation that have to be accommodated.

When a service is sold to a customer, the network devices and associated logical attributes must be installed or allocated, interconnected, configured, activated, and tested. This is the service provisioning and activation process. Any specific attributes associated with these components must also be tracked, for example, data speed, and calling features. Tracking what has been allocated to a customer and being able to trace the path from the customer premise is critical to managing and maintaining the service. A new switch can differ considerably from the previous switch in terms of components, labels, functionality, and terminology.

One of the most complex aspects of an OSS is billing. It is complex because rating calls (determining the class of call and associated billing rate) accurately can be a logical nightmare because a caller can theoretically call from anywhere in the world to anywhere in the world at anytime. The second is that the United States has divided its geographical area into LATAs (Local Access Transport Areas) over which a call is considered a long distance call. Unfortunately, LATAs cross state boundaries which make determining the type of call more difficult. Then, there are the message unit charges for local calls that extend over certain distances (zones) from the caller's central office. A call may come from a ship at sea, an airplane, a hotel, a prison, a pay phone, an educational institution, and so forth, all of which are rated differently. The billing system must not only determine what type of call was made but also what plan a customer has and how the charge must be computed, for example, was it a week day or weekend day, after 9:00 pm, over 1,000 minutes of usage, and so forth? This data is derived from the Call Detail Records (CDRs) captured at the central office switch, and a new switch may differ significantly in terms of the data it captures about calls and the format of the data as well as the procedures needed to bring this data into a billing system.

In order to configure and activate services for a customer, local loops must be acquired and installed, devices such as switches, multiplexers, routers, and customer premise interface access devices must be configured by setting device parameters to meet the attributes of the services purchased; and databases must be updated for porting numbers, 800 numbers, caller ID, 911, and so forth. For example, a last mile DSL provider of voice and data may need to access and configure the following devices to activate the service for a customer:

1. Interface Access Device (IAD) at the customer premise
2. DSL Multiplexer at the Local Service Office
3. ATM Router at the Central Office
4. Internet Router at the Central Office
5. Switch at the Central Office
6. Internet server
7. VPN server
8. Voice Mail server

Entering any local telecommunications market is not a simple thing to do, and a change in the central office switch may cause unforeseen problems (The Competitive Telecommunications Association, 2005).

Technology Components

Many of the OSS components run on systems with clustering capabilities, a database management system, and application software written in a programming language such as C or C++. The system architecture is usually client server where the desktop client uses TCP/IP over an Ethernet network. The Server CPUs are networked for high availability and reliability with multiple network connections. The network disk storage is usually RAID 5 or higher to guarantee data integrity. The database is replicated to ensure a fault tolerant data environment. A hot backup or a cluster is used to guarantee continuous operation. A disaster recovery plan and associated resources are in place. The internal network has redundant paths between remote offices and the OSS system location as well as the disaster recovery location.

Much of the OSS software commercially available does not scale, is not reliable, and is not flexible which, in turn, can cause a new company to struggle with commercial OSS software. It is important to balance what services and functions the Back-office system will provide (Tombes, 2003).

The basic system design for Broadstreet's OSS is shown in the diagram in Figure 2. It is obvious that the OSS is comprised of many different DBMSs, software packages, hardware platforms, operating systems, and networking components. The introduction of a new switch has side effects on many of these components.

Scalability and Reliability

To be successful, a telecommunications company needs to acquire customers and lots of them. The capital, circuit, and labor cost for a telecommunications company is very high and therefore the

Figure 2. OSS software architecture

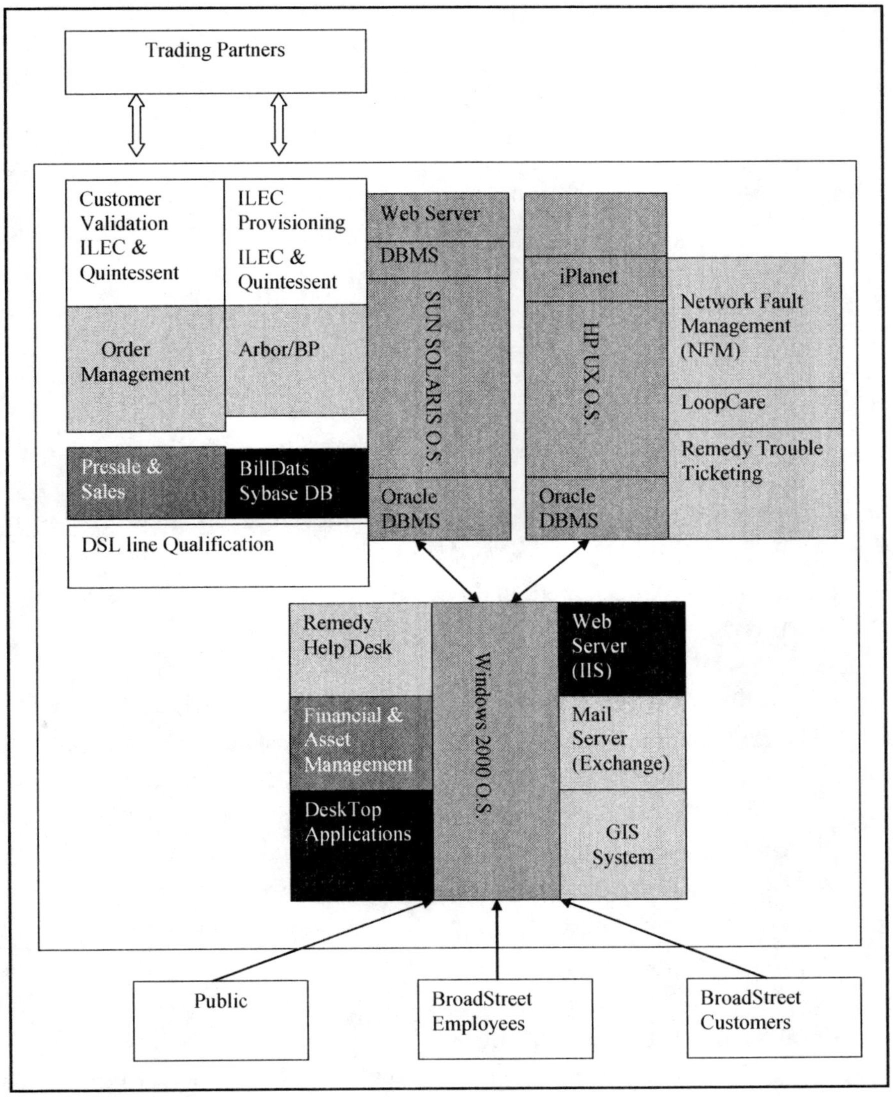

need to utilize the available capacity to produce revenue is essential for survival. The OSS, like the customer service network, must be highly reliable. The OSS must be able to scale with the business and must be available at all times. The scaling can only be accomplished by using efficient application software, database management systems, operating systems, and hardware as well as adequate network bandwidth for users of the OSS.

Management

The management of such a diverse set of technology resources and the people who design, implement, operate and use them requires a formal set

Figure 3.

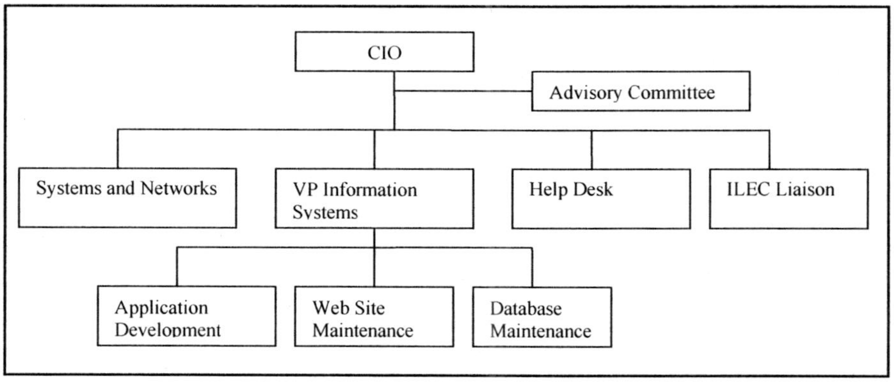

of policies and procedures such as strict change control and a team of talented individuals who are not only dedicated but loyal, motivated, and able to withstand high levels of stress. The difficult management issue related to personnel is that it is very rare to find all these traits in a single individual. The management approach at Broadstreet was to have a detailed model of the systems and processes needed to design, code, test, install, implement, and operate either purchased or developed software. The management challenge was to make sure everyone involved understood the technologies, the business model, the business rules, the technology models, the policies, procedures, and to implement systems within time and budget constraints. The IT department was organized around major functionality components as shown in Figure 3.

Advisory Committee

The advisory committee was composed of representatives from each of the major components of the organization such as sales, marketing, finance, provisioning, engineering, human resources, and so forth. This committee met once a week and was provided updates on progress, asked for advice on implementation and operational issues, and helped defined requirements for functionality, interfaces, and interrelationships. There was an attempt to use sound software engineering principles based on those in publications such as Thayer (2005).

The CIO was part of an executive management committee that met each week to discuss issues, schedules, plans, and resource allocation. The CIO expressed concern with many of the delays and mid-stream changes that marketing and sales advocated since it not only caused delays, consumed unplanned-for resources, and required reworking or acquiring new software, but it also caused morale issues among the IT and engineering staff who seemed to never get anything completed. Typically, the CTO and engineering were aligned with the CIO in objecting to requested changes or lack of specificity in requirements for new features, functionality, or services. The question typically posed by these concerned individuals was: Will this change significantly to improve the service or the revenue stream? The answer was usually, "we don't know for sure but it will make us look good." For example, marketing wanted to significantly change the Web site with more graphics and animation so they could compete for an award for the most attractive Web site. The CIO objected because it would take personnel resources away from more critical software development areas such as billing, provisioning, sales force automation, and so forth. The project was approved anyway. This was a typical pattern at the management meetings.

The fact that sales personnel did not have any customers to call on meant that they had plenty of time to think up interesting data analysis, data presentation, order entry, customer care, and product and service packaging schemes which impacted many parts of the back-office system. The software was under continual enhancement and revision, which made it difficult to complete software to meet the unmet functional requirements of the originally designed OSS and information systems.

Eventually, the CIO was only invited to management meetings where there were technical issues that needed his expertise, and requests for changes, enhancements, and additional functionality were sent to him via e-mail or paper documents. The CIO then implemented an on-line service request system that not only guided the requestor through a set of questions that detailed the requirements for the requested change or enhancement but also required schedule, cost, and benefit estimates. This quickly reduced the number of requests.

When it was announced, after nine months of system development, that the current PathStar central office switch was being replaced by the 5ESS switch, everything came to a halt, and the CIO was invited to all the management meetings once again because the central office switch impacted nearly everything. The CIO and CTO presented a plan for accommodating the new switch into the physical network and the software environment. The impact on the physical network was much less than the impact on the software since the 5ESS switch was not only a well known device but the network engineers were experienced in its installation, configuration, usage, and provisioning. That was not the case for the software engineers who now had to reexamine nine months of software development and software configurations for purchased packages. This also meant that the professional services personnel who had helped configure purchased software had to be called back.

Technology Resources and Constraints

The major constraints for overcoming the introduction of a new switch were time, budget, and finding highly competent personnel and consultants (professional service personnel) who had the interests of the organization's success as a priority rather than their own personal benefit. Budgetary constraints placed limits on hiring more high priced personnel, equipment with capacity for the future, and software packages to perform every needed function.

Every individual in the organization had either a laptop or desktop computer. All sales personnel had laptop computers and most management personnel had both a laptop and a desktop computer. Each of the six regional offices had two servers, namely, a primary and a backup, with a DS1 channel back to the headquarters in Pittsburgh. Headquarters had two servers for the billing, provisioning, and ILEC interface systems and two servers for network monitoring and diagnosis. There were also two systems running purchasing, accounting, human resources and other business functions, two systems running order entry, customer relations, marketing, helpdesk, inventory control, GIS, Web Site, and interfaces to the provisioning, order management, billing, trading partners (for ordering lines, 800 numbers, etc.), and 911 centers and one e-mail server. The servers were all connected via an Ethernet network using Ethernet switches as well as routers for the regional offices. The Internet connection was a DS1 line connected via a router to the internal network and was available to every user.

The OSS and other application software was a combination of purchased packages and in-house developed packages. There were four different database management systems involved with the purchased software. The constraints associated with purchased software are simply those of control over how they function and the interfaces they provide. Some software systems provide configuration model parameters but once they are established and used to initialize the system, they cannot be easily changed. In many cases, the configuration must be done by professional services personnel from the provider since they provide little or no documentation on how to configure the software.

Since more full-time personnel could not be hired, the IT department had to approach the new switch problem using existing personnel, professional service personnel from the providers of the purchased software, and some contract programmers. The problem with contract programmers is that they typically have a long learning curve concerning the application's policies, procedures, and business rules. Professional service personnel also have some learning time about the particular configuration parameters used at an installation. Their services are also quite expensive and the budget for this new development was not extensive.

Organizational Concerns

The concerns that the organization had as the OSS was developed and put into operation were inadequate documentation, failure to follow standards, reliability, stability, configuration limitations, limited integration of components, and adequate functionality to provide flow-through provisioning. Of course, the integrity of data and the security aspects of the OSS were major concerns as well. Hiring experienced IT personnel in adequate numbers who had a wide range of skills was a primary concern. The most personnel IT ever had during the two and a half years of Broadstreet's existence was nine people serving over 180 personnel at six locations. This was due to budgetary constraints as well as the inability to find highly qualified, experienced, and motivated personnel.

With the introduction of a new switch, it was necessary to evaluate its impact on all the Back-office (OSS) software already developed as well as the physical network and determine what changes were necessary. It also was necessary to determine the impact on software not yet completely developed. This caused major concerns among everyone including Broadstreet personnel, the venture capitalists, and the financier, Lucent Technologies.

The major concerns revolved around resources and schedules. It was quite clear that the 12 months originally estimated in the business plan to start selling services to customers was never going to be met. The question was whether the resources that remained would be enough to allow Broadstreet to start realizing revenue, not only to meet its financial covenants, but to remain a viable company capable of meeting its operating cost obligations. It was estimated by the CTO and his network engineers that once the 5ESS switches were on-site it would take about 30-45 days to get them installed and configured and about 30-45 days before the 5ESS switches could be delivered to Broadstreet.

The software effort was estimated to take much longer than the switch installations since marketing and sales now had a much wider array of services than the PathStar switch offered, and they wanted to incorporate them into the products and services offered. Marketing and sales estimated it would take about 30 days to redesign their service and product packages. The CIO estimated it would take about 30-45 days to evaluate the impact the new switch had on existing purchased and developed software and another 10 days to evaluate the impact of the new switch on the redesign of software not yet developed. The CIO would then be in a position to provide a time and resources estimate for changing the existing software and completing the development of the unfinished software. Asked to give a ballpark estimate of time, the CIO estimated it would take another four to six months to complete the Back-office system if adequate resources were available.

Challenges/Problems Facing the Organization

The major challenge facing Broadstreet at the point when a new switch was introduced was the impact it had on operational functionality such as processing orders; provisioning customers; installing devices; configuring devices and software; testing devices and software; activating devices and software; collecting and processing call detail records; calculating and sending bills; monitoring and controlling the physical networks; answering help desk calls; paying personnel; paying taxes; paying for purchases; reconciling reciprocal billing with trading partners; distributing 911 address changes; acquiring 800 numbers; acquiring and managing telephone numbers; managing and distributing IP addresses; as well as acquiring, processing, and presenting sales and marketing data.

Another problem that became evident early in the development of the OSS was that the purchased software configured by the vendor's professional service personnel did not always work correctly because the model that the software was built around was not based on DSL technology where 1 line is used for multiple services, for example Voice and Data but was based on the typical 1 line 1 service model. This, of course, caused a number of last minute changes to make it work correctly. This, in turn, affected many other software packages that either used the data or that provided data feeds to such a package. The vendor, of course, recommended that more professional service personnel needed to be brought on board that had special, niche expertise needed to configure the system. The concern was whether the underlying models for the OSS components were configured incorrectly and were producing incorrect data and interfaces to other system components.

In evaluating the OSS purchased software it was discovered that the product/system documentation was poorly done, was incorrect in many instances, and outdated. Broadstreet had sent IT personnel to training classes on all the purchased software so that the IT personnel who attended the classes would be able to support, maintain, and change the configuration of the system. It was discovered that the training, for the most part, was superficial and introductory. In fact, the trainers, in some cases, were not technically trained but simply followed a lesson plan with canned examples that teach the students how to navigate menus and complete data fields for the simplest cases. The concern was that reliance on a vendor's professional service personnel was not only expensive but unreliable.

Since the OSS must interface with many external systems for exchanging Call Detail Records, 911 data, 800 numbers, porting numbers, SS7 services, and so forth, it was important that standards were followed so that data exchanges occurred flawlessly. It became evident during the evaluation for the new switch that participants in the industry compromise the standards to meet their legacy systems or for other unexplained reasons. The concern was whether billing data, customer data, 911 and other crucial data were being transferred correctly so as to avoid loss of service or liability issues.

The ability to account for every CDR in the billing process without losing potential revenue is critical. Billing is complex because all calls must be typed/classed and rated correctly and unfortunately there are hundreds of call types. The concern was whether or not call records for the new switch were formatted and data elements defined in the same way as the PathStar switch so that calls could be rated and billed correctly.

The new 5ESS switch had a much greater capacity than the PathStar switch. The OSS, like the customer service network, must be highly reliable. The OSS must be able to scale with the business and must be available at all times. The concern was whether or not the systems would scale with the capacity of the new switch.

The requirement for a Web site that allows marketing and sales to advertise products and services, provide customer care, take orders online, track the status of orders, track the status of installation, and allows human resources to provide information to employees and prospective employees was also a major initiative for the company. Marketing wanted changes to the Web site almost every other day to incorporate a great new marketing idea, which consumed valuable resources that were needed for more problem and operationally oriented functions. The concern was that the IT resources consumed by Marketing and Sales reduced the IT resources available for other aspects of the OSS that were in need of changes.

Hiring the right kind of IT personnel and enough of them was a major concern of the organization. Because of limited budget, there were not enough resources to hire another 5-10 personnel and because of the need to have personnel with multiple areas of expertise and with the appropriate experience, work ethic and motivation, it was difficult to find personnel to complete the needed work on the OSS.

Of course, the overwhelming challenge was to recover from the change in the central office switches so that Broadstreet could begin acquiring customers, realizing revenue, and meeting its covenants with its funding partners and financier.

Possible IT Options

The options identified by the IT department when a decision was made to replace all the central office switches with a different switch after 75% of the system had been developed and implemented were:

1. Make a case to management for keeping the existing central switches for a period of time and only installing the new switches in new central offices, thus preserving the back-office systems that had already been developed and would permit acquiring customers, offering services, and realizing income from the existing central offices while performing a redesign, reconfiguration, and rewrite of systems for the new switch which would not only be placed in new central offices but eventually replace the old switches
2. Evaluate the impact that the new switch would have on the existing back-office system and develop a new plan for retrofitting the back-office systems already developed by either:
    a. using only existing IT resources at Broadstreet
    b. using existing IT resources plus professional services from vendors
    c. using existing IT resources and vendor professional services personnel plus outsourcing work to a professional programming services company
    d. outsource the entire back-office application to an Application Service Provider (ASP) firm that supports DSL network technology and the new switch technology until the IT department can redesign, reconfigure, rewrite, and implement a back-office system
    e. partner with an ILEC or other CLEC who has a back-office system using the new switch (5ESS) (which nearly all of them have) while the IT department can redesign, reconfigure, rewrite, and implement a back-office system

REFERENCES

Bhandari, N., & Mania, V. (2005). Business process outsourcing for telecom service providers. Retrieved April 29, 2006, from http://www.dmreview.com/whitepaper/WID527.pdf

Emre, O. (2001). Delivering converged services. Retrieved April 29, 2006, from http://infocus.telephonyonline.com/ar/telecom_delivering_converged_services/index.htm

Garifo, C. (2001). A pain in the ASOG: Managing changes in ordering guidelines presents challenges. Retrieved April 29, 2006, from http://www.xchangemag.com/articles/171back1.html

Jethi, R. (2005). Getting what you wish for: New OSS keeps Northpoint on top Of DSL demand. Retrieved April 29, 2006, from http://www.xchangemag.com/articles/0a1sec8.html

McDermott, M. F. (2002). *CLEC: An insiders look at the rise and fall of competition in the local exchange competition.* Rockport, ME: Penobscot Press.

New Paradigm Resources Group, Inc. (2002). *Measuring the economic impact of the Telecommunications Act of 1996: Telecommunication capital expenditures.* Chicago: Author.

Perez, F. (1994). The case for a deregulated free market telecommunications industry. *IEEE Communications Magazine, 32*(12), 63-70.

Telcordia Technologies, Inc. (2000). *COMMON LANGUAGE(R) general codes—telecommunications service providers IAC Codes, exchange carrier names, company codes - Telcordia and region number: BR-751-100-112.* Morristown, NJ: Author.

Thayer, R.H. (2005). *Software engineering project management* (2nd ed.). Alamitos, CA: IEEE Computer Society Press.

The Competitive Telecommunications Association. (2005). *Entering the local market: A primer for competitors.* Retrieved April 29, 2006, from http://www.comptelascent.org/public-policy/position-papers/documents/CLECPrimerReport.pdf

Tombes, J. (2003, October). Cooking up OSS: Balancing your Back-office diet. *Communications Technology*. Retrieved May 8, 2006, from http://www.ct-magazine.com/archives/ct/1003/1003_oss.html

APPENDIX A. TECHNOLOGY ARCHITECTURE

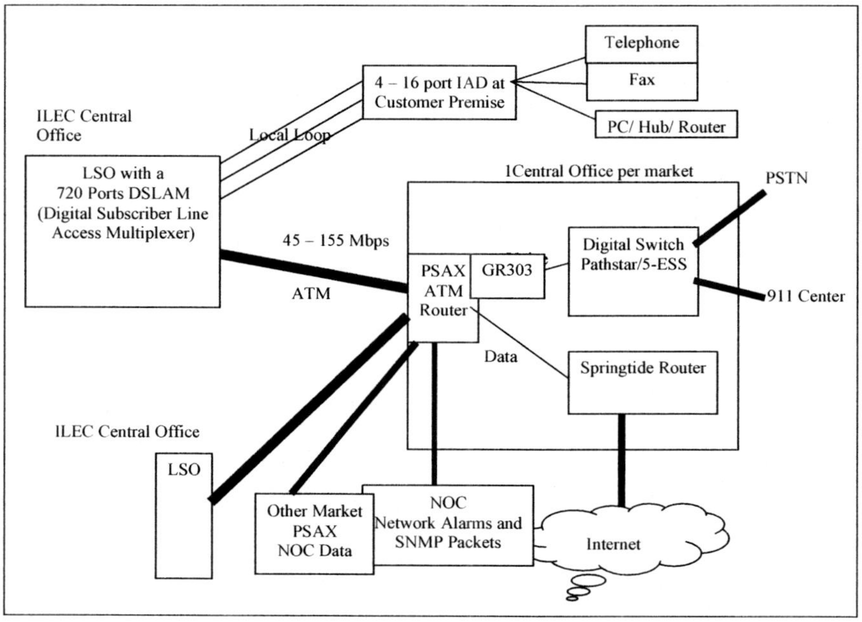

Glossary of Terms

ATM—Asynchronous Transfer Mode: A high-speed small packet based method of transferring digital data between two digital devices such as a DSLAM and a router.

DSLAM—Digital Subscriber Line Access Multiplexer: a device that receives data from many devices over many telephone lines and transmits them in a sequential manner over a single high-speed communication line to another switching center for transfer the PSTN or Internet.

GR303—Generic Requirement-303: A Telcordia standard interface to a Class 5 telephone switch from a digital loop carrier such as DSL. This is the primary interface to the telecommunications central office switch from the outside world. In order to connect directly to the PSTN, IP phones and IP telephony gateways must adhere to GR-303.

IAD—Interface Access Device: connects the devices at the customer site to the telephone line by transforming the data when necessary to a form and format compatible with the DSLAM.

LSO—Local Serving Office: A switching center where local loops connect customer telephones, fax, PC, and so forth to the central office switch.

NOC—Network Operations Center: the place where all network components are monitored and troubleshooting of network malfunctions takes place.

PSAX—PacketStar Access Concentrator: Acts as a concentrator and router for digital data using an incoming ATM format

PSTN—Public Switched Telephone Network: the Plain Old Telephone network that connects calls on a worldwide basis.

SNMP—Simple Network Management Protocol: The software standard used to detect network related errors and report them to the NOC.

Springtide Router—A device used to route packets of data through an Internet Protocol network.

APPENDIX B

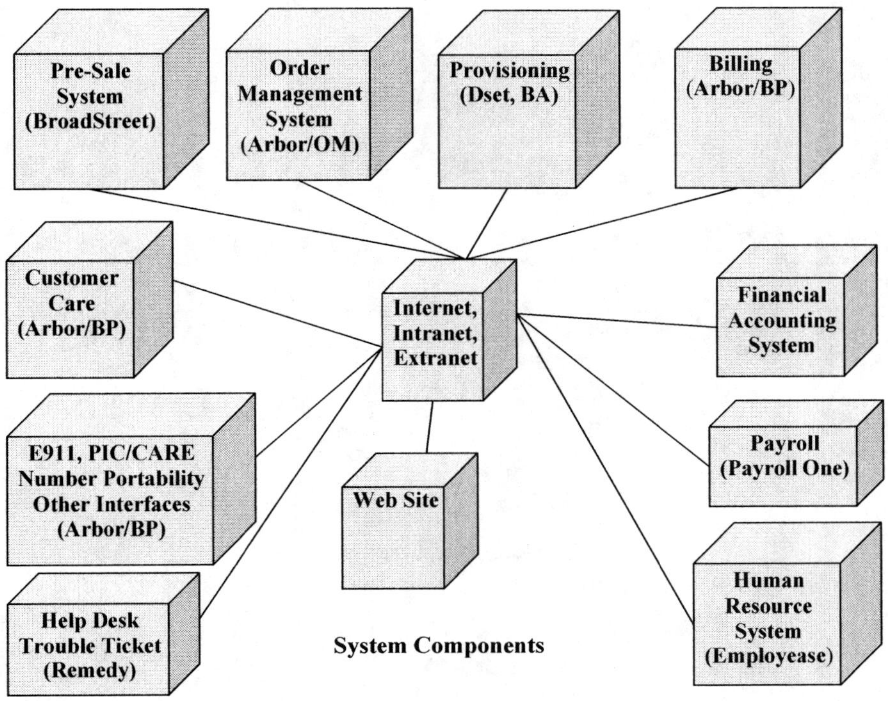

System Components

This work was previously published in International Journal of Business Data Communications and Networking, Vol. 4, Issue 1, edited by Jairo Gutierrez, pp. 30-51, copyright 2008 by IGI Publishing, formerly known as Idea Group Publishing (an imprint of IGI Global).

Chapter VI
A Component-Based Data Management and Knowledge Discovery Framework for Aviation Studies

M. Brian Blake
Georgetown University, USA
Center for Advanced Aviation System Development, The MITRE Corporation, USA

Lisa Singh
Georgetown University, USA

Andrew B. Williams
Spelman College, USA

Wendell Norman
Center for Advanced Aviation System Development, The MITRE Corporation, USA

Amy L. Sliva
Georgetown University, USA

ABSTRACT

Organizations are beginning to apply data mining and knowledge discovery techniques to their corporate data sets, thereby enabling the identification of trends and the discovery of inductive knowledge. Since traditional transaction databases are not optimized for analytical processing, they must be transformed. This article proposes the use of modular components to decrease the overall amount of human processing and intervention necessary for the transformation process. Our approach configures components to extract data-sets using a set of "extraction hints." Our framework incorporates decentralized, generic components that are reusable across domains and databases. Finally, we detail an implementation of our component-based framework for an aviation data set.

INTRODUCTION

Over the past decade, government and industry organizations have enhanced their operations by utilizing emerging technologies in data management. Advances in database methodology and software (i.e., warehousing of transaction data) has increased the ability of organizations to extract useful knowledge from operational data and has helped build the foundation for the field of knowledge discovery in databases (KDD) (Fayyad, Piatetsky-Shapiro, & Smyth, 1996; Sarawagi, Thomas, & Agrawal, 2000; Software Suites supporting Knowledge Discovery, 2005). KDD consists of such phases as *selection*, *pre-processing*, *transformation*, *data mining*, and *interpretation/evaluation*. Selection involves identifying the data that should be used for the data mining process. Typically, the data is obtained from multiple, heterogeneous data sources. The pre-processing phase includes steps for data cleansing and the development of strategies for handling missing data and various data anomalies. Data transformation involves converting data from the different sources into a single common format. This step also includes using data reduction techniques to reduce the complexity of the selected data, thereby simplifying future steps in the KDD process. Data mining tasks apply various algorithms to the transformed data to generate and identify "hidden knowledge." Finally, the area of interpretation/evaluation focuses on creating an accurate and clear presentation of the data mining results to the user.

Excluding the data mining phase, where there are a plethora of automated algorithms and applications, the other phases are mostly human-driven. Data experts are required to complete the tasks related to the majority of steps in the KDD process as explained in the following.

- **Data formatting, loading, cleaning and Anomaly detection.** In the pre-processing phase, data experts must correct and update incorrect data values, populate missing data values, and fix data anomalies.
- **Adding important meta-data to the database.** In the data transformation phase, data must be integrated into a single model that supports analytical processing. This typically involves adding meta-data and converting data sets from text files and traditional relational schemas to star or multi-dimensional schemas.
- **User and tool-generated hints.** In the final phases (i.e., data mining and evaluation), general approaches are needed to assist users in preparing knowledge discovery routines and analyzing results. These general approaches must allow the user to manually specify potential correlation areas or "hints." In the future, the suggestion of new hints may be automated by intelligent software mechanisms.

These human-driven tasks pose problems since the initial data set, which we will refer to as the *raw data*, is large, complex and heterogeneous. Our work attempts to reduce the amount of time required for human-driven tasks in the KDD setting. General reusable components may represent a feasible solution to assist in the execution of the time-consuming processing tasks underlying KDD. In this paper, specific tasks suitable for such components are identified and characterized. In addition, a component-based framework and corresponding process are described to address these tasks.

The paper proceeds in the following section with a discussion of related work with respect to component-based KDD. The paper then introduces the Component-Based Knowledge Discovery in Databases (C-KDD) framework. Subsequent sections provide specific low-level technical details of the C-KDD framework and, in the final sections, the C-KDD is used in an aviation-based study.

RELATED WORK

In practice, the application of KDD to the aviation domain has been done in a limited number of studies. In fact, there are few approaches known by the authors that use data mining techniques in the aviation domain. Earlier aviation studies (Callahan, De Armon, Cooper, Goodfriend, Moch-Mooney, & Solomos, 2001; Nazeri & Jianping, 2002) use static, specialized techniques for aviation security and studies in weather. These earlier approaches do not leverage current data modeling approaches or follow a general purpose design. With respect to KDD-related research, there are many approaches that investigate the data mining phase (Agrawal & Shim, 1996; Netz, Chaudhuri, Fayyad, & Bernhardt, 2001), but few approaches that address the human efforts are particularly using component-based development. Bueno (1998) discusses the benefit of using components to assist with the KDD process. However, similar to other KDD-related research, Bueno focuses on the components for the data mining stage of the process. Bueno does not significantly detail the connection of their components to the underlying database. C-KDD focuses on all KDD stages with an emphasis on the tedious human-driven data collection phase.

Chattratichat, Darlington, Guo, Hedvall, Kohler, and Syed (1999) and Engels (1999) describe component architectures to assist human users in KDD. Chattratichat et al. (1999) address data mining across different organizational regions and Engels (1999) focuses on assisting in the selection of data mining algorithms. Neither project considers support for the human-driven steps required to initially capture the data.

Kim et al. (2002) describes Chamois, a component-based framework implemented by composing a number of third-party software packages. The goal of Chamois is similar to C-KDD; however, the focus is not on the integration of component capabilities at the specification level from the top-down, instead it focuses on building up a framework based on existing applications. The C-KDD approach focuses on the communication channels, particularly from human-to-component. The innovation of C-KDD is the formal data transfer specifications which have not been found in related frameworks supporting the full KDD process.

C-KDD FRAMEWORK

C-KDD is a component-based framework designed to support KDD phases. Five distinct KDD components were chosen to realize the various phases of the KDD process. These KDD components are the *Data Extraction Component*, *Loader Component*, *Data Integration Component*, *Data Mining Component*, and *Hints Component*. The Data Extraction Component and Loader Component combine to extract data from several heterogeneous data sources and populate a central relational data repository. A human, domain specialist identifies common fields between relational entities, and the Data Integration Component transforms these entities into the new component-accessible model. Finally, in the Data Mining, Interpretation, and Evaluation Phases, a human, domain specialist initiates a data mining scenario with the help of the Hints and Data Mining Components. The Hints Component presents the available data model to the domain specialists and accepts the human instructions to submit specialized studies to be executed by the Data Mining Component. The C-KDD framework is illustrated in Figure 1.

In the scope of this paper, we highlight the C-KDD specification techniques that allow human, domain experts to program components and automate portions of the KDD process. The following three sub-sections discuss how domain experts can specify data extraction/loading directives, data transformation instructions, and knowledge discovery instructions.

Component Specification for Data Extraction and Loading

In the selection and pre-processing phases of KDD, various data sources must be identified and the underlying data must be captured in machine-readable format. The approach taken in this work is to perform a component-mediated step that collects heterogeneous data and populates that data in several relational database tables using traditional approaches (i.e., existing data transformation techniques are embedded in component (Chawathe et al., 1994; Nodine, Fowler, Ksiezyk, Perry, Taylor, & Unruh, 1998)).

The C-KDD framework considers several types of data formats. It was not possible in the initial investigations to consider every possible format, but samples of data were taken from the initial deployment domain (i.e., aviation studies). There were several types of structured and semi-structured data considered as listed in the following:

- Delimited Data in Row Format
- Delimited Data in Non-Fixed Row Format
- Relational Database Information
- Name/Value Format
- Mark-Up Format (i.e. XML and HTML)

Figure 1. The C-KDD Framework for KDD

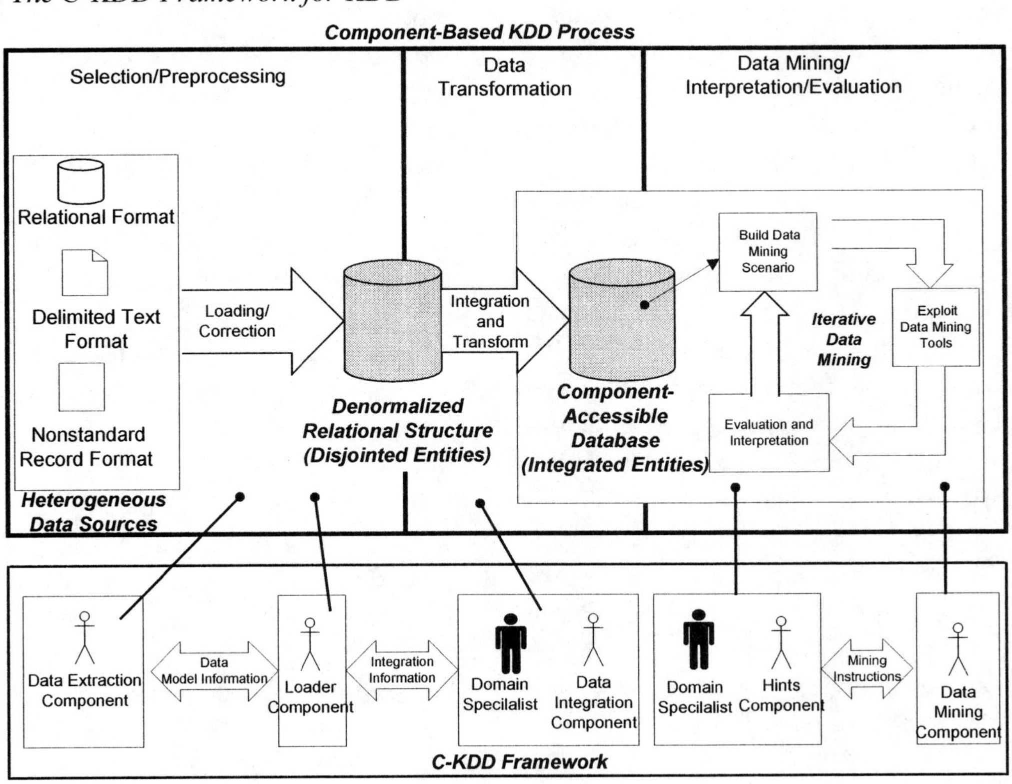

Delimited Data in Row Format, Relational Database Format, Name/Value Format, and Mark-Up Language Format are standard structured formats for capturing data. Using general functions and minimal human intervention, the aforementioned formats are extracted with ease by the Data Extraction Component. Human intervention can be defined in two tasks. The first row in most input files contains the column headings. With respect to human intervention in this case, the human is required to confirm these headings or change the naming, if necessary. In other files, the column headings are not included; therefore, the human user must specify column headings. In mark-up files, XML *element* names are used as headings. In general, for these types of files, the overhead is relatively low, since the user is only required to enter the names or make confirmations with a couple of button presses. There are some exceptional cases, but it was discovered in the aviation domain that this degree of human intervention was sufficient in 17 out of 19 relevant cases.

Exploiting semi-structured data (i.e., Delimited Formats in Non-Fixed Row Format) requires a relatively higher degree of interaction between the human user and the Data Integration Component. The C-KDD framework extends an existing approach referred to as *templating* to describe the semi-structured data schema to the components. With respect to the C-KDD approach, the specification of a template allows the user to *flatten* hierarchical data into row format. The template consists of several aspects. We define a *zone* to be the area of a template identified by delimination parameters (e.g., semi-colon, space, or tab-delimited). Data can be delimited by a string delimiter and also by using fixed column locations. There are specialized attributes that describe the start and end of a zone. Other attributes allow the user to specify a zone by number of columns. It is not in the scope of this paper to describe the templating technique in detail, but technical details can be found in related work (Blake, Singh, Williams, Norman, & Sliva, 2005).

Component-Mediated Processes for Data Transformation

The C-KDD framework includes a Data Transformation Component that generates a generalized, denormalized database. In the C-KDD framework, we introduce a process that generalizes the database schema into a model that C-KDD components can navigate. This is a reproducible process that is valid for any database with entities that share related columns. This process includes the creation of fact tables (Kimball, 1996; Kimball, Reeves, & Ross, 1998) that connect entities. With respect to the C-KDD framework, there is a *CorrelationSpace* entity that links multiple *CorrelationAttibute* entities. The CorrelationSpace table corresponds to the fact table in a star schema. The CorrelationAttribute entity corresponds to the dimension table in the star schema. Example CorrelationAttributes include time and location. Although the results will be discussed in a later section, Figure 2 contains an aviation-based model developed using this approach. There is a CorrelationSpace table, and the AreaSpace and TimeSpace tables are both CorrelationAttributes.

The process for creating the denormalized database in Figure 2 is as follows:

1. Human domain specialist identifies related columns across multiple normalized tables.
2. The domain specialists identifies the specific range of data for the related columns/attributes
   - This specification may consist of a range of time, a specific area, or a range of keyed information.
3. Assuming columns are sorted, C-KDD components create generic CorrelationAttribute tables and preload records for data within that specified range.
   - For example, if the user specifies the time between May and June, the

Figure 2. The ASCEND relational database schema

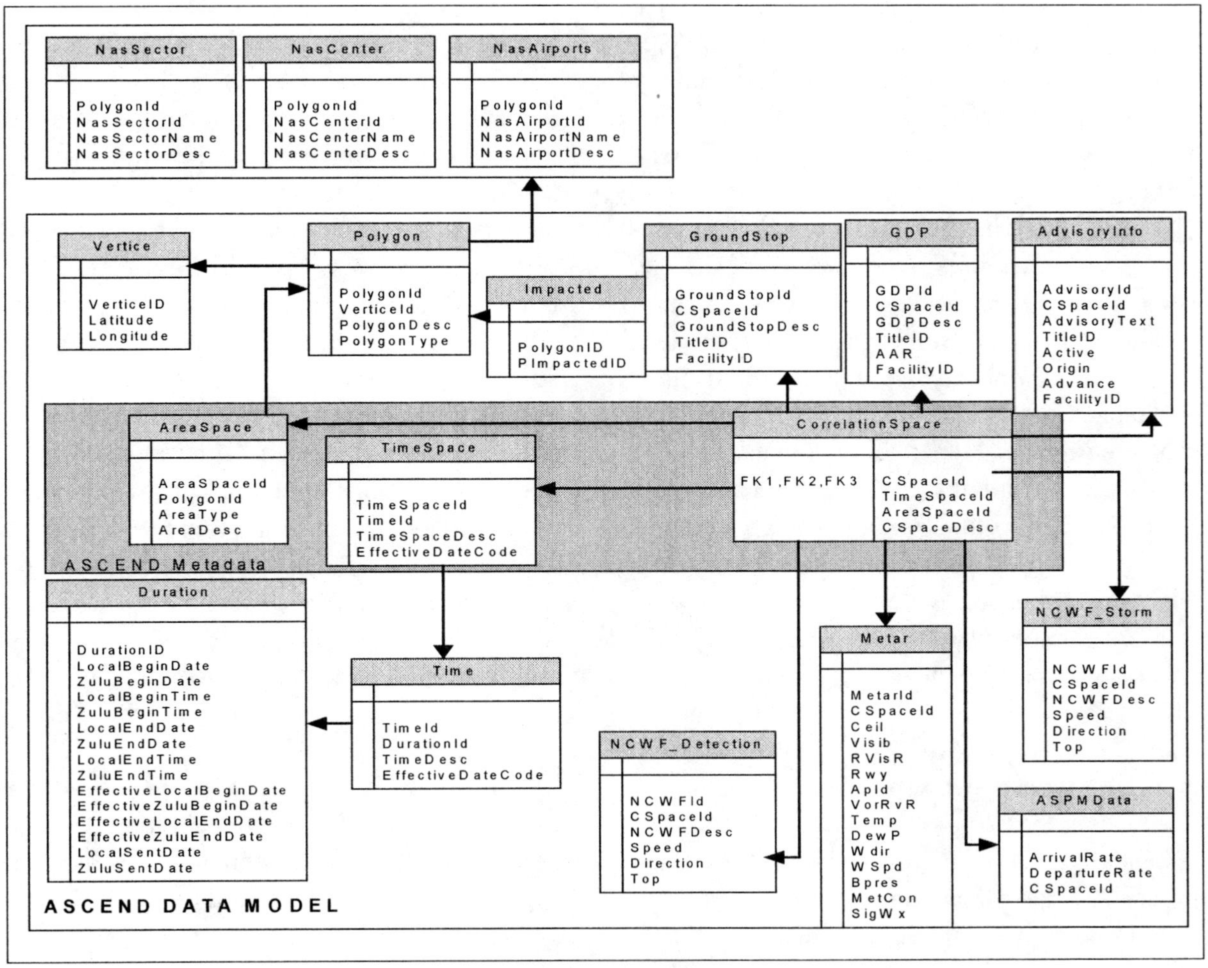

component would build a Correlation Attribute (i.e., TimeSpace) table that has a record for each time increment within that range.

4. Components query the initial data tables and record-by-record create correlation records in the CorrelationSpace entities based on matching the record to the pre-loaded correlation attributes.
   - Considering the earlier example in step 3, the component would extract a record from the original table and search for the time increment corresponding to that record. The two individual records would be merged or "joined" into a single composite record in the CorrelationSpace table.
5. Additional detailed tables can be generated/connected to further describe the correlation attributes. (This is not a component-automated step.)

By using the generic Correlation Space and CorrelationAttribute table structures, component capabilities can be duplicated on any database containing these meta-information tables. The Data

Mining Components are not limited by having to *hard-code* column names and table locations; instead, queries can be managed based on pre-defined conditions. Also, this process essentially pre-loads entity joins into the database model. During loading, the Data Integration Components do not create duplicate correlation identifiers (i.e., CSpaceid in Figure 2), Instead, existing correlation records are reused. In this way, for example, when a weather record has the same correlation conditions (the same time and same airport) as a terminal performance record, both records have the same CorrelationSpace identification. This modeling approach separates the domain-specific information from the database semantics, an approach akin to the separation of concerns in software design. This separation allows the Data Mining Components to be able to perform basic text comparisons without knowledge of domain-specific concerns when generating data sets.

Component-Mediated Data Mining Routines Using Extraction Hints

A major innovation in this work is the formal approach and corresponding user interface design that allows human domain specialists to collaborate with components on data mining scenarios. This approach uses a representation called *extraction hints* as instructions to the Hints Component. Domain specialists, at times, need to determine trends based on a composite list of constraints. In C-KDD framework, these constraints, based on *qualifying events,* are modeled as extraction hints. In the aviation domain, qualifying events can be defined as the combination of weather, processes, and performance conditions (i.e., "Give specified data when the temperature is greater than 70 degrees and the airport arrival rate is less than 15 per hour").

There are two major aspects of the extraction hint that human users can provide. The first aspect, called the *search criteria,* allows the input of a basic constraint. For example, a user might direct the component to explore situations where the cloud ceiling was greater than 1,000 feet and temperature was greater than 90 degrees. This search criteria will specify records from the ASPMData table (Figure 2) where the values of the ceiling column is greater than 1,000 and the temperature column is greater than 90.

Once a search criteria is set, the user can also suggest *information points*. An information point is defined as other information related to the search criteria, as constrained by the correlation attributes (in this case, time and airport). For example, a user may specify an information point as the visibility at the same time and area as the search criteria constraint. A user can also specify an information point for a different location for a different time, perhaps at another airport and, for the time, three hours before the time captured when the search criteria is met. Both the search criteria and information points are composed of the correlation attributes represented in the correlation records.

The Hints Component has a specialized user interface that supports the collaboration of extraction hints between the domain specialists and the Hints Component. At initiation, the domain specialist suggests an initial extraction hint using this interface. Once results have been generated, the Hints Component can suggest a variation of the original hint for further data mining. In future work, the intent is to convert the Hints Component into a software agent to manage new, fully automated data mining scenarios. The Hints Component user interface as customized for the aviation domain is shown in Figure 3.

The search criteria is specified on the left side of the user interface. The *Table* textbox dynamically pulls table names from the C-KDD database. Once the table is chosen, the database columns are dynamically populated in the *Field* textbox. The domain specialist can designate a value (*Value* textbox) using any of the most common relational operators, that are populated in the *Relation* field. The *Preferred Name* textbox allows the user to

Figure 3. User interface for Hints Component

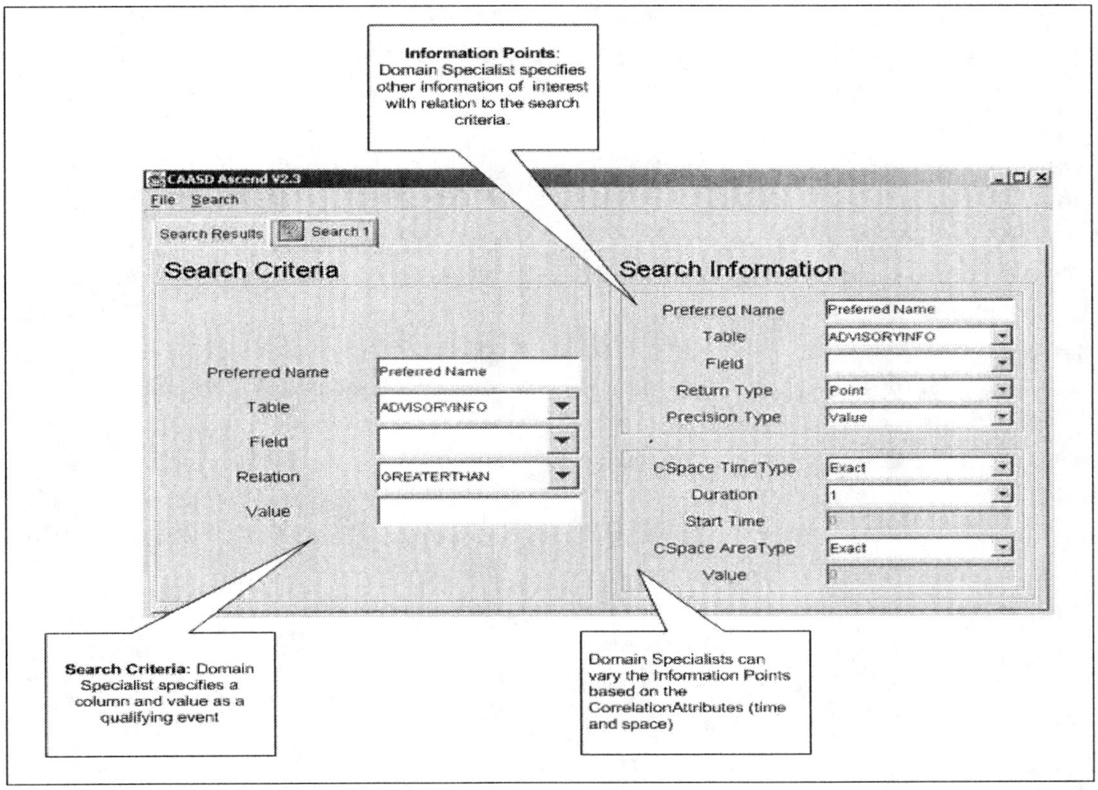

associate a personalized heading for the search criteria. The information point is specified on the right side of the user interface, but was named *search information* for consistency. The domain specialist again can specify a preferred name, table, and field. In addition, the domain specialist can specify if there will be one data return correlated with the search criteria or multiple data returns (*Precision* textbox). The *CSpaceTimeType* and *CSpaceAreaType* textboxes and corresponding textboxes allows the domain specialist to vary the correlation between the search criteria and search information. This user interface has been customized for the aviation domain so, in the case of this illustration, the correlation is based on time and location. However, it should be noted that the correlation attributes are not fixed, but that this implementation is just one embodiment of the approach. The user interface in Figure 3 shows one tab, but multiple tabs can be added on-demand by the domain specialist. Each tab has a new search criteria and information point (search information). Finally, the Hints Component creates an XML message that directs the Data Mining Component on how to extract datasets from the relational database. In the interest of space, there is just an overview of the capabilities of this user interface presented here.

ASCEND: C-KDD FOR ANALYSIS IN THE AVIATION DOMAIN

In evaluating the effectiveness of the C-KDD framework and approaches, the software was customized and deployed in an operational setting and used for an analysis on real data sources. The C-KDD framework and initial prototype was used in a joint project of the Center for Advanced Aviation System Development (CAASD) of The MITRE Corporation and the Department of Computer Science of Georgetown University. This joint project called Agent-Based Software for the Correlation of EN-route Data (ASCEND) was a customization of the C-KDD framework in the aviation domain. The ASCEND software was built predominantly using the C-KDD prototype. An additional goal of the project was to extend the C-KDD components using agent-oriented concepts, but those efforts are out of the scope of this paper.

Initially, the C-KDD software consisted of about 23,000 lines of Java code. The learning components were implemented by integrating the WEKA data mining application and toolkit (WEKA, 2005). The WEKA software was chosen for this research project, since it was also written in Java code and provided run-time program interfaces. The denormalized database was implemented using Oracle9iLite. Oracle9iLite is a fully operational relational database management system that runs on a personal computer. This personal database was chosen to support the portability of the software and the database as one package. This ability made it possible to transfer the entire application and database from machine to machine on one CD. Both the C-KDD database and external databases (as data sources) were accessed using several JDBC interface software. The communication component was implemented using shared object space, specifically Sun Microsystems JavaSpaces implementation. The software for communication was an extension of the agent communication methods developed in earlier work (Blake, 2003b; Blake & Williams, 2003a).

Several software changes were required to customize the C-KDD framework for ASCEND. Several specialized objects were created to converge latitude and longitude values for airports, airspace locations, and weather locations into a common attribute. The user interfaces were

Table 1. Data sources implemented in ASCEND

Data Source Name	Data Format Type	Source of Data	Description
Ground Stop (Order 7210.3S, 2003)	Microsoft Access Database (structured)	Federal Aviation Administration (FAA)	Procedure used to stop operations at a specific airport or group of airports with time and duration
Ground Delay Program (GDP)	Microsoft Access Database (structured)	Federal Aviation Administration (FAA)	Procedure used to control the flow of arrivals into an airport or group of airports with a time and duration
Advisory Information	Delimited Text in *Non-Fixed* Row Format (unstructured)	Federal Aviation Administration (FAA)	Specific information about the status of the National Airspace System
ASPM Data (ASPM, 2005)	Delimited Text in Row Format	Federal Aviation Administration (FAA)	Airport performance metrics
National Convective Weather Forecast (NCWF) - STORM (National Weather Service, 2005)	Delimited Text in *Non-Fixed* Row Format (unstructured)	National Weather Service	Polygons representing areas of convective or inclement weather (prediction)
National Convective Weather Forecast (NCWF) - DETECTION	Delimited Text in *Non-Fixed* Row Format (unstructured)	National Weather Service	Polygons representing areas of convective or inclement weather (current forecast)
METAR	Delimited Text in Row (structured)	National Oceanic and Atmospheric Administration (NOAA)	Weather situation over a particular airport

customized for more specialized extraction hint generation. The sum of software changes was less than 2,000 lines of code which was favorable with regards to the usability of C-KDD.

The ASCEND Database

The ASCEND component-accessible database was populated with several data sources from multiple heterogeneous sources as listed in Table 1. Access to these data sources was provided by The MITRE Corporation. The final ASCEND data model is closely related to the schema used for the C-KDD approach. The final data model is illustrated in Figure 2.

The data sources listed in Table 1 are shown in the data model with direct association to the CorrelationSpace entity. The CorrelationAttribute entities were given more specific names expressed as database views. The names of the two CorrelationAttribute entities are *AreaSpace* and *TimeSpace*. The shaded area represents the meta-data tables added for component navigation by the Data Integration Component. The additional tables in the model are specialized tables added to further describe the Correlation-Attribute entities. The additional specialized Java code described in the previous section extended the C-KDD framework to create and access the additional entities (i.e., Vertice, Polygon, Time, Duration entities).

Verifying C-KDD by Regenerating Known Aviation Rules

C-KDD was evaluated by performing aviation-specific knowledge discovery routines using real operational data. The ASCEND database was populated with 6 months of data from May 2001 to October 2001 using the data sources described in the previous section. This six-month time frame is of great importance to traffic flow management in the United States because the convective weather (i.e., thunderstorms) that occurs during these months is the greatest cause of aviation delay (Convective Weather PDT, 2005). The final database was 600 megabytes in size.

The intention of the experimentation was to verify the correctness of the framework by performing studies that result in known rules. The intention of the first study, involving Los Angeles International Airport, was to verify the C-KDD data mining tools and the second study, involving Atlanta Hartsfield International Airport focused on verifying the data transformation mechanisms. The rules resulting from both studies would indicate trends that support several well-known facts in aviation (i.e., when visibility is bad, aviators are required to use their instruments also certain weather conditions negatively affect the performance of airport departure and arrival operations).

The first study was helpful in evaluating the accuracy of the Data Mining Components by re-generating known rules in real operational data. The chosen experiment was to re-engineer *flight rules*. From an aviation perspective, when visibility is below a certain distance and ceiling (cloud-level) is below a certain altitude, the FAA institutes the instrument flight rules (IFR) restriction as opposed to the visual flight rules (VFR) designation during normal conditions. During IFR, any pilot flying an aircraft should be able to operate the plane using solely the instruments. In the MITRE-maintained METAR data source (National Weather Service, 2005), the MetCon (Flight Rule) field is determined by the value of the Visib (visibility) and the Ceil (Cloud Ceiling) columns. If the Visib field is less than four miles and the Ceil field is less 700 feet, then the MetCon is set to IFR. Otherwise, the MetCon is set to VFR. There are other conditions, but it is not in the scope of this paper to discuss the rules in detail. It is more important to understand that the MetCon column is a known function of the Ceiling and Visibility. The study was executed on one data source (i.e., METAR) to concentrate on the data mining functions. In setting up this experiment, the search criteria was set to Los Angeles airport and the information points were set to most of the METAR attributes (i.e., ceiling, visibility, flight rules, temperature, barometric pressure, wind speed, wind direction, and dew point). The relevant columns were not specified; therefore, a successful experimental result could be shown in two outcomes. The first outcome would be that ceiling, visibility and flight rules should stand out as having the strongest correlation. A second successful outcome would be the determination of correct rules for flight rule designations. The experiment was successful from both aspects. Executing the experiment on all available METAR data for Los Angeles, the decision tree created using Data Mining Component (this component wraps the WEKA software (WEKA, 2005)) represented the known flight rule correlations. There were no significant deviations.

Using the same principle of discovering known rules, the second evaluation experiment was toward the verification of data integration processes and tools. In this experiment, the correlation between wind speed and airport performance was measured. Considering the fact that an aircraft lands against the wind and takes-off against the wind, excessively high wind speed tends to have a strong correlation to poor performance (low arrival/departure rates, high cancellation rates) in airports on a specific day. In this second experiment, we evaluated the affect of wind speed using the METAR data source with the airport performance using the ASPM data source (i.e., integration of separate data sources). As with the first experiment, the study contained many more columns than the columns with known correlation. The experiment was run on data for the Atlanta Hartsfield International Airport (ATL).

For a successful verification in this second evaluation, the expectation was that the high wind speed would result in low departure and arrival counts at Atlanta airport. Again, there were sufficient results to verify the tools ability to integrate the two data sources. The METAR and ASPM data points were reported approximately twice a day during the period of time captured in this study. The Arrival and Departure counts were low in the majority of the cases when the wind speed was high. Several MITRE analysts were consulted to confirm that the results were consistent with the trends of weather-impacted operations at the facilities. The specific aviation-based results of the studies can be found in related literature (Blake et al., 2005).

DISCUSSION

In this paper, we describe a component-based framework that facilitates a human domain specialist in developing and executing KDD routines. This work represents a novel investigation of component capabilities for use in knowledge discovery. The C-KDD framework has been applied to the aviation domain and is currently in use for analysis purposes at The MITRE Corporation. We introduce the notion of a correlation space that enables us to generalize our components related to data preprocessing and data transformation. The innovation in the ASCEND project is the ability of the user to define the data mining scenario with the assistance of a general component-based framework. In addition, this is perhaps the first general-purpose framework that exploits KDD in the aviation domain. This paper presents an experience report of the C-KDD, but in-depth technical details can be found in related literature (Blake et al., 2005).

Although the current implementations have been deployed as a local workstation application, in future work, we plan to distribute the C-KDD engine as a network accessible web service. The underlying technologies (i.e., Java and Jini) would facilitate these future extensions. In other future work, we plan to evaluate the optimality of our correlation modeling approach by recapturing our data model in other more standard formats. With regards to aviation studies, we plan to extend our current studies using a full year of aviation data. Although the May through October period (as analyzed in this work) is most relevant for traffic flow experimentation with respect to delays (Convective Weather PDT, 2005), other times during the year may be more valid for other concerns such as ground operations.

ACKNOWLEDGMENTS

Our thanks to several undergraduate student developers, Lindsay Blohm, Todd Cornett, and Denitsa Apostolava. We thank David Hite, Urmila Hiremath, John Mack, and Joe Hoffman of the MITRE Corporation for their discussions in the evaluation of the framework and the interpretation of the data mining results. This work was produced for the U.S. Government under Contract DTFA01-01-C-00001 and is subject to Federal Aviation Administration Acquisition Management System Clause 3.5-13, Rights In Data-General, Alt. III and Alt. IV (Oct. 1996). The contents of this material reflect the views of the author and/or the Director of the Center for Advanced Aviation System Development. Neither the Federal Aviation Administration nor the Department of Transportation makes any warranty or guarantee, or promise, expressed or implied, concerning the content or accuracy of the views expressed herein.

REFERENCES

Agrawal, R., & Shim, K. (1996). Developing tightly-coupled data mining applications on a relational database system. In *Proceedings of the 2nd International Conference on Knowledge Discovery in Databases and Data Mining*, Portland, Oregon (pp. 287-290). Menlo Park, CA: AAAI Press.

The Aviation System Performance Metrics (ASPM). (2005). Retrieved August 24, 2005, from http://www.apo.data.faa.gov/faamatsall.HTM

Blake, M. B. (2003b). Agent-based communication for distributed workflow management using Jini Technologies. *International Journal on Artificial Intelligence Tools, 12*(1), 81-99.

Blake, M. B., Singh, L., Williams, A. B., Norman, W. N., & Sliva, A. L. (2005). *A component-based framework for data management and knowledge discovery* (Georgetown Technical Report). Retrieved August 24, 2005, from http://daruma.george town.edu/techreports/techrep_data/blake_singh_%20C-KDD2005.pdf

Blake, M. B., & Williams, A. B. (2003a). Developmental and operational processes for agent-oriented database navigation for knowledge discovery. In *15th International Conference on Software Engineering and Knowledge Engineering (SEKE'03)*, San Francisco (pp. 216-223). Skokie, IL: Knowledge Science Institute.

Bueno, J. C. (1998). *KDCOM: A knowledge discovery component framework*. Master's thesis, Spanish Council for Scientific Research.

Callahan, M. B., De Armon, J., Cooper, A. M., Goodfriend, H., Moch-Mooney, D., & Solomos, G. H. (2001). Assessing NAS performance: Normalizing for the effects of weather. In *4th USA/Europe Air Traffic Management R&D Symposium*, Sante Fe, New Mexico. Retrieved August 24, 2005, from http://atm2001.eurocontrol.fr/finalpapers/pap94.pdf

Chattratichat, J., Darlington, J., Guo, Y., Hedvall, S., Kohler, M., & Syed, J. (1999). An architecture for distributed enterprise data mining. In *Proceedings of the 7th International Conference on High-Performance Computing and Networking, Lecture Notes in Computer Science,* Amsterdam, The Netherlands (pp. 573-582). Berlin: Springer-Verlag.

Chawathe, S., Garcia-Molina, H., Hammer, J., Ireland, K., Papakonstantinou, Y., Ullman, J., & Widom, J. (1994). The TSIMMIS project: Integration of heterogeneous information sources. In *Proceedings of the 100th Meeting of the Information Processing Society of Japan (IPSJ)*, Tokyo, Japan (pp. 7-18). Tokyo: IPSJ Press.

Convective Weather Product Development Team (PDT), FAA Aviation Weather Research Program. (2005). Retrieved August 24, 2005, from http://www.faa.gov/aua/awr/prodprog.htm

Engels, R. (1999) .*Component-based user guidance in knowledge discovery and data mining.* Volume 211 Dissertations in Artificial Intelligence-Infix.

Fayyad, U., Piatetsky-Shapiro, G., & Smyth, P. (1996). Knowledge discovery and data mining: Towards a unifying framework. In *Proceedings of the Second International Conference on Knowledge Discovery and Data Mining (KDD-96)* (pp. 82-88). AAAI Press

Kim, W., Park, S., Yong, H., Chae, K., Cho, D., Choi, B., et al. (2002). The Chamois component-based knowledge engineering framework. *IEEE Computer, 35*(5), 44-52

Kimball, R. (1996). *The data warehouse toolkit: Practical techniques to building dimension data warehouses.* New York: John Wiley

Kimball, R., Reeves, L., & Ross, M. (1998). *The data warehouse lifecycle toolkit.* New York: John Wiley.

The National Convective Weather Forecast (NCWF). (2005). Retrieved August 24, 2005, from http://cdm.awc-kc.noaa.gov/ncwf/index.html

National Weather Service METAR/TAF Information. (2005). Retrieved from http://205.156.54.206/oso/oso1/oso12/faq.htm

Nazeri, Z., & Zhang, J. (2002). Mining aviation data to understand the impacts of severe weather on airspace system performance. In *Proceedings of the International Conference on Information Technology: Coding and Computing (ITCC'02)*, Las Vegas, NV (pp. 518-523). Los Alamitos, CA: IEEE Press.

Netz, A., Chaudhuri, S., Fayyad, U., & Bernhardt, J. (2001). Integrating data mining with SQL databases: OLE DB for data mining. In *Proceedings of the International Conference on Data Engineering (ICDE 2001)*, Heidelberg, Germany (pp. 379-387). Los Alamitos, CA: IEEE Press.

Nodine, M., Fowler, J., Ksiezyk, T., Perry, B., Taylor, M., & Unruh, A. (1998). Active information gathering in InfoSleuth. *International Journal of Cooperative Information Systems, 9*(1/2), 3-28.

Order 7210.3S, Facility Operation and Administration, Chapter 5 (Traffic Flow Management). (2003). Retrieved August 24, 2005, from http://www.faa.gov/atpubs/FAC/Ch17/chp17toc.htm

Sarawagi, S., Thomas, S., & Agrawal, R. (2000). Integrating association rule mining with databases: Alternatives and implications. *Data Mining and Knowledge Discovery Journal, 4*(2/3), 89-125.

Software Suites supporting Knowledge Discovery. (2005). Retrieved August 24, 2005, from http://www.kdnuggets.com/software/suites.html

WEKA. (2005). Retrieved August 24, 2005, from http://www.cs.waikato.ac.nz/~ml/weka/

This work was previously published in International Journal of Information Technology and Web Engineering, Vol. 1, Issue 1, edited by G. Alkhatib and D. Rine, pp. 76-90, copyright 2006 by IGI Publishing, formerly known as Idea Group Publishing (an imprint of IGI Global).

Chapter VII
Waleli:
Bringing Wireless Opportunities to Life

Paul Kirwan
Nikos, University of Twente, The Netherlands

Peter van der Sijde
Nikos, University of Twente, The Netherlands

Aard Groen
Nikos, University of Twente, The Netherlands

ABSTRACT

This chapter tells the development story of Waleli, a high-tech company utilizing the latest proven developments in wireless communications to bring innovations to the market. It presents the journey of the firm through the entrepreneurial process, from initial idea right through to value creation. Furthermore, it looks at the roles and origins of network contacts and how these impact the entrepreneur's effort to accumulate the necessary resources to establish and subsequently develop the firm. This chapter adds to our understanding of network activity in the entrepreneurial process and as a real life case has relevance to both potential and practicing entrepreneurs in that it details the trials and tribulations of the process, and academics, as it provides in depth data on the role of the network in the entrepreneurial process.

BACKGROUND

History of the Organization

Waleli[1] is a high tech company which utilizes the latest proven technological developments in wireless communication to bring innovations to the market. It aims to develop wireless communications to solve needs, both between people (mobile applications) and between machines and apparatuses (machine-to-machine, or M2M, communication), in a simple way.

Siete Hamminga, the founding entrepreneur, graduated with a master's degree in industrial engineering and management from the University of Twente in 2001. During his time as a student he ran a one man business, Well-Suited, selling suits to students. He also completed an internship with Unilever in Switzerland and while enjoying the work, this experience reinforced his desire to start his own business. He completed his Master's thesis in Silicon Valley studying the business models of Internet service providers (ISPs). His study period coincided with the collapse of the Internet bubble and his thesis was accordingly entitled "Back to Basics." Inspired by his stay in Silicon Valley, he, along with Roel Pieper (former vice president of Philips, current professor of e-business at the University of Twente and chairman of Favonius Ventures, a venture capital firm specializing in funding software start-ups), and Vincent Kouwenhoven wrote the book *Beyond the Hype* (see Figure 1). In this book, top Dutch executives give their vision on e-business now that the hype is over. Based on these experiences, Siete Hamminga founded Waleli in February 2002.

Armed with the knowledge he obtained in Silicon Valley, Hamminga recognized that current mobile technology ideas were far ahead of consumer needs. The practical idea for the business came from his brother, who posed the question "Why can't a mobile phone be used to answer the doorbell?" To answer this he conducted some formal research, borrowed some money, built the prototype, and launched the "GSM-doorbell."

Figure 1. Beyond the hype

While developing this product he continued to come up with ideas to bring practical solutions to meet needs in a simple way using wireless communication; this became the mission of his company.

Type of Business

Recognizing that current mobile technology ideas were far ahead of consumer needs, Siete founded Waleli on the premise of developing applications for wireless communication which meet needs in a simple way. Siete recognizes the great potential of wireless communication and Waleli has capitalized on this with its first product: the "GSM doorbell." Subsequently, Waleli has extended the market for this product by identifying its usefulness in an alternative industry, that is, health care. Waleli has further increased its product and service offerings to include among others the "Basiqphone" and "MMS-witness" (see Products/Services Provided).

To meet the needs of its customers Waleli engaged with them to transform innovation into concrete products and/or services. Waleli's products have two things in common: they are wireless and they meet needs in a simple way. Further, these applications are thought of and evaluated based on their market potential. Waleli's client portfolio ranges from large corporations through to SMEs; most of these companies have established names and market positions, for example, KPN, Siemens, and JohnsonDiversey.

Waleli views the key to being successful in this market as being able to bridge the technology market gap. Critical within this process is the role of participation, as it allows everyone to excel at what they do best. Also integral to this process is the "art of limitation," defining what a product should not do. To this end, Waleli is a networking organization; it works with specialized development teams and partners to meet the needs of specific project assignments. The innovative nature of the venture's process of turning ideas into applications has been recognized through participation in several innovation competitions where Waleli has won prizes including the LiveWIRE Business Award (www.shell-livewire.com), the Top Three Business Boost, and the Innovation Trophy. Further to their successes with product innovation and building a customer base, Waleli completed a share issue in August 2004 allowing continuing professionalization of the company and in-house innovations. Waleli lists Shell and Roel Pieper among their shareholders.

Siete believes that the IT opportunity based company of the future is small, maneuvers professionally, and has access to efficient resources all over the world. Waleli's international orientation is to grow and remain small at the same time

Table 1. Waleli's major product and service offerings

Products			**Services**		
Product	**What it is/does**	**Industries**	**Service**	**What they do**	**Industries**
GSM-Doorbell	A system that connects a visitor who is ringing the doorbell to the mobile phone of the homeowner	Construction Health Care	Innovation Fource	Using four representatives from the client and four representatives of Waleli, four product ideas are conceived and visualized related to the wireless opportunities of the client's field	All industries with the potential to exploit opportunities in wireless technologies
Switch 21	Product that simplifies the forwarding of telephone calls from fixed lines to mobile telephones	Telecoms	MMS-Witness	Collaboration with the Dutch police force to stimulate the use of digital camera phones for recording criminal activity and also implementing a management system for recording such images	Security Telecoms
Basiqphone	Mobile telephone which only makes and receives calls, features large reading display and touch buttons	Telecoms			

Figure 2. Organizational structure

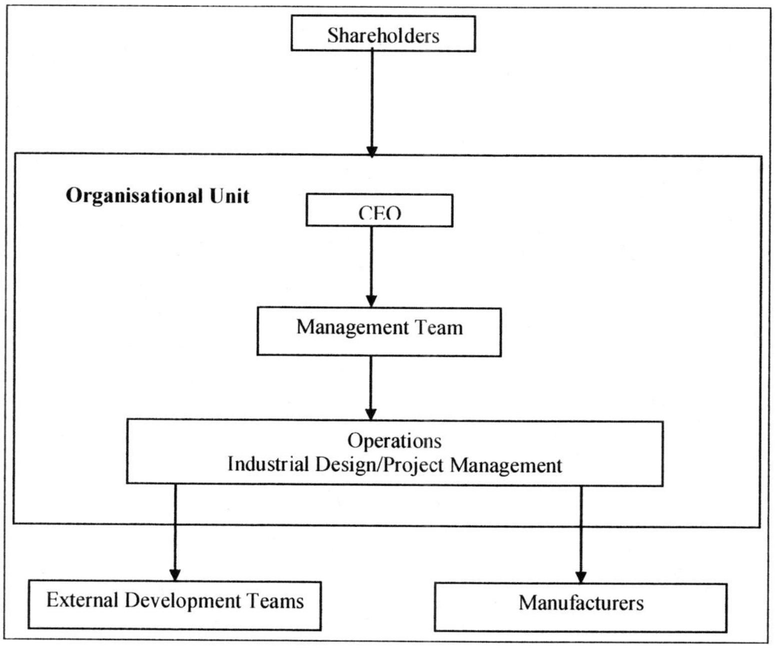

through having a small presence in major international cities, for example, Amsterdam, Berlin, Shanghai, and San Francisco. Currently, Waleli has some products being produced in China, its leaders work with MBA students in Cambridge and have distributors in Belgium and clients in Switzerland, and it is developing new businesses in Germany and Sweden.

Products/Services Provided

As previously mentioned, products of Waleli have two things in common: they are wireless and they meet needs in a simple way. Highlighted in Table 1 is a summary of the major products and services offered by Waleli.

Management Structure

The management structure of Waleli (see Figure 2) is hierarchical; it is led by Siete Hamminga, the founding entrepreneur. The management team consists of Allard van de Rhee as technical director and Hans Joosten, marketing director. Allard van de Rhee started his working career in 1988. Following the fall of the Berlin Wall in 1989, he moved to Eastern Europe to exploit the opportunities within the semipermanent and prefabricated construction market. By 1992 he was in a managerial position and he gained further international experience pursuing business opportunities in Albania when its boarders were opened the following year. In 1995, he returned to the Netherlands and became project leader in a film-set construction firm, which, in 1998, led to him setting up his own enterprise, WharfProductions. In 2003, having sold his share of WharfProductions, he joined Waleli.

Hans Joosten studied business economics at the Catholic University of Brabant. After graduating in 1994, he started working at Nashuatec, a market leader in copier and prints systems. Five years later he introduced a business-to-business formula under the flag of BelCompany, a component of the stock exchange quoted Macintosh. From the middle of 2002 until joining Waleli in the beginning of 2004, he was responsible for the strategic new business team within Nashuatec. Both Allard and Hans are current shareholders in Waleli. There are currently nine operational staff within Waleli, some of whom are focusing on industrial design for multiple projects. On the individual project level, Waleli has one person responsible for each of these. Overall, there are 45 people working on Waleli projects.

Siete views his shareholders as external to the organizational unit; they provide advice and guidance but as the majority shareholder the final decision is his. In addition to the organizational unit there is an external network of firms, including software development teams and foreign manufactures with whom Waleli partner to meet various product requirements.

Siete concedes that this structure would not have been recognizable in daily operation up until recently. In the early development phases, the group was so small and roles were not so well defined that there was no formal organizational structure. However, Siete feels that such a structure is a prerequisite to achieving Waleli's current growth plans.

Financial Status

At foundation, Waleli was valued at one million euros. At the second round of financing in June 2006, the value of the company had almost doubled. Revenues for 2006 are forecast to be over 1.2 million Euro but may be significantly higher (100%) depending on the go ahead of some larger projects. The current challenge is to create cash cows and recurring revenues instead of hourly based project revenues.

Strategic Planning

Waleli has recently begun implementing a growth plan to the year 2009. In this plan Waleli's leaders outline their intentions to develop their organizational structure, international activities, and M2M markets. These events are covered in detail in the Current Challenges section.

Organizational Culture

The organizational culture has always been very entrepreneurial but with the growth plan some organizational structures are becoming more defined and formalized. Nevertheless, Waleli continues to portray a hip and trendy image.

Siete identifies people as being Waleli's greatest asset. However, he has learned some lessons in his human relations policy which has resulted in some changes over the lifetime of the firm. Initially, to keep costs low Waleli hired a lot of graduates and interns. It later became clear that this method cost a lot of time and energy bringing these employees up to speed; therefore, the focus shifted toward young people with relevant work experience and contacts. Siete strives for a results driven culture based on personal responsibility and enthusiasm. The effort is paying off. Waleli's employees do not need to be entrepreneurs; however, they do need to have an entrepreneurial attitude. In some cases, this is stimulated by allocating shares in spin-offs which result from the work effort of employees.

Given the relative size of the company, Waleli spends quite some effort on professionalizing processes like project registration, bookkeeping, public relations, and human relations. This is based on the ambitions and awareness that processes have to be in place before growth absorbs all operational attention.

SETTING THE STAGE

Problem Setting

Entrepreneurs developing new ventures based on IT opportunities are faced with the traditional problems of starting a venture, namely, gathering scarce resources, acquiring knowledge, establishing a reputation, and attracting suppliers, customers, and partners (Autio, Yli-Renko, & Salonen, 1997; Birley & Cromie, 1988; Brush, Greene, & Hart, 2001; O'Farrell & Hitchens, 1988). In many cases, because of the nature of the opportunity, these new ventures are more often operating in international markets. This creates further problems for the starting entrepreneur, as the resources which have to be gathered are sometimes internationally dispersed and they must also prepare both domestic and international sales channels. Further, significant investments must be made to create internationally acceptable goods and services and the pursuit of international trade requires knowledge of international markets and establishing international networks. The resources involved in these activities are significant (Diamantopoulos & Inglis, 1988) and given that high tech new ventures are especially resource poor, lacking the required time, capital, and capabilities to sufficiently prepare international markets (Doutriaux, 1991), investing in international activities may come at the expense of other activities.

Objective

As all starting new ventures are faced with such problems, we describe the entrepreneurial process of Waleli and illustrate how new ventures based on IT opportunities can overcome this paucity of resources through effective networking activities. The entrepreneurial process is a process driven by the entrepreneur, wherein ideas are recognized, prepared, and exploited leading to value creation (Van der Veen & Wakkee, 2006). Waleli, similar to many other IT ventures, is operating in an internationally competitive market. Specifically, this chapter tells the story of the roles and origins of various network contacts and how these impact the entrepreneur's efforts to accumulate the necessary resources to establish and subsequently grow the firm.

CASE DESCRIPTION

Small Beginnings

Siete's ambitions when starting the company were to develop applications for wireless communication which meet needs in a simple way. As previously mentioned, the initial product idea was the GMS-doorbell, a system allowing the occupant of a house to communicate with and open their door through their mobile telephone. Having just graduated from university, Siete Hamminga did not have sufficient resources to launch the business of his dreams. On February 15, 2002, he registered as a "sole trader" at the Chamber of Commerce. Right from the beginning, Siete had a five point strategy for what he had to develop in the coming period: (1) patent, (2) proof of concept, (3) business plan, (4) professional delivery, and (5) cost price analysis. He created a budget estimating the costs of achieving each item on his five point plan. To meet this budget he entered into an arrangement with three private investors, who would lend him €5000 each. In this arrangement, Siete offered three scenarios depending on the development of the company over a 12 month period:

1. If the product idea is sold after a year, then the investors would receive an additional €1,500 on top of their original investment
2. If the company attracted external investment leading to a share issue, then the investor's loans would translate to a 2% stake in the company
3. In the event of the first two scenarios failing, then the investors would forfeit their investment and Siete would not be liable

Siete also registered for a temporary entrepreneurship position[2] (TOP program), which entitled him to an interest-free loan of €12,000 as well as access to a mentor. Part of this program also involved writing a professional business plan.

Siete also set about finding a company that could produce the proof of concept. His modus operandi is that when he is in contact with one party and they inform him that they are not interested, he asks them to suggest a party who might be. Through this method he came in contact with Chess. Chess provides services, products, and solutions in the field of automation for electronic products, development of micro-electronics, embedded software, and business-critical applications. Given his resource constraints, Siete struck a deal with Chess to defer 40% of the development costs for one year, with the stipulation that he would have to pay this 40% if the company ceased to exist. Siete summed up this approach by saying "these were the kinds of deals I made; I made deals everywhere to gather the resources, since I had no money."

Siete also joined several networking organizations to give him some support, both morally and "hopefully" professionally, in the process of starting his own firm. It was at one of these, the De Industrieele Groote Club (IGC), a chic business club in the heart of Amsterdam, that he met with Alexander Berger. During conversation over a beer, it emerged that Alexander ran a graphic design company, G2K. At this time Siete was relying on students from the University of Twente to design his logo, but he was unhappy with their efforts as he felt that it was of a comparable level to what he could produce himself.

Siete explained to Alex the origins of the Waleli brand and his aims for the company. Waleli is the Cherokee Indian name for the humming bird; a bird that characterizes itself by its compactness, velocity, and beauty. It is no larger than a finger bone, it has a velocity of 80 kilometers per hour, yet, despite this, is capable of hovering perfectly still in the air, the wings moving so quickly that not even in slow motion can they be captured with a normal camera. Siete's ambition for Waleli was to be similar to the humming bird, that is, a compact yet mobile company.

Figure 3. Waleli's corporate logo

Alex said that his firm created and designed corporate identities and he offered to undertake this work for Waleli at cost price (see Figure 3). G2K still does this work for Siete today, and, further, he and Alex have become good friends. Reflecting on this scenario, Siete remarked, "At that time it was a really good decision to go for the professional branding and image. Alex pointed out the essence (of this) to me. This was all window dressing to me." Siete didn't want to expend too much energy into these activities, but he changed that strategy within 10 minutes (of talking to Alex) because he realized that "if you start communicating with a hotmail account and a cell phone and print your own business cards you position yourself at the lower end of the market."

The €12,000 from TOP coupled with the investments from the private investors allowed him to build the prototype, buy the patent, create a professional appearance, and fund his own basic costs for the year. In addition to this, Siete realized that he would need an office to work form. He asked Roel Pieper if he could use the attic space in his office premises in Bloemendaal, Amsterdam, free of charge. Roel Pieper agreed to this request, giving Siete not only office space and facilities free of change but also providing him with a professional appearance from the start. This is best illustrated by early visitors complementing the offices and enquiring "Mr. Hamminga how many people are working here?" To which Siete's standard reply was "Around 20 people." This was in fact the truth, however, these people were not employees of Waleli.

During this time, Siete also initiated contacts with Heijmans and Bouwfonds, two of the leading housing construction firms in the Netherlands, with a view to them using the GSM-doorbell in their new building projects. Siete had a friend, who had a friend who worked in Bouwfonds and he used this contact to first pitch the idea. Siete expresses the importance of having a contact name within the company, "You need to have a name within the company and you need to have the name of the person who referred you to them." He also expressed the importance of reusing the contacts you have established when in conversation with other potential customers, for example, when in conversation with Heijmans he told them that Bouwfonds were interested.

At the end of the first year he had produced his "umpteenth" business plan, the prototype had been built, and several project developers were particularly enthusiastic. However, no revenues had been earned. Siete had to go back to the investors and painfully reveal that their money was lost. Despite this setback he remained undeterred; in fact, these experiences increased his belief in the product and he decided to carry on, having learned his first lesson, "It is a challenge to bridge the gap between enthusiasm and commitment... and that takes time."

The Search for External Financing

To move the business forward Siete realized that he needed external financing and decided to conduct a share issue. Roel Pieper had previously indicated that he would invest but Siete did not want to solely rely on this contact and instead he made a list and contacted several leading Dutch entrepreneurs who he thought might be interested in investing. Unfortunately, with these solicitations he failed to find any investment; however, he did receive a copy of one of the entrepreneur's motivational books and the owner of a chain of record stores sent the following amusing response: "Dear Siete, I have made an arrangement with the bank that they don't diversify into selling compact disks and I don't issue any loans."

At the IGC business club Siete also encountered an experienced businessman who expressed an interest in investing in start-up companies. Siete went to the businessman's house to present his business plan; however, this was to yield further disappointment as the businessman commented with a compassionate sigh "opening your door with your mobile is never going to meet a market need. You seem like a likeable guy to me...I'd recommend you look for a decent job as soon as possible."

At a later stage, Siete was able to generate some enthusiasm among four financers. They were even willing to give him an advance payment so that he was able to maintain his livelihood and commitment to the venture. On request, Siete arranged meetings with the management of Nemef and some project-developers so that the potential investors would be able to assess the general opinion about Waleli, that is, to get an idea about his reputation and his chances for success. These meetings were positive, but before the first deposit was made, the investors suddenly pulled out. "*We got carried away by your enthusiasm. We're sorry but Waleli is still only in a development phase and this is too early for us to invest. We think you'll be successful and that we're going to read about you on the front-page of the newspapers,*" remarked one of the potential investors.

Shortly afterwards, Siemens, through their director Kees Smaling, expressed an interest in a plug and play variant of the GSM-doorbell. The importance of being a product developer for a multinational company such as Siemens was not lost on Siete, who remarked "Wouldn't it be fantastic for Waleli as a one man business to acquire Siemens as a client for product development." In June 2003 an offer was eventually tabled but two months later owing to deliberation from their legal department an agreement had yet to be finalized.

Siete then received some encouraging news. He had entered his business plan in the New Venture Business Plan Contest, an initiative of McKinsey & Company organized by "de Baak Management Centre VNO-NCW." The organization called Siete with the specific request to be present during the prize-giving ceremony and not to sit on the balcony. Siete enquired as to whether they ask that of all the participants, and they replied, "Well yes, but you are among the candidates short-listed for the prize." Waleli was among the short-listed candidates; however, it did not win the prize and while the initial good fortune of being nominated was a boost, the experience left Siete despondent.

The situation reached a crisis point. Siete had managed to stretch the restricted starting budget to cover the first 18 months. However, following this, his business and personal finances could not be separated and the demands for payment were piling up to the point where he couldn't take it anymore. Together with his girlfriend he decided to go on vacation to reflect on whether or not he would continue with the business. While on vacation, on the morning of the July 30, Siete received a telephone call from his good friend Daan Westdijk to inform him that there was an article about Waleli's "GSM-doorbell" on the front page of one of the leading Dutch national newspapers, *de Telegraaf* (see Figure 4).

Figure 4. De Telegraaf article

meubel Pagina 26

een afstand van circa twintig kilometer.

Zie verder pag. 6 kol. 1

GSM-DEURBEL

Van een onzer verslaggevers
AMSTERDAM, woensdag

Een gsm-deurbel moet uitkomst bieden aan mensen die niet thuis zijn en toch willen weten wie er thuis voor de deur staat.

Wordt er aangebeld, dan gaat het mobiele toestel van de huiseigenaar rinkelen.

Neemt deze de telefoon op, dan kan er via de intercom worden gecommuniceerd.

Als er sprake is van 'goed volk', kan de bewoner desgewenst via een pincode op zijn toestel de deur openen.

De noviteit, ontwikkeld door het Nederlandse bedrijf Waleli, zal waarschijnlijk eind dit jaar voor het eerst worden toegepast.

Marketingmanager Erwin Evers van ABN Amro-dochter Bouwfonds bestudeert de mogelijkheden om de toepassing standaard, of als optie, in toekomstige nieuwbouwprojecten aan te bieden.

(advertentie)

Voordeel

The Tide Turns

Siete decided that he was only willing to stop the enterprise if he had the feeling that he had tried everything. He formulated a list of five things to do before allowing himself to quit and he set about these on his return, but not before sending a copy of the newspaper article to one of the potential investors who had previously backed out.

Siete also sent a candid e-mail to Siebren de Vries, the founder and owner of Chess telling him he was considering his options for the business (see Table 2). Siebren was at that time on a sailing holiday with his family but recognizing Siete's cry for help he answered, "If you drive to the harbor in Hindelopen, I'll meet you there." Siebren encouraged him and said that he should not stop now after coming so far. He offered him a short tem contract to investigate the M2M market possibilities for Chess. This helped relieve the immediate pressure.

In the meantime Siete realized that Waleli could not continue as a one-man band. On September 16, 2003, Siete received a job application letter from an enthusiastic young man, Daan van Kampen, who had read the article in the Telegraaf and was enquiring about the employment vacancies on the Waleli Web site. In reality, finances were not available to recruit new employees; Siete had only put the vacancies on the Web site to create a more professional profile for those looking. Siete reacted honestly to Daan (see Table 3), who still chose to come for an interview. A short time later, Daan started an internship with Waleli earning €350 per month and with the intention of applying to other companies while gaining some experience.

Table 2. Siete's cry for help

Mail to Siebren de Vries (founder and owner of Chess). Date: July 1, 2003. Time: 10.22 a.m.
Dear Siebren, I would gladly appreciate if, in the coming period, we could organize a private and confidential meeting to discuss the progress of my initiatives and your possible role in this. Your possible role could be from one of the following options: (1) to invest a small quantity of risk capital as an informal investor, (2) to employ us as a client working to make your Bluetooth-vision concrete <u>Current Status</u> • There are seven developers who are interested in implementing the system as standard in new building projects • "Het Bouwfonds" (market leader) wants to use the application of the system in a press campaign this Summer • Siemens have requested and received a tender for the development of a plug and play version on the basis of DECT • The Executive Board of Isolectra have written to inform us that they want to incorporate the product in their portfolio • The market leaders in door (Krego) and lock (Nemef) manufacturing are interested in the product concept and want to market it as a total concept also incorporating their electronic lock • The LTO expressed an interest in marketing the product to their 90,000 farmers and market gardeners • The current patent is too broad and will have to be adopted to new insights • Products which bear technological resemblance are being developed at this time • Within Waleli two new product concepts (concerning Wifi and Bluetooth) have originated which are being developed with others Despite these developments all is not well with Waleli. In spite of the enthusiasm from the market and from some major players, a serious problem exists bridging the gap between generating "interest" and actual cash flow. My gut feeling tells me that Siemens will not pursue the offer. Also, discussions with external financiers have yielded positive feedback with respect to belief and trust in the entrepreneur but not commitment to actual deals. At the moment my business and personal finances have been stretched to the limit, forcing me to consider changing direction. Following Steps There are four possible solutions: 1. To raise some limited bridging capital from one or two informal investors at short notice 2. To undertake part-time consultancy/project management activities to generate cash flow for Waleli 3. To generate direct cash flow by taking a part-time job 4. To leave Waleli dormant and look for a full-time job Best regards, Siete Hamminga

Table 3. Siete's correspondence with Daan van Kampen

Mail from Daan Kampen. Date: September 16, 2003. Time: 3:49 p.m.
Dear Mr. Hamminga, Some time ago I read an article in the *Telegraaf* about the GSM doorbell, which I read with great interest. I tracked down your Web site and have read about your business. Your vision and also the connection between machines and mobile telephones offering many opportunities appeals to me. I have recently graduated from the University of Leiden, where I completed the Master program on Media Technology. At the moment I am looking for a job that is related to my studies. After reading your Web site I thought that is where I want to work. It would fit perfectly to my training to work in a small company with such an innovative atmosphere and I think it would be fantastic to work in such a team. I have a great affinity for technology. I have even persuaded my parents to purchase a home automation system. Although this system does not work at present with the mobile telephone its use in this system is possible. And the GSM door bell is course a beautiful supplement to such a system. On your Web site it stated that in January you were looking for: - Marketing & sales manager - Financial manager - Management assistant/secretary I take it that these functions have been filled in the meanwhile. Therefore, I wonder if there are other job openings within your company. With Kind Regards, Daan van Kampen
Mail to Daan Kampen. Date: September 16, 2003. Time: 2.11 p.m.
Dear Daan, Firstly, thank you for your interest in Waleli. I have received your mail and CV and it looks good. I shall be honest and frank in my reply. Waleli is still in the starting phase, it is not more than a one man business with a passionate and ambitious entrepreneur behind it. The appointments were never filled as there were merely "window-dressing." BUT, … A lot has happened since then. The GSM doorbell is set to fly, we have signed a partnership with Chess and this week I am getting things started with Siemens. Things are beginning to really happen and I am beginning to think about hiring people. From this perspective I would like to meet you with a view to a possible working relationship. If you are still interested given of the new information which I have outlined above, then with this mail I invite you to visit me at Bloemendaal. In the coming time this is possible on Monday the 23rd, Thursday 25th or Friday 26th of September. Kind regards, Siete Hamminga

At the same time Allard van de Rhee and Siete got to know each other through a mutual acquaintance. Siete had mentioned in conversation with this acquaintance that he was looking for someone like minded who understood the fundamentals of entrepreneurship, and she recommended Allard. Allard had just sold his holding interest in WharfProductions and was considering something totally new. They met on the 29th of September and decided to begin working together gradually, with Allard eventually becoming a shareholder and taking on the position of technical director.

Three days before beginning the cooperation with Allard, Siete had brought the situation with Siemens to a head. Siete felt that the uncertainty had to come to an end, in essence, "it was now or never." He organized a meeting with Kees Smaling during which he said, "Kees, you want to pay me to think of opportunities but at the same time you are paying your lawyers to think in problems. I think we need to put a bomb under them to push the situation forward. The following week the contract was finally signed. With the contract secured Siete had the courage to ask the question that had been on his mind for a long time: "why did you choose Waleli when you have your own research and development department?" The answer was simple: "if we do it ourselves it takes four times longer, it costs five times as much and just before it is ready the project gets killed." Through this experience Siete learned the justification for Waleli's existence.

The following month Waleli won the regional final of the LiveWIRE Young Business Award and with it an invitation to compete in the national final, taking place at the head office of Shell, in the Hague. LiveWIRE is an international Shell-program that has been running in the Netherlands, in conjunction with Syntens, the innovation network for entrepreneurs, since 1997. The competition helps technically innovative entrepreneurs with the development of their company, offering personal advice and a professional network. LiveWIRE aims to increase the entrepreneur's chances of success, while adding to the knowledge base of individual companies. Winning the regional final gave Waleli some prestige and helped raise the company's profile among potential investors. Besides Roel Pieper, who always expressed an interest in investing, now Menno Witteveen van Haagveste and André Smit from Shell were also enthusiastic about investing. In early December 2003 Siete met with two further potential investors, this meeting went particularly well and the men were expressing an interest. Siete asked them if they wanted to see the business plan, to which they replied "there would want to be something seriously wrong with it for us not to come on board."

The New Start

On December 9, 2003, the rental contract for the office premises at Keizersgracht 203, the former business premises of Adam Curry (media and podcasting entrepreneur), was signed. Waleli acquired the photo copier machine belonging to Adam Curry in exchange for a GSM-doorbell. They further acquired free desks and chairs through contacts of Daan and Allard and these were immediately resold to acquire capital for new modern office furniture. Waleli were fortunate in that Aquadia, the former ground floor tenants at Keizersgracht had recently gone bankrupt, allowing Siete to cheaply purchase flash desks, chairs, and shelving from this company. The office space was thus brought into the style of Waleli; later this was to become known as "waleli-ization." The office premises were rented allowing for the potential growth of Waleli but in the meantime this left them with more space than was required. Waleli again saw an opportunity in this situation and decided to sublet the office space to other starting entrepreneurs. This action not only allowed Waleli to have their own office space rent free but it also increased the presence within the premises, giving clients the impression that Waleli was a larger company than it actually was. There was now a successful aura surrounding Waleli.

On January 21, 2004, Waleli also won the national final of the LiveWIRE Young Business Award, beating off competition from over 300 other starting entrepreneurs. In the same period, Hans Joosten contacted the company about potential job opportunities. Hans had many years of sales experience working within Nashuatec, a global leader in the field of document management solutions, services, supplies, and consultancy, but he wanted to be an entrepreneur. However, he didn't want to "start from scratch," rather, he wanted to partner with a starting enterprise. Allard, Hans, Siete, the five "informals" (informal investors) had a meeting to discuss the relevant inputs and division of shares.

All in all these affairs were concluded on May 27, 2004, and at the same time it was decided to set up a limited company; thus Waleli B.V. (limited) was born. In the period that followed, Siete went to meet Gert and Bart, two of the original informal investors, for coffee. On behalf of Waleli he returned their investment, including the bonus. They reacted with both astonishment and delight and promised to be available for business support in the future if the need would arise.

Waleli had morphed from a one man business to an entrepreneurial team and the next challenge was to broaden the scope of the company from that of a single product to having a developed product portfolio. The existence of the venture would no longer be dependent on the success of the GSM-doorbell. The search then began to develop further applications for wireless communication which meet customer needs in a simple way.

Broadening the Scope/In Search of Opportunities

Waleli always had the intention and ambition of serving both the consumer and M2M markets. Their first M2M project, which was also their first international activity, was with JohnsonDiversy in Switzerland. Siete contacted an old colleague of his at JohnsonDiversy, which had previously been a subsidiary of Unilever. Soon they found a M2M opportunity to work on. JohnsonDiversy have large cleaning machines, the activities of which were logged twice a year onto a laptop. By implementing an M2M solution, the data is now collected daily, allowing the client to lease the machines per hour while also charging per hour used, thus increasing revenues and efficiencies.

To generate short-term cash flow, Waleli began operating their consultancy services and the "Innovation Fource" (see the section on Products\ Services Provided). The Innovation Fource is a clever strategy,as it is not only beneficial to the client but it also provides further opportunities for Waleli. At the end of each session the client has four concrete and visualized product ideas, the ownership of which is retained by the client for a 12 month period; after this time has elapsed, ownership reverts to Waleli. Further, Waleli offers their clients the opportunity to work together in the product development and bringing the product to market and in these instances the client company retains ownership of the ideas.

In the summer of 2004, Waleli developed the "21Switch" (see Table 1 for more information). This switch allows forwarding of telephone calls from fixed telephone lines to any other phone. The system was previously available but uptake was low because it was a pay service and it required a series of actions to activate it. Waleli simplified this process so that it can be done at the flick of a switch. Waleli pitched this idea at KPN and by Christmas they had initiated a trial run with some 3,000 households. Depending on the outcome of the trial, Waleli might need to supply KPN with some 4.2 million switches to be distributed free to KPN's customers.

Fortunately for Siete, he had already begun a relationship with a manufacturer who is capable of meeting such demand. When carrying out his original search for someone who could produce the "21 Switch," another starting entrepreneur mentioned a Dutch manufacturer in China. Siete coincidently met the daughter of the company's owner, having already received the quote for the initial production. Siete used this opportunity to find out more information about the father and it transpired that he has seven factories, employing some 5,000 people, in China. Siete concluded, "Our relationship with this production company is becoming a very interesting asset for Waleli."

Waleli continued to churn out ideas through the Innovation Force and also through the new product development team throughout the course of the following year, including, among others, the Basiqphone and the MMS-Witness Collaboration with the Dutch police force. Waleli continued pursuing leads in the M2M market but unfortunately these can have a long time delay before projects are realized, and this was especially true for Waleli, as they have insufficient resources to devote to these activities. By this time Waleli had grown to a business with 13 employees, many external development teams, some of which are located at Keizersgracht, and a valuable association with the Chinese based manufacturer. Despite this growth, Siete's ambitions for the company have not yet been realized and for this to happen, some challenges must first be overcome.

CURRENT CHALLENGES

Organizational Challenge

Through its Innovation Fource and internal innovation activities, Waleli has a new business idea funnel which creates multiple opportunities for the firm. However, this also creates the challenge of managing the process of going through the funnel so that they are not focusing on 20 projects at the same time and spending too much time on particular projects. To this end Waleli leaders developed a business model to evaluate their product portfolio, that is, to stop, kill, or recycle projects, and also to see if the resource allocation matches the potential projects revenues (see the project assessment model below). An important issue here is creating recurring revenues because Siete recognizes that Waleli will never reach its growth potential by only doing development projects and sending invoices for the hours worked, as Siete remarked "Then it is a capacity company and you won't ever get big." For Waleli to grow, it needs intellectual property, so it can do royalty agreements or margins on its own products *or* they get shares in exploitation companies, for example, an inventor comes to them and says "I have this great product idea and it fits well with the vision and competencies of Waleli. Can you help me to develop the product and put it to the market?" Waleli says "Yes, we can" and instead of charging €200,000, they charge €100,000 and take a percentage stake in the idea. Only through these mechanisms will Waleli ever grow to meet the growth expectations of the shareholders. It is a daring strategy in the sense that it has a relatively high-risk profile but given the criteria for stopping and starting projects the outlook on recurring revenues is a very important one.

The rationale behind the project evaluation model is to match resource allocation with the potential of the project. To do this Siete defined criteria to judge their current project portfolio; projects with greater potential and that have a greater synergy with Waleli's image and mission should get a greater resource allocation than those that have less potential. Siete commented "This is logical and makes good business sense; however, in practice, it was not always the case." The model (see Figure 5 has four quadrants, which are explained as follows:

1. **Runners:** Are those projects that have a paying client behind them, that is, they are currently cash generators. 70% of Waleli's resources will be allocated to such projects.
2. **Applicants:** Are those projects which might become a runner. Contacts have been es-

Figure 5. Waleli's project assessment model

Runners (70%)	**Applicants (25%)**
Receptionists (5%)	**Mummies (0%)**

tablished but final contracts have yet to be finalized. These projects will absorb 25% of the resource allocation.

3. **Receptionists:** Are those projects that might go live but Waleli is not going to put too much further energy into them. Waleli will remain by and large passive, as Siete feel that they have done everything to acquire the project and now it is up to the client. Waleli's only action will be to phone the client once a month to see if there is any progress on their part. These projects are allocated 5% of Waleli's resources.
4. **Mummies:** Are those remaining projects in Waleli's portfolio, they are used as examples but are inactive. These projects include projects that were both successful or unsuccessful and no resources are allocated to these projects.

Siete has presented a growth plan (2005-2009) to the shareholders in which the project appraisal scheme and the growth paths are important elements. In order for Siete to act on a high level and have less time spent on operational requirements, he needs to have people around him who are able to work independently and see the necessity of creating revenue streams from all available opportunities. Up until now Waleli has been very reactive with its recruiting policies. In essence, its leaders have never actively recruited for a particular position; rather, potential employees have approached them. Siete remarked "People have always approached us and when they are enthusiastic, we say great, come on board. We don't pay too much but you are more than welcome."

On a positive note, this has resulted in Waleli having a very enthusiastic and young team. Conversely, this has also resulted in poor role definition because it has not been defining roles and then hiring the people to fill them; rather Waleli has been taking on staff and then examining what the staff could do. Another drawback of having such a young, albeit enthusiastic, team is that these employees lack a prior reference point for what it is like to work in an actual company. To illustrate this point Siete recounted a recent conflict with respect to the use of Microsoft's MSN™ in the office. Siete was aware that MSN™ was been used by the office staff to chat with friends during work hours; however, he saw too many windows open at one time and felt it was being over used, so he asked his employees to curtail their use of the program. This did not happen, which lead to Siete uninstalling MSN™ from all the computers. The employees responded saying "We are a hip company, we are a cool company in the heart of Amsterdam, we should be allowed to use MSN™." In return Siete asked them to name one company that takes itself seriously that allows their employees to use MSN™ during office hours. Siete concluded, "They understood my position but they have a lot to learn."

From the MSN™ experience, Siete established new criteria for the hiring of future employees. First, new employees will have to have prior work experience. "Very basically, they will need to know what it is like to work, especially in a professional environment." Second, Siete wants future new employees to bring in a relevant network of their own. Siete explained the benefits of this policy as follows, "If you have been a product manager at Philips you know how certain things work, you probably bring in your own network, so it gives Waleli a flying start." Thus, Waleli is changing its recruitment policy from a very responsive or reactive approach to a more focused approach. That is also one of the main reasons it has written the growth plan: to establish where Waleli wants to be in 2009. Part of this process entails evaluating who it already has employed and what tasks they perform, establishing what job openings need to be filled and then evaluating what kind of people they need.

Internationalization Challenge

The current challenges include going international with the 21Switch product. Waleli has already made contact with, among others, Deutsche Telecom, British Telecom, and Telia (Sweden), and these contacts are close to fruition. To help with international efforts, Waleli has entered into an arrangement with Lawrence Masle. Lawrence Masle is an American with a very strong corporate track record; he was vice president of Corporate Communication and Business Development with Ericsson in Sweden. He then went to London to set up an Ericsson office there and following this he went to work for Philips in the Netherlands. Lawrence is currently self employed and teaches MBA classes at a university in Belgium.

Lawrence Masle had read several articles on Waleli and on reading the third one he thought it would be interesting to meet the founder. Siete and Lawrence had several lunch meetings and they decided to start working together. Siete told him on their first meeting that given his curriculum vitae he couldn't afford to employ him, but Lawrence said he was not looking for a job. They settled on an agreement where Siete pays Lawrence a fixed amount every month to cover general expenses. Lawrence recognizes the potential in Waleli and he wants to be involved so much so that he also wants to invest in the company and Waleli is planning a new share issue in the coming time in which Lawrence will participate. Their first project together was to jointly go to Stockholm, where Lawrence introduced Siete to his network, that is, the people who used to work for him there. While there they also met the former CEO of Ericsson, Mr. Sven Christian Nielsen, who is currently on the board of Telia, and through him Waleli began discussions for introducing the 21Switch to the Swedish market.

In a previous shareholders' meeting, Siete informed the shareholders that he wanted to expand the company to give it more of an international orientation. To do this Siete planned to focus on getting smaller offices operational in other countries. The shareholders agreed that it was a great idea but they had their reservations, saying "Let's work on Waleli in the Netherlands first" and also questioning "Who is going to do that, when your time is required here?" So Siete changed this strategy and said "Tickets rather than offices." Let us first focus on developing markets in North Western Europe and finding the right person for this task. Those things collided: the right person, someone who had experience at the highest corporate levels and was internationally orientated, came along at the right time and Lawrence is now conducting the international business development.

Market Challenge

Waleli's ambition is to have a strong presence in both the consumer products market and the industrial applications for the M2M market. Currently, Waleli's activities in developing its market for M2M applications is constrained by a lack of resources, especially in relation to the time available to dedicate to these activities. It only has one live project, namely JohnsonDiversy, dealing with the M2M market. Siete's aim is to embark on the acquisition of M2M projects creating a luxury problem, whereby when they acquire projects they can employ staff to work on these activities.

To develop his M2M aspirations, Siete attended a networking meeting where he hoped to find small development companies with whom he could jointly develop M2M opportunities and possibly take-over, or at least bring under the flag of Waleli. To his surprise, however, he found that the participants at this meeting were all actually potential clients, so Siete took the opportunity to discuss potential business opportunities and made a good head start. The event itself was organized in part by a competitor of Waleli but as a paying guest Siete had no scruples about using this opportunity to pursue acquisitions. In fact, all the companies were responding so positively that Siete took the list of all the companies, including those who did not attend but had expressed an interest in doing so, and called these companies, saying how they missed each other but that he was curious as to what they were doing in this area.

Waleli currently has a list of some 10 projects, half of which arose from the M2M networking event, which have already been discussed with various firms. Siete realizes that they won't acquire all of these projects but hopes to be successful with between one third and one half of these. In the growth plan he outlined plans to hire a senior M2M person. The ideal candidate would be a seasoned business person with the relevant experience and, of course, the relevant network for developing this market. This position will be the top earning position in the firm and there will be a dedicated budget for undertaking these activities. Siete hopes to have acquired some projects so that the day the new employee takes the position Siete can say "Well, I've already acquired three projects, the management of these and the relevant extended development teams will take 60% of your time, while the other 40% will be for acquisitions."

All the projects on the M2M list have also been evaluated using their new business model and there are three projects there Waleli is actively pursuing. These are all projects with larger firms, as the potential upswing from conducting business with such firms is higher. Also, through his experiences of dealing with potential M2M projects, Siete has learned a valuable business lesson, namely, when you put so much time and effort into a project and it doesn't work out, you should look and see who else could use it, that is, offer it to the competitor. Siete remarked that this is an obvious business practice in which they did not previously engage but they are going to more actively pursue this strategy in the future.

REFERENCES

Autio, E., Yli-Renko, H., & Salonen, A. (1997). International growth of young technology-based firms: A resource-based network model. *Journal of Enterprising Culture*, *5*(1), 57-73.

Birley, S., & Cromie, S. (1988, September). *Social networks and entrepreneurship in Northern Ireland.* Paper presented at the Enterprise in Action Conference, Belfast, Northern Ireland.

Brush, C.G., Greene, P.G., & Hart, M.M. (2001). From initial idea to unique advantage: The entrepreneurial challenge of constructing a resource base. *Academy of Management Executive, 15*(1), 64-80.

Diamantopoulos, A., & Inglis, K. (1988). Identifying differences between high- and low-involvement exporters. *International Marketing Review 5,* 52-60.

Doutriaux, J. (1991). High-tech start-ups, better off with government contracts than with subsidies: New evidence in Canada. *IEEE Transactions on Engineering Management*, *38,* 127-135.

O'Farrell, P.N., & Hitchens, D.M. (1988). Alternative theories of small-firm growth: A critical review. *Environment and Planning, 20*(2), 1365-1383.

Van der Veen, M., & Wakkee, I. (2006). Understanding the entrepreneurial process. In P. Davidsson (Ed), *New firm startups* (pp. 27-65). Cheltenham, UK: Edward Elgar Publishing Ltd.

http://www.shell-livewire.com

http://www.utwente.nl/top

http://www.waleli.com

ENDNOTES

1. For information on Waleli, see www.waleli.com The information is the present tells the story of development until May 2006.
2. For information on the TOP program, see www.utwente.nl/top/.

This work was previously published in Cases on Information Technology Entrepreneurship, edited by J. A. Medina-Garrido, S. Martinez-Fierro, and J. Ruiz-Navarro, pp. 148-170, copyright 2008 by IGI Publishing, formerly known as Idea Group Publishing (an imprint of IGI Global).

Chapter VIII
Designing a Resilient and High Performance Network

Abid Al Ajeeli
University of Bahrain, Bahrain

Yousif Al-Bastaki
University of Bahrain, Bahrain

ABSTRACT

This article describes the detailed configuration and LAN infrastructure design at the University of Bahrain (UOB). The article describes the configuration based on the new setup and migration requirements and indicates how the design satisfies those requirements. The article explains the detailed configuration of the design process of the distribution layer switches and shows how these switches can be configured in the final implementation. The article also discusses the modifications that occurred during the implementation/migration phase. The design of the network at UOB campuses incorporates resiliency into the network core in order to manage problems effectively. This will enable user access points to remain connected to the network even in the event of a failure. This incorporation aims to provide services and benefits to users without impediments.

INTRODUCTION

A computer network is simply a system of interconnected computers. The article emphasizes the design and implementation of one type of computer network—the local area network (LAN). It is a group of computers and associated devices that share a common communication line or wireless link and typically share the resources of a single processor or server within a limited geographic area. Usually, the server has applications and data storage that are shared by multiple computer users.

Major local area network technologies include Ethernet, Token Ring, and Fiber Distributed Data Interface (FDDI). Ethernet is by far the most commonly used LAN technology. A number of corporations use the Token Ring technology. FDDI is sometimes used as a backbone LAN interconnecting Ethernet or Token Ring LANs.

Another LAN technology, ARCNET, was the most commonly installed LAN technology and is still used in the automation industry.

At UOB, a suite of application programs is kept on the LAN server. Users who need an application frequently can download it once and then run it on their local hard drives. Users in each department can order printing and other services as needed through applications run on the LAN server. A user can share files with others at the LAN server, and a LAN administrator maintains read and write access. A LAN server also may be used as a Web server.

The proposed network design has a wireless LAN that is sometimes preferable to a wired LAN, because it is cheaper to install and to maintain. The implementation supports a Resilient Packet Ring (RPR), a network topology developed as a new standard for fiber optic rings. The Institute of Electrical and Electronic Engineers (IEEE) began the RPR standards (IEEE 802.17) development project in December 2000 with the intention of creating a new Media Access Control layer for fiber optic rings (IEEE, 2000).

In order to provide readers with more information on network models, we sketch the interfaces, which identify seven layers of communication types as in Figure 1.

Figure 1. The OSI network model

Application
Presentation
Session
Transport
Network
Data Link
Physical

Each layer depends on the services provided by the layer below it down to the physical layers, which define network hardware such as the network interface cards and the wires that connect the cards.

The motivation is to design and build a network infrastructure to solve scalability issues for a network to scale for future upgrade leveraging high-speed gigabit infrastructure with the ability to grow to 10 gigabits per second. The result is that UOB has a network ready for future technologies such as IPv6 and MPLS. Another motivation is the manageability power added to the network with a enterprise management system, which will result in better productivity of the IT staff, faster fault isolation and response, as well as proactive management planning for future applications.

The rest of the article is organized as follows. The second section provides background details on the UOB network organization; related work to high performance and resiliency has been discussed in the third section. The fourth section discusses the physical connectivity and topology of the network at UOB; the fifth section provides discussion details of WAN connectivity. The IP addressing and VLANs—layer 3 design—is outlined in the sixth section. The seventh section discusses network management issues at UOB; and the eighth section gives a brief conclusion with some discussion of future research directions in this area.

BACKGROUND

The Internet is a network of networks, and it is not a network of hosts. For example, the UOB network is connected to one of the Batelco networks, which has a backbone that is connected to other networks. This concept is outlined in Figure 2.

This research study describes a project that aims to design and implement a LAN infrastructure for the UOB upgrading from the legacy ATM network to a state-of-the-art gigabit and

Figure2. View of Internet-connected networks

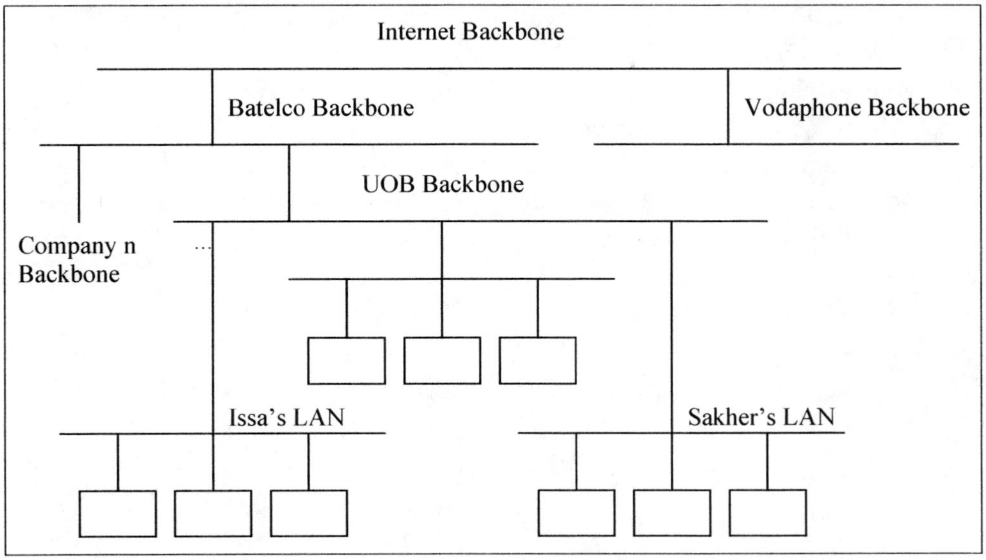

a 10/100 desktop network in the two university sites, Sukhair and Issa-town. The university's two campuses, located near the capital of Manama, offer a full range of study in subjects ranging from electrical engineering to business and education. UOB also is undergoing rapid growth, having added several institutes and expanded from four to nine colleges in the past four years. During this time, enrollment has doubled from 10,000 to 20,000 students, with continued growth projected well into the future.

To facilitate this expansion, UOB originally invested more than $1.3 million to network its campuses with Asynchronous Transfer Mode (ATM) technology from Nortel Networks. Recently, however, increased traffic seriously decreased network throughput, hampering administrative and e-learning applications. Students had difficulty registering online; grades could not be submitted and processed in a timely fashion; and access to e-libraries was sporadic. Moreover, the Nortel ATM infrastructure was incapable of creating additional virtual LANs (VLANs) and otherwise scaling to satisfy UOB's rapidly escalating needs, rendering the network obsolete in just four years.

For these reasons, we started our system by planning, designing, and implementing a network topology that provides the highest application performance and network availability. Furthermore, the system is based wholly on (3Com) core and edge switches technology. The aim is to support local switching on modules, reducing congestion and enabling intensive applications to run faster. Multiple VLANs segregate network traffic and ensure maximum efficiency. Hardware routing reduces packet loss and latency time, allowing flawless transfer of mission-critical data as well as enterprise resource management applications and videoconferencing.

There are a total of 19 Intermediate Distribution Interface (IDF) wiring closets in Sukhair and 18 in Issa-town; each terminates the end user connections. There is one new switch (3Com) in each IDF, which works as an aggregator to the existing L2 switches that provide user connectivity. The switch also provides connectivity to some users at 10/100/1000 Mbps (in some sites, its 10/100 Mbps using 4400 switches).

Two core switches (3Com 7700R) are to be used at the core of the network, one as a core of Sukhair site and the other as an Issa-town core switch. Both will provide connectivity between the wiring concentrator and server farm switches. Whenever available, the wiring closets with 4,924 switches are connected to the core with two gigabit links as link aggregates to the related core.

All of the equipment is managed by a management system, the 3Com Network Director (3ND).

RELATED WORK

Networks have proliferated in many walks of life and have become an integral part of the corporate world. Ubiquitous computing and Internet-capable cellular phones have provided people with the ability to remain connected with individuals, even if they are away from a wired office environment. Numerous researchers continue to develop and to outline new design technologies, including software, hardware, routers, and countless other products (Netcraft, 2004; Tanenbaum, 1996). With wide use of computers in the corporate world, the speed and convenience of using them to communicate and to transfer data have altered forever the landscape of how people conduct business.

Xerox Corporation originally created Ethernet in 1973. Originally, it ran at 3 Mbps and was called X-Wire. In 1979, Xerox and Digital Equipment Corporation joined forces along with Intel Corporation to standardize and to promote the use of a 10-Mbp version called DIX. In 1983, the Institute of Electrical and Electronics Engineers (IEEE) approved the first 802.3 standard, which, for the most part, was the same technology as the DIX standard.

The deployment of networked service applications, such as collaborative tools and LAN-based videoconferencing, created the need for a cost-effective and high-performance network infrastructure (Netcraft, 2004). This led to the advancement of network technologies from ATM to Gigabit Ethernet (GbE) (Chowdhry, 1997), which is the latest in a line of successful Ethernet technologies. It is very similar to its predecessors, but it is faster. Data travels across GbE at 1000 Mbps, which is 100 times faster than the original Ethernet.

Intel (2005) discussed the emergence of multi-gigabit Ethernet and showed how it enabled data centers to accept more transactions and to provide faster responses. To take full advantage of this network capacity, Intel advised data center managers to consider the impact of high-traffic volume on server resources. A compelling way to efficiently translate high bandwidth into increased throughput and enhanced quality of service is to rely on Intel I/O Acceleration Technology (Intel, 2005).

Network availability plays a crucial role in the design of network systems today, especially given the popularity of distributed applications facilitated by the Internet. The all-IP network essentially acts as the system bus to these applications. An article published by ZNYX Networks (2001) claims that a preferable design would include an automated system that circumvents faults in order to reduce challenges facing system designers.

As applications became more network-intensive and as user connection speeds grew, network administrators recognized the need to connect switches to each other at higher speeds in order to prevent congestion on the interswitch links. Initially, this was done with technologies such as Fiber Distributed Data Interface (FDDI) and later with ATM.

In 1992, work began on a higher-speed version of Ethernet, keeping the same basic characteristics of the original but working at 100 Mbps. This was originally designed not to replace the backbone technologies of FDDI and ATM but to give servers a higher-speed connection to the network. It became very popular as a backbone technology, however, because it did not require any kind of translation or conversion of format. It was Ethernet all the way from the desktop to the server and made network administration much easier (Gigabit, 1999; Krunz, 1998).

Nortel Networks (2004a) has engineered a network infrastructure to provide sufficient bandwidth resources to end users. The network was designed for the mission-critical function of an information transfer system. The resiliency network aimed to ensure that those resources are consistently available. Resiliency and redundancy, however, are not the same thing. Redundant networks often have two of every network element. Resilient networks are comprised of network devices that provide reliable failover mechanisms, either within the device or by working in concert with other network elements, so that all network devices can be utilized simultaneously.

A redundant network, as defined by Nortel, is not always the most resilient. Redundant network elements can increase network complexity and can be expensive to implement. A truly resilient network provides the maximum amount of network uptime without requiring an entire duplicate network. As a result, Nortel designed a healthcare network infrastructure with a network resiliency in order to maximize network uptime. Healthcare organizations, furthermore, seek proper implementation and operation in order to ensure sustainable operational performance of the network. In this case, a framework for planning, implementing, operating, and maintaining such a network must be incorporated (Nortel Networks, 2004b).

The article produced by Cisco Systems and Network Appliance (NetApp) highlights key market drivers for IP-connected network-attached storage and elaborates on some common deployment scenarios in customer environments (Ahmed, Godbole, & Vishwanathan, 2005). Faster online response for users is one of the key drivers for customer retention and loyalty in an e-business environment. For example, the eight-second-rule states that if a Web page does not completely load within eight seconds, customers might not return to the Web site, which leads to lost revenue. While performance typically is addressed at several levels in the infrastructure design, networked storage, in the meantime, plays an important role in meeting overall performance goals (Ahmed et al., 2005; Houdt & Blondia, 2000).

IDC (2000) confirmed that networked storage grew from $6 billion in 1999 to approximately $32 billion in 2004 (CAGR = 51.3%), while direct-attached storage shrunk from $14 billion in 1999 to approximately $9 billion in 2004 (CAGR = -10%).

Liu and Yang (2004) claim that the widespread use of the Internet and the maturing of digital video technology have led to the increase in various streaming media applications. In order to support multimedia communication, it is necessary to develop routing algorithms that make decisions based on multiple Quality of Service (QoS) parameters achievable. However, the problem of QoS routing is difficult, because finding a feasible route with two independent path constraints is NP-complete problem (Koyama et al., 2004).

Liu and Yang (2004) confirmed that QoS routing algorithms for broadband networks must be adaptive, flexible, and intelligent for efficient network management. Some researchers proposed a multi-purpose optimization method for QoS routing based on Genetic Algorithm (GA). The simulation results (Koyama et al., 2004) show that the proposed method has a good performance and, therefore, is a promising method for QoS routing.

Hadzic and Szurkowski (2005) discussed Ethernet as deployed in metropolitan area networks (MANs) as a lower-cost alternative to SONET-based infrastructures. MANs usually are required to support common communication services such as voice and frame relay based on legacy synchronous TDM technology in addition to asynchronous packet data transport. The article by Hadzic and Szurkowski (2005) addressed the clock synchronization problem that arises when transporting synchronous services over an asynchronous packet infrastructure such as Ethernet.

A novel algorithm for clock synchronization is presented, which combines time-stamp methods used in the network time protocol (NTP) with signal processing techniques applied to measured packet interarrival times. The algorithm achieves the frequency accuracy, stability, low drift, hold-over performance, and rapid convergence required for viable emulation of TDM circuit services over Ethernet (Liu & Yang, 2004; Koyama et al., 2004; Hadzic & Szurkowski, 2005).

Resilient links protect network against an individual link or device failure by providing a secondary backup link that is inactive until it is needed. A resilient link comprises a resilient link pair that contains a main link and a standby link. If the main link fails, the standby link immediately takes over the task of the mail link. Herbert (2003), in his article "Issues in Resilient Network Design," outlined the misleading possibilities of packet flooding. The packet flooding effect can contrive to mislead network administrators and engineers about the true utilization levels on the switches and cause unnecessary hardware upgrades to take place.

PHYSICAL CONNECTIVITY AND TOPOLOGIES

The University of Bahrain is located at two different sites: Sukhair and Issa-town. The Sukhair site consists of many buildings spread over 10 km. Issa-town has more than 40 detached buildings.

Topology Diagrams

A network topology is the method in which nodes of a network are connected by links. A given node has one or more links to others, and the links can appear in a variety of different shapes. The simplest connection is a one-way link between two devices. A second return link can be added for two-way communication. Modern communica-

Figure 3. Network topology examples

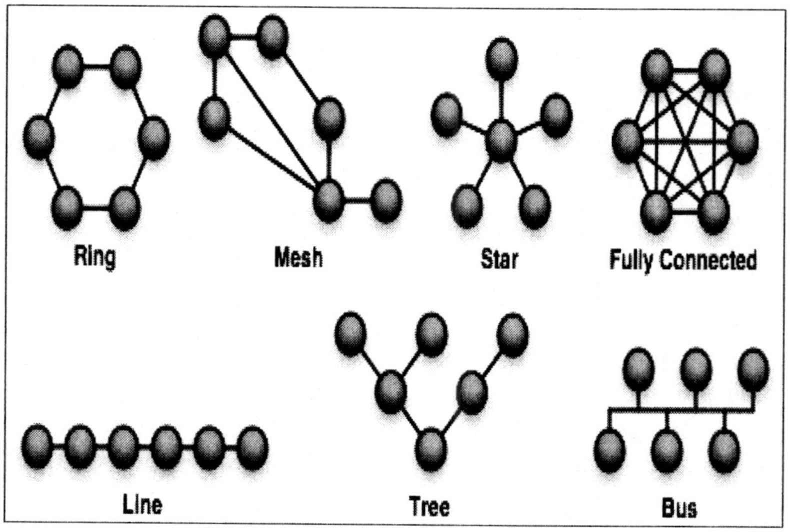

Figure 4. UOB—3Com Network—Sukhair site

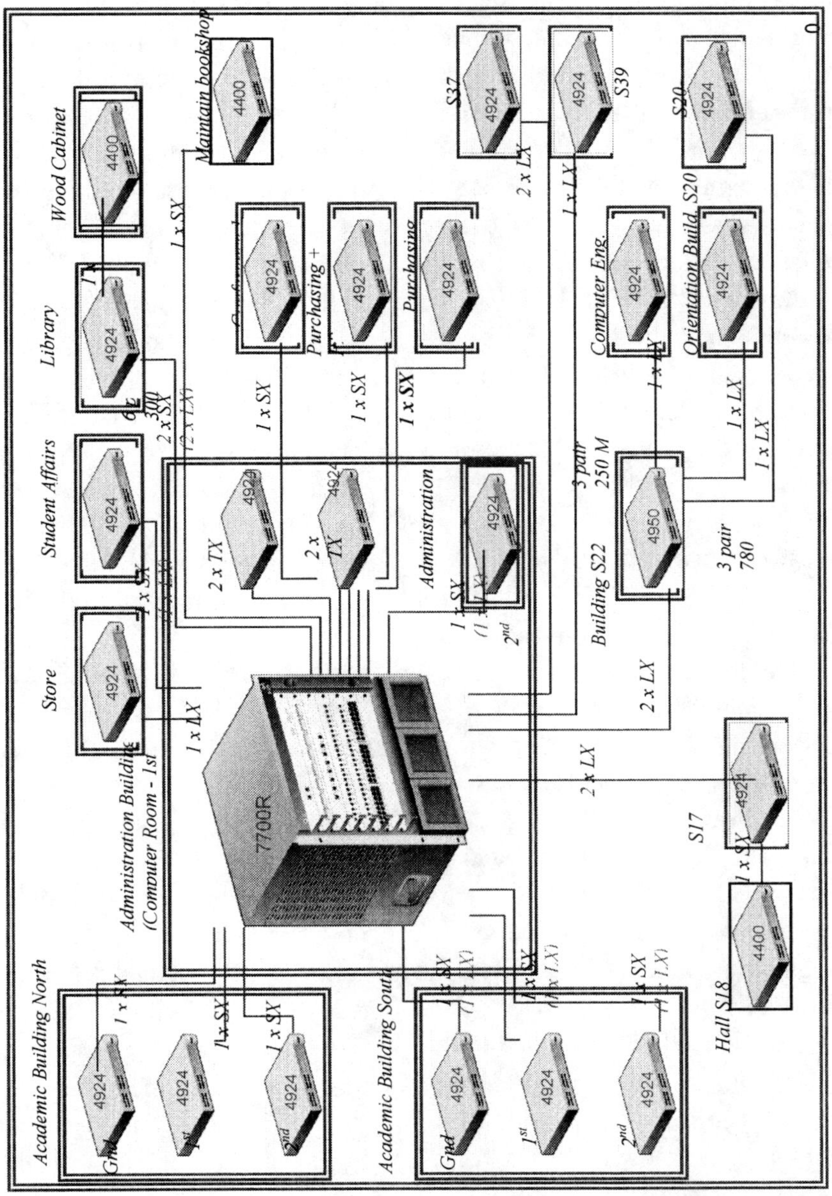

tion cables usually include more than one wire in order to facilitate this, although very simple bus-based networks have two-way communication on a single wire.

Network topology is determined only by the configuration of connections between nodes; it is, therefore, a part of a graph theory. Distances between nodes, physical interconnections, transmission rates, and/or signal types are not a matter of network topology, although they may be affected by it in an actual physical network. Figure 3 shows a number of topologies (Wikipedia, 2005).

Real-life topology applications consist of hybrid network topology. Hybrid networks use a combination of two or more topologies in such a way that the resulting network does not have one of the standard forms. For example, a tree network connected to a tree network is still a tree network, but two star networks connected together exhibit hybrid network topologies. A hybrid topology always is produced when two different basic network topologies are connected.

The design of the network at UOB also is based on hypoid topologies. The hybrid topologies are required because we deal with different network technologies, different network architecture, and different distribution switch layers with different speeds and load sizes.

UOB Network Interconnectivity

Connectivity mechanisms between various UOB sites are outlined in Figure 4 and Figure 5. The figures show, for example, the interconnectivity between the IDF wiring closets and the Switch 7700R cores in both sites (Sukhair and Issa-town) as well as the server farm closet.

The server farm aggregation is designed and implemented using two Giga ports per 4924 switch to provide redundant connectivity for servers. The server aggregators are linked to the Switches 7700R core with 2Gbps.

In Sukhair, we designed the network to be 99 VLANs. Each VLAN has 253 nodes. The VLANs distribute traffic load using either 1 Gigabit links or aggregated links. In Issa-town, the network design follows similar principles adopted in the Sukhair site, in which connectivity uses single links or aggregated links as in buildings 15, 28, and the library.

Wherever two links are connected, the traffic between edge stacks and core is utilizing two aggregate links to the core. Core switches in Sukhair and Issa-town are connected to each other via ATM links on Nortel switches.

Sukhair Campus Switches Setup

Each building at the Sukhair site has one or more 4924 distribution layer switches acting as an aggregator for current building edge switches. As in Figure 4, Sukhair core terminates connections of different buildings at the Sukhair campus. Today, most of these buildings are connected with 1 Gigabit link. Some buildings with high-density users act as distribution links for two or more buildings. High-density buildings are connected by dual links as an aggregated link.

Sukhair 4924 distribution switches are set to have all front ports untagged in user VLAN. They also are used to terminate users and user edge switches—Nortel, 3Com, and Cisco. The Gigabit uplink from the back is tagged on the Interconnect VLAN. VLAN and IP design will be described in detail in the fifth section.

Issa-town Campus Switches Setup

Each of Issa-town's buildings has a 4924 distribution switch that acts as an aggregator for current building edge switches and terminates some power users. As Figure 5 shows, Issa-town core terminates connections of different buildings in the Issa-town campus. Most of these buildings are connected by 1 Gigabit link. Some of the high-density users' buildings are connected by dual links as an aggregated link or acting as distribution switches for more than one building

Similar to Sukhair, the Issa-town core switch has been equipped with dual switch fabric for redundancy as well as 2 × 20 port 1000 BaseX and 1 × 20 port 10/100/1000 for servers and routers connectivity.

Issa-town 4924 distribution switches are set to have all front ports untagged in user VLAN and shall terminate users and user edge switches—Nortel, 3Com, and Cisco. The Gigabit uplink from the back is tagged on the Interconnect VLAN. VLAN and IP design will be described in detail in the fifth section.

Figure 5. UOB network based on 3Com Giga Ethernet solution — Issa-town

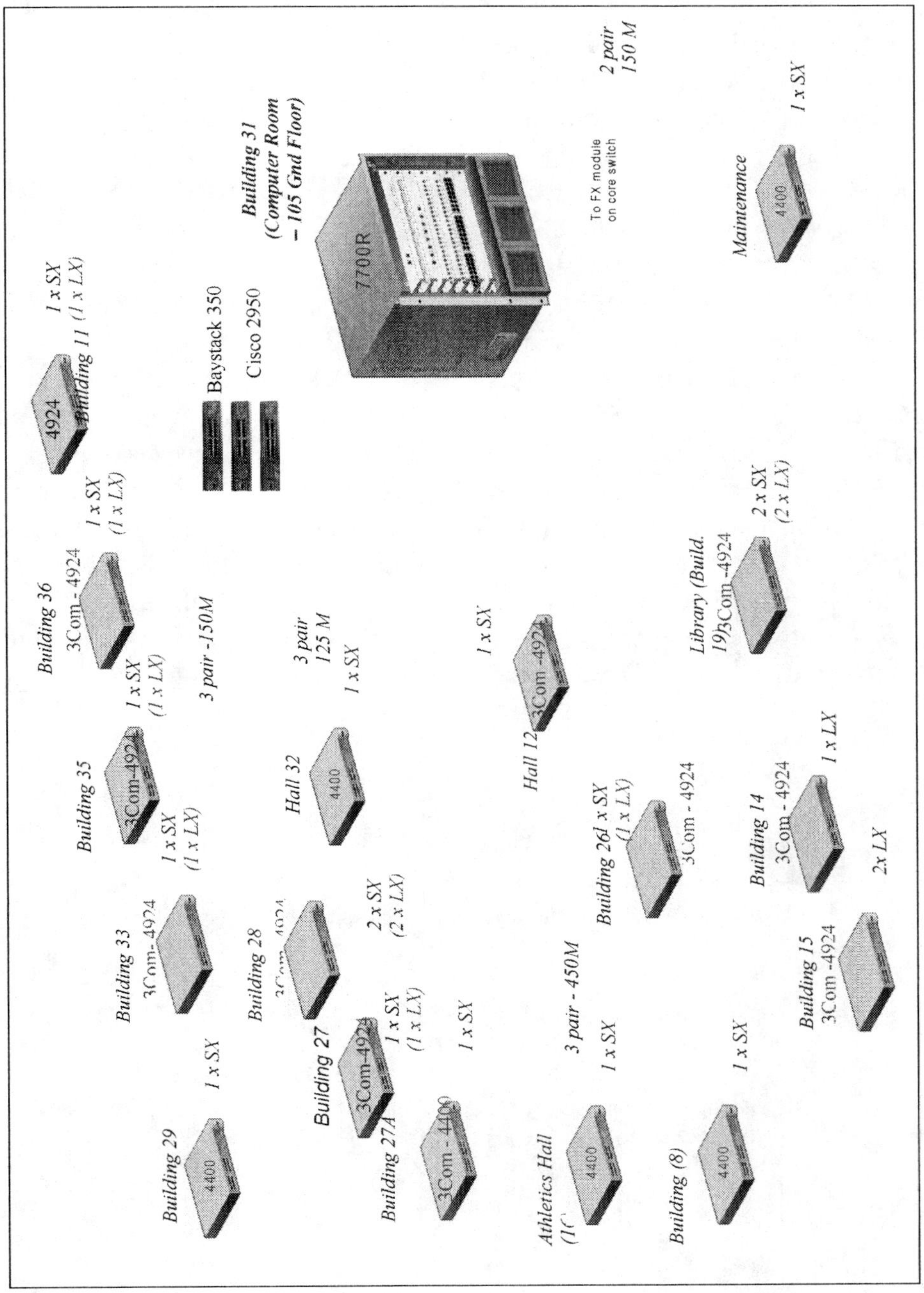

WAN INTERCONNECTIVITY SETUP

Wan Interconnectivity includes an intercampus link between Issa-town and Sukhair as well as Internet connectivity. The intercampus link uses ATM to connect the two campus locations via Batelco at OC3 (155 Mbps speed). As the connection is established through Nortel C5000 and C1400 switches, 3Com network design keeps the connectivity via WAN VLAN connects directly to Nortel on both sides. WAN VLAN is an isolated network on VLAN 1 on each 4007R that connects at 100 Mbps speed to the C5000 on both sides. A Gigabit speed also can be used instead. However, the limitation will be in Batelco connection to 155 Mbps. Figure 6 illustrates WAN connectivity based on ATM.

The WAN link is divided into 2 VCIs (Virtual Circuits); one carries Normal Intercampus data, and the other carries data with regard to the registration and Oracle access VLAN. Filtering is implemented on the C5000 on both sides in order to classify and to manage routing through the two VCIs.

IP ADDRESSING AND VLANS—LAYER 3 DESIGN

In order to understand the modification carried out on the IP schema, a brief description is introduced next. TCP/IP (Transport Control Protocol/Internet Protocol) is the language of the Internet. Agents can perform assigned tasks on the Internet when they learn to speak TCP/IP. A host, for example, that has TCP/IP functionality (such as Unix, OS/2, MacOs, or Windows NT) easily can support applications that use the network.

IP is a network layer protocol that allows the host to actually talk to each other. Such things as

Figure 6. UOB network WAN connectivity based on ATM

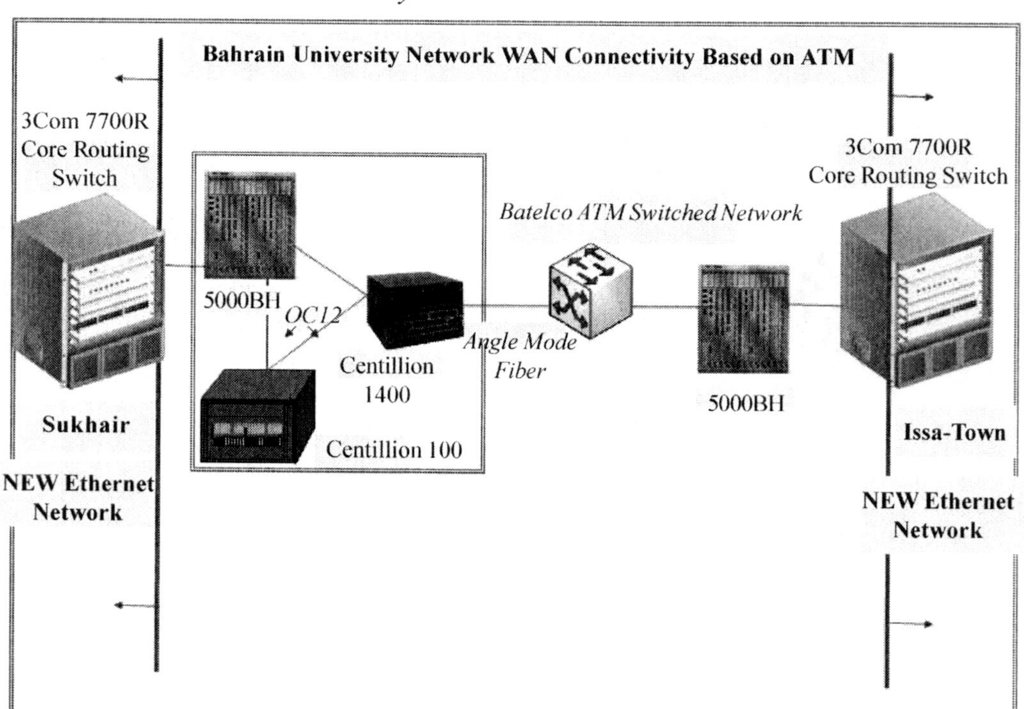

carrying datagram; mapping the Internet address (e.g., 10.2.3.4 to a physical network address such as 08:00: 69:0a:ca: 88); and routing ensures that all of the devices that have Internet connectivity can find the way to each other (CISCO SYSTEMS, 2005).

In our design, some changes have been applied to the IP scheme in both the Sukhair and Issa-town sites. The key characteristic is to have the IP scheme reflect the building number as conveniently as possible. All user VLANs/subnets are in the range of 192.168.x.0, where *x* is the building number. Some buildings have names, not numbers, or have extended numbers such as A20, S18, and in this case, the IP scheme of these buildings will not reflect the geo-location.

As the design employs Layer 3, additional VLANs are used as Interconnect VLANs. Interconnect VLANs are used to carry the RIP routing information. The Interconnect VLANs are in IP range of 172.16.x.0, and they only exist between 7700 Core and 4924 Layer 3 Distribution.

The distribution switches update the cores through the communications with routing information protocol. However, alternative default route has been implemented in each 4924 distribution switch to the core.

The IP addressing of the Interconnect VLAN carries the same third octet number in the native user VLAN. However, LAN ID is the third octet number plus 1000 (e.g., Building 39 has a user VLAN **(39)** 192.168.**39**.0, and its interconnect VLAN **(1039)** to the core is 172.16.**39**.0).

At the Issa-town campus, the VLANs start with 100 in order to avoid repetitive VLANs across the two campuses (e.g., Building 11 has a VLAN 111). Also, for interconnect, VLANs start with 1100; therefore, the associated interconnect VLAN to the same Building 11 will be 1111.

Figure 7. VLANS layer 3 design

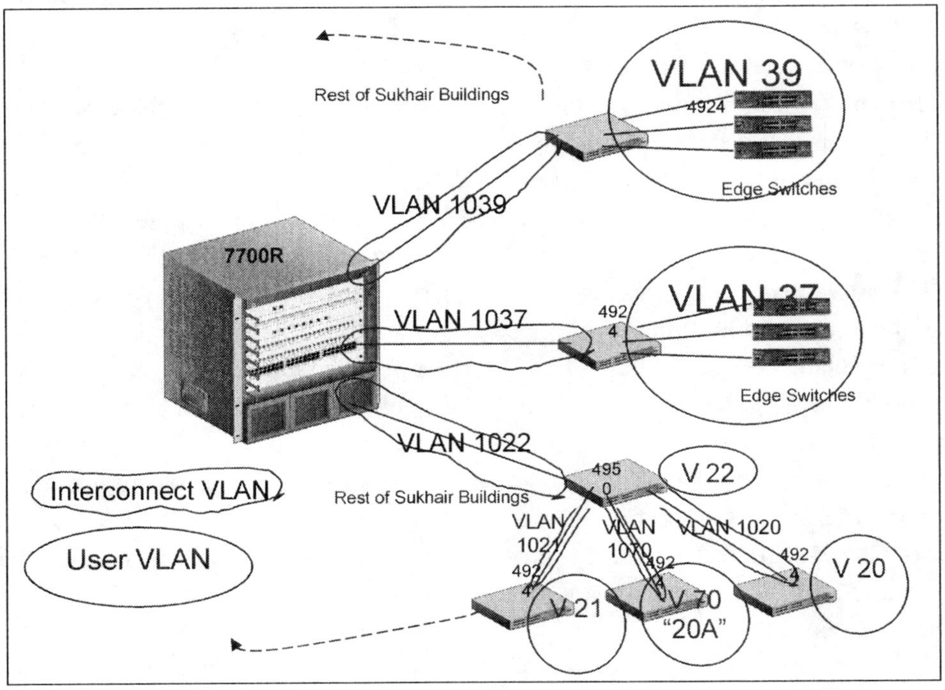

The Interconnect VLANs have only two IP addresses each.

The Routing Information Protocol (RIP) is only enabled on these two interfaces.

VLAN Interface in 7700R	VLAN Interface in 49xx Distribution
172.16.x.1 1	72.16.x.254

For Users of VLANs, the 4924 distribution switches act as a Gateway L3 border. The 4924 addressing for VLAN users (VLAN Interface L3 address) would be 192.168.x.**1**.

Figure 7 shows RIP routing at the Sukhair site. Issa-town RIP routing also will follow a similar concept.

Servers reside in VLAN 100 and VLAN 19 Public VLANs. They utilize 4924s as distributed server VLANs and route through the 7700. In addition to the server VLANs, the 7700 core will have two server VLANs including the following:

- VLAN 3 Oracle VLAN
- VLAN 4 DHCP VLAN

Both servers utilize the UTP 10/100/1000 modules.

There is only one exception for Building 22, which is a distribution for Buildings S20, S21, and S20A. Therefore, the RIP domain has been extended to carry Layer 3 information across the 4950 to the 4924s.

The Issa-town IP scheme follows the same concept with a small change of adding a 100 to the building number. Some buildings with high density have more than one VLAN. These buildings are not numbered in sequence and have been set after consultation with Issa-town network administrator.

The network is primed to support planned improvements with a minimum additional cost when network upgrading is needed. However, in addition to converging voice and data, the design process took into consideration future expansions, so we planned 3Com wireless systems to connect students in dormitories, cafeterias, and meeting rooms.

The administration and management of network processes within an organization can dramatically reduce cost and increase efficiency of the network. Management can properly support applications ranging from registration to instructional material delivery planning. This can help to show a significant saving in teacher and administrative time and can benefit the university, as staff can spend less time on routine tasks and more time helping the university to reach its key objectives (Barrett & King, 2005). A security plan must be put into place, and users need to be educated and trained. Access control lists confine users to areas of the network for which they are authorized, and layer 3 protocol authentication prevents unauthorized users from entering the network, ensuring that sensitive students' data and payroll are safe from prying eyes.

As UOB is growing rapidly, the new design will sufficiently be able to extend for new requirements in the near future. The design addresses capacity/performance and protocols. In terms of capacity, the new network supports upgrade capacity in certain ways: ability to move forward to 10 Gigabit in the Network Core and ability to scale for 100% upgrade capacity within the current equipment. The building block approach opens the limit for scalability with multi-tier network architecture. In terms of applications, UOB is planning to gradually renew and update its infrastructure in order to support new educational-based real-time applications such as videostreaming, IP telephony, e-learning, and so forth. As networking technology changes in order to cope with new application demands, the new network supports current and near future technologies such as IPv6 and MPLS for open Internet world, Internet 2.

NETWORK MANAGEMENT

All of the network devices should be managed from a single console, which should display graphical topology information and provide details of switch configuration, status, and link utilization. The used network management tools provide proactive management for the whole network with comprehensive real-time as well as historical reporting mechanisms.

3Com 3ND is used to manage the network. The Network Management workstation is supplied by ABK as per the specification shown in Figure 8. The recommended specification is advisable. The PCs are stand-alone and are used for the sole purpose of the network management.

Network management also includes grouping and grouping policies and methods for keeping the network running smoothly and efficiently. Knowing what to look for and how to fix it is a necessary skill (Curtin, 2004).

The design of the network at UOB campuses provided a manageable solution to network growing problems by integrating network resilience into core networking devices. This resiliency integration into the network core enables user access points to remain connected to the network even in the event of a failure. By ensuring the availability of the network, core converged applications can provide the services and benefits for which they were designed without impediments (Nortel Networks, 2004b).

Having the new network equipment will extend high-speed Internet everywhere and increase global outreach in order for students to access the Internet for education. The network is now ready for e-learning application, including video and audio multicast with real-time application-aware network devices.

CONCLUSION AND RECOMMENDATIONS

This article described our experience in designing and implementing LAN infrastructure at the University of Bahrain. The research showed how the switching mechanisms were set up at both campuses. The server farm aggregation was implemented by using two Giga ports per 4924 switches. This approach was aimed to provide redundant connectivity for servers.

This article also emphasized the importance of network resiliency, which is a multi-layered process that allows UOB services to support current applications while providing a solid foundation for their future network growth. Working toward five 9s resiliency starts in the network core with network hardware elements, design, and protocols working together to ensure reliability. By ensuring the maximum amount of resiliency in the most heavily used area of the network, network managers can feel confident that applications like IP telephony, multicasting, and collaboration

Figure 8.

	Minimum	Recommended
IBM PC or compatible	500MHz Pentium III	2.4GHz Pentium IV
RAM	512MB	1GB or more
Free hard disk space	2GB	4GB
Graphics capabilities	1024x768/256 colours	1024x768 /64K colours
CD-ROM drive	Yes	Yes
Network adapter card	Yes	Yes

tools can provide the network resources when and where they are needed.

There are two issues that need to be tackled in future upgrading.

1. **RIP Routing Update.** In post-implementation testing, an issue of RIP updates from 7700 are inconsistent, and sometimes information may be lost. This has been experienced with VLANs 10xx, where xx is from 24 to 99. This drawback has been dealt with in the current design by implementing a static default router in the 4924 switch to forward the traffic to the 7700, which is an alternative to RIP routing. We recommend fixing this issue in the upcoming release of the 7700 distribution layer switch.
2. **Intercampus Routing.** The intercampus routing over Batelco ATM utilizes the legacy old equipment of Nortel C5000 and C1400 switches. Upgrading these switches to ATM routers (3Com 6000) is necessary in order to provide a framework for end-to-end 3Com environment, fully manageable environment, and the ability to set end-to-end QoS and filtering.

REFERENCES

Ahmed, I., Godbole, R., & Vishwanathan, S. (2005). *An open standards approach to network-centric storage.* Retrieved July 2005, from http://www.netapp.com/library/tr/3121.pdf

Barrett, D., & King, T. (2005). *Computer networking illuminated.* Jones and Bartlett Publishers.

Chowdhry, P. (1997, December 15). Can gigabit ethernet and RSVP beat ATM? *PC Week Magazine.* Retrieved from http://www.zdnet.com/pcweek/reviews/0908/08rsvp.html

Cisco Systems. (2005). Retrieved June 2005, from http://www.cisco.com

Curtin, M. (2004). *Introduction to network security.* Retrieved December 2004, from http://www.interhacknet/network-security

Gigabit. (1999). *IEEE and gigabit ethernet alliance announce formal ratification of Gigabit ethernet over copper standard* (Press Release). Gigabit Ethernet Alliance.

Hadzic, I., & Szurkowski, E. S. (2005). High-performance synchronization for circuit emulation in an ethernet MAN. *Journal of Communication and Networks, 7*(1), 1-12.

Herbert, J. (2003). *Issues in resilient network design.* International Network Services Inc.

Houdt, B. V., & Blondia, C. (2000). Performance evaluation of the identifier splitting algorithm with polling in wireless ATM networks. *International Journal of Wireless Information Networks, 7*(2), 2345-2355.

IDC. (2000). *Disk storage system forecast & analysis (1999-2004).* Retrieved from http://www.idc.com

IEEE. (2005). Retrieved July 2005, from http://searchsmb.techtarget.com/gDefinition/0,294236,sid44 _gci214428,00.html

Intel. (2005). *Accelerating high-speed networking with Intel® I/O acceleration technology.* Intel Corporation. Retrieved January 2006, from http://www.microsoft.com/whdc.winhec/partners/intel05_IOAT.mspx

Koyama, K., Barolli, L., Capi, G., Apduhan, B. O., Arai, J., & Durresi, A. (2004). An efficient multi-purpose optimization method for QOS routing using genetic algorithm. *Journal of Interconnection Networks, 5*(4), 409-428.

Krunz, M. (1998, Fall). *ECE 564: Broadband networks and multimedia communications* [Lecture notes]. University of Arizona.

Liu, F., & Yang, C. (2004). Proxy design for improving the efficiency of stored MPEG-4 FGS video delivery over wireless networks. *Journal of Communication and Networks, 6*(3), 280-286.

Netcraft. (2004). *June 2004 Web Server Survey.* Retrieved from http://news.netcraft.com/archives/2004/06/06/june_2004_web_server_survey.html

Nortel Networks. (2004a). *Designing a resilient network.* Retrieved from http://www.nortel.com/products/01/passport/8600_rss/collateral/nn107680-031804.pdf

Nortel Networks. (2004b). *High-availability network for the healthcare industry.* Retrieved from http://www.nortel.com/corporate/success/ss_stories/collateral/nn107403-021804.pdf

Tanenbaum, A. S. (1996). *Computer networks* (3rd ed.). Prentice Hall.

Wikipedia. (2005). *Network topology.* Retrieved January 19, 2006, from http://en.wikipedia.org/wiki/Network _topology

ZNYX Networks. (2001). *Network level resiliency for high availability (HA) in ethernet networks.* Retrieved from http://www.znyx.com/products/software/openarchitect/OAHA/OA_HA_30page_001_web.pdf

This work was previously published in International Journal of Business Data Communications and Networking, Vol. 2, Issue 2, edited by J. Gutierrez, pp. 37-54, copyright 2006 by IGI Publishing, formerly known as Idea Group Publishing (an imprint of IGI Global).

Chapter IX
Collaboration Challenges in Community Telecommunication Networks

Sylvie Albert
Laurentian University, Canada

Rolland LeBrasseur
Laurentian University, Canada

ABSTRACT

This article reviews the literature on networks and, more specifically, on the development of community telecommunication networks. It strives to understand the collaboration needed for innovative projects such as intelligent networks. Guided by a change management framework, collaboration within a community network is explored in terms of the formation and performance phases of its development. The context, content, and process of each phase is analyzed, as well as the interaction of the two phases. User involvement and technology appropriation are discussed. Collaboration challenges are identified and linked to the sustainability of the community network. Policy makers are presented with a model that gives some insight into planning and managing a community network over time.

INTRODUCTION

Collaboration in networks and managing performance across organizations has gained the attention of researchers (Huxham & Vangen, 2000). Our comprehension of collaborative networks[2] has progressed substantially over a couple of decades (Oliver & Ebers, 1998), but it lacks integration (Ebers, 2002).

Collaborative networks cover a range of purposes such as innovation requiring heavy investment in R&D, international ventures, and the delivery of public services like health and education. This article is focused on telecommunication networks that operate within a physical and shared community space. The more ambitious community networks aim to become "intelligent" communities with broad participation and significant impact on the local social and economic development. To understand them as a dynamic phenomenon, a framework is needed that can accommodate and organize the conceptual pillars of organizational environment, structure, culture, leadership, and management. Pettigrew (1992, 1987) offers such a framework, and Ebers (2002) and LeBrasseur et al. (2002) demonstrate its effective application.

Organizations in all sectors have become more interested in inter-organizational collaboration to encourage synergy, innovation, and economic development. Although there are many pockets of successful collaborative efforts, there is a continuing need to identify the challenges and opportunities inherent to community networks. With this focus, this article is divided into four main sections. First, collaborative networks are defined and described, and community telecommunication networks and their potential for supporting intelligent communities are analyzed. Second, key collaboration challenges that impact on the development of a community network are introduced. Third, the literature is reviewed and organized according to the context, content, and process involved in these community networks during their two phases of development–formation and performance. The collaboration challenges present in each phase of development are explored, including challenges that the users experience. Fourth, the article concludes with policy implications for network planners.

TELECOMMUNICATION NETWORKS AS AN EXAMPLE OF COLLABORATION

Collaboration is the pooling of resources (e.g., information, money, labour), by two or more stakeholders or partners[3], to solve a set of problems, which neither can solve individually (Gray, 1985). It involves an interactive process whereby organizations, using shared rules, norms, and structures, act or decide on issues related to a problem domain (Wood & Wood, 1991). The intentional goal-oriented collaborative arrangement that emerges is that of a network (Poyhonen & Smedlund, 2004).

Networking represents a particular form of organizing or governing exchange relationships among organizations and is an alternative to markets and hierarchies (Ebers, 2002, p. 23). Network partners maintain their autonomy and retain residual property rights over their resources that have been pooled to achieve mutually agreed outcomes (Bailey & McNally-Koney, 1996; Brown et al., 1998; Gray & Hay, 1986; Huxham & Vangen, 2000; Oliver & Ebers, 1998). The principal coordination mechanisms for allocating resources are negotiation and concurrence. Informal social systems, rather than bureaucratic ones, coordinate complex products or services and reduce uncertainty (Jarillo, 1988; Jones et al., 1997).

Networks have gained in importance over the last two decades. For the private sector, globalization and the speed of change have encouraged collaborative efforts. For government, downloading[4] since the 1990s has forced new ways to view management of programs and services for resource maximization (Bradford, 2003; Bailey et al., 1996). Municipalities and regions have also demonstrated an increased interest in collaboration efforts and network development to attract new opportunities and maintain their competitive advantage. Collaborative networks typically increase the scale and visibility of program efforts, increase support for projects, and leverage capital to enhance feasibility, speed, and effectiveness (O'Toole, 1997). Synergy is achieved through improved resource management and intensive exchanges on specific projects.

To achieve synergistic gains and programming enhancements from sharing resources, risks, and rewards, stakeholders need to shift their focus toward collaborative rather than competitive advantage (Lowndes & Skelcher, 1998). Too often in the past, public sector organizations built independent silos and their private sector counterparts viewed potential partners as competitors rather than collaborators. Public policies dealing with ambitious or complex issues, like community sustainability, are likely to require networked structures that allow for the pooling and mobilization of resources from both private and public sectors within a government policy initiative (O'Toole, 1997).

Community telecommunication networks reflect the trend in western society away from bureaucratic government to network governance (Sorensen, 2002): the latter delivers more services efficiently with less risk and uncertainty (Considine & Lewis, 2003; Jones et al., 1997). Stakeholders and collaborators include municipalities, health, education, social services organizations, and private sector organizations. These networks are part of a wider agenda to increase the country's capability for the knowledge-based economy.

There are several kinds of community networks (Gurstein, 2000; Pigg 2001), ranging from those serving a restricted membership (usually called private networks) to those serving a broader segment of the community or region. A private network may, for example, link several schools and/or municipal sites, and members would include the students, administration, and staff of these organizations. In contrast, a community network is built on a platform that gives broad access to community citizens, businesses, and agencies; it encourages many stakeholders to become a user and service provider. These stakeholders may come together simply to upgrade an aging infrastructure, especially when market forces cannot be relied upon to meet community needs or to co-build economic foundations. Leading communities strive to build partnerships and synergy to overcome barriers to access, job creation, and innovation (Agres et al.,1998; Eger, 2001; Tan, 1999; Industry Canada, 2002a, 2002b). Community networks facilitate information dissemination, discussion, and joint activity by connecting neighbours, creating new opportunities, and empowering residents, institutions, and regions (Carroll & Rosson, 2001; Igbaria et al., 1999; Canadian National Broadband Task Force, 2001).

A community network has four basic components: a telecommunication infrastructure with broadband capability, applications or content, devices (such as computers, cellular telephones, i-pods, and blackberries), and users. The development of a community telecommunication network typically occurs through a governing board representing the needs of users, which is supported by a small management structure (e.g., executive committee and network manager). The network relies on information and communication technologies (ICTs) and allows the community to import and export knowledge, encourage innovation, and overcome distance. The opportunities for economic and social development are contingent on attracting many users and creating a culture of "digital" use. The network must fulfill user needs and be attentive to their requirements, which may include a fair price, access to computers and the Internet, and training and education.

Infrastructure investment in the telecommunication network aims for the widest possible coverage of the community and region, with the constraint of reasonable cost. Investment also tries to ensure that users have access devices; some users have modest means, and schools and other organizations may have budget constraints. On the human resources front, technical training of local staff may be required to install the infrastructure and devices, and provide support to users. Organizations may need to re-design processes in order to meet the changing needs of their supplier and distribution partners, and to offer services online to end-users. The transformation effort may also require promotion campaigns to attract both individual and organizational users. These many resource challenges imposed on the community require a collaborative effort to pool resources and find innovative solutions.

A community network has users at the individual, organizational, and community levels of human activity and endeavours. Individuals or end-users use the network to communicate with friends, play games, access information, obtain training, and occasionally innovate. Organizations are often stakeholders and use the network for a wide variety of purposes (Waits, 2000), including providing information and services, and selling online. They are intermediate users (Williams et al., 2005) and are the drivers of the development of the network. These organizations are the channels through which collective innovation is exercised and community change takes place (de la Mothe, 2004; Rycroft, 2003). At the community level, these individuals and organizations create aggregate demand and use of the network, and determine the sustainability of the network. The challenge is to create a culture of "digital" use that is integrated into the broader culture that is shared by community members.

During the development of the network, user involvement can be traced through participation in articulating a "digital" vision for the community, in the purchase of access devices and services that will connect users to the network (e.g., telephone, cable, wireless, computers, and Internet), and in the utilization of applications being made available through these access devices. Users may also be involved in creating employment by innovating on the network configuration, and in helping to create a culture of use by providing additional attractive applications.

Good governance requires legitimacy with an appropriate range of stakeholders, and involves building cohesion and commitment. Relationships are voluntary, and network survival depends upon the collective will and commitment of the stakeholders. The intentionally-planned network takes on a collaborative structure composed of local residents, non-governmental organizations, private sector businesses, and government. The stakeholders create a product that reflects the concerns, priorities, and aspirations of the local population. If the infrastructure, devices, and applications meet the needs of the users, a culture of "digital" use emerges as an organic extension of existing community ways and practices. Without broad participation, the network is likely to reflect narrow interests and weaken the community's social sub-system, which in turn will limit the economic success of the network.

A sustainable community telecommunication network makes consistent and positive contributions to the economic and social development of the community (ITU, 2003), thereby enhancing the community's capital base. In large measure, these positive outcomes depend upon the collaboration of partners. They also reinforce the efforts invested in collaboration. Networking allows individuals, professionals, and entrepreneurs to access information and knowledge, learn about a wide range of issues, recognize opportunities, and achieve innovative products and services (Suire, 2004; Martin & Matlay, 2003; Corbett, 2002; Ardichvili & Cardozo, 2000; Kickul & Gundry, 2000). Whether a community network realizes its potential depends upon how well it is developed.

The above discussion portrays the formal structure of a community network as a fluid organization composed of volunteers with the purpose of facilitating the community's transition and participation in the information society. However tempting, this viewpoint is non-critical in nature; it ignores the community context and processes by which the network emerges (Pigg, 2001; Day 2002).

COLLABORATION CHALLENGES FOR COMMUNITY NETWORKS

Communities around the world have demonstrated that transformation is possible using network technology. For example, Sunderland (UK) reduced unemployment from 30% to 4% by moving from a shipbuilding and coal industrial base to a knowledge and technology economy. Similarly, Spokane Washington (USA), once a railroad town reliant on natural resources, dramatically improved the fortunes of its downtown by installing the highest density of broadband in the country. In Tianjin (China), a major push on broadband connectivity was accompanied by rapid user growth, from 20,000 to 2,700,000 in two years. Their stories make ample reference to the intensive collaboration of many actors, but the patterns of influence are not well articulated.[5] Bell (2001) compared six urban communities noted for their telecommunication achievements and identified two effective patterns of collaboration: (1) a comprehensive and formal plan, and (2) a coherent pattern of individual initiatives. Similarly, Williams et al. (2005) reviewed numerous ICT initiatives, both small and large, and emphasized the overlapping nature of the planning, implementation, and use stages of development. These patterns are explored under the phases of network development section of this article.

Individuals and organizations involved in the creation of a community network face four collaboration challenges:

1. Defining and agreeing on the transformation effort (includes vision, transformation, and planning).
2. Assembling and mobilizing resources (includes interdependence, tasks, and structure).
3. Assembling and mobilizing trust (includes prior experience, communication, and distribution of power among collaborators).
4. Balancing leadership and collaborative management (includes the broadening requirements of the network, user appropriation, and power).

These challenges are tied to the coherence and adaptability of the network, and specifically to the dynamic relationship between the formation and performance phases of its development. Collaboration is inter-woven in each of these challenges. Network sustainability is achieved by collaboration efforts that evolve during the network's development.

PHASES OF DEVELOPMENT OF A COMMUNITY NETWORK

We propose that network development takes place in two phases that are iterative in nature. Phase 1, the formation of the community network, is marked by the emergence of a leader and/or a board of directors, to respond to environmental pressures. These pressures may occur as a result of globalization and the need to remain competitive in the face of other communities or regions. It may occur as a result of downsizing or social development pressures (e.g., lack of medical practitioners, youth out-migration). The broad goals of the network are developed, including a representation of the future user. Phase 2, network performance, involves the concrete objectives and steps that the board takes to achieve the community goals that were agreed upon and the measures taken to attract and retain users. User involvement can and should take place in both phases of development.

Smaller communities need collaborators to solve a wide variety of challenges including infrastructure availability. Larger communities tend to have more resources and thus need collaboration to resolve economic and social pressures rather than infrastructure issues. In this second phase, the network can develop a culture and structure that gives meaning and coherence to a variety of projects. Some communities are more liberal and hands-off, allowing the private sector and citizens to develop content and opportunity. Others intentionally plan a vision of community transformation based on an improved telecommunication infrastructure. Phase 1 depends highly on leadership dynamics whereas Phase 2 is closer to managerial dynamics but with a distinctive collaborative flavor. These two phases are interdependent over time in that formation sets the stage for performance, and performance impacts on the board and leadership dynamics. Positive outcomes at the performance phase consolidate the dynamics of the formation phase; negative outcomes challenge the board and leadership and initiate a re-formation phase. This iterative process was demonstrated in the feedback loop identified by Arino and de la Torre (1998) and Thomas (1993).

Because networks are fluid in nature (pooling from existing resources, changing membership, and varied timelines) and focused on both results and relationships, two interactive phases are considered sufficient. The two phases are supported by case studies of strategic alliances (Doz, 1996) that found that successful partners actively exchanged information, re-evaluated the project (in terms of efficiency, equity and adaptability), and consequently readjusted the initial conditions of their cooperation. They are also consistent with the ICT social learning findings of Williams et al. (2005).

Formation Phase

The push and pull factors in the environment impact on the community members and prompt them to consider uniting their forces to address the issue or opportunity that has been identified. Under the leadership of a visionary, and through ample interpersonal communication, a group is assembled that represents the initial membership of a potential network. If a consensus on vision and goals is attained, the group becomes the founding board of a network and plans for the performance phase. The principal outcome is a collaborative base on which to build the network. Table 1 provides an overview of the critical factors present in the formation phase.

Context of Formation

The outer context or environment includes factors such as economic, political, culture, demographics, funding opportunities, pressures from government agencies, and technology innovation trends (Agres et al., 1998; Bailey & McNally-Koney, 1996; Igbaria et al., 1999; Keenan & Trotter, 1999; and Tan, 1999). Global competitiveness and turbulence are the underlying catalysts for creating networks for organizations, communities, and governments (Poyhonen & Smedlund, 2004; Scheel, 2002).

Interdependencies exist because organizations possess or control vital resources (material, human, political, structural, or symbolic) and thus are the source of environmental pressures for one another (Wood & Wood, 1991). Organizations seek to reduce these pressures and manage the interdependencies by gaining control over crucial resource supplies. The network form, as opposed to markets and hierarchies (e.g., vertical integration), provides a neutral space within which organizations can meet to explore solutions and synergies.

Table 1. Influencing factors at the formation phase

Context	Content	Process
• Economy • Social/cultural • Political • Urbanization • Funding • Technology • Globalization & competition • Cost Benefit/ Synergy	• Vision • Power • Board Membership • Concept of Sustainability • User representation	• Values • Expectations • Goals • Planning • Leadership (transformational, visionary)

International bodies such as the World Bank (1999), the United Nations (1998), and OECD (1997) have adopted the paradigm of the information society as a guide to many of their development policies. According to Castells (1996, 1997, 1998), ICTs have produced a network society in which mobilizing knowledge and information have become more important than mobilizing physical resources. He argued that both organizations and individuals can benefit from electronic networks; they support the development and dissemination of knowledge and information, and facilitate innovation. Castells warns that these changes are accompanied by growing wealth disparities, social fragmentation, and dislocation. Governments are addressing these concerns, in part, by financially supporting the creation of community networks with broad accessibility. Locally, these new opportunities are often communicated through the chamber of commerce and other economic development agencies to mobilize or inspire stakeholders into action.

Communities come in all sizes and density, and all are influenced by the urbanization trend. Rural settings are witnessing the exodus of their youth and an erosion of their economic base as cities attract both talent and investment, including initiatives in telecommunications (OECD, 2004). Recent studies of Canadian rural communities concluded that ICTs can act as enablers of community building and development processes (New Economy Development Group Inc., 2001; Canadian Advisory Committee on Rural Issues, 2004). Given that the Internet is content-rich, offers online education, facilitates social networking, and offers a platform for the creation of new enterprises and the expansion of existing ones, the viability of the digital community network becomes crucial for the future of small communities.

When governments create generous programs to create community networks (e.g., Brown et al., 1998), communities are pressured to apply for capital funds even when they may not have the organizational and resource capacity to sustain the network. Smaller communities have relatively fewer and less diverse resources and a push-style policy may be the only way to spur action. Another example of a push factor is when a major telecom firm seeks a community partner for a demonstration project, or when the private sector chooses to make an investment to upgrade its infrastructure. The telecom supplies the ICTs, but the community stakeholders still need to demonstrate and plan on how the technology can be applied to personal and organizational purposes. Often, infrastructure is built and languishes until there are other pressures in the environment of the community, such as closure of a major employer or the arrival of a strong champion. At other times, communities struggle with the lack of open access that inhibits economic development and competition. Pushing for open access can discourage the involvement of incumbent carriers, at least at the onset. The key here is to evaluate how context issues can stimulate communities into action toward their transformation.

Content of Formation

Stakeholders need to find the community vision attractive and see a benefit for themselves and for their organization. When the problem is broad in scope and complex, such as economic development, it requires a larger variety of stakeholders with legitimate interest to devise solutions and bring sufficient resources to bear. Stakeholders must have the right and the capacity to participate, and include organizations with legitimate power as well as those who will be affected by the network.

Collaborative action necessarily involves interdependence between individuals and organizations (Ouchi, 1980) and can yield both intangible (e.g., image of citizenship) and tangible benefits (e.g., cost reductions and additional revenues). Interdependence is strongly linked to the vision of the network and the factors motivating stakeholders. It allows for an exchange among stakeholders that is built on trust, and an understanding of mutual benefit or advantage. According to Olk and Young (1997), the more ties an organization has to others in a network, the less likely is it to act opportunistically. Blois (1990) argued that collaborators should engage in bargaining on who will accept responsibility for certain elements of the transaction costs. They must come to the table understanding their role and develop a level of interdependence and mutual benefit in order to sustain the network effort.

The economic and social exchanges that take place are mediated by mutual trust. Ring (2002) distinguishes between "fragile" and "resilient" trust. The former is typical of opportunistic settings such as markets and involves the formal processes of negotiation, transaction, and administration. In contrast, the latter is the foundation of successful networks and is based on the informal processes of sense-making, understanding, and commitment. However, prescribing resilient trust does not ensure that is takes place. Ring proposed that it will emerge when the participants have a shared history of experience and when reputations for reliability are well established. On the other hand, Doz (1996) has documented the role of trusted intermediaries in helping other participants to shift gradually from fragile to resilient trust. We conclude that if a community has rich social relations, it can establish resilient trust early, but that parachuting in partners and stakeholders makes fragile trust more likely. However, if trusted intermediaries become involved, they can build the level of trust within the network.

There is a need for legitimate authority, credibility, and multiple memberships (Bailey & McNally-Koney, 1996; Gray & Hay, 1986) if a sustained transformation is to occur. Jones et al. (1997) have argued that networks need to restrict membership access and choose its members according to their reputation and status. Important stakeholders may choose to join a network or a project because of the presence of other members. A smaller number of leaders may allow the network to realize quick wins, reduce coordination costs, and improve interaction frequency. One could argue that success will breed success—trust will increase, motivation will increase, and faster output can be generated. This view is less applicable to community networks where innovation, legitimacy, and broad reach is critical and leads to a large membership and numerous exchanges. Therefore, a smaller, more restricted network may be mobilized quickly and act efficiently, but be less effective in producing varied output. The larger network may slow the pace of change, but may be important enough to attract accomplished leaders. Structure issues become important in managing a larger group of stakeholders and are discussed in the performance phase of the network.

Another content issue is sustainability. Stakeholders want to know, "How much will it cost," but few ask "How will the network become sustainable in the long-run?" Sustainability is a function of revenues (stemming from the use of the infrastructure and its applications), and the costs of the network (human resources, equipment, and materials). There are opportunities for synergistic gains when partners chose to purchase as a group, or share the operating costs. At the formation phase, the concept of sustainability is hazy, but becomes clearer as projects develop during the performance phase. Nevertheless, the board must carefully address the sustainability issue early to ensure that it becomes incorporated into their common frame of reference.

In the formation stage, the planning includes an explicit model of future users, their communication needs, and their likely use of the telecommunication network. Williams et al. (2005, p. 112, Figure 5.2) identify ways for direct involvement of users, such as user panels, market research, and trials. They also identify sources of indirect evidence about users through information on demand and markets for similar products, and competitive offerings. With the additional input of board members who understand their community, a representation of the users is developed. This user-centered approach is helpful in guiding the design of the system and identifying training and promotion requirements. However, Williams et al., emphasize its limitations and the design fallacy that it breeds: "the presumption that the primary solution to meeting user needs is to build ever more extensive knowledge about the specific context and purposes of an increasing number and variety of users in the technology design" (p.102). The idea of perfect user representation ignores the reality that users are active agents and appropriate the technology later, primarily in the performance phase of network development.

Communities would be wise to involve users in all facets of their formation stage, but users are often thought of as passive participants that can be surveyed for the eventual purchase of devices or services at the performance stage. Yet, users have concerns over ownership, access and distribution of information, privacy, security, and copyrights (Agres et al., 1998), and most of these issues need consideration early on. However, the design fallacy mentioned above emphasizes the limitations of comprehensive user involvement in the formation phase.

Process of Formation

Leaders and champions can enhance or constrain the development of a community network (Industry Canada, 2002; Jones et al., 1997; Huxham & Vangen, 2000). Leaders tap into the collective awareness of the community stakeholders and mobilize the initial change efforts by supplying a vision and practical steps to realize it (Bailey & McNally-Koney, 1996; Roberts & Bradley, 1991). Sustaining collaboration depends on the emergence of a common view of the community and shared organizational values. Leaders and champions play a role in consolidating and expanding the collaborative initiatives, but a wider involvement is needed to foster innovation. It is important to have a community cross-section of members as well as individuals with sufficient power to rally other stakeholders. The parties must freely participate, knowing and agreeing on who is involved and in what capacity (Glatter, 2004; Roberts & Bradley, 1991); prior experience and trust facilitate the membership drive.

Network goals are created, implemented, evaluated, and modified through purposeful social construction among network stakeholders and partners (Van de Ven & Poole, 1995; Ring & Van de Ven, 1994). Network effectiveness may be defined as the harmonization, pursuit, and attainment of the goals sought by the various stakeholders and partners. With diverse stakeholders, it becomes difficult to satisfy all parties equally; therefore, managing expectations and potential conflicts help to maintain the social cohesion of the network. Members will likely persist so long as they can positively identify with the intermediate and long term outcomes, whether they are social or economic in nature.

According to Hardy and Phillips (1998), when individuals come to share a vision of the issues and the solutions, they become stakeholders and begin to create a collective identity with mutually agreed upon directions and boundaries that, in time, may become a permanent network. The catalyst is a transformational leader who encourages collaboration as a means to create synergy for innovation, growth, or to protect against future turbulence. Engaging the stakeholders in a planning exercise can address their many concerns; tasks and roles can be organized and assigned within the network to fit their expectations. Because work is complex and elaborate in networks, planning and coordinating task-specialized activities is required (Roberts & Bradley, 1991). However, planning follows the visioning that the leader has enacted.

Challenges in the Formation Phase

Defining and Agreeing on the Transformation Effort

It is argued that a multi-sectoral and multi-organizational network is needed for a transformation to an intelligent community. The wide variety of stakeholders impact the style of leadership and structure needed for joint initiatives. The leader (or leaders in the case of shared roles) shares a vision of a desirable future and initiates a flexible structure that can accommodate differences in orientation (profit versus not for profit), time horizons (short versus long term), and civic engagement (self versus community focus). Given the diversity of stakeholders, the visioning must be consistent and persuasive, but large enough in scope so that stakeholders can personalize the vision to suit their personal and organizational interests. Key activities include:

- Utilizing context issues to create a sense of urgency and sell the concept of the community network.
- Identifying solutions to problems and synergistic opportunities.
- Preparing a plan for producing meaningful and motivating outcomes.

Agreeing on the vision depends on the availability and abilities of the local leader. Individuals with strong communication skills, an established reputation of trustworthiness, an ability to deliver on promises made, and conceptual skills to craft a vision are in short supply. While large communities have a greater pool of candidates, small communities may have to draw more on external talent and work hard on establishing trustworthiness.

Assembling and Mobilizing Resources

The community network depends upon its board to acquire the physical, financial, and organizational resources that make a broadband network functional. Collaboration among stakeholders and partners facilitates the pooling of their resources. Choosing board members should flow from resource requirements and the likelihood that the stakeholders recruited or volunteering are favorably disposed to sharing with other organizations. Community citizenship of board members channels the resources to create and enhance the network. Key activities include:

- Assembling the representatives of a variety of public and private sector organizations to form the board, including both small and large stakeholders.
- Mobilizing the resources controlled by board members and reaching out to obtain vital resources from the environment.

Too many resources may harm the development of the network if the board lacks the capability to make good use of them. Waste would damage the network's reputation and make future resource acquisitions more difficult. Likewise, too few resources can harm the network because the scope of activities would be narrow and appeal to only a small segment of the community's population. A narrow focus would appear self-serving and lack broad legitimacy.

Assembling and Mobilizing Trust

For the board to be effective in creating and enhancing the network's resource base, its members must trust each other so that extensive sharing becomes possible. When stakeholders engage in

Table 2. Influencing factors at the performance phase

Context	Content	Process
• Structure • Roles • Trust • Power of stakeholders • Interdependence & Culture	• Goals • Achievement/output • Innovation	• Team management • User appropriation • Communication

joint efforts and initiatives, they are putting the community first and themselves second, making them vulnerable to exploitation by less citizen-minded organizations. When trust exists on the board, stakeholders can tolerate some exposure. Therefore building and maintaining trust in a realistic manner is essential to the network's resource base and projects. Key activities include:

- Assembling the board membership on the basis of reputation, prior experience, and diversity of stakeholders.
- Creating a shared vision that reflects the underlying values of community spirit.
- Distinguishing between fragile and resilient trust, and building the latter.

Building and maintaining resilient trust is at the core of the inter-dependent culture that emerges in the network. When a transformational vision is complemented with solid resources and trust, the community network has met the challenges of the formation phase of its development and is ready to shift into the performance phase.

Performance Phase of Development

The performance phase of network development is centred on concrete projects that require the pooling of resources by its members. The resources may be tangible (finances, staff secondment, office space, and equipment) and intangible (time, information, influence, and reputation) in nature. Pooling is facilitated by both the culture and structure of the network in which horizontal interactions, exchanges among equals, are based on trust. These resources are organized and controlled to attain the project objectives, and the management style is collaborative and accountable to the membership of the network. Pursuing these objectives gives collaborators opportunities to learn how they can make the network function effectively. In the short term, the level of attainment of the project's objectives dominates; small wins and their public recognition are important to confirm the value of the network (Bouwen & Taillieu, 2004). Effective project management is needed. In the long term, the board focuses on the level of attainment of the broad goals of the network. To ensure that the projects and the general management of the network are aligned with the original vision and goals, effective leadership is required. Table 2 provides an overview of the critical factors in the performance phase.

CONTEXT OF PERFORMANCE

The interdependence of members within a community network is reflected in both its structure (O'Toole, 1997) and culture. Structure requires careful attention because a poor structure—one that gives too much power to one partner or that does not embody the values of stakeholders—will affect the performance and longevity of the collaboration.

Poyhonen and Smedlund (2004) and Nooteboom (1999) identified three network structures: vertical, horizontal, and diagonal. The latter consists of firms and organizations from several different lines of business. A diagonal structure is appropriate for community networks because it includes as many collaborators as possible to create synergy and innovation within and between sectors; transformational, as opposed to incremental change, is facilitated. The success of collaborative networks is contingent on managing the ambiguity, complexity, and dynamics of the structure. It becomes more important in the performance phase because it must sustain an action plan and organize resources to carry it out. However, a telecommunication network is developed to resolve dynamic context issues and can only do so within a process of continuous improvement. A rigid structure that minimizes innovation diminishes the network's sustainability. Though difficult to assess, the effectiveness of the structure can be judged by its internal coherence and fit with the culture of the network. This puzzle, identified by Bailey & McNally-Koney (1996), needs a solution that retains the fluidity of communications and decision-making, while providing for a framework for productivity and sustainability.

Collaboration is associated with incremental innovation when partners share on several levels: a larger purpose, explicit and voluntary membership, an interactive process, and temporal property (Roberts & Bradley, 1991). Hardy and Phillips (1998) pointed out that more powerful stakeholders may force collaboration on weaker players to control them. Consequently, there is a lessening of the level of interdependence and common vision. Weaker stakeholders are bound to minimize their participation and find excuses to exit the network when they are being coerced. Though asymmetrical power is a likely reality, leaders that seek innovation must put less emphasis on control and more on incentives and opportunities.

Creating a culture of collaboration gives coherence to the stream of actions that builds the community network. Collaboration is described as a relational system of individuals within groups in which individuals share mutual aspirations and a common conceptual framework (Bailey & McNally-Koney, 1996). Individuals are guided by their sense of fairness and their motives toward others (caring and concern, and commitment to work together over time). Through communication and shared experiences, they create a system of shared assumptions and values, and accepted approaches and solutions to problems, including collective sanctions, to safeguard exchanges and reinforce acceptable behaviors (Jones et al., 1997). Sanctions may include exclusion from certain benefits (present or future) and opportunities (participation in projects), and as a last measure forced exit (temporary or permanent) from the network.

Collaborators often choose to stay in a poorly performing network based on the strength of their social ties. However, if they conclude that they can meet all of their needs outside of the network, they may view the network as superfluous (Brown et al., 1998). Linkages or interdependence must be solid and intentional (Bailey & McNally-Koney, 1996) and may be a strong indicator of sustainability (Olk & Young, 1997). Conversely, Brown et al. (1998) identified that greater resource interdependence makes successful partnerships more difficult to achieve. In order to find common ground and encourage persistence, the reasons for enhancing an interdependence need to be emphasized, and stakeholders must want to belong and believe in the vision.

Content of Performance (Specific Projects)

The content of performance includes a wide variety of projects that meet the goals of the network, including the needs of stakeholders and users. Among them are projects to launch or upgrade an infrastructure, acquire devices to deliver applications, develop content for the network, and promote the network to potential users. The outcomes include cost savings to deliver services, revenues from users, and additional capability for the social and economic development of the community.

Waits (2000) described collaborative networks in terms of their pursuits:

- Co-inform: actions to identify members and impacts, promote a heightened awareness of the issues, and improve communication among the members.
- Co-learn: educational and training programs sponsored by the network.
- Co-market: collective activities that promote member products or services abroad or domestically.
- Co-purchase: activities to strengthen buyer supplier linkages or to jointly buy expensive equipment.
- Co-produce: alliances to make a product together or conduct R&D together.
- Co-build economic foundations: activities to build stronger educational, financial, and governmental institutions that enable them to compete better.

Some of these pursuits appear easier to realize and only require fragile trust (co-inform and co-learn). They are more likely to give quick "small wins." Others may be challenging and require resilient trust (co-market, co-purchase, and co-produce); their success will take more time but are more highly valued. Co-building economic foundations appeals less to self-interest and more to a communal interest, and depends on a broad vision that will lead to a series of concrete actions and sustained effort. Waits' objectives are compatible with each other, but have different time horizons and commitments. The strength of the formation phase influences the commitment of stakeholders in the development phase. In particular, a strong collaborative climate encourages them to be patient and willing to invest additional time and resources to achieve long term goals.

PROCESS OF PERFORMANCE

Leaders require managerial sophistication to recognize appropriate circumstances and tools for collaboration (Glatter, 2004). In networks, collaboration depends upon an ongoing communicative process (Lawrence et al., 1999). Roles and responsibilities are negotiated in a context where no legitimate authority is necessarily recognized (Glatter, 2004; Lawrence et al., 1999; Lowndes & Skelcher, 1998). Like in partnerships, there is concern for trust, politics, emotions, and results. Furthermore, leaders require an understanding of user appropriation of the digital network to effectively channel the collaborative efforts.

Du Gay et al. (1997) describe the appropriation of technology as an active process in which users make choices around the selection and local deployment of the technological components, and create meaning and sense of the technology. Appropriation has both a technical and cultural side. In this spirit, Williams et al. (2005) have argued that user appropriation has two distinct but inter-related processes: innofusion (users adjust and innovate to improve the usefulness of the technology) and domestication (users adapt the use of the technology to integrate it meaningfully in their activities). When both processes are fully engaged, the community may be said to have a "digital" culture that sustains the network.

The pace of change within the network must be properly managed. Effective use of communication will allow collaborators to react and contribute. Because of large boards and membership and turnover in representation, some collaborators may not know everyone or their status. Indeed, some may be confused over the degree of autonomy they have in making decisions for their organization (Huxham & Vangen, 2000). Changes in government mandates and organizational priorities create uncertainty as collaborators plan and structure the network. Communication and recognition of accomplishments become important to keep everyone focused.

The board's effectiveness in tackling problems within their community as well as within their respective organizations will directly influence the achievement of the intelligent community objectives. Leaders need to guide the board and create bridges with important outside players. They must align the requirements of their own organization with the vision of the intelligent community initiative for success; they must create a high performance team environment (Albert, 2005; Wheelan, 1999; Smith, 1994). This standard is not easily achievable, especially for a volunteer board with diverse membership and affiliation.

Challenges in the Performance Phase

Continuing Challenges from the Formation Phase

The consensus on the community vision that was created in the formation phase needs to be reinforced. The leader can remind stakeholders of the urgency to capture opportunities, but must incorporate measures for sustaining collective efforts. Key transformation activities include:

- Expanding planning and monitoring projects and measures of performance.
- Marketing the network concept to mobilize and gain the support of the wider community and further engage the stakeholders.

In terms of resources, the community network continues to depend upon its board to acquire resources to develop, acquire, and develop applications to attract numerous users. Key activities include:

- Modifying board membership to improve the resource base of the network as projects change over time.
- Engaging both small and large partners for innovation to create new resources.
- Creating a small management structure for the performance phase of the network.

As for trust, the performance phase requires continuing sharing of resources in the face of uncertain outcomes. Key activities include:

- Applying different trust standards as the situation warrants.
- Encouraging the broad sharing of resources instead of specialized contributions.

Resilient trust can block new stakeholders and partners from joining the network; they may have key resources but be deemed untrustworthy. In such a case, the network requires the flexibility to resort to fragile trust with its emphasis on formal agreements and contracts. The reverse situation can also damage the network, when fragile trust dominates relationships. While formal contracts increase accountability of the parties, they are narrow in scope and participation is contingent on self-interests being satisfied. Community considerations remain secondary. In time and through active leadership, these new members may buy into community citizenship through association and success.

Balancing Leadership and Collaborative Management

Both the formation and performance phases of development have their champion. The leader dominates the formation (and re-formation) phase through visioning, planning, and attracting and retaining stakeholders with key resources and disposed to collaborate. The manager guides and maintains the performance phase, and ensures that both tangible and intangible benefits are created for the stakeholders and the community.

The "collaborative" manager is needed to reinforce the user appropriation by supporting the innofusion and domestication in which users engage. By encouraging the involvement of intermediaries (e.g., Chamber of Commerce, owner of a cybercafé, entrepreneur who wants to keep control), the network manager allows the network to evolve along lines that reflect the different groups and segments in the community's population (Williams et al., 2005).

Formal planning becomes less important, as a pattern of coherent projects becomes established. At the same time, these intermediaries (or small groups of individuals in large networks) interact to keep the board informed and reinforce their individual efforts. By working together, they ensure that the vision of the network creates a coherent set of initiatives and projects, and opportunities and issues relevant to the board meetings are identified. Key activities include:

- Encouraging innovation and proper planning to achieve the transformation effort.
- Reaching out to intermediaries to broaden user involvement.
- Ensuring that the vision that binds the board members remains true to the community values as the network develops and expands.
- Confronting head-on the need to modify the board composition to respond to internal or external factors.
- Managing projects with a blend of fragile and resilient trust, the former with binding contracts and the latter with negotiation and concurrence.
- Choosing projects that are likely to succeed and that are valued by the stakeholders.
- Building and maintaining redundant communication systems, both formal and informal, to reflect the culture of inter-dependence that binds the stakeholders of the network.

The network can be damaged by a dominant leader or manager who insists on being involved at all times and on controlling the process, whether at the board or project level. This situation emerges when there is a failure to share multiple roles and to act as a team. The lack of experienced persons may push one individual to assume both the leadership and managerial role; this solution ensures positive momentum, but may block future sharing of roles as the incumbent becomes entrenched. Similarly, the abundance of strong and experienced personalities facilitates the sharing of roles, but may slow down momentum as too many persons insist on prominence. Developing a team spirit among members of the board and management should be encouraged as early as possible in the network's development (Albert, 2005).

Collaboration Challenges for Users

At the formation stage, the infrastructure and applications are planned and guided by a vision. Stakeholder requirements are addressed in the planning of the network through the methods of user representation. At the performance stage, when the network is functional, the users actualize the network in both expected and emerging ways. A community network is validated by the applications it makes available to its users, and the extent to which the users actually use them. Furthermore, the design features of the telecommunication network influence the collaboration opportunities that the network creates. When the network design enhances collaboration, it has succeeded in creating effective socio-technical patterns (Huysman & Wulf, 2005; Evans & Brooks, 2005).

Challenges for Individual Users

IT and a community network challenge the individual because they put into question existing ideas and routines, and add knowledge and skill requirements. Being open to change means making efforts to understand and use the network. The younger generation makes more use of the internet than the established older generation for social contact and is likely to push for internet connection in the home (Bernier & Laflamme, 2005; Crowley, 2002). The older adults are more likely to be introduced to ICT changes in the workplace. Age aside, the Internet facilitates the local-global link through which knowledge and expertise from around the world can be channelled to community members (Stevenson, 2002). Creative individuals can interact to exchange expertise and create innovations (e.g., open source development), and are motivated by reputation and recognition built into the Web site (Fischer et al., 2004). To generate ideas, group support systems that ensure anonymity appear more effective (Pissarra & Jesuino, 2005). In general, the individual must learn to assess the trustworthiness of the Internet information sources (Franklin, 1999; May, 2002) and assume risks when making transactions online. Similarly, participating in virtual communities and discussion forums challenges the individual to change roles from spectator to contributor (Ginsberg, 2001) and activist.

Challenges for Organizational Users

Organizations that are stakeholders in the community network need to share their "network" vision with their board members, managers, employees, and organizational partners within their supply chains and customer/client networks. Key individuals likely were involved in the network formation stage to ensure that the design of the systems would support expected transactions and activities. At the performance stage, each organization is challenged to mobilize its ICTs, skill base and network use, and do so in dialogue and coordination with their organizational networks. Internally, this means empowering employees and lower levels of management through information systems and decision-making authority. Externally, this refers to the network of relations and the integration of the organizational and community networks. Failure to have extensive collaboration diminishes the benefits that the community network can deliver to stakeholders. Knowledge sharing (Van den Hooff et al., 2004) and knowledge management (Ackerman & Haverton, 2004) are useful frameworks for channelling this collaboration. In addition, involvement can include intra-preneurship (Von Oetinger, 2005) and joint ventures supported by collaborative groupware (McKnight & Bontis, 2002). The organization can also reach out to innovators and entrepreneurs in the community, who view the network as their business platform, and initiate partnerships. The above array of activities pushes leaders and senior managers to adopt an organizational model that incorporates trust.

Challenges for the Community

As the community network is fully implemented, the stewardship vision (Block, 1993) incipient in the formation phase must be reinforced by extending inclusiveness to all segments of the local population, imagining a broad culture of use, and providing for economic development with a digital component. Community leaders should have concrete programs to diminish access barriers such as network connectivity at a reasonable cost (or at no cost for public terminals) and to training and education. Adoption of the network will vary across socio-economic dimensions, and programs are needed that are adapted to specific groups such as youth, seniors, and the non-profit and small business sectors. Developing and implementing these programs can take place with community stakeholders in collaborative projects. An innovation culture (Martins & Terblanche, 2003), linked to the network, can be encouraged.

A culture of "digital" use is emerging in many communities; the Internet and its many activities are being integrated into everyday routines of social communication, work, and play (Bernier & Laflamme, 2005; Crowley, 2002; Wellman et al., 2001). In contrast, civic participation has had less success. The evidence indicates that internet use reinforces civic participation and makes it more sophisticated, but does not increase the levels of activity (Shah, 2002; Uslaner, 2004; Warkentin & Mingst, 2000; Wellman et al., 2001). Pigg (2001) has argued that networks can be designed to enhance civic participation, but so far, these designs have failed to incorporate the nature of participation. The designs typically focus on customer services and support instead of sharing of information, ideas, and knowledge to influence civic decisions. With a customer focus, the civic authorities may increase the satisfaction of its citizenry, whereas a participation focus obliges the authorities to share decision-making powers and accept more uncertainty in the process and outcomes.

CONCLUSION

A community network faces four inter-related collaboration challenges during its development that are tied to transformation, resources, trust, and management. When these challenges are met, the network will have a solid culture and structure of interdependence, and the flexibility to change over time. The network will maintain a positive momentum that is constructive and manageable, and lead to medium and long-term sustainability. When these challenges are not met adequately, the pace of change will be either too slow or too fast, or blocked at some point in time. Sustainability of the network will be compromised unless the underlying issues are addressed.

These four challenges are anchored in the proposed network development model where formation and performance phases, and adaptation through reformation are critical for the sustainability of the community network. Policy makers and change agents among the stakeholders of community networks are well advised to shape their interventions with the aim of establishing and maintaining positive momentum, while paying continued attention to issues of visioning, resources, trust, leadership, and management. They would do well to expand their views of technology development to include user appropriation and the challenges that users face. They must accept the uncertainty that is inevitable with user involvement to support the goal of network sustainability.

REFERENCES

Ackerman, M., & Haverton, C. (2004). Sharing expertise: The next step for knowledge management. In M. Huysman & V. Wulf (Eds.) *Social capital and information technology* (Chapter 11). Cambridge, USA and London, England: The MIT Press.

Agres, C., Edberg, D., & Igbaria, M. (1998). Transformation to virtual societies: Forces and issues. *The Information Society, 14*(2), 71-82.

Albert, S. (2005). Smart community networks: Self-directed team effectiveness in action. *Team Performance Management, 1*(5), 144-156.

Ardichvili, A., & Cardozo, R. N. (2000). A model of the entrepreneurial opportunity recognition process. *Journal of Entreprising Culture, 8*(2), 103-119.

Arino, A., & de la Torre, J. (1998). Learning from failure: Towards an evolutionary model of collaborative ventures. *Organizational Science, 9*(3), 306-325.

Bailey, Darlyne, & McNally-Koney, K. (1996). Interorganizational community-based collaboratives: A strategic response to shape the social work agenda. *Social Work, 41*(6), 602-610.

Bell, R. (2001). *Benchmarking the intelligent community—a comparison study of regional communities.* The Intelligent Community Forum of World Teleport Association.

Bernier, C., & Laflamme, S. (2005). Uses of the Internet according to type and age: A double differentiation. [Usages d'Internet selon le genre et l'age: Une double differenciation] *The Canadian Review of Sociology and Anthropology/La Revue Canadienne De Sociologie Et d'Anthropologie, 42*(3), 301-323.

Block, P. (1993). *Stewardship—Choosing service over self-interest.* San Francisco: Berrett-Koehler Publishers.

Blois, K. (1990). Research notes and communications—transaction costs and networks. *Strategic Management Journal, 11*, 493-496.

Bouwen, R., & Taillieu, T. (2004). Multi-party collaboration as social learning for interdependence: Developing relational knowing for sustainable natural resource management. *Journal of Community & Applied Social Psychology, 14*, 137-153.

Bradford, R. (2003). Public-private partnerships? Shifting paradigms of economic governance in Ontario. *Canadian Journal of Political Sciences, 36*(5), 1005-1033.

Brown, M., O'Toole, L., & Brudney, J. (1998). Implementing information technology in government: An empirical assessment of the role of local partnerships. *Journal of Public Administration Research and Theory, 8*(4), 499-525.

Canadian National Broadband Taskforce. (2001). *Report of the national broadband taskforce: The new national dream: Networking the nation for broadband access.* Ottawa, Canada: Industry Canada.

Canadian Rural Partnership. (2004, October). *Report of the advisory committee on rural issues.* Paper presented at the Third National Rural Conference, Red Deer, Canada.

Carroll, J. M., & Rosson, M. (2001). Better home shopping or new democracy? Evaluating community network outcomes. *3*(1), 372-377.

Castells, M. (1996). *The rise of network society, vol. 1 of the information age: Economy, society and culture.* Oxford: Blackwell.

Castells, M. (1997). *The power of identity, vol. 2 of the information age: Economy, society and culture.* Oxford: Blackwell.

Castells, M. (1998). *End of millennium, vol. 3 of the information age: Economy, society and culture.* Oxford: Blackwell.

Caves, R. (2001). E-commerce and information technology: Information technologies, economic development, and smart communities: Is there a relationship? *Economic Development Review, 17*(3), 6-13.

Considine, M., & Lewis, J. (2003). Networks and interactivity: Making sense of front-line governance in the United Kingdom, the Netherlands and Australia. *Journal of European Public Policy, 10*(1), 46-58.

Corbett, A. (2002). Recognizing high-tech opportunities: A learning and cognitive approach. *Frontiers of Entrepreneurship Research* (pp. 49-60).Wellesley, MA: Babson College.

Crowley, D. (2002). Where are we now? Contours of the internet in Canada. *Canadian Journal of Communication, 27*(4), 469-508.

Day, C. (2002). *The information society—a sceptical view.* Malden, MA: Blackwell Publishers.

De la Mothe, J. (2004). The institutional governance of technology, society, and innovation. *Technology in Society, 26*, 523-536.

Doz,Y. (1996). The evolution of cooperation in strategic alliances: Initial conditions or learning processes? *Strategic Management Journal, 17*, 55-83.

Du Gay, P., Hall, S., Janes, L., Mackay, H., & Negus, K. (1997). *Doing cultural studies: The story of the Sony, Walkman,* London and New Delhi: Sage.

Ebers, M. (2002). *The formation of inter-organizational networks*. Oxford: Oxford University Press.

Eger, J. (2001, November). *The world foundation for smart communities.* Retrieved January 28, 2003 from www.smartcommunities.org

Evans, J., & Brooks, L. (2005). Understanding collaboration using new technologies: A structural perspective. *Information Society, 21*(3), 215-220.

Fischer, G., Scharff, E., & Ye, Y. (2004). In M. Huysman & V. Wulf (Eds.). *Social capital and information technology* (Chapter 14). Cambridge, MA and London: The MIT Press.

Franklin, U. (1999). *The real world of technology.* Toronto: House of Anansi Press.

Ginsburg, M. (2001, November). *Realizing a framework to create, support, and understand virtual communities.* Maastricht, Holland: Infonomics.

Glatter, R. (2004). Collaboration, collaboration, collaboration: The origins and implications of a policy. *MiE, 17*(5), 16-20.

Gray, B. (1985). Conditions facilitating interorganizational collaboration. *Human Relations, 38*(10), 911-936.

Gray, B., & Hay, T. (1986). Political limits to interorganizational consensus and change. *The Journal of Applied Behavioral Science, 22*(2), 95-112.

Gurstein, M. (2000). *Community informatics: Enabling communities with information and communications technologies* (Introduction, pp. 1-29). Hershey, PA: Idea Group Publishing.

Hardy, C., & Phillips, N. (1998). Strategies of engagement: Lessons from the critical examination of collaboration and conflict in interorganizational domain. *Organizational Science, 2*, 217-230.

Hock, D. (2000). Birth of the chaordic age, *Executive Excellence, 17*(6), 6-7.

Huxham, C., & Vangen, S. (2000). Ambiguity, complexity and dynamics in the membership of collaboration. *Human Relations, 53*(6), 771-805.

Huysman, M., & Wulf, V. (2004). *Social capital and information technology.* Cambridge, MA and London: The MIT Press.

Huysman, M., & Wulf, V. (2005). The role of information technology in building and sustaining the relational base of communities. *The Information Society, 21*(2), 81-89.

Igbaria, M., Shayo, C., & Olfman, L. (1999). *On becoming virtual: The driving forces and arrangements* (pp. 27-41). New Orleans, LA: ACM.

Industry Canada. (2002a, April 4). *Fostering innovation and use.* Retrieved July 30, 2002 from http://broadband.gc.ca/Broadband-document/english/chapter5.htm

Industry Canada. (2002b, April 4). *Smart communities broadband.* Retrieved July 12, 2002 from http://smartcommunities.ic.gc.ca/index_e.asp

ITU (International Telecommunications Union). (2003). *World summit on the information society* (pp. 1-9). Retrieved from www.itu.int

Jarillo, C. (1988). On strategic networks. *Strategic Management Journal, 9*(1), 31-41.

Jones, C., Herterly, W., & Borgatti, S. (1997). A general theory of network governance: Exchange conditions and social mechanisms. *Academy of Management Review, 22*(4), 911-945.

Keenan, T., & Trotter, D. (1999). The changing role of community networks in providing citizen access to the Internet. Internet Research. *Electronic Networking Applications and Policy, 9*(2), 100-108.

Kickul, J., & Gundry, L. (2000). Pursuing technological innovation: The role of entrepreneurial posture and opportunity recognition among internet firms. In *Frontiers of Entrepreneurship Research,* MA: Babson College.

Lawrence, T., Phillips, N., & Hardy, C. (1999). Watching whale watching. *The Journal of Applied Behavioral Science, 35*(4), 479-502.

LeBrasseur, R., Whissell, R., & Ojha, A. (2002). Organizational learning, transformational leadership and implementation of continuous quality improvement in Canadian hospitals. *Australian Journal of Management, 27*(2), 141-162.

Lowndes, V., & Skelcher, C. (1998). The dynamics of multi-organizational partnerships: An analysis of changing modes of governance. *Public Administration, 76*, 313-333.

Martin, L., & Matlay, H. (2003). Innovative use of the Internet in established small firms: The impact of knowledge management and organizational learning in accessing new opportunities. *Qualitative Market Research, 6*(1), 18-26.

Martins, E. & Terblanche, F. (2003). Building organisational culture that stimulates creativity and innovation. *European Journal of Innovation Management, 6*(1), 64-74.

May, C. (2002). *The information society—A sceptical view.* Cambridge, UK: Polity Press.

McKnight, B. & Bontis, N. (2002). E-improvisation: Collaborative groupware technology expands the reach and effectiveness of organizational improvisation. *Knowledge and Process Management, 9*(4), 219-227.

New Economy Development Group Inc. (2001). *Sustainability project on sustainable communities.* Paper presented at the Canadian Rural Partnership. Rural Research and Analysis, Government of Canada.

Nooteboom, B. (1999). Innovation and inter-firm linkages: New implications for policy. *Research Policy, 28*(8), 793.

OECD. (1997). Organisation for economic cooperation and development. *Towards a global information society.* Paris: OECD.

OECD. (2004). Organization for economic cooperation and development. *Information and communication technologies and rural development.* Paris, France: OECD Publication Service.

Oliver, A., & Ebers, M. (1998). Networking network studies: An analysis of conceptual configurations in the study of inter-organizational relationships, *Organization Studies, 19*, 549-83.

Olk, P., & Young, C. (1997). Why members stay in or leave an R&D consortium: Performance and conditions of membership as determinants of continuity. *Strategic Management Journal, 18*(11), 855-877.

O'Toole, L. (1997). Treating networks seriously: Practical and research-based agendas in public administration. *Public Administration Review, 57*(1), 45-52.

Ouchi, W. (1980). Markets, bureaucracies, and clans. *Administrative Science Quarterly, 1*, 129-141.

Pettigrew, A. (1992). The character and significance of strategy process research. *Strategic Management Journal, 13*, 5-16.

Pettigrew, A. (1987). Context and action in the transformation of the firm. *Journal of Management Studies, 24*(6), 649-670.

Pigg, K. (2001). Applications of community informatics for building community and enhancing civic society. *Information, Communication & Society, 4*(4), 507-527.

Pissarra, J., & Jesuino, J. (2005). Idea generation through computer-mediated communication: The effects of anonymity. *Journal of Management Psychology, 20*(3/4), 275-291.

Poyhonen, A., & Smedlund, A. (2004). Assessing intellectual capital creation in regional clusters. *Journal of Intellectual Capital, 5*(3), 351-365.

Ring, P. (2002). Processes facilitating reliance on trust in inter-organizational networks. In M. Ebers (Ed.), *The formation of inter-organizational networks* (pp. 113-45). Oxford, England: Oxford University Press

Ring, P., & Van de Ven, A. (1994). Developmental processes of cooperative interorganizational relationships. *Academy of Management Review, 19*, 90-118.

Roberts, N., & Bradley, R. (1991). Stakeholder collaboration and innovation: A study of public policy initiation at the state level. *Journal of Applied Behavioral Science, 27*(2), 209-227.

Rycroft, R. (2003). Technology-based globalization indicators: The creativity of innovation network data. *Technology in Society,* 25(3), 299-317.

Scheel, C. (2002). Knowledge clusters of technological innovation systems. *Journal of Knowledge Management, 6*(4), 356-367.

Shah, D. (2002). Nonrecursive models of internet use and community engagement: Questioning whether time spent online erodes social capital. *Journalism & Mass Communication Quarterly,* 79(4), 964-987.

Snow, C. & Thomas, J. (1993). Building networks: Broker roles and behaviours. In P. Lorange, B. Chakravarthy, J. Roos, & A. Van de Ven (Eds.), *Implementing strategic processes: Change, learning and co-operation* (pp. 217-38). Oxford: Blackwell.

Sorensen, E. (2002). Democratic theory and network governance. *Administrative Theory & Praxis, 24*(4), 693-720.

Stevenson, T. (2002). Communities of tomorrow. *Futures, 34*(8), 735-744.

Suire, R. (2004). Des réseaux de l'entrepreneur aux ressorts du créatif Quelles stratégies pour les territoires? *Revue Internationale PME, 17*(2), 123-143.

Tan, M. (1999). Creating the digital economy: Strategies and perspectives from Singapore. *International Journal of Electronic Commerce, 3*(3), 105-22.

United Nations. (1998). *Knowledge societies: Information technology for sustainable development.* Report prepared by R. Mansell & U. Wehn. Oxford: United Nations Commission on Science and Technology for Development/Oxford University Press.

Uslaner, E. M. (2004). Trust, civic engagement, and the Internet. *Political Communication, 21*(2), 223-242.

Van de Ven, A. & Poole, M. (1995). Explaining development and change in organizations. *Academy of Management Review, 20*(3), 510-540.

Van den Hooff, B., de Ridder, J. & Aukema, E. (2004). Exploring the eagerness to share knowledge: The role of social capital and ICT in knowledge sharing. In M. Huysman, & V. Wulf (Eds.), *Social capital and information technology* (Chapter 7). Cambridge, USA and London, England: The MIT Press.

Von Oetinger, B. (2005). From idea to innovation: Making creativity real. *The Journal of Business Strategy, 25*(5), 35-41.

Waits, M. (2000). The added value of the industry cluster approach to economic analysis, strategy development, and service delivery. *Economic Development Quarterly, 14*(1), 35-50.

Warkentin, C., & Mingst, K. (2000). International institutions, the state, and global civil society in the age of the World Wide Web. *Global Governance, 6*(2), 237-257.

Wellman, B., Haase, A. Q., Witte, J., & Hampton, K. (2001). Does the internet increase, decrease, or supplement social capital? Social networks, participation, and community commitment. *American Behavioral Scientist, 45*(3), 436-455.

Wheelan, S. (1999). *Creating effective teams: A guide for members and leaders* (p. 154). Thousand Oaks, CA: Sage Publications.

Williams, R., Stewart, J., & Slack, R. (2005). *Social learning in technological innovation—Experimenting with information communication technologies.* Cheltenham, UK and Northampton, USA: Edward Elgar.

Wood, D., & Wood, G. (1991). Toward a comprehensive theory of collaboration. *Journal of Applied Behavioral Science, 27*(2), 139-162.

World Bank (1999). *World development report 1998/99: Knowledge for development.* New York: Oxford University Press.

Zollo, M., Reuer, J., & Singh, J. (2002). Interorganizational routines and performance in strategic alliances. *Organizational Science, 13*(6), 701-713.

ENDNOTES

1 The authors acknowledge the helpful comments of the reviewers. By addressing their concerns and suggestions, this article found a better balance between organizational and involvement issues.

2 Multi-organizational collaboration, partnerships, and networks are considered interchangeable terms and refer to a variety of organizations collaborating for a common purpose. "Collaborative network" is proposed as an inclusive alternative.

3 A stakeholder is defined as an organization that contributes programs and services to the network. A partner is one that makes a financial contribution to the overall project.

4 The term downloading has become a popular expression in Canada as a result of higher levels of government shifting responsibility for programs to lower levels of government. Municipalities have inherited a number of costs and responsibilities previously held by the province and the province has inherited responsibilities previously held by the federal government.

5 These communities have been highlighted at the annual conference of ICF (Intelligent Communities Forum).

This work was previously published in International Journal of Technology and Human Interaction, Vol. 3, Issue 2, edited by B. C. Stahl, pp. 13-33, copyright 2007 by IGI Publishing, formerly known as Idea Group Publishing (an imprint of IGI Global).

Chapter X
A Generic Framework for Bluetooth Promoted Multimedia on Demand (BlueProMoD)

Panayotis Fouliras
University of Macedonia, Greece

Nikolaos Samaras
University of Macedonia, Greece

ABSTRACT

In recent years many technologies have converged to integrated solutions and one of the hottest topics has been the deployment of wireless personal area networks (WPANs). In this article we present a generic architecture scheme that allows voice and other real-time traffic to be carried over longer distances. The proposed scheme is a novel framework that combines a wired backbone network including Bluetooth access points (APs) with the mobile Bluetooth-enabled devices of the end users. This scheme is called Bluetooth Promoted Multimedia on Demand (BlueProMoD). BlueProMoD is a hybrid network and provides free-of-charge communication among customers, multimedia advertisements, as well as location-based and other value-added services.

INTRODUCTION

The concept of personal area network (PAN) is relatively new. A PAN, basically, is a network that supports the interoperation of devices in personal space (Elliott & Phillips, 2003). In this sense, it is a network solution that enhances our personal environment, either work or private, by networking a variety of personal and wearable devices within the space surrounding a person and providing the communication capabilities within that space and with the outside world (Prasad & Munoz, 2003). A wireless PAN (WPAN) is the natural evolution of this concept, where all participating devices communicate wirelessly. Furthermore, a WPAN is a network that moves with a person, linking all the devices carried by the person with each other, as well as any devices that are met along the way.

Since a WPAN has by definition a limited range, compatible devices that are encountered along its path can either link to it or leave it when they go out of its range in a flexible and secure way.

The limited range of a WPAN offers additional advantages such as low-emitted power (thus reducing potential health risks), lower power consumption (hence longer battery life), and lower probability of interference from other WPANs as well as the possibility of location-based services (LBSs). Nevertheless, the core of a WPAN is the wireless technology employed. Nowadays there are many such technologies to choose from. Each one offers specific advantages and disadvantages, which should be taken into consideration before deciding on the most suitable for a particular service or environment.

Of all current wireless technologies Bluetooth is the most promising and employed for many real-life applications. Applications using Bluetooth have become important in hot spots such as at hotels, shopping malls, railway stations, airports, and so forth. Bluetooth is a well-established communications standard for short distance wireless connections. A wide range of peripherals such as printers, personal computers, keyboards, mouse, fax machines, and any other digital device can be part of a Bluetooth network.

Bluetooth has many advantages: (1) low cost, (2) considerable degree of interference-free operation, (3) speed, (4) appropriate range, (5) low power, (6) connectivity, (7) provision for both synchronous and asynchronous links, and (8) wide availability in mobile phones, PDAs, and other devices. Bluetooth is usage-scenario driven, in the sense that its design points were optimized to satisfy established market needs (Bisdikian, 2005). Such usage scenarios are headset to mobile phone connectivity (hands free); mobile device to computer synchronization; digital camera to printer connection for printing; and so forth. More sophisticated applications in diverse areas have been investigated such as hotel services (electronic door locks, check-in/out) in Starwood Hotels and portable patient monitoring in hospitals so that recovering patients are not confined to their rooms (Dursch, Yen, & Shih, 2004). Another interesting application of Bluetooth technology took place at the CeBIT 2001: Guiding services for finding the shortest path to a particular exhibitor in the hall as well as additional exhibitor information services were implemented (Kraemer & Schwander, 2003).

The core of Bluetooth technology is based on the IEEE 802.11 standard and it is a wireless system for short-range communication. This standard defines the protocol for two types of networks; client/server and ad-hoc networks. Bluetooth supports both point-to-point and point-to-multipoint connections. Both Bluetooth and most of IEEE 802.11x share the same 2.4 GHz industrial, scientific and medical, license-free frequency band. Compared with other systems operating in the same frequency band, the Bluetooth radio typically hops faster and uses shorter packets.

In this article we propose a generic architecture scheme that allows voice and other real-time traffic to be carried over longer distances, while simultaneously showing how the providing organization can experience sufficient revenues in order to finance and maintain the necessary infrastructure. The proposed scheme is a novel framework that combines a wired backbone network including Bluetooth access points (APs) with the mobile Bluetooth-enabled devices of the users. The end result is a hybrid network offering free voice and other communication in return for short, specifically targeted multimedia advertisements and tracking information on behalf of the stores or branches operating at a large shopping center or complex. Location-based and other services are also envisaged as a natural side effect. An additional advantage is that the user perceives such capabilities as part of the services offered by his/her respective WPAN.

This article is structured as follows. In the following section we give a detailed overview of the most important technical characteristics of

Bluetooth, such as hardware, connectivity, security, error control, interference, and health issues. The proposed generic framework (BlueProMoD) is described next, where we present all the critical components. Finally, conclusions together with future work are presented.

BLUETOOTH DESCRIPTION

In this section we briefly discuss some important characteristics for the Bluetooth technology. The basic form of a Bluetooth network is the piconet. This is a network with a star topology with up to eight nodes participating in a master/slave arrangement (one master and up to seven slaves, see Figure 1a). More specifically, the master is at the center and transmits to the slaves; a slave can only transmit to the master, provided it has been given prior permission by the master. This protocol allows both asynchronous and isochronous services to be realized. The communications channel is defined by a pseudorandom sequence of frequency hops over 79 frequency sub-bands 1 MHz wide, in the 2.4 GHz band, ranging 2.402 - 2.480 GHz. There are three main Bluetooth versions according to the transmission rate (see Table 1).

There are also three classes of antenna power emission and thus possible ranges, namely Class 1 (100 mW, 100 meters), Class 2 (2.5 mW, 10 meters), and Class 3 (1 mW, 1 meter). Typical mobile phones operate at Class 2. Due to the extremely low power emission, investigators agree that Bluetooth products have not been identified as posing any health risks (Erasala & Yen, 2002). Table 2 presents all the possible modes of a device using Bluetooth technology.

A piconet is created in two stages, the first of which is optional: In the first stage, a device inquires for other devices in its vicinity; if a device responds to the inquiry message it transmits its own Bluetooth address among other information. A Bluetooth address is a unique 48-bit integer derived by the manufacturers from the same address space and authority as the IEEE 802 MAC addresses.

In the second stage, a device (the master) that wants to communicate with another device must perform paging. The paged device (the slave) responds by notifying the paging device that it is ready to communicate with it. In this case the paging device assigns a so-called active member

Table 1. The main Bluetooth versions

Transmission rate (gross)	Bluetooth version
1 Mbps	1.0/1.1/1.2
3 Mbps	2.0/2.1 (+ EDR)
480 Mbps	"Seattle" (draft status)

Table 2. Possible modes of a device under Bluetooth

Mode	Description
STANDBY	The device is not connected in a piconet; it listens for messages every 1.28 seconds over 32 hop frequencies.
PAGE/INQUIRY	The master sends 16 identical page messages to the slave in 16 hop frequencies, repeating this process if there is no response from the slave.
ACTIVE	This represents the actual data transmission.
HOLD	The device conserves power by not transmitting any data.
SNIFF	The slave does not take any active role in the piconet, but only listens at a reduced power consumption level; though this level is higher than Hold.
PARK	The slave is synchronized to the piconet, but is not part of the normal traffic.

Figure 1. Examples of a Bluetooth: (a) piconet, (b) scatternet

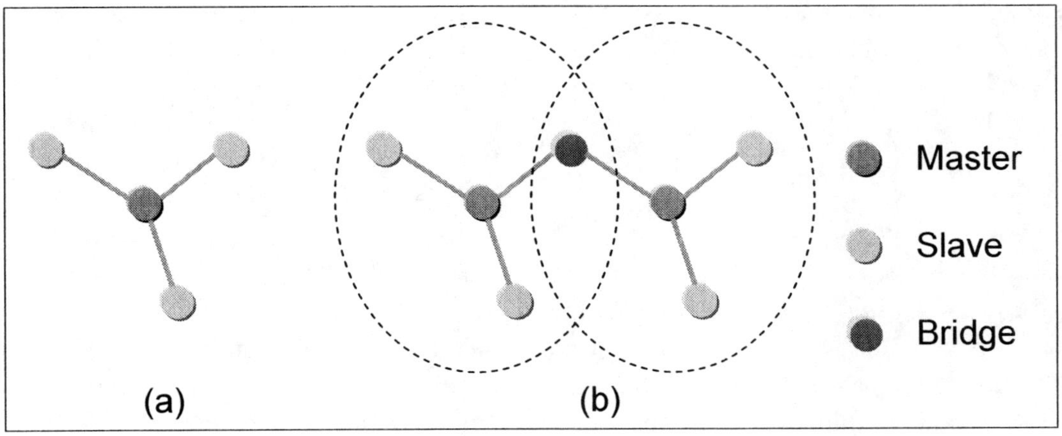

address to the paged device. This is a relative address, which is only of local significance.

In a sense, the limit of seven slave nodes per piconet can be overcome whenever an eighth active node appears: The master takes a node and places it in "park" mode. In this state, the particular node relinquishes its active member address which is given to the newcomer. This feature allows up to 256 nodes to participate in a piconet, but only up to seven slaves to actively exchange data with the master at a time. Communication links may be of two types: *synchronous connection-oriented* (SCO) and *asynchronous connectionless* (ACL). SCO links are typically used for voice communication, whereas ACL links for data. The former are strictly real time, hence lost or damaged packets are never retransmitted over SCO links.

Bluetooth scatternets can also be formed by the connection of two or more piconets. In this case they can have entirely or partially overlapping frequencies, apart from the ongoing communication with their respective members. Evidently one or more nodes must assume the role of a bridge between adjacent piconets (see Figure 1b). As expected, a bridge node operates using the store-and-forward paradigm in order to forward a packet from one piconet to an adjacent one. It is possible for a bridge node to be the master in one of the piconets, but in general this is a slave node. Communication is first initiated by the master of the first piconet, which sends a packet to the bridge node where it is stored; whenever the master of the second piconet allows it the bridge node sends the stored packet to the master. Given that the two piconets are not coordinated in any way (in effect the two masters), the overall waiting time at the bridge node during switch over can be quite long. The Bluetooth specification switch-over does not define any particular mechanism, protocol, or algorithm in order to guarantee some minimum time for multi-hop communication in scatternets.

In order to ensure that different hardware implementations are compatible, Bluetooth devices use the host controller interface (HCI) as a common interface between the host and the core. Logical link control and adaptation protocol (L2CAP) is built on top of HCI, which basically segments and reassembles large data packets over Bluetooth baseband connections. Higher level protocols such as service discovery protocol (SDP), RFCOMM (serial port emulation), and telephony control protocol (TCS) are built over L2CAP. Applications are built over them using the services provided. To give the reader some rough background about other aspects of the Bluetooth communication, such as security, interference, and real-time traffic, we present them briefly next.

Security

A challenge-response mechanism is employed for authentication using Bluetooth address as the public key and a 128-bit integer during device initialization as the private key. In addition, another 128-bit random number is used for each new session. Encryption is also included to maintain link privacy. As a result, devices can be classified as "trusted" and "untrusted," and services as "requiring authorization and authentication," "requiring authentication only," and services open to all devices.

In order to reduce security risks, as well as accelerate device discovery, it has recently been proposed to use limited duration visual tags provided by a central trusted authority for specific devices, so that users can quickly identify the one of interest among many available and connect to it (Scott, Sharp, Madhavapeddy, & Upton, 2005). Nevertheless, security problems have been identified with Bluetooth. In a recent article (Shaked & Wool, 2005) an implementation of a passive attack is described under which a 4-bit PIN can be cracked in less than 0.3 seconds on a Pentium III 450 MHz computer.

Interference

Frequency hopping and the short range of a piconet are the deciding factors for experiencing minimal interference problems under Bluetooth. This minimizes interference not only among Bluetooth devices or piconets, but also interference from other types of devices such as microwave ovens, mobile phones, and baby monitors.

The case of coexistence of Bluetooth with wireless LANs in particular has been studied extensively due to the popularity of IEEE 802.11x devices, which happen to use the same frequency band. Nallanathan, Feng, and Garg (2006) present a complete integrated analysis taking various scenarios into consideration. A Bluetooth piconet in a multi-piconet environment is first examined

Figure 2. Example of 3-slot packet from a slave after a 1-slot packet from the master

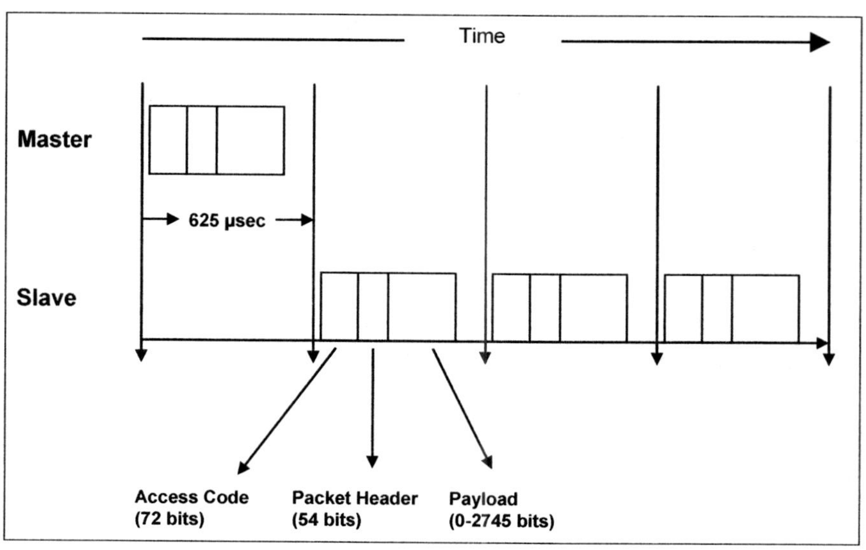

for all three types of packets (that is, 1-, 3-, and 5-slot packets), concluding that 1-slot packets are the best for high density interference, 3-slot packets for moderate, and 5-slot packets for low density interference (see Figure 2). In the mixed environment of Bluetooth and IEEE 802.11b, the latter suffers more from interfering Bluetooth devices than the other way around; in fact IEEE 802.11b successful reception packet rate increases when Bluetooth uses 5-slot packets.

Real-Time Traffic over Bluetooth

Real-time traffic is of particular importance in present-day networks, since it constitutes a significant proportion of user sessions—especially voice due to its limited bandwidth requirements. As noted earlier, Bluetooth supports voice traffic over SCO links by design, using constant bit rate (CBR) at 64 kbps per channel. In order to support this rate and the expected quality of service, a strict timing scheme has to be observed, which may lead to severe deterioration of the asynchronous links performance. Furthermore, only up to three SCO links are supported per piconet.

Given these limitations it is not surprising that researchers tried to calculate the actual SCO link performance. The next step was to use the similar experience from voice-over-IP (VoIP) in order to explore the possibility of using ACL links for voice traffic. In Misic, Misil, and Chan (2005) a detailed analysis is presented under which the presence of a single SCO link reduces the bandwidth available to ACL links to two thirds of the original; with two SCO links, the ACL bandwidth is reduced to half of the original; and with three SCO links the ACL bandwidth is practically zero. The authors evaluated their scheme, which uses multi-slot ACL packets to carry CBR voice traffic, concluding that it is more efficient.

Zeadally and Kumar (2005) used the audio/video distribution transport protocol (AVDTP) specification standard. The results showed that even AVDTP itself can offer 48.2% performance improvement when the optimized version is used over the default version. A more recent study (Razavi, Fleury, Jammeh, & Ghanbari, 2006) considered MPEG-2 (video) traffic coming from an IP network to a Bluetooth master of version 2.0, supporting 2-3 Mbps. Their simulation results lead to similar conclusions. It is, therefore, clear that no complete scheme exists that is able to support simultaneous voice traffic for maximum size piconets (that is, all seven slaves carrying voice traffic) over ACL links. Such a scheme would greatly enhance the usefulness of Bluetooth, especially in view of the fact of the high gross data rates of version 2.0 and possibilities of interconnectivity with IP backbone networks.

DESCRIPTION OF THE BlueProMoD GENERIC FRAMEWORK

In this article a framework for the deployment of a novel usage model is presented. The proposed model offers users free of charge voice, location-based, and other services in return for specifically targeted short advertisements on their mobile devices, based on a Bluetooth hybrid network. The whole network environment is run by a single managing authority typically in a large shopping mall or organization (for example, a university campus with short distances between buildings). Revenues to support the system functionality come from the advertisements of participating resellers as well as value-added services, such as the one coming from the post processing of users movement tracking.

The BlueProMoD generic platform requires four types of hardware communicating entities: (1) the server(s), (2) the Bluetooth APs, (3) the interconnection network between the APs and the server, and (4) the Bluetooth-enabled user devices. In its basic form, the user first downloads the client application from specific APs at the reception area. The client application distinguishes BlueProMoD APs from all possible surrounding

Figure 3. BlueProMoD architecture

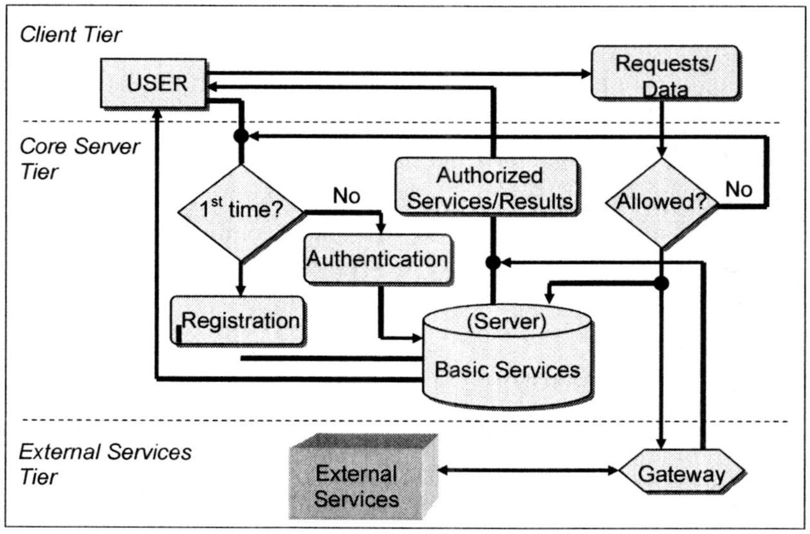

Bluetooth devices and selects one of them for connection, typically the one with the strongest signal, after proper registration or authentication. From this time onwards the user is able to use all free services offered by the system. The mandatory services provided by all variations of BlueProMoD are: User Reception, SMS, and MMS over Bluetooth, Voice over Bluetooth, Users' Directory, and User's Location. The latter is the foundation upon which LBSs may be built, among which route finding and m-advertisements from shops are the most important. The User Reception service is obviously essential since it allows a user to connect to the system.

BlueProMoD is a three-tier architecture, an overall description of which is shown in Figure 3. These are the Client, the Core Server, and the External Services tier.

Client Tier

The client tier is composed of a Bluetooth-capable cell phone or PDA running the appropriate client application, which allows the user to connect to any available Bluetooth AP, offering the Reception Service. The client application is downloaded and installed only at specific reception areas. For this reason first-time users can seek assistance from experienced personnel (for example, at the help desk or information points). At the same time, this arrangement places fewer burdens on the overall system since long or unsuccessful loading efforts are limited to areas that do not need to constantly interact with the rest of the system. The whole process mentioned previously is more secure and reduces user frustration from unsuccessful or long loading times.

After registration and authentication, the user is presented with a list of available services. This step is important since it is the only way that the system can ask personal information from the user and identify him/her every time he/she returns for another visit. Since Bluetooth devices are widespread it is quite natural that many such devices may exist in the same area apart from client devices and Bluetooth APs. In order to improve user satisfaction and reduce client connection time, the client application automatically stores user identification information on his/her device that is made directly or indirectly available (for example, via a signed ticket for higher security) upon request from a Bluetooth AP.

The client application searches for Bluetooth devices offering the Reception Service (hence with a specific UUID). To reduce overall system workload even further, apart from user information, the client application sends a specific ID which indicates to the Bluetooth AP that the sender is a client application. In this way, only clients can connect to Bluetooth APs and only Bluetooth APs can connect to clients.

Typically there are at least four types of clients: (1) administrator, (2) employee, (3) retailer, and (4) other. The first class is reserved for the system administrators and the second for privileged users who must have more rights than other users (for example, unlimited talking time and no advertisements); retailers who may pay for the system are assigned to the third class with almost the same rights as the second class; ordinary users obviously belong to the final class. There are three basic states at which a user may appear in the system. These states are Ready, Busy, and Disconnected (see Figure 4).

The first state is entered after the user connects to the system; the user remains in this state throughout the connection with the exception of the Busy state. The Busy state is entered when the user is in the middle of a phone call. The last state represents a user who has logged out or disconnected from the system for a specific timeout period. These states are also necessary for the Users' Directory service to be up-to-date. In this way system resources are not wasted unnecessarily (for example, if a user A knows that user B is busy talking, he/she will not try to call him/her for a while unless it is really urgent).

Connection time can be further improved once a client device has connected once and remained in the vicinity of the same Bluetooth AP. The Bluetooth standard allows up to 255 devices to remain in the PARK mode, waiting to "wake up" and become one of the seven slave nodes that are actively communicating with the master. In our case, this implies that the device discovery time can be eliminated in subsequent communication sessions between the user and the particular Bluetooth AP.

Core Server Tier

The core server tier is composed of the processes involved in order to provide the basic services of BlueProMoD. These processes are: User Reception; short message service (SMS) and multimedia message service (MMS) over Bluetooth; voice over Bluetooth; users' Directory; and user's location. Upon running, the client application first determines the appropriate Bluetooth AP to be connected with. After a network connection is es-

Figure 4. The three basic states of a user

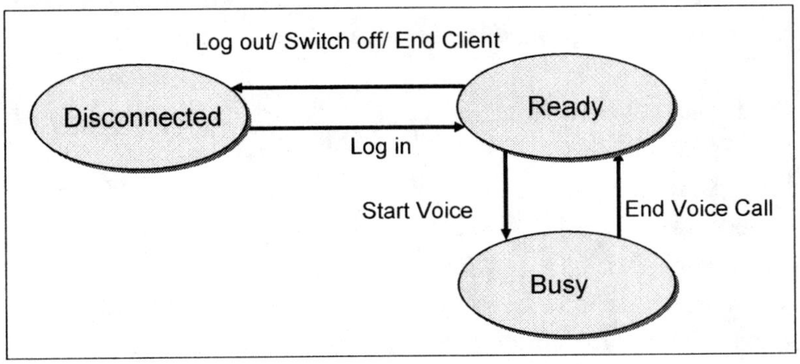

tablished with the system for the first time, the user is asked several questions in order to be registered with the system. After this is completed, the user logs in to the system and his/her account is updated, reflecting his/her voice credits, IP address, position, and so forth. All these separate actions are performed by the User Reception as the main service that interacts with the Users' Directory and User's Location services. From then onwards, the User Reception service is chiefly responsible for periodically checking whether a user is still connected to the system, since it is possible that he/she disconnects in an abrupt way.

The rest of the services run as processes that mostly sit, passively waiting for an event either from a user (for example calling or being called) or from the external services tier (an advertisement generated by a retailer). Since these services require at least a BlueProMoD server, the Bluetooth APs, and the interconnection network, it is important to turn our attention to these components and their interaction.

Bluetooth APs Placement

Bluetooth APs must be appropriately placed so that there is a complete coverage of the targeted area. In order to reduce unnecessary load as well as annoyance to the users, the APs only recognize and generate connections with client devices that run the particular client application; other Bluetooth devices (for example hands free) as well as other Bluetooth APs cannot interfere since they have a different than expected service ID. This arrangement, however, covers only one of the considerations during the deployment of the APs.

The Bluetooth APs must cover all areas traversed by potential users. Since Class 2 devices have a typical range of 10 meters one could assume that placing APs at approximately 20 meters apart would solve the coverage problem. Unfortunately this is not that simple. If there are many obstacles that absorb part of the signal, more APs are needed, placed at a high point in relation to these obstacles (for example top of shelves). Large areas with no columns in the middle also cause a problem, typically solved by placing APs on the ceiling. Nevertheless the effective range of the particular APs is now reduced. For ceilings that have a height approximately equal or larger than the typical AP range of 10 meters, the straightforward solution is to create appropriate booths at the floor for the APs to be housed, although it is not always possible to do so.

Another consideration in placing Bluetooth APs is the coverage of congested areas. By "congested" we mean areas where there are likely to be more than seven users that want to use the voice service simultaneously. Typically, this is the reception area as well as coffee shops and any other area where users can sit and relax. A Bluetooth AP without obstacles can cover an area of $\pi R^2 \sim 300\ m^2$, where only up to seven clients can be active in the respective piconet. Such an area could have as many as 100 users who may wish to place or receive voice calls simultaneously in the worst case. Although they can all be accommodated in the PARK mode, they cannot be active in the same piconet. Obviously all such areas require more Bluetooth APs so that the waiting time for the voice service can be minimized. Other considerations may also prove important, such ease of service and replacement for defective APs, security, and cost.

Interconnection Network

In terms of network infrastructure BlueProMoD requires at minimum a WPAN for each user and Bluetooth AP, as well as a backbone network. Although an IP-wired network can be used for this purpose, different packetization schemes and other problems could tempt designers to opt for a scatternet-based backbone. Considerable research effort has been placed on the development of scatternet and routing schemes to this end. Pabuwal, Jain, and Jain (2003) have developed a complete architectural framework, implementing among others several ad-hoc routing protocols, such as destination sequence distance vector (DSDV), clusterhead switch routing (CGSR), and cluster based routing protocol (CBRP). Although the authors report that they built a complete multimedia application to run over it, no details are given about its performance under different scenarios.

The work in Duggirrala, Ashok, and Agrawal (2003) provides a more detailed scheme called BEAM, under which Bluetooth bridges can effectively negotiate their participation in neighboring piconets. They report that slave piconet nodes are better for the bridge role, as well as that such nodes should be assumed as carrying additional traffic apart from the transit one between neighboring piconets. Although their proposal is interesting it requires Bluetooth manufacturers to incorporate their framework in their products. However, the most important result for our discussion is that bridge nodes can delay traffic significantly since they are slaves and need to use store-and-forward by default.

For this reason as well as the complexity in the formation and routing in scatternets, and security reasons described earlier, we believe that IP-wired networks are better candidates for backbone networks. In addition, such networks offer much higher, interference-free bandwidth at a smaller cost, as well as easier software development, since programming for communicating IP nodes using the sockets API is a de facto standard. The Bluetooth APs must support the PAN profile (the LAN profile is deprecated in Bluetooth v2.0). Typically, L2PCAP is used to establish an ACL connection between a Bluetooth AP and a client device. IP data transfers are facilitated using two possible methods. Under the first (and older) method, IP runs over PPP, which, in turn, runs over RFCOMM. Under the second method, IP runs over BNEP, which offers a layer-2 protocol type similar to Ethernet. Newer and more expensive cell phones support the PAN profile, whereas cheaper and older cell phones support the former method. In all cases, a Bluetooth AP essentially plays the role of a bridging network device between the two different network technologies.

In terms of device connectivity, there are at least two possible paths. Under the first, each Bluetooth AP is an intelligent device, which can directly translate between Bluetooth and IP. In this sense, it can be a dedicated device or a USB dongle connected to a small computer. Unfortunately, the USB standard specifies that the maximum distance from a computer to a full-speed device is 5 USB hubs (at 5 meters each) and an additional 5-meters cable, yielding a total of 30 meters. Obviously, such an arrangement is appropriate only for congested areas in the best case. Vendors, however, have come up with various cost-effective solutions that address this limitation. Some sell USB extenders, which can run reliably at distances of up to 50 meters from the attached computer. Others propose a hybrid approach where Ethernet cables are used to connect suitable USB hubs at 12 Mbps per port; up to five such hubs may be connected to a single computer in a transparent way for existing applications. Such a computer would obviously have to host a proxy ARP, router and—preferably—DHCP servers (depending upon the client connection type/profile) and would connect with the BlueProMoD server. Yet another proposal essentially incorporates a miniature computer at the end of an Ethernet cable with a Bluetooth interface on the other end. In all cases, there are several ways to address the problem of interfacing the Bluetooth APs to the backbone.

The Server

The server is responsible for all of the basic available services as well as other administrative work such as client activity logging and connection to external services. The primary service offered is the reception service which works closely with the users' directory and location services. Client activity logging is also performed on this server, with the relevant data periodically transferred to the External Services tier, so that only recent activity is maintained; older activity is moved out for post processing (for example marketing research). Since BlueProMoD is intended for large organizations, at least two more issues are involved: scalability and reliability.

The server must be able to connect and support all Bluetooth APs. Each Bluetooth AP should not only support its piconet of up to seven active nodes, but also all of the nodes in PARK mode. All these possible user connections are dynamic in nature and could overflow a central DHCP server. Hence, a DHCP server must exist either at each Bluetooth AP, or more preferably on a pool of Bluetooth APs covering neighboring areas. Such a hierarchy reduces the DHCP server load, avoids the single point of failure, and facilitates user roaming as discussed later.

The second issue regarding the BlueProMoD server is reliability of service. Since user authorization/authentication and activity logging takes place at this server, a back-up BlueProMoD server is required. In order for the server transition to take place gracefully, the back-up server must be kept at strict synchronization with the primary. At the same time, for security reasons, no other node can communicate directly with it when the primary server operates normally.

Location-Based Services

LBSs are an emerging technology in wireless telecommunications. One of the most powerful ways to personalize mobile services is based on location. Mobile nodes or even PANs can be combined with a suitable backbone network in order to determine their location and advertise their presence in the overall system. This basic capability has led to the development of numerous services, such as service route finder; access to catalogue data, area guide, and advertisement of goods offered by local retailers; tracking of vehicles and people; or even mobile gaming (Hand, Cardiff, Magee, & Doody, 2006). Some of them are very essential and will be offered by the BlueProMoD.

However, for such services to be effective, the location of the mobile node must be adequately fine grained. For example, an accuracy of 100 meters for route finding in a national highway may be adequate, but is unacceptable in a commercial mall environment. By design, Bluetooth has a typical range of 10 meters (Class 2). However, Bluetooth signal strength is adaptable in the effort to conserve power consumption; a node tries to reduce its signal strength to the minimum required for communication with its peers. Hence, the location accuracy can be further improved if Bluetooth signal strength measurements are taken into account (Peddemors, Lankhorst, & De Heer, 2003). Nevertheless, there are two inherent Bluetooth characteristics that must be taken into account: First, the link quality is relative since it is measured in a manufacturer-specific way. Second, Bluetooth device discovery is typically performed at regular intervals of 20 seconds; this can be a problem in the case of a fast moving node. In the case of a pedestrian walking in a large building complex such as a shopping center, hotel, hospital, or university, however, the previous characteristics do not represent a major problem.

External Services Tier

This is composed of one or more servers, external to the basic BlueProMoD, offering all the non-basic services of BlueProMoD, as well as a DBMS for the advertisements, accounting and Internet services. For example, under a certain variant of BlueProMoD it is possible for some clients to be offered VoIP over Bluetooth, in order to call any telephone outside the area supported directly by BlueProMoD. Nevertheless, the advertisements (announcements in the case of universities or other nonprofit organizations) through targeted MMSs and LBS such as route finding remain some of the most important services, since they represent a significant portion of the system revenues. These should be sent sparingly to the appropriate target user groups to have maximum effect. For example, sending an advertisement for children shoes to a single man is different from advertising a popular children's toy to a mother in her early 30s just before Christmas.

Apart from the marketing issues, which are outside the scope of this article, there are several technical issues involved that belong in this tier. Free voice calls and MMS are expected to be the main attraction of the system, apart from route or goods finding. Instead of creating multiple versions of Web pages to be displayed on various device classes (for example, cell phones, PDAs, portables, info-kiosks, etc.), the content creator could follow certain guidelines to prepare a single version; upon request, the particular Web page could then be sent to a gateway, which would translate it to the appropriate format before sending it to the requesting user. This gateway also plays the role of a buffer between the core server and external services tier (see Figure 3).

Handover

An important part of the system functionality is the handover mechanism, when a client moves from one Bluetooth AP to another. When a user uses his/her cell phone for voice communication, he/she is used to speaking while moving at a normal walking speed; hence, he/she expects to receive an unbroken voice communication service. One of the major limitations of the Bluetooth standard is that it does not provide any mechanism for managing micro-mobility. In Kraemer and Schwander (2003), the inquiry procedure for the determination of the locations of user mobile devices and the connection to the nearest AP was measured to be up to 5 seconds. Furthermore, it is possible for a user that remains stationary to lose his/her connection to the respective Bluetooth AP for a short period of time. What happens when the user tries to reconnect? If he/she is in the middle of a voice call, reconnects, and his/her device assumes a different IP it appears as if the user lost the previous session.

The solution to the latter problem is relatively straightforward taking into account a particular feature of DHCP. More specifically, a client that has recently acquired an IP address can reconnect asking to be assigned the same IP address. If this IP address has not been assigned to anyone else, it is possible for the DHCP server to grant such a request. In our case, let us assume that the IP address space is large enough and each new connection is granted one of the not recently used IP addresses; there is a good chance then that a recently disconnected client will be granted the old IP address. This approach works well and places minimum burden on the network and the DHCP server. Furthermore, users in PARK mode at the same Bluetooth AP appear connected constantly with the same IP address initially assigned to them.

Such an approach also works for roaming users if a DHCP server is responsible for a pool of neighboring Bluetooth APs. Such users typically walk at a slow pace when they talk and tend to remain in the same general area. Hence, the only disruption will be short and depend almost entirely on the adopted Bluetooth hand-off mechanism. In this way more complex forms of addressing the roaming user problem (for example mobile IP) are avoided. Voice over Bluetooth is the only form of user communication requiring special consideration; all other forms of communication are nothing but a series of short TCP/UDP transactions not significantly affected by any change in the IP address of communicating parties. Finally, communications among users is only allowed on a user-to-user basis, so that malicious users cannot easily overload the system.

CONCLUSION

Bluetooth consists of the most promising and well-established wireless technology for a vast number of applications. In this article we have presented a generic framework, called BlueProMoD, based on Bluetooth technology. The main aim of the proposed framework is to support large commercial enterprises (or organizations) at the retail level. BlueProMoD uses all the advantages of Bluetooth in order to provide free-of-charge communication and other services to simple users in return for limited and centrally controlled advertisements from the local retailers, who will have to pay for the system (hence promote it). However, user activity logging is envisaged as the main source of revenue, since valuable marketing information may be extracted from it. The advantages of deploying and supporting the BlueProMoD-based system are straightforward, which combined with its cost effectiveness make it an important tool in such environments.

A future work is to build a prototype as well as conduct extensive simulations in order to determine specific costs and technical profiles that will demonstrate these advantages quantitatively.

REFERENCES

Bisdikian, C. (2005). A framework for building Bluetooth scatternets: A system design approach. *Pervasive and Mobile Computing, 1,* 190-212.

Duggirrala, R., Ashok, R. L., & Agrawal, D. P. (2003). BEAM: A framework for bridge negotiation in Bluetooth scatternets. In *12th ICCCN Conference* (pp. 339-344).

Dursch, A., Yen, D. C., & Shih, D. (2004). Bluetooth technology: An exploratory study of the analysis and implementation frameworks. *Computer Standards & Interfaces, 26,* 263-277.

Elliott, G., & Phillips, N. (2003). *Mobile commerce and wireless computing systems.* Addison-Wesley.

Erasala, N., & Yen, D. C. (2002). Bluetooth technology: A strategic analysis of its role in 3G wireless communication era. *Computer Standards & Interfaces, 24,* 193-206.

Hand, A., Cardiff, J., Magee, P., & Doody, J. (2006). An architecture and development methodology for location-based services. *Elsevier Electronic Commerce Research and Applications, 5,* 201-208.

Kraemer, R., & Schwander, P. (2003). Bluetooth based wireless Internet applications for indoor hot spots: Experience of a successful experiment during CeBIT 2001. *Computer Networks, 41,* 303-312.

Misic, J., Misil, V. B., & Chan, K. L. (2005). Talk and let talk: Performance of Bluetooth piconets with synchronous traffic. *Ad Hoc Networks, 3,* 451-477.

Nallanathan, A., Feng, W., & Garg, H. K. (2006). Coexistence of wireless LANs and Bluetooth networks in mutual interference environment: An integrated analysis. *Computer Communications, 30,* 192-201.

Pabuwal, N., Jain, N., & Jain, B. N. (2003). An architectural framework to deploy scatternet-based applications over Bluetooth. *IEEE International Conference on Communications, 2,* 1019-1023.

Peddemors, A. J. H., Lankhorst, M. M., & De Heer, J. (2003). Presence, location and instant messaging in a context-aware application framework. In *4th International Conference on Mobile Data Management* (LNCS 2574, pp. 325-330). Springer.

Prasad, R., & Munoz, L. (2003). WLANs and WPANs towards 4G wireless. *Artech House.*

Razavi, R., Fleury, M., Jammeh, E. A., & Ghanbari, M. (2006). An efficient packetization for Bluetooth video transmission. *Electronic Letters, 42*(20), 1143-1145.

Scott, D., Sharp, R., Madhavapeddy, A., & Upton, E. (2005). Using visual tags to bypass Bluetooth device discovery. *ACM Mobile Computing and Communications Review, 9*(1), 41-53.

Shaked, Y., & Wool, A. (2005). Cracking the Bluetooth PIN. In *3rd International Conference on Mobile Systems, Applications and Services* (pp. 39-50).

Zeadally, S., & Kumar, A. (2005). Design, implementation, and evaluation of the audio/video distribution transport protocol (AVDTP) for high quality audio support over Bluetooth. *Computer Communications, 28,* 215-223.

This work was previously published in Journal of Information Technology Research, Vol. 1, Issue 1, edited by Mehdi Khosrow-Pour, pp.1-13, copyright 2004 by IGI Publishing, formerly known as Idea Group Publishing (an imprint of IGI Global).

Section III
Tools and Technologies

Chapter XI
Using Mobile Phones and PDAs in Ad Hoc Audience Response Systems

Matt Jones
University of Waikato, New Zealand

Gary Marsden
University of Cape Town, South Africa

Dominic Gruijters
University of Cape Town, South Africa

ABSTRACT

This chapter investigates how to create ad hoc audience response systems using nonspecialist devices. The chapter revolves around two case studies: one involving the use of mobile phones, and the other based on PDAs. Both case studies are carried out in tertiary education institutions, showing how these devices can be used to facilitate audience participation using devices that students might, themselves, bring to lectures. Both are evaluated from the perspective of the student and the educator, using a mixture of observational and interview-based techniques.

INTRODUCTION

Anyone who has given a talk or lecture to a large audience will be well acquainted with the uncomfortable silences, embarrassed glances, and nervous shuffling that greet requests for audience participation. This anecdotal evidence is supported by survey findings presented by Draper and Brown (2004), indicating that if a lecture class is asked for a verbal response, 0% to 3.7% of students are likely to respond: even for the less exposing, "hands-up" response style, the participation rate might also be a low 0.5%-7.8%.

Not all audiences are so shy, though. In the late 1990s, the television game show, "Who Wants to Be a Millionaire?" attracted large, viewing numbers throughout the world. As part of the game format, the contestant could "ask the audience," getting each member to answer the multichoice question using a handset.

Draper and Brown have taken similar handsets out of the TV studio and into the classroom. In Draper and Brown (2004), and an earlier paper (Draper, Cargill, 2002), they present pedagogic motivations for their work, which we share, and will not elaborate on here, beyond noting the value of interactivity and engagement between the learners (students) and the learning-leader (lecturer).

In a long-term, extensive study, summarized in Draper and Brown (2004), the personal response system they used for multiple-choice questions (MCQs) was seen as being of benefit: for example, 60% of 138 first-year computer students rated the system "extremely" or "very" useful; and, similar responses were seen in other disciplines as varied as medicine and philosophy. Handsets are also likely to increase the participation levels: when asked whether they would work out an answer if asked to vote using the system, between 32%-40% agreed.

Of course, specialized handsets have many advantages such as providing simple, direct ways for students to respond (they just press a button): however, there are some drawbacks, including large costs involved in providing handsets ubiquitously, for every student and every lecture; organizational-overheads (e.g., handing out and collecting handsets); and, the impoverished range of responses possible (a single selection for MCQ use).

Inspired by Draper and Brown's experiences, we sought to address these sorts of drawbacks by using a technology that most students now carry with them to every lecture—the mobile telephone. We were interested in whether the pervasiveness and easy familiarity students have with this technology would allow it to serve as a replacement for the purpose-built handsets. Furthermore, we wanted to explore the possibilities beyond MCQs such as students sending free-text questions or, perhaps suggestions and comments to the lecturer. Although other researchers have considered the use of mobile phones in a university setting, for example (Cheverst et al., 2003), we believe this to be a novel application.

Mobile phones are becoming increasingly sophisticated, with a number of current models, sometimes termed "smartphones," providing the sorts of functionality, such as web browsing and document editing, and wireless connectivity, like Wi-Fi and Bluetooth, as well as conventional mobile telecom networking, seen on the handheld personal digital assistants (PDAs). In light of these technological advances, we developed MISPE—the mobile information sharing in the presentation environment, to explore future interaction possibilities for audiences.

The use of personal technologies, like advanced mobile phones and PDAs, has the potential to help all students play a more active role in their education experiences. For people in developing countries though, for example those in South Africa or India, the mobile is a "bridging technology" that can span the digital divide (Marsden, 2003). In these contexts, access to traditional information technology is limited: meanwhile, in South Africa, for instance, over 40% of the population owns a cell phone (rising to 70% for Europe). Staggeringly, over one billion people worldwide own a GSM handset!

In this chapter, we present our experiences in terms of two case studies: the first involves the use of mobile phones to enable the audience to give real-time feedback and responses; the second considers the role of an ad hoc network consisting of the audience's personal technologies, and the lecturer's computer, using MISPE. We discuss both technology issues such as infrastructure requirements and limitations, as well as others relating to the users' experience.

CASE STUDY: TEXT MESSAGING

While the specialized handset studies provided us with a very useful set of functional and non-functional possibilities, we decided to also run some sessions bringing together a group of eight experts in both human-computer interaction and education (all of which were also lecturers), to brainstorm requirements. In the process, we developed scenarios such as this one:

Dr. Monday begins her lecture on advanced linguistic analysis to 300 first-year students. "Before we go any further, are there any questions about last week's topic? Send me a text now from your mobile phone to 444." After a minute, Dr. Monday checks the computer display and sees there are 25 questions, listed in the order they arrived: she can reorder the list alphabetically and by size of message as well. She selects one of the questions to answer.

Later in the lecture, Dr. Monday wants to test the students' understanding of "focus." "Here's a quick quiz," she says. "If you think focus is related to the subject, text 1 to 444; if you think it is related to the topic, text 2; and if you think it is related to the verb, text 3 to 444." Moments later, Dr. Monday can display a bar chart showing the students what the most popular choice was. "Most of you are wrong," she says, wryly, "The correct answer is 2—the topic."

Several times in the lecture, Monday asks the students to text their current "happiness level": "send a text message to 444 now to show how well you understand the lecture so far," she says, "enter H followed by a number from 0 to 9, where 0 is the worst." She can view the changing level of "happiness" over time as a line graph.

After the lecture, Monday returns to her office, and can access all the questions sent by students: she can also review the bar charts for each multiple-choice question, and see the "worm" trace plotted over time. All this information helps her review the lecture content, and plan for next week's session.

Such discussions clarified some of the additional forms of interactivity mobiles might provide over specialised handsets:

- Allowing multiple responses to an MCQ, for example, "choose 2 of the 5 features listed below."
- Parameterised responses, for example, "text your answer (1-5) and how confident you are in your answer (0-100%)."
- Open-ended "conversations" between the lecturer and audience.
- As an active, lecture-experience feedback device.

Pilot-Study System

Before building a full-scale system tailored specifically to the lecture-context, we decided to acquire a third-party, commercial text-polling system to first explore the issues and feasibility of our ideas. The software chosen was the SMS PollCenter by Code Segment. (For information and a demonstration, see http//www.codesegment.com/) The system runs on a PC (we ran it on a laptop in the field studies), and also requires a mobile phone to be connected to the computer via a serial cable, so that sent text messages can be gathered. MCQ results can be displayed in a range of forms such as bar chart and a pie chart. The "SMS Chat" facility displays incoming texts in a scrolling whiteboard format. Software such as this has been used commercially to provide audience response facilities in a range of situations, including television programmes and conferences. Figure 1 illustrates the system in use.

Left-hand image shows a mobile phone being used to send user's response to the lecturer-posed MCQ: background shows lecturer and live

Figure 1. Pilot system use

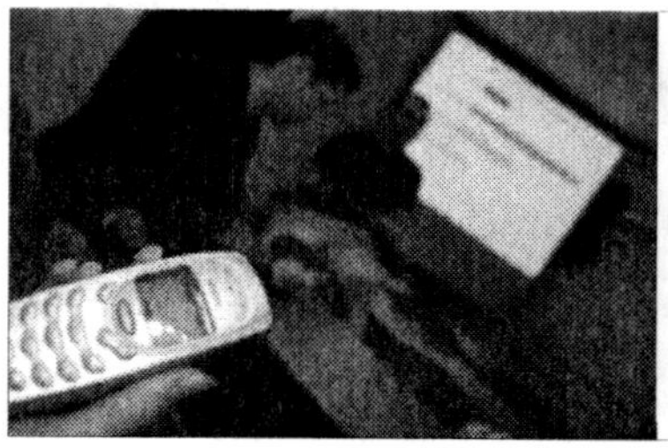

results chart summarizing audience's overall response. Right-hand image shows free-form question: "How do I write a function in C++?" being entered on mobile phone: when sent, it is displayed in the system's "SMS Chat" window (which can be displayed to just the lecturer or the entire audience).

FIELD STUDIES

Initial Experience Gathering

In the first deployment of the system, we studied its use over six, 1-hour sessions spread over 2 months. Our aim was to gather impressions in a range of contexts, so we chose situations with different characteristics, and used the system in a variety of ways (Jones & Marsden, 2004). Three courses were involved:

- *A*: first-year programming class run in New Zealand (NZ).
- *B*: first-year programming class run in South Africa (SA).
- *C*: a fourth-year human-computer interaction class in South Africa.

For courses *B* and *C*, we carried several trials, each separated by around a week. During each session, researchers set up and operated the system for the lecturer: they also observed the class interaction, and were involved in interviewing students at its end. In classes *A* and *C*, the authors were the lecturers—we wanted to experience the system from the front, as it were: two other lecturers were involved in presenting class *B*. Figure 2 shows the system in use in the first-year programming class in Cape Town.

In each session (e.g. session 2), there was one or more uses of the system (e.g., 2.1, 2.2). Questions were either factual (based on lecture content), or personal (eliciting subjective opinion). Text messages sent were either single selections relating to an MCQ, or free text (chat style). Messages/poll results were either fully visible (results shown during polling and dynamically updated), partially

Figure 2. The pilot system in action at the University of Cape Town. Lecturer is discussing results of an MCQ poll (shown on the RHS display; the poll question is displayed on the LHS screen).

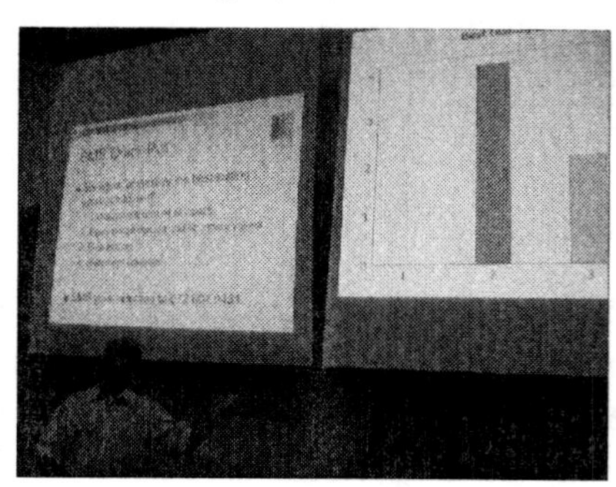

visible (final results shown at end of polling), or hidden (only the lecturer saw the messages).

A summary of each session, and use of the system within them, is shown in Table 1, along with data on the number of text messages received during each use. While this table gives some raw indications of interactivity, it is worth highlighting some of the specific behaviours and effects we noticed. First, 19% of all logged responses to MCQ style questions were in a form that was not recognized by our answer matching filters: for example, in Session 2.1, the students were asked to enter a single integer, but one sent "Turn 72 degqees" (sic). Second, on average, 10% of respondents sent more than one message in response to a question (either resending their initial response, or changing their vote). Third, in SA, 6% of all messages were spam (e.g., "Let the universe decide SMS "oracle" to 34009"); no spam was received in NZ. Fourth, in most of the MCQ cases, as the lecturer discussed the results of the poll chart, additional messages would arrive—sometimes this was a mobile telephone network effect (5%-10% of messages were delayed), but there was also evidence of a "playfulness" as students attempted to "disrupt" the lecturer by altering the results.

At the end of each session, we asked for volunteers to remain behind and give feedback on the system. Overall, we spoke to around 50 people in this way. Views were consistent, in that students liked the idea of the approach (it gave them more of a role in the lecture, changed the pace of the session, etc.); strongly preferred the MCQ style of interaction over the chat scheme (as texting a freeform question could take too long, and the display of comments to the whole class could be distracting); but, they had concerns over the cost of sending messages (over and over again we were told "if sending a message was at a reduced rate, or free, I'd use it a lot more").

We also discussed the experience with the class *B* lecturers. They were less enthusiastic and more cautious about the scheme than the students. Their main concerns were the potential negative impacts of the technology on the "natural" flow of the lecture, and the need for more flexibility in the software to respond dynamically.

Table 1. Summary of sessions and system use

Session/ system use	Course	Question type	Response elicited	Visibility	# people in class	#unique respondents (% of total)
1	A	factual	MCQ	Full	155	35 (23%)
2.1	B	factual	MCQ	Full	180	32 (18%)
2.2	B	personal	chat	Full	180	16 (9%)
3.1	B	personal	MCQ	Partial	150	17 (11%)
3.2	B	factual	MCQ	Partial	150	10 (7%)
4.1	C	personal	MCQ	Full	40	15 (38%)
4.2	C	personal	chat	Full	40	3 (1%)
5.1	C	factual	MCQ	Full	40	6 (15%)
5.2	C	personal	chat	Hidden	40	3 (1%)
6.1	C	personal	MCQ	Full	33	10 (30%)
Mean					**101**	**15 (15%)**

Longitudinal Study

Following this probing set of sessions, we carried out a more focused, longer trial during another course, using the most successful of the methods: multiple-choice questions where results are visible to all.

Over the first 5 weeks of a first-year programming course, one of us used the system to support lecture materials. The course included two lectures every week, and during one of the lectures, students were presented with an MCQ that they could answer using their mobile phone. On average, the number of people attending the lecture was 112 (which represented around 50% of those enrolled in the course). Table 2 presents the data for each weekly session.

During the sessions, we further observed the impact of the system on the lecture experience. Even though the numbers of people responding to the MCQ were low, there was a noticeable effect on the entire audience. The approach helped to strengthen the rapport between the lecturer and the students.

To gain a further perspective on the relative usefulness of the new system, we also deployed two other methods of gathering audience feedback during weeks two to five of the course. Each week, in the lecture that did not include the mobile MCQ system, we distributed post-it notes to each student, and asked them to write an anonymous, short comment, question, or suggestion to the lecturer. These were then collected up after the lecture when the students had left the room.

In addition, we used a Web-based polling system accessible to all the students in the course via their own or university computer. The same MCQ that had been asked in the class lecture via the mobile system was presented. Students could answer the question, and then the current results (showing the frequency of people selecting each choice) were then displayed. We recorded the number of unique respondents one week after each question was posed.

Both of these more conventional methods achieved higher participation rates than our new approach: on average, both achieved around 27% of the total number of possible respondents, where the total number in the post-it note case was the number of attendees in the lecture, and in the Web poll context, the total number of enrolled students.

Discussion

The results suggest that using the handsets to SMS responses to MCQs could improve the level of participation: in the initial study we saw a response rate of 7%-38% (much higher than that predicted by Draper and Brown for "hands-up"). The system was most successful when the results were always on display to the students (from the

Table 2. Week-by-week response rates to MCQ used in first-year programming class

Week #	people in class #	unique respondents	response rate
1	110	16 1	5%
2	105	5	5%
3	110	19 1	7%
4	110	12 1	0%
5	126	8	6%
Mean 1	**12 1**	**2**	**11%**

start to the end of the poll): we discovered that students liked watching their messaging change the display dynamically.

Even when the messaging rate was low, the technique appeared to have a positive impact on the lecture experience: the sessions became more participative, with the lecturer engaging the students in a discussion of the poll results, for instance.

While a novelty effect might well have been in play, in the initial study the response rate seen in 6.1 (30%) compares favorably with that of the earlier session for that class (4.1 (38%)), even though the second session took place approximately 1 month after the earlier one. In the second study, as the weeks went by, there were fluctuations in response rate, but we did not detect a steadily decreasing pattern of enthusiasm. Given Draper and Brown's experience, we predict the enthusiasm for the approach will grow, particularly if charging issues can be resolved (e.g., by providing free texting for students).

The "chat" form of interaction was disappointingly received in the initial study (and we did not proceed with it in the second study). However, we intend to explore this form further with a tailored system, as its potential was undermined by the constraints of the pilot system (e.g., lack of filtering or censoring facilities for the lecturer). Another area for potential was discovered in the form of interesting emergent "community" behaviour when the chat screen was visible to all students: as well as communicating with the lecturer, students posed questions to *each other,* and received replies from within the audience. While there is much exciting work on mobile communities for noncollocated people, this experience suggests there is some useful work to be done on supporting *immobile* mobile communities, such as crowds in football stadia.

Unlike when using specialized handsets in closed networks, designers of mobile phone-based response systems will have to accommodate "invalid" input, both from the users and spammers. In setting up software to process student MCQ responses, for instance, the aim should be to accommodate the variety of answer messages likely to be sent (e.g., "1," "one," "the first choice").

While the more conventional feedback methods used in the second study led to greater participation, they did not, however, foster higher in-class interaction.

CASE STUDY: PDAS

One of our motivations for using mobile phones as the basis of an audience response system, rather than the purpose-built handsets seen elsewhere, was to consider the richer forms of interaction they might facilitate. The previous case study, for instance, illustrated some potential roles of text messaging.

The study also showed, though, that entering responses, particularly free text questions or selections, can be problematic due to the impoverished text-entry facilities of conventional handsets. PDAs and advanced phones provide more sophisticated input and output facilities to the user—stylus-based handwriting recognition and larger, higher resolution displays, for example.

To consider the potential for these emergent, more advanced sorts of personal technology, we built the mobile information sharing in the presentation environment that connects students and their PDAs to the lecturer, with their laptop machine, in a lecture setting. In the sections that follow, we focus on describing the usage scenario and evaluation of the approach: a detailed discussion of the architecture and implementation can be found in Fry, Gruijters et al. (2004).

Usage Scenario

Students arrive at a lecture with their PDA or smartphone—these devices are the "clients" of the system. The lecturer turns on his or her laptop, and uses it to run the "server" application.

Students then connect to this local, lecture server, wirelessly, the system supporting both Wi-Fi, and the shorter-range Bluetooth protocol.

The class begins, and as the lecturer presents the materials to the students in the form of a slide presentation, their devices receive the slides. Once the slide has arrived on a student's PDA, they may annotate it with pertinent information in much the same manner as they would usually annotate hard-copies of slides. At any point, the student can also write a question that is directly displayed on the lecturer's laptop.

As the lecture proceeds, the lecturer presents questions—eliciting free-form text answers—and MCQ polls for the audience to respond to, watching the results appear in his/her slideshow in real time (see Figure 3 for an example).

Once the class is over, the lecturer can save all information created during the class (student answers, student questions, and voting results) enabling them to reflect on the impact of the presentation on its audience.

The image on the left shows the display on the student PDA. This screen is generated automatically, without the lecturer having to explicitly place controls on the screen. Instead, the lecturer works through a dialog box configuring the type of question they require. The software then renders the question in the most appropriate way for the handheld device. The image on the right shows the results received from the PDAs. The lecturer's software is a shell wrapped around PowerPoint™, with audience response slides being created automatically.

Related Work

The Pebbles project has done a substantial amount of work in the areas of collaborative environments involving handheld computers and other devices (Myers, 2001). Among the Pebbles applications is the Slideshow Commander, which allows educators to run a PowerPoint™ presentation from his/her laptop while controlling the presentation from a PDA connected via a wireless link to the laptop. The PDA provides thumbnails of the slides available, and allows the educator to move freely between slides using their PDA.

A system that allows multiple people to share an interactive whiteboard using PDAs is described in Myers, Stiel et al. (1998). This system allows multiple contributors to take turns to use a pen or drawing device on a virtual whiteboard, using PDAs to wirelessly access this single whiteboard display. PebblesDraw, as the system is named, allows users to take turns drawing on their PDAs, and having their contribution appear on a single, central whiteboard or display.

While these systems are useful in an educational context, they supplement a very small proportion of the educator's overall workflow. MISPE differs from the systems developed by

Figure 3. MISPE client (PDA) and server (laptop) example views

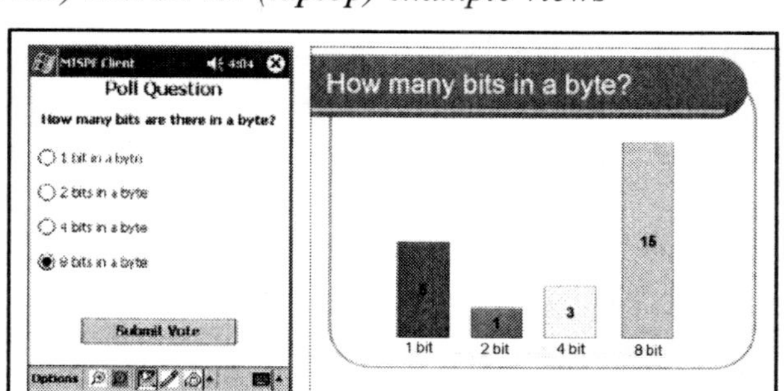

Pebbles by designing and evaluating a system that addresses more of an educator's workflow than just presenting information.

In terms of authoring presentation material, Hexel, Johnson, (2004) describe a tool that allows presenters to create materials that can be customised for specific members of an audience. This tool provides a server that delivers customised presentation data, based on the specifics of the audience members, with relation to language, culture, or disabilities. The tool provides no authoring capabilities in terms of questions or voting. In addition, the system provides only one-way communication from lecturer to audience member.

The work by Hexel et al. (2004), and that reported in Webster and Ho (1997), provides evidence that the sorts of features seen in MISPE may enhance the educational experience. This work suggests that learners experience higher engagement in multimedia presentations which (1) are more challenging, (2) provide more opportunities for feedback, (3) allow more presentation control, and (4) vary in multimedia features.

MISPE uses ad hoc networking. Ad hoc networks are networks that are able to exist without a specific network infrastructure, and without a fixed network topology (Buszco et al., 2001; Doshi, Bhandare, 2002). They also do not require a central, authoritative body. This makes them suited to a highly mobile environment where network nodes may come and go as they please.

Evaluation

An initial, small-scale, user-centred evaluation of MISPE has been carried out to assess its usefulness and usability in real lecture settings.

Method

Two lecturers volunteered to use the system during a lecture. The class size for both lectures was small: five in the first lecture and six in the second. The mean age of the student participants was 18, and all reported having moderate levels of computer experience (the course they were participating in was computer-related).

Researchers attended both sessions to carry out naturalistic observations: that is, to record impressions of how the system impacted on both the lecturer and students during the lecture itself. The observers recorded any unusual uses of the systems, as well as common trends of normal use. They were asked to collect data on the features most used, comments made, and overall user reactions to the system. They were also asked to note how the students' interaction with the technology affected their participation in the lecture: were they able to effectively make annotations, and ask and answer questions, while still listening to the lecturer, or did they fall behind, losing the flow of the content?

Observations of the educators specifically noted the normal teaching style and actions of the educator, as well as when and how the educators interacted with the system.

At the end of both sessions, the lecturer and participants were questioned about the usefulness and usability of the system.

IMPRESSIONS AND FEEDBACK

Lecturers' Perspective

Overall, the response of the two lecturers was enthusiastic: they were keen to use the technology again. In terms of the audience response features provided, they rated the facilities that allowed them to pose questions, and to carry out in-class votes, as the most useful.

In terms of student submitted comments and questions during the lecture, however, the usefulness of the system was hampered by the limited way the system accommodated the dynamic, "performance" nature of lectures. The technology, then, was seen to disrupt the natural flow of the lecture: lecturers would stop interacting with the class for periods of between 10-15 seconds, as they focused on their computer in an attempt to select a comment to discuss, or a student's answer to display to the rest of the class.

The immobile nature of laptop also caused problems: the lecturers often moved around the lecture room during the class. While they stood next to the computer when they wished to control the slideshow, they would often move away from the system to talk about the new slide displayed. This made it difficult for them to view any audience feedback—spontaneous comments or questions—that occurred as they were speaking.

Students' Perspective

The students felt the system led to them being more engaged during the lecture. Being able to ask and answer questions, and to receive feedback using their own device, made them feel more personally involved. Most students used the system to submit many comments and questions during the lecture: they were uninhibited as, in contrast to conventional verbal question asking, they felt their comments would be less disruptive.

There were, though, two usage trends that may lead to a negative impact on the lecture experience. First, students "played" a lot with the system when the lecturer talked for a long while about a slide's content. When they became bored, that is, the system became a distraction. Second, despite the more flexible, easier to use text input methods seen on the PDA, compared to those of the mobile phones in the first case study, the authoring of slide annotations and questions still took too long, causing students to fall behind, and to lose the context of what the lecturer was discussing.

Discussion

As in the texting case study, the use of MISPE provides some evidence that personal technologies can enhance the audience's participation. From the lecturer's point-of-view, though, we need to design a better way for them to control and interact with the system when they are in full-flow of a lecture performance. Specifically, the system should accommodate the lecturer's mobility: as in the SlideShow commander (Myers, 2001), it would seem important that the lecturer has a handheld device that they can use to orchestrate the slideshow and audience participation. Simple interactions with this control device, for example, one-handed button presses, need to provide them with direct access to, say, incoming student comments.

The frustration observed by students as they wrote or answered free-text questions could be overcome by providing a much more lightweight way of authoring. In the present version of the system, submissions have to be entered as textual strings, either by the student tapping on the letters of an onscreen keyboard, or by using the handwriting recognition facilities of the device. A faster approach would be to allow users to use sketching facilities: a scrawled note or comment can be created with the image being sent the lecturer without any preprocessing.

CONCLUSION

Personal technologies—mobile phones, both conventional ones, and the increasingly sophisticated smartphones, along with wireless-capable handheld computers—offer the potential for increasing audience participation in lecture settings.

Unlike the special-purpose handsets used in other trials, these devices offer practical benefits (such as lower cost set-ups, flexibility in deployment, and richer forms of audience response) as well as less tangible impacts on the audience experience arising from the personal nature of the device itself. People relate to these technologies they own and carry everywhere in engaging ways.

In the text-messaging studies, we saw higher participation rates than might be experienced using traditional verbal or hand-show methods. While the response rate was not overwhelming, with, on average, 15% of the audience directly taking part, the impact on the overall lecture experience was significant. In our studies, students had to pay the cost of every message sent, and they indicated that if the service was free, they would more readily respond using it.

In the PDA trial, where all student messages were free, we saw a much higher level of question answering and comment giving. In the next several years, most mobile phones will be equipped with Wi-Fi and Bluetooth capabilities, so that messaging sending costs can be eliminated.

Lectures are often dynamic, lively performances with their own rhythms and flow. Lecturers and audiences will want to use a range of participation methods, both technologically-based and traditional: one moment, the lecturer will ask for a show of hands; the next for questions to be sent to a shared, digital whiteboard.

There is a need, then, for any audience response system to fit within this ecology of resources. Too often, technology fails to accommodate the context, fails to fit in: instead of playing its part in the ecology, it devours, in the process destroying the user experience. In our case studies, we saw some examples of the impact of suboptimal contextual design, but also suggested ways of improving later prototypes.

ACKNOWLEDGMENT

Thanks to Hussein Suleman and Donald Cook, who set aside time in their lectures. Dave Nichols and Mike Mayo helped with the NZ observations, and the Waikato HCI group worked on scenarios.

REFERENCES

Buszko, D., Lee, W., & Helal, A. (2001). Decentralized ad hoc groupware API and framework for mobile collaboration. *Proceedings of the 2001 International ACM SIGGROUP Conference on Supporting Groupwork* (pp. 5-14). ACM Press.

Cheverst, K., Dix, A., Fitton, D., & Rouncefield, M. (2003). Exploring the utility of remote messaging and situated office displays. *Proceedings of Mobile HCI 2003* (pp. 336-341). Springer.

Doshi, S., Bhandare, S., & Brown, T. X. (2002). An on-demand minimum energy routing protocol for a wireless ad hoc network. *Mobile Computing and Communications Review, 6*(3), 50-66.

Draper, S. W., & Brown, M.I. (2004). Increasing interactivity in lectures using an electronic voting system. *Journal of Computer Assisted Learning, 20*, 81-94.

Draper, S. W., Cargill, J., & Cutts, Q. (2002). Electronically enhanced classroom interaction. *Australian Journal of Educational Technology, 18*(1), 13-23.

Fry, B., Gruijters, D., & Reid, S. (2004). *MISPE - Mobile Information Sharing in the Presentation Environment.* Technical report CS04-22-00. Cape Town: University of Cape Town, Department of Computer Science.

Hexel, R., Johnson, C., Kummerfeld, B., & Quigley, A. (2004). PowerPoint™ to the people: Suiting the word to the audience. *Proceedings of the Fifth Conference on Australasian User Interface* (pp. 40-56). Dunedin, NZ: ACM Press.

Jones, M., & Marsden, G. (2004). Please Turn ON your mobile phone: First impressions of text-messaging in lectures. *Proceedings of the 6th International Symposium on Mobile Human-*

Computer Interaction (Mobile HCI '04) (pp. 436-440). Glasgow, UK: Springer.

Marsden, G. (2003). Using HCI to leverage communications technology. *Interactions, 10*(2), 48-55.

Myers, B. (2001). Using hand-held devices and PCs together. *Communications of the ACM, 44*(11), 34-41.

Myers, B. , Stiel, H., & Gargiulo, R. (1998). Collaboration using multiple PDAs connected to a PC. *Proceedings of the 1998 ACM Conference on Computer Supported Cooperative Work (CSCW)* (pp. 285-294). Seattle: ACM Press.

Webster, J., & Ho, H. (1997). Audience engagement in multimedia presentations. *ACM SIGMIS Database, 28*(2), 63-77.

This work was previously published in Audience Response Systems in Higher Education: Applications and Cases, edited by D. A. Banks, pp. 359-372, copyright 2006 by IGI Publishing, formerly known as Idea Group Publishing (an imprint of IGI Global).

Chapter XII
An Evaluation of U.S. City Government Wireless Networks for Mobile Internet Access

Ben Coaker
Whiting-Turner Contracting Company, USA

Candace Deans
University of Richmond, USA

ABSTRACT

The purpose of this chapter is to provide guidelines for city governments considering implementing large-scale wireless networks to provide Internet access for their citizens and businesses. Case studies of cities in the United States that have implemented wireless networks will be evaluated in the context of opportunities and potential challenges. Some key considerations discussed in this chapter involve free versus fee-based models, security considerations, conflicts with local telecommunications companies, and network support. Opportunities to benefit police and emergency services are examined in terms of potential benefits as well as considerations of security in mission critical situations. Strategy guidelines will be presented as a means for providing structure to this decision-making process.

INTRODUCTION

The spectrum used for wireless technology is under FCC control, but is not charged a usage or license fee. This simple fact has many cities looking at the opportunity of providing wireless access to local residents and businesses. Some municipalities have a business model that provides this service for free, while others are considering the options for charging some type of usage fee. Many issues come into play as city government administrators and boards evaluate the opportunities and potential problems associated with taking on the responsibility of implementing and maintaining a wireless network that provides residents with new conveniences and flexibility afforded by mobile Internet use. There is also potential benefit to businesses to enhance visibility and marketing efforts at a cost less than telecommunications companies or other profit-oriented businesses might provide.

Evaluation of long-term benefits for the municipality is essential and yet difficult to evaluate given the fast changes in technological developments. Should city governments even get into this business of mobile Internet access? What are the decision variables? This chapter will help simplify the decision process by providing analysis of city case studies and the current state-of-the-art in terms of benefits and potential drawbacks.

The major objective of this chapter is to evaluate the options for U.S. city governments considering the implementation of large-scale wireless networks to provide mobile Internet access to residents and businesses. Several cities within the United States have already successfully implemented these systems for their residents. Case studies of cities that have implemented wireless networks will be examined to provide insights into the opportunities and potential challenges that are unique to government entities. Revenue generation is a key issue in the overall strategy and decision process. This chapter will provide a discussion of the pros and cons of implementing wireless networks in the context of unique issues faced by city governments.

BACKGROUND

Currently, there is very little research and literature addressing the issues of wireless Internet implementation for government entities. The technology is relatively new in this environment and cities have only recently begun experimentation in this arena. Most of the literature currently available is in trade publications and press releases that address current issues of interest to business and government leaders. The focus of this chapter is on city governments in the United States. Similar trends and issues are emerging in other parts of the world as well.

A major issue surrounding municipal ownership of wireless networks involves competition with the telecommunications companies. Most telephone and cable companies oppose community efforts to offer wireless Internet and in many cases have lobbied to stop municipal Internet zones (DeGraff, 2005). Those favoring municipalities in these efforts believe legislation will ultimately impact America's ability to compete globally. It is essential that all Americans have access to Internet services in order to benefit from the higher standard of living afforded by the Internet. Those without access will be left behind in the rapidly evolving marketplace. According to the National Telecommunications and Information Administration, 40% of Americans do not have dial-up access to the Internet at home and 80% do not have high speed access. One out of four Americans do not use the Internet at all. This places the U.S. behind many countries in Europe and Asia. Those without Internet access cannot benefit from online advertisements for job openings and other available information on the Internet. Those who lack Internet access are typically low income, minority, less educated, and unemployed. Affordable or free Internet access could provide these members of society with the benefits of participating in an Internet-based society (DeGraff, 2005).

Income is a major factor in this divide between the Internet haves and the have-nots. Charges run on average \$40 to \$60 a month and prices continue to rise. Cable and phone companies claim that municipal Internet service is unfair competition. They argue that cities could still provide access to community centers, schools, and libraries through their services. Public interest groups argue that communities have a right to build their own Internet networks to offer more services at a lower cost. School districts could save considerably on current costs for high-speed Internet access (DeGraff, 2005).

Some argue that cities could better spend their money on teaching literacy and computer skills since these are the issues at the root of the digital divide. They argue that broadband Internet access is different than power, telephone, and other "utilities". The costs are astronomical and failures are also high. Building a redundant competitive broadband system may deter the private sector from building new systems or upgrading systems already in place. The private sector can provide higher quality and more secure systems (Effros, 2005).

As more cities investigate this alternative, the debate will likely become more heated as both sides move to protect what they perceive as their best interests. Closing the digital divide is the ultimate goal but finding the best solution remains the challenge for all parties involved.

TYPES OF WIRELESS TECHNOLOGY

There are two main types of wireless technology that are available to municipalities that are interested in implementing a large-scale wireless network. These available technologies are local area networks (LANs) using Wi-Fi technology or metropolitan area networks (MANs) using Wi-Max technology. Each of these technologies has benefits as well as weaknesses associated with them.

Wi-Fi technology utilizes the IEEE 802.11 standard. This standard operates on three different levels. 802.11b was the first standard released and provides transfer speeds up to 11 mbps and operates in the 2.4 GHz range. 802.11a was released several years later and increased transfer speeds up to 54 mbps; however 802.11a operates in the more expensive 5.0 GHz range. Recently, 802.11g has been introduced as a cost effective alternative. 802.11g operates in the 2.4 GHz range keeping it relatively cheap, but still provides transfer speeds up to 54 mbps (Senn, 2004). While Wi-Fi technology is relatively inexpensive, the main weakness of Wi-Fi is the fact that the signal starts to degrade after approximately 100 feet. This gives Wi-Fi an effective operating area of about 31,000 ft^2, which is a little more than one city block. Using Wi-Fi technology, the municipality will have to put one node on approximately every corner in the covered area. Not every node will need to be hard wired to an Internet server, but a large portion will. Wi-Fi performance starts to deteriorate as more and more users access the system. In order to counter the problem of lesser service, several Wi-Fi nodes will need to tie directly into an Internet server, giving the wireless network several access points needed to cope with the anticipated volume.

A new development in wireless technology is the use of MANs, using Wi-Max technology. Wi-Max technology falls under the IEEE 802.16 specifications. Wi-Max technology operates in much the same way cellular technology operates. Instead of having an effective operating distance of 100 feet as in Wi-Fi, Wi-Max can operate up to 30 miles under ideal conditions. Wi-Max operates in one of two ways. The first is through line of site from one tower to another. A steady stream of data is beamed from one tower directly to another, up to 30 miles under ideal conditions. The effective distance can be affected depending on weather conditions and other obstacles that may be in the way. The second way Wi-Max operates is through non-line-of-sight, similar to the way Wi-Fi works. When Wi-Max is not relying on line of sight, its effective distance is cut to about a 5 mile radius, which translates to approximately 25 mi^2. Under this scenario, a city could cover its entire corporate limits with four to six towers strategically placed to provide maximum coverage. Also, since the line of sight aspect of Wi-Max can carry such high capacities, very few towers (possibly only one) will need to be hardwired into an Internet server.

When deciding whether to use Wi-Fi or Wi-Max technologies for its wireless Internet network, each municipality has to make some basic decisions which will dictate how to proceed. First, the local government needs to determine if the existing network of cell phone towers would provide adequate Wi-Max coverage should the cell phone companies allow the municipality to attach a Wi-Max transmitter atop the towers. Second, if towers are not available, or coverage would not be adequate, what would the local reaction be to putting up more towers? Would more towers cause an eyesore to the existing layout of the covered area? Third, would it be more expensive to erect Wi-Max transmitter towers or place a Wi-Fi transmitter node on every city block corner?

CITY GOVERNMENT WIRELESS NETWORK INITIATIVES: CASE STUDIES

The following case studies represent first initiatives by cities in the United States to implement city-run wireless networks. Each city faces different opportunities and challenges but the solutions are similar. The main issue involves whether to offer the service for free and if not what to charge for the service. In some cases, cities have found sponsors to help defray the costs while in other cases the citizens cover the entire costs through monthly fees. The challenge is to find the long-term solution that best meets the needs of the particular situation at hand. The lessons learned by these cities will provide valuable insights for other cities evaluating the potential for their own citizens.

Philadelphia, Pennsylvania

According to the 2000 census, there are approximately 1.5 million people living in the Philadelphia county area. Philadelphia also saw 25.5 million visitors (business and leisure) in 2004, (www.gophila.com). Therefore, on any given day there are more than 2 million people within the Philadelphia metropolitan area. These population facts accompanied by the fact that Philadelphia is a major business center and tourist destination along the east coast, makes this city a perfect test bed for low cost wireless Internet access. Philadelphia has decided to install a city-wide wireless network based on current Wi-Fi 802.11b standards. In July of 2004, Mayor John F. Street appointed an independent council to research, recommend, and coordinate the implementation of a low cost wireless Internet network. This council, the Wireless Philadelphia Executive Committee (Wireless Philadelphia, 2005), is acting in an advisory/advocacy role in the process of developing a wireless community network by instituting community outreach programs, communications with the press, and participation in meetings and conferences (www.phila.gov/wireless).

With the creation of Wireless Philadelphia, the city now has a full-time organization whose sole purpose is to oversee the implementation of a wireless network throughout the city and coordinate all the resources needed to maintain this network. The sole purpose of Philadelphia Wireless can be broken down into three main functions: provide a forum, recommend policy, and identify barriers along with strategies to overcome them.

According to its Web site, www.phila.gov/wireless, Philadelphia Wireless will provide a forum that will enable potential future users to provide feedback on various issues of concern. This forum also allows Philadelphia Wireless to explore any and all emerging technologies that may enhance and improve performance on the wireless network once it is in place or before implementation is complete. Another aspect of the forum to Philadelphia Wireless is the promotion of third-party development of research, development, and use of mobile mesh networks. Inclusion of a third party, or multiple parties, will help defray the cost of the wireless network as well as the burden of maintaining and servicing this new network. By using the open forum to explore concerns, new technologies, and partnering scenarios, Philadelphia Wireless allows for a higher volume of information to be examined as well as multiple viewpoints to be presented. With all the information gathered from the forum, Philadelphia Wireless will then be able to make recommendations on policy.

The second function of Philadelphia Wireless is to provide recommendations on different policy areas. These policy areas include fee structure, roles and responsibilities of personnel, level of service, and privacy and security issues. Some recommendations of Philadelphia Wireless already include a fee structure that provides some level of free wireless Internet access to everyone within Philadelphia city limits with a tiered system of payments for more advanced access. Preliminary fees for advanced access are $10/month for qualified residents and less than $20/month for non-qualifying residents and businesses. Since Philadelphia is a large city with infrastructure in place, wireless nodes will be placed on city streetlights and traffic lights. Locating the nodes in these places will allow for connection to a pre-existing source of power as well as a grid network that is already in place. According to a press release by Philadelphia Wireless, no city money will be used to install and/or maintain the wireless network. Third-party personnel will place and connect the nodes as well as maintain and service the nodes as appropriate. The level of service provided initially will be free access in some parks and public spaces (Philadelphia Wireless Press Release dated October 3, 2005) with an opportunity to expand service areas as more nodes are installed. Privacy and security issues will be handled by the individual service providers and the end users. Each service provider and end user will be responsible for protecting their data and abiding by all applicable laws.

The third function to Philadelphia Wireless will be to identify barriers and develop strategies to overcome these barriers. One such barrier is the resistance from telecommunication companies. As more and more cities are starting to look into the benefits and opportunities of providing free and/or low cost wireless Internet access to the population, telecommunication companies are providing more and more resistance through political lobbying techniques. For example, in November of 2004, Pennsylvania Governor Ed Rendell signed a law that disallowed any local government from providing broadband Internet access "within the service territory of a local exchange telecommunications company". The law also states that the only way a municipality can deploy its own network is if the government first offered the local telecommunication company the opportunity to deploy its own network. This law was largely backed by telecommunications giant, Verizon (Chen, 2005). Philadelphia was allowed to circumvent this roadblock by brokering a deal with the governor and Verizon.

The most important fact of Philadelphia's initiative to provide wireless Internet access to its population is that no money will come from city coffers for installation, maintenance, and upgrades of the wireless network. According to a press release from Philadelphia Wireless, (October 3, 2005) Philadelphia Wireless and Earthlink have entered into the final stages of contract negotiations for Earthlink to "...develop and build the nation's largest municipal Wi-Fi broadband network." According to the press release, the agreement between Earthlink and Philadelphia Wireless specifies that Earthlink will "...finance, build, and manage the wireless network, and provide revenue-sharing fees to support Wireless Philadelphia." While providing a great service to the population such as free and/or low cost wireless Internet access is important, more important is the need to relieve financial stress on the municipality. In today's economic times, everybody is trying to reduce budgets and cut spending. Therefore, finding a means to increase services to city residents and visitors without spending more money is the goal of every local and state government in the United States. According to the press release (October 3, 2005), the agreement between Philadelphia Wireless and Earthlink specifies that Earthlink will provide a 135mi^2 city-wide Wi-Fi mesh network that shall provide the following:

- Low cost wireless high-speed Internet access.
- Hot spot access providers will posses roaming capabilities.
- Free access in some parks and public spaces.
- Occasional users and visitors will be able to obtain daily and weekly access.
- Small businesses will be able to connect to the wireless network and use it as a T-1 alternative.
- Future expansion of wireless network into different areas of Philadelphia.
- Future implementation of emerging technologies as they become economically viable.

According to Philadelphia Wireless's Web site, www.wierlessphiladelphia.org, "Philadelphia's goal is to become the number one wireless city in the world and intends to set the standard by which wireless accessibility is measured." This statement, along with the recent progress made by Philadelphia Wireless, shows the aggressive nature that Philadelphia is pursuing in wireless technology in order to provide a better experience for its citizens and visitors. Estimates of installing Wi-Fi transmitters in Philadelphia are 10 units per day with approximately 8 to 16 units needed per square mile.

Alexandria, Virginia

Alexandria, Virginia is another locality that is pioneering the municipality supplied wireless Internet access. However, Alexandria's plan differs from Philadelphia's plan in two important ways. Alexandria is planning on offering wireless Internet access at no charge and only in a small portion of the town as well as all the public libraries. Alexandria's wireless plan, appropriately named Wireless Alexandria, is to provide free wireless Internet access in a small portion of Old Town Alexandria, specifically the eight block zone from Washington street to the Potomac River along King Street (Gowen & MacMillan, 2005). This portion of Alexandria is a vibrant area that attracts tourists and residents alike due to the large number of shops and restaurants in the vicinity. There are several positives as well as negatives associated with a wireless network of this size.

Wireless Alexandria has all the usual benefits that come with a municipal government providing wireless Internet access. The wireless footprint provided can easily be enlarged due to the fact that the local government owns several buildings in the area as well as all the street and traffic light poles on which to mount the wireless receivers. Emergency officials can utilize the wireless network while within the area receiving the wireless signal. Tourists and residents alike can check and send e-mail or surf the Internet while enjoying a nice day in an outdoor café or while taking in the sites alongside the Potomac River. Due to the fact that the network is confined to a small area, Alexandria has been able to implement the network at a relatively low cost of approximately less than $14,000 initial investment and about $650 per month for the T1 Internet connection (Fifer, 2005). One other benefit unique to the situation in Alexandria is the high tourism traffic in this area. Parking control can be greatly enhanced due to the wireless network. For instance, instead of having traffic officers concentrate on reading meters all day, these same meters can be monitored automatically in a central location, and when the meter expires a notification can be sent to any officer in the vicinity to go and ticket the parking meter violation. Another possible usage for Alexandria's wireless network would be for trash cans to sense when they are getting full and notifying the sanitation department automatically (Walsh, 2005). This section of Alexandria has many points of interest, whether it is shops, restaurants, or historical landmarks. A free wireless network will allow users to learn about the history while actually standing in front of the point of interest, or peruse the dinner menus of several different restaurants while walking down the street. The most important benefit of Alexandria's wireless network is the way the network is set up; there is no competition between the telecommunication companies and the local government providing free wireless Internet access.

While there are many benefits to Alexandria's wireless network, there are also some negative aspects as well, many common to free networks. First, the wireless network that Alexandria has put into place is an unsecured network. An unsecured network is one that has no encryption software in place guarding the information that is being sent back and forth along the network. Alexandria is making a concerted effort to let users know that Wireless Alexandria is an unsecured network and no financial transactions or any type of communications with sensitive information should be conducted while connected to Alexandria's wireless network. Alexandria's wireless Internet is set up so that the only reliable signal will be available out doors along the eight block area between Washington Street and the Potomac River, along King Street. Finally, since this is a free service, users are not going to be able to get customer support 24 hours a day.

Alexandria is avoiding the wrath of the large telecommunication companies that some other cities are coming into contact with as they try to implement wireless technology for public use. However, it should be pointed out that in large cities such as Philadelphia, where the local government is trying to install a service that will take the place of the private Internet service provider, Alexandria is not trying to replace the services provided by the telecommunication companies. Alexandria's main purpose for the wireless network is to provide services for emergency personnel and other city services in a strategically located area. The fact that the residents and tourists are able to "piggy back" onto this wireless network is merely a beneficial secondary opportunity. The fact that this network is available for outdoor use only, is an unsecured network, and lacks 24-hour customer support can almost guarantee that the telecommunication companies in Alexandria will see no noticeable drop in customers due to the implementation of a free wireless Internet access zone in Old Town Alexandria.

New Orleans, Louisiana

In late August of 2005, Hurricane Katrina slammed ashore near the outfall of the Mississippi River as a category 3 hurricane. This storm caused considerable damage far and wide; however the city that received the most attention was New Orleans. The storm surge from the hurricane caused New Orleans' extensive levee system to fail and the waters of the Gulf of Mexico came flooding into the city streets. What followed over the course of the next week or so can only be described as chaos. Emergency services broke down, communication networks throughout the city were non-existent, and many lives were lost. During the confusion, the difficulty in getting accurate information in a timely manner caused many delays for much needed emergency supplies and services. The response time was further delayed due to lack of communication in the next few weeks. Some of this lack of communication was due to the fact that phone and power lines were down over a very substantial area. Another source for lack of communication was the fact that information coming out of New Orleans was so disjointed, that it was very difficult to put the big picture together because there was no way to organize the information and distribute it to the responsible parties quickly enough. Once the storm waters subsided and the rebuilding efforts got under way, New Orleans officials began to investigate a means for speeding the recovery process and giving incentives for displaced residents to return to the streets of New Orleans.

On November 29, 2005, city of New Orleans officials announced plans to implement a free wireless Internet access network initially in the central business district and the French Quarter. New Orleans is in a unique situation right now. Hurricane Katrina has caused massive damage to the public and private utilities as well as the structures in the New Orleans area. Many other cities that have tried to implement a wireless Internet network have done so as a single project, not as a large-scale rebuilding effort. New Orleans has an opportunity to not only replace the infrastructure that was damaged or destroyed by Hurricane Katrina, but upgrade the infrastructure to a higher level of service. The purpose behind New Orleans' wireless Internet network is two-fold. First, the wireless network will aid in the recovery effort by giving the police, emergency services, and aid workers a communications network to utilize in order to make the rebuilding process more efficient. By using the wireless network, city officials can assess damage and log their findings while still on site. Aid workers will be able to communicate with their parent organization and order more supplies efficiently using the wireless Internet network. Police and emergency services will be able to use the wireless Internet network to access databases on current activity as well as coordinate responses to deal with any situation more efficiently than ever before. Second, the wireless network will hopefully jump start the economy in New Orleans. With free wireless access, more tourists are likely to come back to New Orleans quicker than if the network was not in place. Residents will have an incentive to come back home and try to rebuild what they lost in the storm. But the most important result from the city of New Orleans implementation of a free wireless Internet network is the fact that this kind of initiative has never before been deployed by a city trying to recover from a natural disaster. By moving forward and implementing a city-owned free wireless Internet access network, the city of New Orleans is showing that they are not only trying to replace what was lost to Hurricane Katrina, but city officials are trying to make New Orleans a better place than it was before the storm ravaged this world famous city. However, not everyone's reaction to New Orleans' announcement was positive. Just hours after the city of New Orleans made the announcement to implement a city-owned free wireless Internet network, BellSouth officials withdrew a building they had donated for the police to use as their new headquarters (Krim, 2005).

Cities all over the United States are starting to look into the opportunities available stemming from a free or low cost wireless Internet network. Obviously there are several benefits that can come out of a city taking on this kind of initiative; however, the biggest roadblock is coming from the telecommunications companies. When Philadelphia implemented their wireless Internet network, the governor had to get involved and state legislation was created just for that instance. The BellSouth building damaged by Hurricane Katrina only had a flooded basement. Since Katrina, New Orleans headquarters police forces have been scattered throughout the city staying wherever there is space, such as hotels, precinct stations, and other makeshift locations. The idea was to renovate the BellSouth building and use that as the new police headquarters building. This is just one example of the many negative responses that other telecommunication companies have had whenever a city decides it wants to implement a city-owned, free, or low-cost wireless Internet network.

Other Cities

The three previous case studies represent the most prevalent examples of city government implementations of wireless networks in the U.S.

Other cities have implemented similar wireless systems as those discussed. The opportunities and challenges fit similar trends as the three cities examined. Scottsburg, Indiana is an example of a town that was losing jobs to nearby Louisville because of its lack of affordable high-speed Internet access. The mayor addressed the issue head on by providing Wi-Fi access to the city's 6,000 residents who pay $35 a month for the city-run wireless service. Chaska, Minnesota offers a broadband network that costs its citizens $16 a month and provides for more effective communication with the police as an added benefit. St. Cloud, Florida deployed wireless Internet in order to attract more people to its downtown area. Plans are in place to expand free Internet access to the entire city rolling the costs into the general budget (DeGraff, 2005). Although the reasons and opportunities vary from city to city, the advantages are outweighing the costs as more cities decide to move in this direction.

Similar Applications

Large-scale wireless networks have been implemented by entities besides cities. One example is universities that are rolling out campus-wide wireless networks. One university in particular is the University of Richmond. The University of Richmond deployed a wireless Internet network with the intent to provide the freedom for students and professors the ability to teach and learn wherever they choose. The University of Richmond's wireless network uses the 802.11g IEEE standards, coupled with a 10 gigabyte wired network in order to supply users with fast reliable service. The network consists of 560 wireless access points and covers all the buildings on campus as well as many outdoor areas. In order to supply the bandwidth needed for an academic setting, each wireless access point is hard wired to a "data cabinet" which is then connected to the University's Internet backbone. In order to provide security to wireless users, the University of Richmond is using the Wi-Fi Protected Access (WPA) security standard. The security provided as well as ample bandwidth truly makes the University of Richmond's wireless network a success story as well as a good model for other academic institutions and small localities. For more information see http://is.richmond.edu/network/wireless.htm.

STRATEGY FRAMEWORK

When a city government decides to look at the possibility of providing wireless access to a certain area, there are some critical decisions that need to be made before pursuing any type of implementation. First and foremost, the city needs to research state laws to ensure that there is nothing on the books that would not allow a local government to provide a wireless network in direct competition with the area telecommunication companies. Case in point is the situation in Philadelphia, Pennsylvania, where Verizon lobbied the Governor of Pennsylvania to enact a law that would make it illegal for Philadelphia to provide wireless Internet access without first giving the local telecommunications company an opportunity to do the same (see the Philadelphia case study earlier in the chapter). Philadelphia was able to overcome this road block because the governor made a deal with Verizon to make Philadelphia an exception. If there are no legal barriers already in place then the municipality should make every effort to get state and federal government official's support before the telecommunication companies can lobby and change their minds. Once the decision has been made to provide wireless Internet access, there are two more key decisions the municipality must entertain that will dictate future actions. The first key decision is whether to provide wireless Internet to a small portion of the city or provide access throughout the city limits. The second key decision is whether to provide wireless Internet access for free or for some type of usage fee.

When deciding to provide the network for free or for a nominal user fee, the municipality must weigh such factors as maintenance, upgrades, implementing new technologies, and financial strain or benefit on the local government's budget. If the network is provided for free, then there will need to be some source of income other than user fees in order to pay for the deployment, maintenance, and future upgrades for the wireless network. Some cities are able to do this by creating a non-profit organization that runs the wireless network (i.e., Wireless Philadelphia) from funds through grants, donations, and other sources. If the municipality decides to offer wireless Internet access for some type of user fee, there is a large opportunity to leverage the wireless network into a revenue generating operation. The technology is already available and inexpensive, while the installation of the network can be done quickly with sections of the network becoming operational as soon as transmitters are put into place.

The good thing about wireless technology is that it is immediately operational in its effective area as soon as the transmitter is installed. Because of this fact, the municipality does not have to decide on whether to roll out full-scale city-wide access or small pockets of wireless Internet access at the onset of the initiative. Several small wireless Internet access points can be set up throughout the city initially in public spaces such as parks and libraries and then expanded as demand increases and funds become available. Even if the plan is to implement wireless Internet access city wide, this can be done quickly.

Opportunities

The ability to access the Internet anywhere in the city carries great opportunities that cities are eager to tap into. Wireless access city-wide can benefit police and emergency service workers (Malykhina, 2005). Police can use the wireless access to hook up to databases that store information on known criminals, sex offenders, and get up-to-date information on traffic situations as well as other cruiser locations. Firefighters and other emergency personnel can use wireless access to retrieve patient information from local hospitals to better serve their patients upon arrival at the location. Using the wireless network to enhance emergency services is a benefit of the wireless network, however there still needs to be a backup plan in the case of a disaster. The wireless network depends on a power source in order to be operative. In the case of a disaster, power may not be available and therefore an alternate means of communication needs to be readily available. Cities providing wireless access can use this as an incentive for businesses to come back to the downtown areas and revitalize areas that have seen an egress of residents in the past few decades. Another opportunity would be for tourist attractions to make their Web sites interactive for tourists on the property. Since municipalities already own the street lamps and all the government office buildings in a downtown area, they will be able to avoid the leasing fees that most telecommunication companies have to pay in order to put up transmitters and receivers (Malykhina, 2005). Local businesses within the wireless zone will be able to use this feature in their marketing plan in order to attract more customers or get customers to stay longer. Along with the further development of wireless technology, more and more cities will be looking into this service as options become faster, cheaper, and more reliable. However, even with all these opportunities, there are still several roadblocks that need to be overcome in order to consider wireless access a viable project to undertake.

Challenges

Whenever a governmental entity takes on a new project of any scale there are several challenges it must overcome. The major challenge for municipalities looking into the prospect of providing free or low-cost wireless Internet access is the reaction of the telecommunication companies. Some telecom companies fear that with more and more cities looking into the ability to provide wireless access, their customers will continue to dwindle. Others, however, feel that government-provided wireless access will be of such quality that a large population will continue to use the telecom's services (Reardon, 2005). Cities providing wireless access to its residents are faced with the daunting task of building this network from the ground up. Starting from scratch requires solutions to such problems as:

- Level of customer support to provide.
- Maintenance of the system and who will provide this service.
- How to pay for the system (bill residents, bill businesses, increase local taxes, get a major corporation to sponsor, or some combination of these options).
- Implementation delays of needed upgrades or initial roll-out of the system due to the cumbersome process of city government operations.
- Interference with neighboring and/or pre-existing networks.
- Providing the high quality the customers of telecommunication companies have come to expect while still offering the service for free or at a low cost.
- Having enough capacity once the wireless signal reaches the router with the actual wired connections.

Each municipality will deal with these challenges in different ways, but these are the issues that will play a big role in the decision of whether to move forward with implementation.

Another challenge that each municipality must overcome is network security. Wireless information security is vital to the continued use of wireless technology. If data transferred wirelessly is not protected properly, then the data can be stolen or corrupted. Knowing the intended uses of your wireless network will allow you to map out a proper plan to provide adequate security. The level of security should depend on the intended use of the network. For municipal-owned networks, the level of security provided is a function of costs. If the network is intended to replace personal, private networks, then the municipality should be prepared to provide adequate security to allow sensitive data to be transmitted as well as financial transactions to be performed. If the network is intended to be a supplemental network, then users should be made aware of the intended level of security to be provided. In order to provide security, the municipality must first determine the intentions of the network.

REVENUE GENERATION

One of the big issues in this debate is whether to provide Internet access as a free service to everyone or as a means to generate revenue. This decision needs to be made at the onset of the project. A free service may attract residents and businesses while additional revenue may be necessary to provide value-added services. If the municipality decides it wants to use the wireless network to generate revenue, then the equation is fairly simple. The installation cost is low when divided out on a per capita basis. Therefore, the city will be able to charge an equally low fee in order to recoup its costs, cover the network's overhead, and generate revenue at some point in the future. If the municipality wants to provide the network for free to attract business then the situation becomes more complicated. Many cities that are providing or getting ready to provide free wireless access have found some companies to sponsor the initiative. Companies such as Cisco Systems, Dell, IBM, Microsoft, and SAP have done this (Malykhina, 2005). Once the system is in place, these cities need to find a way to fund the maintenance, customer support, and other overhead for the network. These miscellaneous expenses can be put into the budget in a variety of places and funded by the increased (hopefully) income from sales and other taxes that the wireless network will bring in by providing a more appealing environment for certain types of shoppers, businesses, and tech savvy residents.

FUTURE TRENDS

Wireless technologies will continue to evolve offering more alternatives and opportunities to cities and private companies. One of these trends is open mesh network technology. Mesh technology allows nodes in the network to send data over multiple routes to multiple neighbors. This results in redundant connections with many paths for transport of data. Data that is sent directly to a neighbor node in the LAN without accessing the Internet travels at faster speeds. Mesh technology also allows for decentralization. There is no need for a central server or central administration for the network. The best path is chosen by each node based on signal strength and shortest distance. This technology is suitable for larger outdoors LANs (Watkins, 2005).

Technological advancements will continue to impact the decision-making process and in some cases change the course of events in unexpected ways. It is essential to keep abreast of technological trends and remain adaptable as new advances provide opportunities that were not before possible.

An emerging technology that will affect future wireless networks is the advancement of Wi-Max to construct wireless MANs. The ability of Wi-Max to operate in two separate modes makes it a perfect candidate for future wireless uses. As mentioned before, Wi-Max can operate through line of sight from tower to tower, as well as an access point in much the same way as Wi-Fi, but with a much larger broadcast area. These two modes of operation will enable Wi-Max to be a viable alternative. The fact that it can carry more data than a Wi-Fi signal will alleviate the bandwidth problem, and become a viable alternative to expensive T1 lines. An indication that Wi-Max is a technology for the future is the fact that it has standards for both fixed wireless (desktop and laptop computers), as well as mobile devices (cell phones, PDAs, and blackberries).

As businesses continue to become more "mobile" in their operations and work environments, these trends will flow into homes and impact personal expectations. As the world becomes more connected and mobile, expectations for ordinary citizens and demand for mobile Internet access will increase. Cities who are already leaders in this arena will likely reap the benefits of being early movers in the midst of complex technological advancements. Decisions made today will influence future opportunities and direction regardless of the outcome of this debate.

CONCLUSIONS

Given the emerging trends as discussed in this chapter, it has become essential that city governments develop a strategy for wireless network deployment. The issues should be addressed head on and discussed openly. The decision of whether to move or not move in the direction of implementing a city-run wireless network is not the key issue. Competition today dictates that city governments be pro-active in this arena and at least debate the issues and determine a plan of action. It is no longer acceptable to be unaware of these trends.

The mobile wireless Internet revolution is only in its infancy. Much experimentation and trial and error is underway and much remains to be determined in terms of long-term direction and outcomes. Decisions made today that will impact access to the Internet by all sectors of society will have lasting and important implications not only for businesses and governments but for all citizens as participants in the evolving global digital future.

REFERENCES

Chen, M. (2005). *Philly to defy Telecom giants, set up public wireless network.* Newstandardnews.net. Retrieved December 10, 2005, from www.newstandardnews.net/content/index.cfm/items/1658

DeGraffK. (2005). Community wireless networks: Why not? *Mobile Government*, June, 7-9.

Effors, S. (2005). Community wireless networks: Why? *Mobile Government,* June, 19-21.

Fifer, C. (2005). *City launches "Wireless Alexandria".* Alexandriava.gov. Retrieved December 15, 2005, from www.alexandriava.gov/fyi_alexandria/sept_05/fyi_alexandria4.html

Gowen, A., & MacMillan, R. (2005). The Web is in the air. *Washington Post.* Retrieved December 15, 2005, from www.washingtonpost.com/wp-dyn/content/article/2005/06/09/AR2005060901770_pf.html

Krim, J. (2005). Angry BellSouth withdrew donation, New Orleans says. *Washington Post.* Retrieved December 21, 2005, from www.washingtonpost.com/wp-dyn/content/article/2005/12/02/AR2005120201853_pf.html

Malykhina, E. (2005). Square off over Wi-Fi in the town square. *Information Week, 26*(September), 47-53.

Reardon, M. (2005). *The citywide Wi-Fi reality check.* News.com. Retrieved December 10, 2005, from http://news.com.com/The+citywide+Wi-Fi+reality+check/2100-7351_3-5722150.html?tag=st.num

Senn, J. A. (2004). *Information technology, principals, practices, opportunities.* Upper Saddle River, NJ: Pearson Education, Inc.

Walsh, T. (2005). When trash talks: Embracing wireless technology. *Government Computing News.* Retrieved December 15, 2005, from www.gcn.com/vol1 no1/daily-updates/35466-1.html

Watkins, S. (2005). Open mesh. *Mobile Government,* June, 27-28.

Wireless Philadelphia. (2005). *Wireless Philadelphia enters final contract discussions with Earthlink.* Wirelessphiladelphia.org. Retrieved December 10, 2005, from www.wirelessphiladelphia.org

This work was previously published in Electronic Government: Concepts, Methodologies, Tools, and Applications, edited by A. Anttiroiko, copyright 2008 by Information Science Reference, formerly known as Idea Group Reference (an imprint of IGI Global).

Chapter XIII
Information-Theoretic Methods for Prediction in the Wireless and Wired Web

Dimitrios Katsaros
Aristotle University of Thessaloniki, Greece

ABSTRACT

Discrete sequence modeling and prediction is an important goal and challenge for Web environments, both wired and wireless. Web clients' data-request forecasting and mobile location tracking in wireless cellular networks are characteristic application areas of sequence prediction in such environments. Accurate data-request prediction results in effective data prefetching, which combined with a caching mechanism can reduce user-perceived latencies as well as server and network loads. Also, effective solutions to the mobility tracking and prediction problem can reduce the update and paging costs, freeing the network from excessive signaling traffic. Therefore, sequence prediction comprises a very important study and development area. This chapter presents information-theoretic techniques for discrete sequence prediction. It surveys, classifies, and compares the state-of-the-art solutions, suggesting routes for further research by discussing the critical issues and challenges of prediction in wired and wireless networks.

INTRODUCTION

The proliferation of wireless cellular networks and the penetration of Internet services are changing many aspects of Web computing. Constantly increasing client populations utilize diverse devices to access the wired and wireless medium, and various heterogeneous applications (e.g., traffic- and weather-condition broadcasting, streaming video) are being developed to satisfy the eager requirements of the clients. In this environment, seamless and ubiquitous connectivity as well as low client-perceived latencies are two fundamental goals. The first goal calls for smart techniques for determining the current and future location of a mobile node, and the second goal calls for efficient and effective techniques for deducing future client requests for information pieces.

Both of the aforementioned problems are related to the ability of the underlying network to record, learn, and subsequently predict the mobile user's behavior, that is, its movements or its information needs. The success of the prediction is presupposed and is boosted by the fact that mobile users exhibit some degree of regularity in their movement and/or in their access patterns (Bhattacharya & Das, 2002; Nanopoulos, Katsaros, & Manolopoulos, 2003). This regularity may be apparent in the behavior of each individual client or in client groups. The detection of regularity patterns can lead to drastic improvements on the underlying wireless network's performance. Accurate data-request prediction results in effective data prefetching (Nanopoulos et al., 2003) combined with a caching mechanism (Katsaros & Manolopoulos, 2004; Vakali, 2001) can reduce user-perceived latencies as well as server and network loads. Also, effective solutions to the mobility tracking and prediction problem can reduce the update and paging costs, freeing the network from excessive signaling traffic (Bhattacharya & Das, 2002).

These issues had been treated in isolation, but pioneering works (Bhattacharya & Das, 2002; Vitter & Krishnan, 1996) are paving the way for treating both problems in a homogeneous fashion. They exhibited the possibility of using methods that have traditionally been used for data compression (thus characterized as information-theoretic) in carrying out prediction. The unifying principle is that they model the respective state space as finite alphabets comprised of discrete symbols. In the mobility tracking scenario, the alphabet consists of all possible sites (cells) where the client has ever visited or might visit (assuming that the number of cells in the coverage area is finite). In the request-prediction scenario, the alphabet consists of all the data objects requested by the client plus the objects that might be requested in the future (assuming that the objects come from a database and thus their number is finite).

A smart network can record the movement (request) history and then construct a mobility (data-access) model for its clients. The history refers to the past, but the model is probabilistic and extends to the future. As uncertainty is inherent in mobile movements or requests, we can consider the requests to be the outcome of an underlying stochastic process, which can be modeled using established information-theoretic concepts and tools (Misra, Roy, & Das, 2004; Vitter & Krishnan, 1996).

In our earlier work, reported in Nanopoulos et al. (2003), we described a framework that was able to embrace some algorithms that had been presented in the context of Web prefetching. That framework was able to present those algorithms as variations of the standard PPM (prediction by partial match) technique. The emphasis of the framework was on differentiating between the techniques based on whether they record contiguous subsequences or noncontiguous subsequences. From that work, it was clear that the discovery of noncontiguous subsequences required considerable computational effort and could be realized only through off-line algorithms.

Extending this work, this chapter provides a unifying framework for all the methods, which deal with the issues of location tracking and prediction and request forecasting using known information-theoretic structures, not only the PPM structures; the framework treats them as (variable or fixed-length) Markov chains and presents the different families of methods, categorizing the state-of-the-art algorithms into their respective families. It mainly deals with the discovery of contiguous subsequences, although it can relatively easily be extended to include noncontiguous subsequences. An important objective of the chapter is to include in the presentation not only the algorithms that are familiar in the wireless-communications community, but also techniques that have been developed in other disciplines, like computational biology, machine learning, and the

World Wide Web, in order to achieve cross-discipline understanding and the proliferation of ideas. The purpose of the categorization is to reveal the shortcomings and advantages of each method and to identify routes for further research. Closely related to our work is that reported in Begleiter, El-Yaniv, and Yolan (2004), which, although it has a more narrow scope, examining only online prediction methods, it gives a completely different notion for the variable-length Markov chain, defining it as a combination of various Markov chains that are of different length.

The rest of the chapter is organized as follows. The next section describes in mathematical terminology the problem of discrete sequence prediction. Then the chapter surveys the different families of Markov predictors, and then provides a qualitative comparison of them. Next, we present a new online prediction algorithm that does not belong to any of the families, though it combines many of the merits presented by each family. Finally, we discuss some fields for further research, and then conclude the chapter.

THE DISCRETE SEQUENCE PREDICTION PROBLEM

In quantifying the utility of the past in predicting the future, a formal definition of the problem is needed, which we provide in the following lines (Feder, Merhav, & Gutman, 1992; Merhav & Feder, 1998). Let $\mathbf{\Sigma}$ be an alphabet consisting of a finite number of symbols $s_1, s_2, \ldots, s_{|\Sigma|}$, where $|\cdot|$ stands for the length or cardinality of its argument. A predictor, which is an algorithm used to generate prediction models, accumulates sequences of the type $\boldsymbol{\alpha}_i=\alpha_i^1, \alpha_i^2, \ldots, \alpha_i^{ni}$, where $\alpha_i^j \in \mathbf{\Sigma}$ for all i, j and n_i denotes the number of symbols comprising $\boldsymbol{\alpha}_i$. Without loss of generality, we can assume that all the knowledge of the predictor consists of a single sequence $\boldsymbol{\alpha}= \alpha^1, \alpha^2, \ldots, \alpha^n$. Based on $\boldsymbol{\alpha}_i$, the predictor's goal is to construct a model that assigns probabilities for any future outcome given some past information. Using the characterization of the mobility or request model as a stochastic process $(X_t)_{t \in N}$, we can formulate the aforementioned goal as follows.

Definition 1 (Discrete Sequence Prediction Problem). At any given time instance t (meaning that t symbols $x_t, x_{t-1}, \ldots, x_1$ have appeared, in reverse order), calculate the conditional probability:

$$\overline{P}[X_{t+1} = x_{t+1} \mid X_t = x_t, X_{t-1} = x_{t-1}, \ldots],$$

where $x_i \in \mathbf{\Sigma}$ for all $x_{t+1} \in \mathbf{\Sigma}$. This model introduces a stationary Markov chain since the probabilities are not time dependent. The outcome of the predictor is a ranking of the symbols according to their $\overline{P}$. The predictors that use such kind of prediction models are termed Markov predictors.

Depending on the application, the predictor may return only the symbol(s) with the highest probability, that is, implementing a most-probable prediction policy, or it may return the symbols with the *m* highest probabilities, that is, implementing a top-*m* prediction policy, where *m* is an administratively set parameter. In any case, the selection of the policy is a minor issue and will not be considered in this chapter, which is only concerned with methods for inferring the ranking.

The history $x_t, x_{t-1}, \ldots$ used in the above definition is called the context of the predictor, and it refers to the portion of the past that influences the next outcome. The history's length (also, called the length, memory, or order of the Markov chain or predictor) will be denoted by l. Therefore, a predictor that exploits l past symbol will calculate conditional probabilities of the form:

$$\overline{P}[X_{t+1} = x_{t+1} \mid X_t = x_t, X_{t-1} = x_{t-1}, \ldots, X_{t-l+1} = x_{t-l+1}].$$

Some Markov predictors fix, in advance of the model creation, the value of l, presetting it in a constant k in order to reduce the size and complexity of the prediction model. These predictors and the respective Markov chains are termed fixed-length Markov chains or predictors of order k. Therefore, they compute probabilities of the form:

$$\overline{P}[X_{t+1} = x_{t+1} \mid X_t = x_t, X_{t-1} = x_{t-1}, \ldots, X_{t-k+1} = x_{t-k+1}].$$

where k is a constant. Although it is a nice model from a probabilistic point of view, these Markov chains are not very appropriate from the estimation point of view. Their main limitation is related to their structural poverty since there is no means to set an optimized value for k.

Other Markov predictors deviate from the fixed-memory assumption (Buhlmann & Wyner, 1999) and allow the order of the predictor to be of variable length, that is, to be a function of the values from the past.

$$\overline{P}[X_{t+1} = x_{t+1} \mid X_t = x_t, X_{t-1} = x_{t-1}, \ldots, X_{t-l+1} = x_{t-l+1}],$$

where $l=l(x_t,x_{t-1},\ldots)$.

These predictors are termed variable-length Markov chains; the length l might range from 1 to t. If $l=l(x_t,x_{t-1},\ldots) \equiv k$ for all $x_t, x_{t-1}, \ldots$, then we obtain the fixed-length Markov chain. The variable-length Markov predictors may or may not impose an upper bound on the considered length. The concept of variable memory offers richness in the prediction model and the ability to adjust itself to the data distribution. If we can choose in a data-driven way the function $l=l(\cdot)$, then we can only gain with respect to the ordinary fixed-length Markov chains, but this is not a straightforward problem.

The Markov predictors (fixed or variable length) base their probability calculations $\overline{P}$ on counts of the number of appearances of symbols after contexts. They also take special care to deal with the cases of unobserved symbols (i.e., symbols with zero appearance counts after contexts), assigning to them some minimum probability mass, which is acquired from the respective mass of the symbols already seen. For the location-tracking and request-prediction applications, though, the predictors usually adopt a nonprediction approach for the unobserved events and do not apply any smoothing mechanism because the possible alternative symbols may be quite large. Therefore, for the rest of the chapter, we will not deal with the zero-frequency problem and will not adopt smoothing in the presented examples.

FAMILIES OF MARKOV PREDICTORS

We explained earlier how Markov predictors create probabilistic models for their input sequence(s). To realize these models, they need a data structure, a dictionary to keep track of the contexts of interest, and some counts used in the calculation of the conditional probabilities $\overline{P}$. The preferred choice for this task is the use of digital search trees (trees). The root node of the tree corresponds to the null event or symbol, whereas every other node of the tree corresponds to a sequence of events; the sequence is used to label the node. An invariant for the trees is that no two children of a father node may have the same label. In the rest of the chapter, we will consider a Markov predictor to be equivalent to its respective tree. Each node is accompanied by a counter, which depicts how many times this event has appeared after the sequence of events corresponding to the path from the root to the node's father that has been observed.

For our convenience, we present some definitions useful for the sequel of the chapter. We use the sample sequence of events $\boldsymbol{\alpha}$=aabacbbabbacbbc. The length of $\boldsymbol{\alpha}$ is the number of symbols it contains, that is, $|\boldsymbol{\alpha}|=15$. We term that the maximal prefix of a (sub)sequence, say, acb, is the (sub)sequence without its rightmost symbol, that is, ac; the maximal suffix of the (sub)sequence acb is the (sub)sequence without its leftmost symbol,

Figure 1. A PPM Markov predictor for the sequence aabacbbabbacbbc

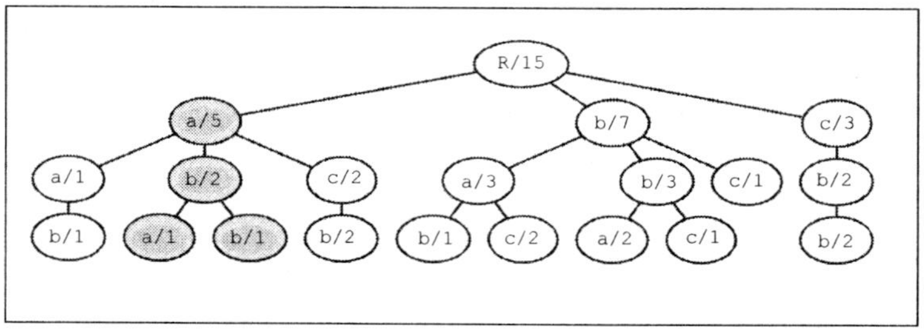

that is, cb, whereas a *suffix* of the acb comes out of acb by removing 0, 1, ..., |abc| symbols from the left of acb. The null sequence denoted as R is a suffix of any sequence and it holds that |R|=0.

The appearance count of subsequence **s**=ab is E(**s**)=E(ab)=2, and the normalized appearance count of **s** is equal to E(**s**) divided by the maximum number of (possibly overlapping) occurrences a subsequence of the same length could have, considering **a**'s length, that is, $E_n(\mathbf{s}) = E(\mathbf{s})/(|\mathbf{a}|-|\mathbf{s}|+1)$. The conditional probability of observing a symbol after a given subsequence is defined as the number of times that symbol has shown up right after the given subsequence divided by the total number of times that the subsequence has shown up at all, followed by any symbol. Therefore, the conditional probability of observing the symbol b after the subsequence a will be denoted as $\overline{P}(b|a)$ and is equal to $\overline{P}(b|a)= E(ab)/E(a)=0.4$.

The generic procedure for deciding which subsequences will be inserted into the tree is specific to each family of Markov predictors and will be described in the next subsections. For purposes related to the clarity of presentation and comparison, we will build the respective tree of each family considering as input the sequence aabacbbabbacbbc. We will present the construction of each family's tree as simple as possible, omitting any optimizations, and we will assume that the input is given beforehand, although some predictors, that is, the online ones, do not demand the whole input to be known in advance.

The Prediction-by-Partial-Match Scheme

The prediction-by-partial-match scheme is based on the universal compression algorithm reported in Cleary and Witten (1984) and constructs a prediction model for an input sequence as follows. It assumes a predetermined maximal order, say, k, for the generated model. Then, for every possible subsequence of length of 1 up to k+1, if it has never been encountered before, we determine the node whose label is the maximal prefix of the considered subsequence. We create a new node under that node. The label of the new node is the length-1 suffix of the considered subsequence, and the new node's counter is initialized to the value of 1. If the considered subsequence has been encountered before, then the counter of the respective node is incremented by 1. Although this description implies that the whole input sequence is known in advance, the method works in an online fashion by exploiting a sliding window of size k+1 over the sequence as it grows symbol by symbol. The PPM predictor for the sample sequence aabacbbabbacbbc is depicted in Figure 1.

Upon completion of the construction phase, we can compute the probability of a symbol σ to appear after a context s by detecting the sequence sσ as a path in the tree emanating from the root, provided that |sσ| ≤ k. The conditional probability of sσ is computed as the ratio of the node counter corresponding to sσ divided by the counter corresponding to σ. Therefore, having built the predictor of Figure 1, we can use it to carry out symbol prediction for a progressing sequence of events as follows: We determine the maximum context with length less than or equal to k that appears as a path in the tree, and compute the conditional probabilities of all symbols to appear after this context. For instance, adopting a most-probable prediction policy, the predicted symbol for the test context ab is a or b, and its conditional probability is 0.50 for either of them (see the gray-shaded nodes in Figure 1).

The maximum context that the PPM predictor can exploit in carrying out predictions is k, though all intermediate contexts with length from 1 to k-1 can be used since they have already been stored into the tree. This model is also referred as the all-kth-order PPM model because it encodes a set of PPM predictors whose order ranges from 1 to k. The interleaving of various-length contexts does not mean that this scheme is a variable-length Markov predictor (although sometimes it is referred to as such) because the decision on the context length is made beforehand and not in a data-driven way.

Apart from this basic scheme, a number of variations have been proposed that attempt to reduce the size of the tree by pruning some of its paths or suffixes of some paths based on statistical information derived from the input data. They set lower bounds for the normalized appearance count and for the conditional probabilities of subsequences, and then prune any branch that does not exceed these bounds. Characteristic works adopting such an approach are reported in Chen and Zhang (2003), Nanopoulos et al. (2003), and Deshpande and Karypis (2004). Their basic motivation stems from the assumption that the pruned states add very little to the prediction capability of the original model, and thus they could be eliminated without sacrificing significantly its effectiveness. The validity of this assumption cannot be justified and, in any case, it strongly depends on the input data distribution. Apparently, these schemes are offline, making one or multiple passes over the input sequence in order to gather the required statistical information.

Application Fields

The PPM scheme was the first compression algorithm that was used in carrying out prediction in wired networks (Fan, Cao, Lin, & Jacobson, 1999; Palpanas & Mendelzon, 1999). Earlier, it had been exploited for the same task in databases (Curewitz, Krishnan, & Vitter, 1993). Although the currently implemented approaches, for example, in the Mozilla browser, implement link prefetching, the sophisticated procedure of the PPM could provide significant benefits.

The Lempel-Ziv-78 Scheme

The Lempel-Ziv-78 Markov predictor, LZ78 for short, is the second scheme whose virtues in carrying out predictions were investigated very early in the literature (Bhattacharya & Das, 2002; Krishnan & Vitter, 1998; Vitter & Krishnan, 1996). The algorithm LZ78 (Lempel & Ziv, 1978) arose from a need for finding a universal variable to the fixed-length coding method and constructs a prediction model for an input sequence as follows. It makes no assumptions about the maximal order for the generated model. Then, it parses the input sequence into a number of distinct subsequences, say, $\mathbf{s}_1, \mathbf{s}_2, ..., \mathbf{s}_x$, such that for all j, 1≤j≤x, the prefix of subsequence $\mathbf{s}_j$ (i.e., all but the last character of $\mathbf{s}_j$) is equal to some $\mathbf{s}_i$, for some 1≤i≤j. The discovered subsequences are inserted into a tree in a manner identical to that of the PPM scheme. In addition, the statistics regarding the number of appearances of each subsequence are stored into the nodes of the tree.

As the process of incremental parsing progresses, larger and larger subsequences are inserted into the tree, allowing the computation of conditional probabilities of increasingly larger subsequences, thus exploiting larger contexts. The

Figure 2. (Left) An LZ78 Markov predictor for the sequence aabacbbabbacbbc. (Right) An LZ78 predictor enhanced according to Bhattacharya and Das (2002)

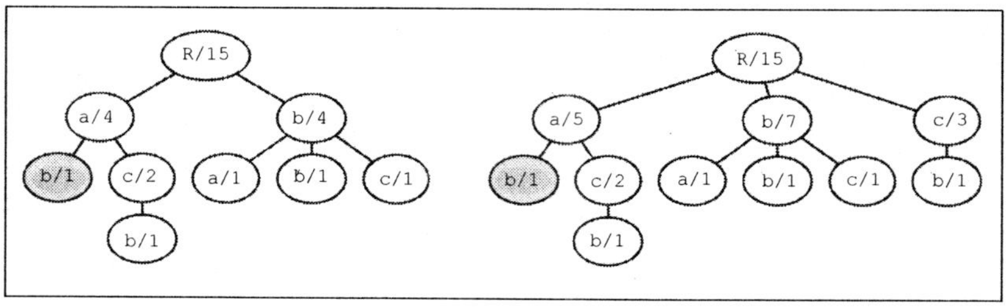

LZ78 predictor for the sample sequence aabacbbabbacbbc is depicted in the left part of Figure 2. The computation of conditional probabilities takes place in a manner completely analogous to that of PPM. However, LZ78 for this example is not able to produce a prediction for the test context ab (i.e., there is no subtree under the gray-shaded node).

Apparently, the LZ78 Markov predictor is an online scheme, it lacks administratively tuned parameters like lower bounds on appearance counts, and it is a characteristic paradigm of a variable-length Markov predictor. Although results do exist that prove its asymptotic optimality and its superiority over any fixed-length PPM predictor, in practice, various studies contradict this result because of the finite length of the input sequence. Nevertheless, the LZ78 predictor remains a very popular prediction method. The original LZ78 prediction scheme was enhanced in Bhattacharya and Das (2002), and Misra et al. (2004) in a way such that apart from a considered subsequence that is going to be inserted into the tree, all its suffixes are inserted as well (see right part of Figure 2).

Application Fields

Apart from the traditional use of the LZ78 algorithm in data-compression areas, recently it has found important application in problems related to location management in mobile networks. It has been proposed as a tool to reduce the communication overhead of the messages that are exchanged between the network and the roaming client (Misra et al., 2004; Roy, Das, & Misra, 2004). However, the applicability of the prediction algorithm is not confined to situations in which the alphabet is easily recognized; in the case of the wireless network, the alphabet consists of the cell IDs. We can have more general situations in which the application defines the symbols of the alphabet. The LZ78 algorithm has been used to track and predict the position of the inhabitants in smart-home applications (Das, Cook, Bhattacharya, Heierman, & Lin, 2002) in order to provide control over home devices. In this context, the house is modeled as a collection of nonoverlapping areas, which are later mapped into a symbolic map corresponding to the neighborhood information for each area. These notions are depicted in Figure 3 and Figure 4. Once we have done this mapping, the application of the prediction algorithm is straightforward.

The Probabilistic Suffix Tree Scheme

The probabilistic suffix tree predictor, PST for short, was introduced in Ron, Singer, and Tishby (1996), and it presents some similarities to LZ78 and PPM. Although it specifies a maximum order for the contexts it will consider, it is actually a

Figure 3. The areas of a smart home

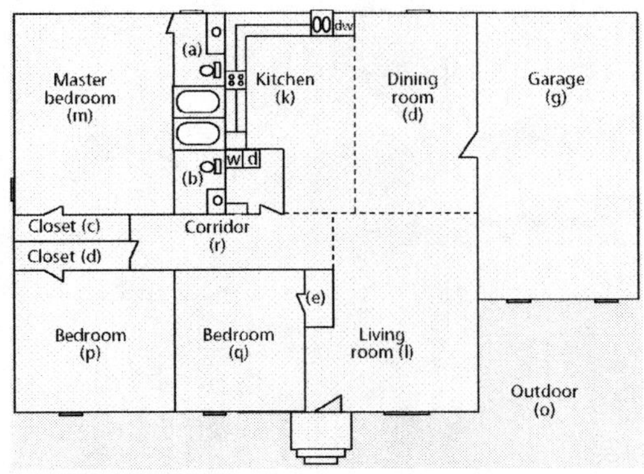

Figure 4. The symbolic representation of the example smart home

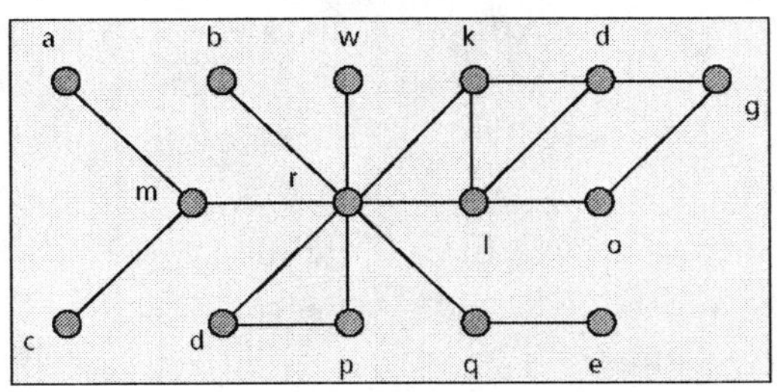

variable-length Markov predictor and constructs its tree for an input sequence as follows. The construction procedure uses five administratively set parameters: k, the maximum context length; a P_{min} minimum normalized appearance count for any subsequence in order to be considered for insertion into the tree; r, which is a simple measure of the difference between the prediction capability of the subsequence at hand and its direct father node; and γ_{min} and α, which together define the significance threshold for a conditional appearance of a symbol. Then, for every subsequence of length of 1 up to k, if it has never been encountered before, a new node is added to the tree, labeled by this subsequence in reverse symbol order provided that a set of three conditions hold. To exhibit the conditions, suppose that the subsequence at hand is abcd. Then, this subsequence will be inserted into the tree of the PST if

1. $E_n(abcd) \geq P_{min}$, and
2. there exists some symbol, say, x, for which the following relations hold:

a. $$\frac{E(abcdx)}{E(abcd)} \geq (1+\alpha)\gamma_{min}$$

b. $$\frac{\tilde{P}(x|abcd)}{\tilde{P}(x|abc)} \geq r \text{ or } \leq 1/r \equiv \frac{E(abc)}{E(abcd)} * \frac{E^*(abcdx)}{E(abcx)} \geq r \text{ or } \leq 1/r$$

In addition, the node corresponding to the considered subsequence stores the (nonzero only) conditional probabilities of each symbol to appear after the subsequence. Obviously, the labels and statistics of each node of a PST differ from those of a PPM or LZ78 scheme. The PST predictor with

Figure 5. A PST Markov predictor for the sequence aabacbbabbacbbc

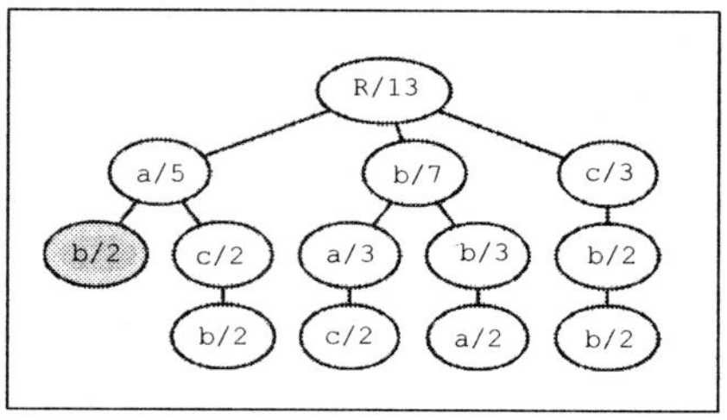

the following set of parameters k=3, P_{min}=2/14, r=1.05, γ_{mim}=0.001, a=0 for the sample sequence aabacbbabbacbbc is depicted in Figure 5. Apparently, PST is a subset of the baseline PPM scheme when k is the same.

Upon completion of the construction phase, we can compute the probability of a symbol σ to appear after a context s by reversing s, and, starting from the root, can detect either the node whose label equals the reversed sequence or the deepest node whose label is a prefix of the reversed sequence. However, PST for this example is not able to produce a prediction for the test context ab (i.e., there is no subtree under the gray-shaded node).

Application Fields

Apart from this basic scheme, a number of variations have been developed, the most important reported in Apostolico and Bejerano (2000), which provided improved algorithms, that is, linear algorithms for the procedures of learning the input sequence and making predictions. Other approaches adapt the technique to specialized contexts, such as computational biology (Bejerano & Yona, 2001; Largeron-Leteno, 2003). Currently, this scheme has not been applied in online problems such as location prediction, but it could be effectively employed in the request-prediction scenario under the assumption that the request pattern of the application does not change dramatically. It is relatively stable for large time intervals.

The Context-Tree Weighting Scheme

The context-tree weighting Markov predictor (Willems, Shtarkov, & Tjalkens, 1995), CTW for short, is based on a clever idea of combining exponentially many Markov chains of bounded order. The original proposition dealt with binary alphabets only, and its later extensions for multialphabets (Volf, 2002) maintained this binary nature. For this reason, we will first describe the CTW Markov predictor for binary {0,1} alphabets and then give the most interesting and practical extension.

The CTW assumes a predetermined maximal order, say, k, for the generated model, and constructs a complete binary tree T of height k, that is, a binary tree in which every nonleaf node has two children and all leaves are at the same height k. An outgoing edge to the left-side children of T is labeled 0, and an outgoing edge to the right-side children of T is labeled 1. Each node s in T is associated with the sequence corresponding to the path from this node to the root. This sequence is also denoted as s. We can find to which input subsequence it corresponds by reversing it. The left and right children of node s are denoted as 0s and 1s, respectively.

Each node s maintains two counters, a_s and b_s, that count the number of 0s and 1s, respectively, that followed context s in the input sequence so far. Initially, all counters are set to 0. Then, we scan the input sequence by considering all subsequences of length k and for each subsequence, we update the counters of the nodes along the path defined by this subsequence. Additionally, each context (node) s maintains, apart from the pair (a_s, b_s), two probabilities, P_e^s and P_w^s. The former, P_e^s, is the Krichevsky-Trofimov estimator for a sequence to have exactly a_s 0s and b_s 1s, and it is computed as

$$\frac{\frac{1}{2}*\frac{3}{2}*\cdots\frac{2a_s-1}{2}*\frac{1}{2}*\frac{3}{2}*\cdots\frac{2b_s-1}{2}}{1*2*3*\cdots*(a_s+b_s)},$$

with $P_e^s(0,0)=1$, $P_e^s(1,0)=1/2$ and $P_e^s(0,1)=1/2$. The latter probability, P_w^s, is the weighted sum of some values of P_e, and it is computed with the following recursive formula:

$$P_w^s=\begin{cases}P_e^s & \text{for } |s|=k\\ \frac{1}{2}P_e^s+\frac{1}{2}P_w^{0s}P_w^{1s} & \text{for } 0\le|s|<k\end{cases}$$

With P_e^R and P_w^R, we denote the Krichevsky-Trofimov estimate and the CTW estimate of the root, respectively. We can predict the next symbol with the aid of a CTW as follows. We make the working hypothesis that the next symbol is a 1, and we update the T accordingly, obtaining a new estimate for the root $P_e^{'R}$. Then, the ratio $P_w^{'R}/P_w^R$ is the conditional probability that the next symbol is a 1. If the next event is indeed a 1, we need not do any update to T; otherwise, we restore the previous values of the tree and perform the update that corresponds to appending a 0 to the input sequence. The CTW predictor for the sample binary sequence 010|11010100011 is depicted in Figure 6. The first three binary digits (at the left of |) are used to create a context for the sequence.

Figure 6. A CTW Markov predictor for the binary sequence 010|11010100011

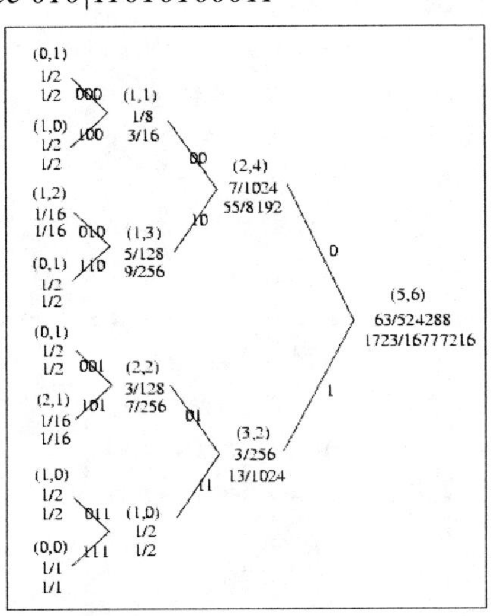

For the case of nonbinary alphabets, Volf (2002) proposed various extensions. We present the decomposed CTW, DeCTW for short, as the best compromise between method efficiency and simplicity. First, we assume that the symbols belong to an alphabet Σ with cardinality $|\Sigma|$. We consider a full binary tree with $|\Sigma|$ leaves. Each leaf is uniquely associated with a symbol in $|\Sigma|$. Each internal node v defines the binary problem of predicting whether the next symbol is a leaf on v's left subtree or a leaf on v's right subtree. Then, we attach a binary CTW predictor to each internal node. We project the training sequence over the relevant symbols (i.e., corresponding to the subtree rooted by v) and translate the symbols on v's left (respectively, right) subtree to 0s (respectively, 1s). After training, we predict the next symbol σ by assigning each symbol a probability that is the product of binary predictions along the path from the root of the binary tree to the leaf labeled by σ. A diagram of the DeCTW is depicted in Figure 7.

Application Fields

The inherent binary nature of the CTW prohibits its wider applicability. However, it has been successfully applied to some problems related to improving the performance of computer architecture. In particular, Federovsky, Feder, and Weiss (1998) described a direct application of the CTW method to the branch-prediction problem, which is the problem of assessing whether the program under execution will follow a branch or not. They looked at the program as a binary source that generates a binary sequence in which 1s correspond to taken branches and 0s corre-

Figure 7. A sketch of the DeCTW Markov predictor for the sequence aabacbbabbacbbc

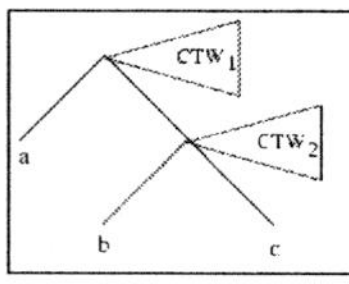

predictor	sequence it models
CTW_1	aaxaxxxaxxaxxxx
CTW_2	bcbbbbcbbc

spond to not-taken branches. Furthermore, they model this binary symbol source using the CTW and perform branch prediction by blending the individual prediction models. Several efforts have been done toward alleviating the binary nature of the CTW and extending it for multialphabets. Of particular importance is the work by Sadakane, Okazaki, and Imai (2000), which provided a very simple and practical implementation along with a technique for combining the prediction strength of PPM and CTW (Okazaki, Sadakane, & Imai, 2002).

COMPARISON OF PREDICTION SCHEMES

In the preceding section, we surveyed a number of Markov predictors. Implicitly or explicitly, they are all based on the short-memory principle, which, simply stated, says that the (empirical) probability distribution of the next symbol, given the preceding sequence, can be quite accurately approximated by observing no more than the last k symbols in that sequence.

Although this principle appears to be simple, the complications it introduces for the prediction algorithms are far from being simple. The algorithms are faced with the problems of selecting an appropriate value for k, which in general depends on the actual values of these most recent symbols. In absence of any other information, some methods fixed in advance the value of k (e.g., PPM, CTW). Such policies mainly suffer from the following drawback. If the value of k is too low and thus too general to capture all the dependencies between symbols, then the prediction efficiency of the respective model will not be satisfactory. On the other hand, if the value of k is too large, then the model will overfit the training sequence. Therefore, variable-length Markov predictors (e.g., LZ78, PST) are most appropriate from this point of view. This was the motivation for subsequent enhancements to PPM and CTW so as to consider unbounded-length contexts, for example, the PPM* algorithm (Cleary & Teahan, 1997).

On the other hand, variable-length predictors face the problem of which sequences and of what length should be considered. PST attempts to estimate the predictive capability of each subsequence in order to store it in the tree, which results in deploying many tunable parameters. LZ78 employs a prefix-based decorrelation process, which results in some recurrent structures being excluded from the tree, at least at the first stages. This characteristic is not very important for infinite-length sequences, but may incur severe performance penalty for short sequences or for sequences in which the patterns appear only a limited number of times; for instance, the pattern bba is missing in both variants of LZ78 of Figure 2.

Despite its superior prediction performance (for instance, see Effros, 2000), PPM is far less commonly applied than algorithms like LZ78. In practice, the LZ78 schemes are favored over PPM algorithms for their relative efficiencies in memory and computational complexity. While the LZ78 predictors can be implemented with O(n) memory and complexity, straightforward implementations of some PPM variants require worst

Table 1. Qualitative comparison of discrete sequence-prediction models

Prediction Method		Overhead			Drawback
Family	**Markov Class**	**Training**	**Parameterization**	**Storage**	
LZ78	variable	online	moderate	moderate	misses patterns
PPM	fixed	online	moderate	large	fixed length high complexity
PST	variable	Off line	heavy	low	parameterization
CTW	fixed	online	moderate	large	binary nature

case $O(n^2)$ memory and complexity to process a data sequence of length n (Cleary & Teahan, 1997). The high computational complexity of PPM algorithms remains an impediment for their more widespread use.

Finally, particular features encountered in each algorithm may make them less appealing for some applications. For instance, the CTW, due to its coupling with binary alphabets, is not the preferred choice for applications regarding multialphabets. Off-line schemes, for example, those in Deshpande and Karypis (2004) and Ron et al. (1996), are not appropriate when request prefetching must be performed by mobile clients.

Table 1 summarizes the Markov-predictor families and their main limitations and advantages. While the considered features and metrics provide a general guideline for algorithm evaluation, the choice and performance of a specific model largely depends on the application.

A SUFFIX-TREE-BASED PREDICTION SCHEME

The suffix-tree-based prediction scheme, STP for short, is a new prediction algorithm not belonging to any of the aforementioned families, and it is described in Katsaros and Manolopoulos (2005). It works as follows. It finds the largest suffix of the input sequence s_1^n—let us call it ss_1^n—whose copy appears somewhere inside s_1^n. Then, it takes a suffix of ss_1^n (the length of this suffix is a parameter of the algorithm) and locates its appearances inside s_1^n. The symbols that appear after the appearances of it are the candidate predictions of the algorithm. The final outcome of the prediction algorithms is the symbol that appears the most times. In pseudocode language, the algorithms are presented in Figure 8.

To explain how the STP algorithm works, we present a simple example.

Example. Suppose that the sequence of symbols seen so far is the following:

s_1^{24}=abcdefgabcdklmabcdexabcd. The largest suffix of s_1^{24} that appears somewhere in s_1^{24} is ss_1^4=abcd. Let α=0.5. Then, sss_1^2= cd. The appearances of cd inside s_1^{24} are located at the positions 3, 10, 17, and 23. Therefore, the marked positions are 5, 12, 19, and 25. Obviously, the last one is not null since it contains the symbol we want to predict. In the general case, all marked positions will contain some valid symbol. Thus, the sequence of candidate predicted symbols is e, k, e. Since the symbol that appears most of the time in this sequence is e, the output of the STP algorithm, that is, the predicted symbol at this stage, is e.

Figure 8. The STP algorithm

Algorithm *STP*
// Current sequence is $s_1^n = s_1, s_2, \ldots, s_n$.
// Predict the symbol after s_n.
begin
STEP 1.
Find the *largest* suffix of s_1^n, whose copy appears somewhere inside s_1^n.
Let this suffix be named ss_1^l. Its length is l and starts at
the position i in s_1^n, i.e.,
$ss_1^l = (s_{n-l+1}, s_{n-l+2}, \ldots, s_n) = (s_{n-i-l+1}, s_{n-i-l+2}, \ldots, s_{n-i})$.
STEP 2.
Take a suffix of ss_1^l of length k with $k = \lceil \alpha * l \rceil$, where α is a parameter.
Let this suffix be named sss_1^k, where $sss_1^k = (s_{n-k+1}, s_{n-k+2}, \ldots, s_n)$.
Suppose that ss_1^l appears m times inside s_1^n.
Each such occurence defines a *marker* and the m positions after
each market are called *marked positions*.
STEP 3.
The predicted symbol is the symbol that appears
the most times in the marked positions.
(In case of ties, the prediction consists of multiple symbols.)
end

The implementation of the algorithm requires an appropriate data structure to support its basic operations, which are the following: (a) the determination of the maximal suffix (at Step 1), and (b) substring matching (at Steps 1 and 2). These two operations can be optimally supported by a suffix tree. The suffix tree of a string $x_1, x_2, \ldots, x_n$ is a tree built from all suffixes of $x_1, x_2, \ldots, x_n\$$, where \$ is a special symbol not belonging to the alphabet. External nodes of a suffix tree contain information about the suffix positions in the original string and the substring itself that leads to that node (or a pair of indexes to the original string in order to keep the storage requirement linear in the string length). It is a well-known result that the suffix tree can be built in linear (optimal) time (in the string length), and can support substring finding in this string also in linear (optimal) time (in the length of the substring). Therefore, the substring searching operation of our algorithm can optimally be implemented. As for the maximal suffix determination operation, if we keep pointers to those external nodes that contain suffixes ending with the \$ symbol (since one of them will be the longest suffix we are looking for), then we can very efficiently support this operation as well. From the above discussion, we conclude the following: (a) The STP algorithm is online, which means it needs no training or preprocessing of the historical data, (b) the storage overhead of the algorithm is low since it is implemented upon the suffix tree, and (c) the algorithm has only one tunable parameter α, which fine-tunes the algorithm's accuracy.

FURTHER RESEARCH

This section presents a couple of directions that we feel would be significant to explore in future research. The first suggestion concerns the development of a new prediction model, and the second proposes to remove one of the assumptions that lead to the development of the current models.

The classical result about the duality between lossless compression (Feder & Merhav, 1994) and prediction implies that any universal lossless compression algorithm can be used to carry out prediction. Although quite a lot of theoretical lossless compression schemes do exist in the literature, only a few of them have been implemented for practical purposes. This is due to the need for effectively combining prediction efficiency, computational complexity, and low implementation effort. These three dimensions limit the range of possible alternative, practical prediction models. Toward this direction, the Burrows-Wheeler (BW) lossless compression scheme offers significant opportunities (Effros, 2000) for combining the excellent prediction ratios of PPM and the low complexity of schemes based on LZ78. So far, no practical prediction scheme is based on the BW scheme, and a plethora of issues have yet to be considered to propose a practical model based on the BW method.

The cornerstone for building the Markov predictors described in this chapter is the "stationarity" assumption, which implied time-homogeneous transition probabilities. Under this assumption, the tree of each predictor grows node by node, increasing the respective node counters; that is, identical subsequences are aggregated (mapped) into the same node of the tree. If we remove the stationarity assumption, this technique is no longer appropriate. In the simplest case, for a mobile client whose roaming patterns change gradually, the predictors will tend to favor the old habits of the client and will adapt to the changing conditions at a very slow rate. Therefore, the assumption of non-time-homogeneous transition probabilities makes the current predictors inefficient and raises some design challenges for any new scheme that will be designed to address this assumption. As we mentioned, full aggregation is not helpful; partial (controlled) or no aggregation could be considered as well, but in any case, novel prediction algorithms should be designed. The technique reported in Ehrenfeucht and Mycielski (1992) could open some directions for research.

CONCLUSION

Discrete sequence prediction is an effective means to reduce access latencies and location uncertainty in networking applications. Due to the importance of the problem in various scientific fields, for example, machine learning, the Web, and bioinformatics, various methods have been reported in the literature. This chapter serves as a survey in this field, promoting the cross-discipline proliferation of ideas, although it by no means covers all proposed techniques. Important research issues have yet to be addressed, such as predictions for nonstationary sequences. We envision predictive model design as a fertile research area with both theoretical and practical solutions.

ACKNOWLEDGMENT

This research was supported by a Γ.Γ.Ε.Τ. grant in the context of the project Data Management in Mobile Ad Hoc Networks funded by the ΠΥΘΑΓΟΡΑΣ national research program.

REFERENCES

Apostolico, A., & Bejerano, G. (2000). Optimal amnesic probabilistic automata or how to learn and classify proteins in linear time and space. *Journal of Computational Biology, 7*(3-4), 381-393.

Begleiter, R., El-Yaniv, R., & Yolan, G. (2004). On prediction using variable order Markov models. *Journal of Artificial Intelligence Research, 22*, 385-421.

Bejerano, G., & Yona, G. (2001). Variations on probabilistic suffix trees: Statistical modeling and prediction of protein families. *Bioinformatics, 17*(1), 23-43.

Bhattacharya, A., & Das, S. K. (2002). LeZi-update: An information-theoretic framework for personal mobility tracking in PCS networks. *ACM/Kluwer Wireless Networks, 8*(2-3), 121-135.

Buhlmann, P., & Wyner, A. J. (1999). Variable length Markov chains. *The Annals of Statistics, 27*(2), 480-513.

Chen, X., & Zhang, X. (2003). A popularity-based prediction model for Web prefetching. *IEEE Computer, 36*(3), 63-70.

Cleary, J. G., & Teahan, W. J. (1997). Unbounded length contexts for PPM. *The Computer Journal, 40*(2-3), 67-75.

Cleary, J. G., & Witten, I. H. (1984). Data compression using adaptive coding and partial string matching. *IEEE Transactions on Communications, 32*(4), 396-402.

Curewitz, K., Krishnan, P., & Vitter, J. S. (1993). Practical prefetching via data compression. *Proceedings of the ACM International Conference on Management of Data (SIGMOD)* (pp. 257-266).

Das, S. K., Cook, D., Bhattacharya, A., Heierman, E., & Lin, T. Y. (2002). The role of prediction algorithms in the MavHome smart home architecture. *IEEE Wireless Communications Magazine, 9*(6), 77-84.

Deshpande, M., & Karypis, G. (2004). Selective Markov models for predicting Web page accesses. *ACM Transactions on Internet Technology, 4*(2), 163-184.

Effros, M. (2000). PPM performance with BWT complexity: A fast and effective data compression algorithm. *Proceedings of the IEEE, 88*(11), 1703-1712.

Ehrenfeucht, A., & Mycielski, J. (1992). A pseudorandom sequence: How random is it? *The American Mathematical Monthly, 99*(4), 373-375.

Fan, L., Cao, P., Lin, W., & Jacobson, Q. (1999). Web prefetching between low-bandwidth clients and proxies: Potential and performance. *Proceedings of ACM International Conference on Measurement and Modeling of Computer Systems (SIGMETRICS)* (pp. 178-187).

Feder, M., & Merhav, N. (1994). Relations between entropy and error probability. *IEEE Transactions on Information Theory, 40*(1), 259-266.

Feder, M., Merhav, N., & Gutman, M. (1992). Universal prediction of individual sequences. *IEEE Transactions on Information Theory, 38*(4), 1258-1270.

Federovsky, E., Feder, M., & Weiss, S. (1998). Branch prediction based on universal data compression algorithms. *Proceedings of the International Symposium on Computer Architecture (ISCA)* (pp. 62-71).

Katsaros, D., & Manolopoulos, Y. (2004). Web caching in broadcast mobile wireless environments. *IEEE Internet Computing, 8*(3), 37-45.

Katsaros, D., & Manolopoulos, Y. (2005). A suffix tree based prediction scheme for pervasive computing environments. In *Lecture notes in computer science* (LNCS 3746, pp. 267-277). Volos, Greece: Springer-Verlag.

Krishnan, P., & Vitter, J. S. (1998). Optimal prediction for prefetching in the worst case. *SIAM Journal on Computing, 27*(6), 1617-1636.

Largeron-Leteno, C. (2003). Prediction suffix trees for supervised classification of sequences. *Pattern Recognition Letters, 24*, 3153-3164.

Merhav, N., & Feder, M. (1998). Universal prediction. *IEEE Transactions on Information Theory, 44*(6), 2124-2147.

Misra, A., Roy, A., & Das, S. K. (2004). An information-theoretic framework for optimal location tracking in multi-system 4G wireless networks. *Proceedings of the IEEE International Confer-*

ence on Computer Communications (INFOCOM) (pp. 286-297).

Nanopoulos, A., Katsaros, D., & Manolopoulos, Y. (2003). A data mining algorithm for generalized Web prefetching. *IEEE Transactions on Knowledge and Data Engineering, 15*(5), 1155-1169.

Okazaki, T., Sadakane, K., & Imai, H. (2002). Data compression method combining properties of PPM and CTW. In S. Arikawa & A. Shinohara (Eds.), *Progress in Discovery Science* (LNCS 2281, pp. 268-283).

Palpanas, T., & Mendelzon, A. (1999). Web prefetching using partial match prediction. *Proceedings of the 4th Web Caching Workshop (WCW).*

Pitkow, J., & Pirolli, P. (1999). Mining longest repeating subsequences to predict World Wide Web surfing. *Proceedings of the USENIX Symposium on Internet Technologies and Systems (USITS)* (pp. 139-150).

Ron, D., Singer, Y., & Tishby, N. (1996). The power of amnesia: Learning probabilistic automata with variable memory length. *Machine Learning, 25*(2-3), 117-149.

Roy, A., Das, S. K., & Misra, A. (2004). Exploiting information theory for adaptive mobility and resource management in future wireless networks. *IEEE Wireless Communications Magazine, 11*(4), 59-65.

Sadakane, K., Okazaki, T., & Imai, H. (2000). Implementing the context tree weighting method for text compression. *Proceedings of the Data Compression Conference (DCC)* (pp. 123-132).

Vakali, A. (2001). Proxy cache replacement algorithms: A history-based approach. *World Wide Web Journal, 4*(4), 277-297.

Vitter, J. S., & Krishnan, P. (1996). Optimal prefetching via data compression. *Journal of the ACM, 43*(5), 771-793.

Volf, P. (2002). *Weighting techniques in data compression: Theory and algorithms.* Unpublished doctoral dissertation, Technische Universiteit Eindhoven, Eindhoven, The Netherlands.

Willems, F. J., Shtarkov, Y. M., & Tjalkens, T. J. (1995). The context-tree weighting method: Basic properties. *IEEE Transactions on Information Theory, 41*(3), 653-664.

Ziv, J., & Lempel, A. (1978). Compression of individual sequences via variable-rate coding. *IEEE Transactions on Information Theory, 24*(5), 530-536.

This work was previously published in Web Data Management Practices: Emerging Techniques and Technologies, edited by A. Vakali and G. Pallis, pp. 159-178, copyright 2007 by IGI Publishing, formerly known as Idea Group Publishing (an imprint of IGI Global).

Chapter XIV
Empirical Prediction of Computer–Network Evolution

Sami J. Habib
Kuwait University, Kuwait

ABSTRACT

This article presents a computer-aided integration tool, iCAD, that can predict a network evolution. We have used the term a network evolution to mean predicting changes within the physical network topology as time evolves. iCAD is connected to four device libraries, each of which contains a distinct set of network-technology devices, such as Ethernet hubs, ATM switches, IP routers, and gateways. As a network technology changes, each device library is updated. Then, we have plotted the cost and performance changes between the old and recent network technologies, enabling us to predict future changes to a first order. This article presents empirical results from 1999 until 2005 recording the network evolution progress, where the lower and upper bounds of network evolution came out to be 10% to 25% and 57% to 74% respectively in terms of network-design cost reduction.

INTRODUCTION

The maintenance process of a network topology so that it can carry out all its communication tasks effectively and efficiently requires a steady tuning and upgrading. Therefore, how often and how to carry on such a process are questions yet to be answered. In other words, we want to identify a metric for predicting the network evolution and develop a design tool to carry on the prediction. Here we have defined a network-evolution metric analogous to Moore's Law (Moore, 1965), which states that the number of transistors in an integrated circuit doubles every 18 months. We have stated that the cost of network-topology design when considering three types of network technologies is reduced at a lower rate of 10-25% and an upper rate of 57-74% in every 24 months.

The network-topology problem with technology considerations entails the allocation and integration of network devices (such as ATM switch,

Ethernet hub, IP router, and gateway) that enable all clients to communicate and access file servers while minimizing the network-integration cost. Literally millions of possible network topologies can be considered for a given network. It is impossible for a human network designer to examine all possibilities for typical large installations. We have applied computer-aided design (CAD) techniques to design and redesign hierarchical computer networks to determine the network topology and network devices. The problem is known to be NP-complete (requiring an exponential number of steps to find an optimal solution) according to Gerla and Kleinrock (1977). The continual changes in the price, performance, and availability of network devices make the network-design process more difficult. The choices of network devices, which are made at a particular time, may not be appropriate a few months later.

The CAD tool, iCAD, is based on evolutionary approach. The tool creates an initial population of network topologies by selecting all network devices randomly. The software analyzes and evaluates each topology in the current population, selects the most fit topologies, modifies some, and discards the rest. New topologies are created by mutating existing topologies or by crossing over genetic material between two topologies. Then the optimization process continues. After several generations specified by the user, the tool terminates. The iCAD is connected to four device libraries, each of which contains a distinct set of network-technology devices, such as Ethernet hubs, ATM switches, IP routers, and gateways. Each device library contains information about device types and their attributes, such as cost, capacity, number of ports, and type of wire. As the network technology matures, the device library is updated. Thus, we have extended the usage of iCAD in Habib, Parker & Lee (2002) so that it can synthesize networks based on the current price and on the predicted future price. Because of this extension on iCAD, we are able to collect many empirical data regarding the network-design evolutions. In addition, we have insured that the synthesized network, once designed, has been optimized for technologies available at the design time. We have examined the topological differences between the current network technologies and the future technologies. These differences would give guidance as to where the network should be upgraded in the future. Finally, we have upgraded existing design constraints by maintaining as much as possible of its original devices so that the upgrading costs are minimized. Maintaining as much as possible of the original devices is known as the legacy problem.

We have been observing and collecting the changes with the network technology, especially with respect to cost and capacity, for the past six years. Habib and Parker (2002) have recorded an initial study on the network evolution by comparing the outcomes from two sets of device libraries in years of 1999 and 2001. In this article, we have expanded our study by running iCAD with four different sets of libraries that go back to 1999, 2001, 2003, and 2005. We have plotted the cost and performance changes between networks designed with the old and recent network technologies. This enables us to predict future changes to a first order. The network-design cost has evolved at a decreasing rate from 10-25% to 57-74% for the past six years as our empirical data has shown. Over time, with more data points, the predictions will become more accurate.

The rest of the article is organized in four sections. The related work has been examined in following section. In "An Overview of the iCAD Tool," we describe the network design tool that is going to carry out all network evolution studies. The experimental results are presented in the section of the same name, and the section after that contains the conclusion.

RELATED WORK

The automatic-network evolution of physical-network topology has not been well researched and documented yet. However, the solution approaches to the network-topology-design problem are built on a wealth of research that can be classified into that of an abstract approach and that of a concrete approach. We have used the term *abstract* approach to mean determining only a set of links and their capacities to connect a set of nodes without specifying any type of network technology. On the other hand, we have used the term *concrete* approach to convey that at least one type of network technology is utilized to synthesize a network topology. Here we list many published papers that have addressed the design of network topology, either the abstract (Gerla & Kleinrock, 1977; Pierre & Legault, 1998; Saksena, 1989) or the concrete (Cosares, Deutsch, Saniee, & Wasem, 1995; Doshi, Dravida, & Harshavardhana, 1995; Elbaum & Sidi, 1996; Fahmy & Dougligeris, 1995; Gresht, 1989; Mitra, Morrison, & Ramakrishnan, 1998). These papers proposed many models and design methodologies for the network physical topology. We have studied and extended their models to be utilized for the network evolution even though none of these papers considered the network evolution. A few groups of researchers have attempted to examine the network evolution from different prospectives. The first group (Frogner & Cannara, 1999) has described a new way of achieving high accuracy in predicting network performance, where a network predictor is built for QoS networks, and it used a combination of network-component models and probabilistic analysis to predict the network performance. The predictor continuously monitors the quality of service, compares it with the expected or desired values and projects future values. Such a predictor uses the past data and performance measurements to predict the network behavior in the future, and it assumes that the network configuration and the policies are unchanged.

The second group (Vekiarides & Finkel, 1998) presented a NETCAP tool, which performed an analysis of Ethernet local area network (LAN) conditions and predicted the future performance of the LAN based on the prescribed changes. The third group (Beigi & Verma, 2001; Beigi, Calo, & Verma, 2004) of researchers from IBM Watson Researcher Center has presented network policies that enable the network administrator to determine whether adding, deleting, or changing a quality-of-service policy would affect the network performance. Network policies are a set of rules determined by the network administrator, to be used in configuring and sharing network resources between all users. A work by Verma (2002) has proposed to leverage the power of policies to simplify the provisioning and configuration of the different devices within the network. This simplification of the management functions is obtained via two elements of the policy-management tool and the policy architecture, namely centralization and business-level abstractions. *Centralization* refers to the process of defining all the device provisioning and configuration at a single point (the management tool) rather than provisioning and configuring each device itself. *Business-level abstractions* make the job of policy administrator simpler by defining the policies in terms of a language that is closer to the business needs of an organization rather than in terms of the specific technology needed to deploy it.

The last group of researchers from MIT and Cornell University (Nowell & Kleinberg, 2003) has defined and studied a basic computational problem underlying social-network evolution, the entitled link-prediction problem. Their problem states that, given a snapshot of a social network at time t, it seeks to accurately predict the edges that will be added to the network during the interval from time t to a given future time t'. In contrast to iCAD, the above articles analyze a network behavior for given policies, limit to one specific network technology, or predict the network performance. In other words, these papers identify the *need* for network evolution; however, they cannot *carry out* network evolution. With tool, we both identify and carry out the network evolution.

AN OVERVIEW OF THE iCAD TOOL

The prototype network tool consists of three main procedures as shown in Figure 1. An evolutionary approach is used to search the design space for a minimal three-level network cost that meets design and performance constraints. The first procedure generates an initial population of three-level network designs by selecting and integrating all network devices randomly. This procedure runs only one time and it insures that all topologies within the population are valid. The second procedure evaluates the cost and estimates the performance of each topology in the population. The third procedure mutates or interchanges portions of some selected topologies in the current population. The second and third procedures execute until the specified number of generations has been produced. The underlying model on which the iCAD tools are based is a large set of mixed-integer nonlinear constraints that must be satisfied to yield a correct design.

We have modeled a network as a three-level hierarchy (backbone, site, and group). The communication within a *group task* utilizes the lowest-level local area network (LAN) that connects a set of *clients* (workstations). The group tasks communicating within a *site task* use the intermediate-level network that connects a set of group LANs. The *backbone task* uses the highest-level network that connects all sites. An assumption is made that a network enterprise locates its clients in a manner where the most intense communications occur locally; moreover, the communications become less intense as the distances increase between the groups and sites. However, this might not be the case uniformly in

Figure 1. An overview of the iCAD tool

```
iCAD(CLT, CTM, TND, DL, EAP)
    /*CLT: is a table containing the client locations*/
    /*CTM: is a matrix containing the client traffic*/
    /*TND: represent a threshold network delay on the network performance*/
    /*DL: represents the four sets of network device libraries*/
    /*EAP: represents a set of parameters used by the evolutionary technique*/
  Begin
      gen = 0;
      generate_initial_network_population();
      evaluate_network_population();
      while(gen < max_gen)
            begin
                  select_parents_for_next_offsprings();
                  perform_genetic_operations();
                  evaluate_network_population();
                  gen++;
            end
End
```

all network enterprises. In order to perform all the collaborative group and site tasks, all clients need to communicate among themselves, and the network should provide the proper support to accomplish the global task within an acceptable time.

iCAD selects one network device for each network level (group, site, and backbone) and integrates all the selected network devices into a three-level network. We embedded a simple performance method (summing all the average delays of all network devices) within the tool to evaluate the synthesized networks (Bersekas & Gallager, 1992). In our formulation, the fitness function represents the total network cost, consisting of the summation costs of network devices, interconnections, and wires.

The output of iCAD is a population of optimized three-level network topologies. The inputs to the prototype network tool are application inputs and tool inputs. The application inputs include the client location table (CLT), client traffic matrix (CTM), and threshold network delay (TND). The CLT and CTM vary from application to application. The TND is a real value given by the designer to provide an upper bound for the average network delay (AND) of a synthesized three-level network topology. The tool inputs are the evolutionary approach's (EAP) parameters and device libraries (DL). The EAP's parameters refer to the population size (PS), number of generations (NG), crossover rate (CR), and mutation rate (MR). The PS indicates the number of candidate solutions that are used during the evolutionary-optimization process. The NG represents how often the evolutionary-optimization process will run. The crossover operator performs higher order transformations by combining parts from two networks (parents) to create a new pair of offspring. Thus, the crossover rate represents the probability of selecting the parents. The mutation operator performs unary transformations by changing a single feature in the network; therefore, the mutation rate represents the probability of selecting a topology for an alteration.

Description of Device Libraries

iCAD is connected to four libraries containing information about network-device types (ATM, Ethernet, IP router, and gateway) and their attributes such as cost, capacity, number of ports, and type of wire. In this article, we illustrate four sets of network libraries, which are collected and upgraded through the years, from 1999 to 2005. Table 1 summarizes iCAD's first network-library set (LIB99), which reflects the network device's status in the year 1999. The table shows four different network technologies with different ranges in communication bandwidth, number of ports, and costs.

Table 2 summarizes the updated iCAD's network device libraries (LIB01), which reflect the network components' status in the year 2001. This table contains network technologies similar to those shown in Table 1, but with different costs. Also, Tables 3-4 summarize the device libraries (LIB03) and (LIB05) that reflect the network components' status in the years 2003 and 2005 respectively.

From the four tables, we concluded that the prices of all network technologies dropped, but at different rates. Ethernet technology has dropped almost to 50% in cost from 1999 to 2001, while both ATM and IP router have dropped at the same rate of almost 25% in cost from 1999 to 2001. Ethernet technology has continued to drop 30% in cost from 2001 to 2003, but at slower rate than both ATM and IP router, which have dropped at the rate of 50% and 60% respectively. This is due to the maturity of ATM and IP-router technologies, which makes each the ultimate choice for high-speed networks. On the other hand, the Ethernet technology still offers a competitive choice due to legacy and high bandwidth such as GigaEthernet. However, the Ethernet technology

Table 1. LIB99, representing the status of network technology in 1999

Library Number	Network Technology	Number of Components Within Library	Range of Number of Ports	Range of Bandwidth, Mbps	Range of Cost ($)
1	Ethernet	36	2-32	10 – 1000	50 – 1,000,000
2	Gateway	25	2	1.5 - 1350	20 – 12,000
3	ATM	36	5-32	25 – 622	2000 – 250,000
4	IP Router	36	5-20	15 – 3000	1000 – 2,000,000

Table 2. LIB01, representing the status of network technology in 2001

Library Number	Network Technology	Number of Components Within Library	Range of Number of Ports	Range of Bandwidth, Mbps	Range of Cost ($)
1	Ethernet	36	2-32	10 – 1000	20 – 500,000
2	Gateway	25	2	1.5 - 1350	8 – 9,000
3	ATM	36	5-32	25 – 622	1500 – 200,000
4	IP Router	36	5-20	15 – 3000	700 – 1,400,000

Table 3. LIB03, representing the status of network technology in 2003

Library Number	Network Technology	Number of Components Within Library	Range of Number of Ports	Range of Bandwidth, Mbps	Range of Cost ($)
1	Ethernet	36	2-32	10 – 1000	14 – 350,000
2	Gateway	25	2	1.5 - 1350	4 – 4,500
3	ATM	36	5-32	25 – 622	750 – 100,000
4	IP Router	36	5-20	15 – 3000	280 – 560,000

Table 4. LIB05, representing the status of network technology in 2005

Library Number	Network Technol-ogy	Number of Components Within Library	Range of Number of Ports	Range of Bandwidth, Mbps	Range of Cost ($)
1	Ethernet	36	2-32	10 – 1000	10 – 100,000
2	Gateway	25	2	1.5 - 1350	4 – 2,000
3	ATM	36	5-32	25 – 622	600 – 75,000
4	IP Router	36	5-20	15 – 3000	50 – 75,000

is limited to three capacity selections (10Mbps, 100Mbps, and 1Gbps), and the price differences between them are huge. That could make ATM technology or IP router a more attractive choice during 2003, since both ATM and IP router offer more selections with different communication bandwidth. This makes ATM and IP router competitive with Ethernet. In 2005, the cost of all network technologies continued to drop up to 90% from 2003.

Network Cost Model

The network cost model (NCM) consists of the summation of device cost, interconnection cost and wire cost as shown in equation 1. The network-device cost (NDC) reflects all network devices (such as Ethernet hub, ATM switch, or IP router) needed to integrate the three-level network, and the device cost is listed within the device library. The network-interconnection cost (NIC) reflects the cost of protocol translators (gateways) that are needed to enable distinct group and site-network devices to communicate with each other, and distinct site and backbone-network devices to communicate with each other. The network-wire cost (NWC) reflects the cost of wires and cables that are needed to connect all network devices.

$$\textit{Minimize } NCM = NDC + NIC + NWC \qquad (1)$$

Network Performance Analysis

To determine the "goodness of fit" of each solution produced by iCAD, performance should be approximated rapidly during the evolutionary-optimization process. We utilized a network delay estimation that is based on a well-known formulation (Bersekas, 1992). This formulation views the network as a network of M/M/1 queues, where each group, site, and backbone is modeled as an M/M/1 queue. Summing all delays generated by all group tasks, site tasks, and the backbone task represents the average network delay (AND). Thus, a network's AND must satisfy the threshold network delay (TND), which is given by iCAD's user. This formulation has been used by many researchers for its simplicity and fast approximation for an average delay (Elbaum & Sidi, 1996; Gerla & Kleinrock, 1977).

To analyze an M/M/1 queue, we need the mean arrival rate (λ) and the mean service rate (μ). In our case, λ represents the total traffic flow within a task (T_i) while μ represents the allocated network capacity for the network device (ND_i). λ is derived from the client traffic matrix (CTM), but μ is a decision parameter. Equation 2 estimates the total average network delay within the three-level network topology and Γ represents the total traffic within the three-level network infrastructure.

$$AND = \frac{1}{\Gamma}\sum_{i=1}^{N}\left(\frac{\lambda_i}{\mu_i - \lambda_i}\right) + \text{PTO} \qquad (2)$$

The term PTO stands for a *protocol translation (gateways) overhead*, which is the sum of all translation-process times between an Ethernet hub and ATM switch. This PTO occurs if an Ethernet hub and ATM switch are connected directly, for example.

EXPERIMENTAL RESULTS

The network tool iCAD was implemented in C++ on a SUN Blade 100. The software contains about 12,000 lines of code. The lowest network design cost found by the evolutionary approach that satisfied all design and performance constraints is the recommended solution to the problem. In all four experiments, a proportionate selection scheme was used in the evolutionary approach with the following parameters' values: population size (PS) = 250, number of generations (NG) = 3000, mutation rate (MR) = 0.05, and crossover rate (CR) = 0.80. A proportionate selection scheme (Michalewicz, 1994), which is a simple selection method, is used to compare each design's fitness function with the average fitness function of the entire population. The fitness function represents the network cost model (NCM), as stated in Equation 1. If a candidate network's fitness function is less than or equal to the average fitness function of the entire population, then this candidate network is kept for the next generation. Otherwise, the candidate network is selected for redesign. The convergence criterion used in our experiments is to terminate iCAD when the number of generations reaches the limit specified.

A hypothetical three-level network topology is considered in our study of the network evolution, as depicted in Figure 2. The three-level network represents the backbone, intermediate, and local area networks respectively. The network consists of eight local area networks (LAN), four intermediate networks, and one backbone network. The 65 clients' nodes are clustered around the 8 LANs as listed in Table 5. Such a type of network is an example of enterprise networks used by organizations, which have certain network requirements, such as high communication bandwidth, high-transfer rate and low delay bounds.

An enterprise network has rapidly become highly distributed collaborative activities involving teams of people in different locations. We used a typical midsize enterprise network—which contains 4 site tasks, 8 group tasks, and 65 client tasks as shown in Table 5—as our example.

Table 5 provides detailed information about the clustering of clients and groups. In the experiments, we focused on predicting network evolution for a three-level network topology with a local star backbone topology to connect its four sites. A backbone network has three network choices: Ethernet hub, ATM switch and IP router.

The traffic flow is summarized by three parameters for each task: local traffic, outgoing traffic and incoming traffic, all of which are calculated from the client traffic matrix given by the user. The

Figure 2. A typical three-level network enterprise infrastructure

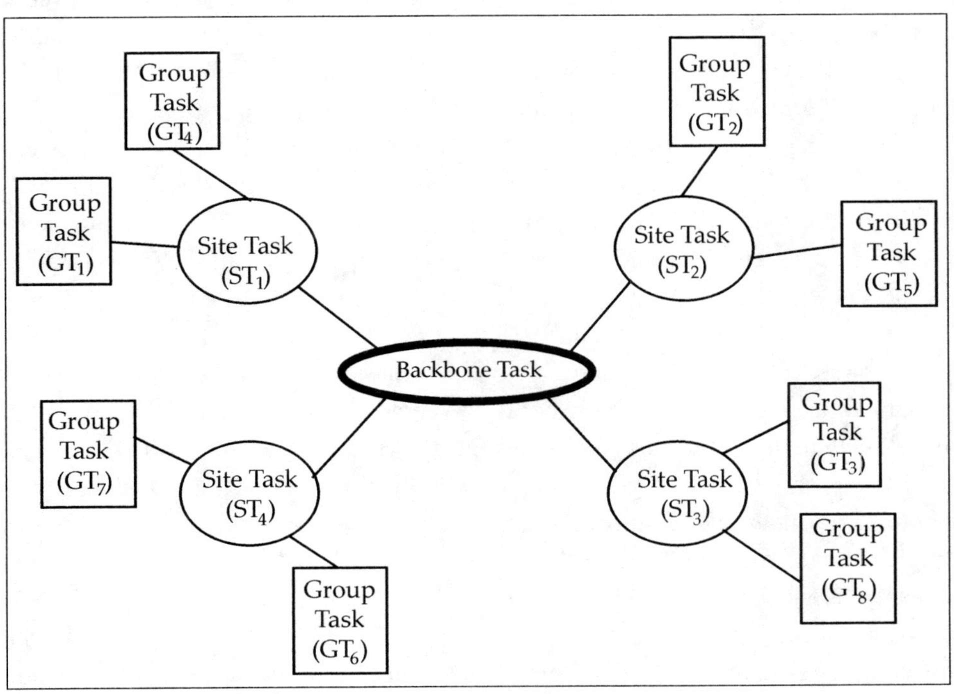

Table 5. Clients and groups clustering information

Site Tasks (ST)	Group Tasks (GT)	Client Tasks (CT)
1	1	1-10
	4	11-15
2	2	16-20
	5	21-30
3	3	31-40
	8	41-50
4	6	51-57
	7	58-65

local traffic represents all the traffic flow within a task. The *outgoing traffic* represents all the traffic flow from a task to all other tasks. The *incoming traffic* represents all the traffic flow coming into a task from all other tasks. Table 6 shows the traffic flow given for the experiments and it is measured in megabits per second (Mbps). The traffic flow within the backbone task can be summarized by one parameter (backbone local traffic, BLT) or site traffic matrix (STM) depending on which topology is used. For the star topology, the backbone local traffic (BLT = 93.125 Mbps) represents all the traffic flow between all sites.

The relationship between the network-cost model (NCM) and the threshold network delay (TND) can be shown in Figure 3. The four curves represent the design cost when iCAD is connecting to LIB99, LIB01, LIB03, and LIB05 respectively. The seven points in each curve represent seven automatically produced network designs by iCAD with TND equal to 5.0, 10.0, 20.0, 30.0, 40.0, 50.0, and 60.0 seconds respectively. Each point in a curve represents an average of ten runs, where each run takes less than three minutes on a SUN Blade 100. The gaps between the four curves represent the shift in the network-design cost. The collected empirical data from 1999 to 2001 has shown a reduction rate in the range of 10.02% to 25.75% while from 2001 to 2003 the reduction rate has been shifted in the rage of 41.51% to 50.59%. The largest shift in the network-design cost occurs between 2003 and 2005, and the reduction rate is in the range 57.08% to 74.70%. We observed the domination of the Ethernet and IP router technologies over the ATM technology in the last few years, where the IP router technology leads in the interconnections among subnetworks, and the Ethernet technology takes over all local area networks (subnetworks).

Also, the plot in Figure 3 illustrates the classical inverse relationship between the network cost and delay. The tradeoff relationship between cost and delay was obvious from 1999 to 2003; however, it was obscure from 2003 to 3005. This is due to reduction in the manufacturing cost of all network technologies. The three-level network design cost is reduced by more than half, from $167,815.00 to $78,960.13, when iCAD is connected to LIB99 and TND increases from 5.0 seconds to 60.0 seconds. When iCAD is connected to LIB01 and TND increases from 5.0 seconds to 60.0 seconds, the three-level network-design cost is reduced by almost half, from $124,609.00 to $64,400.10. When iCAD is connected to LIB03 and TND increases from 5.0 seconds to 60.0 seconds, the three-level network design cost is reduced by more than half, from $72,889.00 to $33,468.10. During 2005, the network-design cost went from $13,885.10 to $22,761.30 after tightening the threshold network delay from 60.0 seconds to 5.0 seconds.

Table 6. Traffic flow for site and group tasks.

Task	Local Traffic (Mbps)	Outgoing Traffic (Mbps)	Incoming Traffic (Mbps)
site task 1	0.0	55.6	2.5
site task 2	0.0	15.0	21.3
site task 3	0.0	15.0	33.8
site task 4	4.2	7.5	35.6
group task 1	27.0	7.5	0.0
group task 2	2.0	2.5	2.5
group task 3	9.0	15.0	7.5
group task 4	8.0	48.1	2.5
group task 5	9.0	12.5	18.75
group task 6	4.2	7.7	16.6
group task 7	5.6	4.0	23.2
group task 8	9.0	0.0	26.3

Figure 3. Network-design cost versus performance over seven years

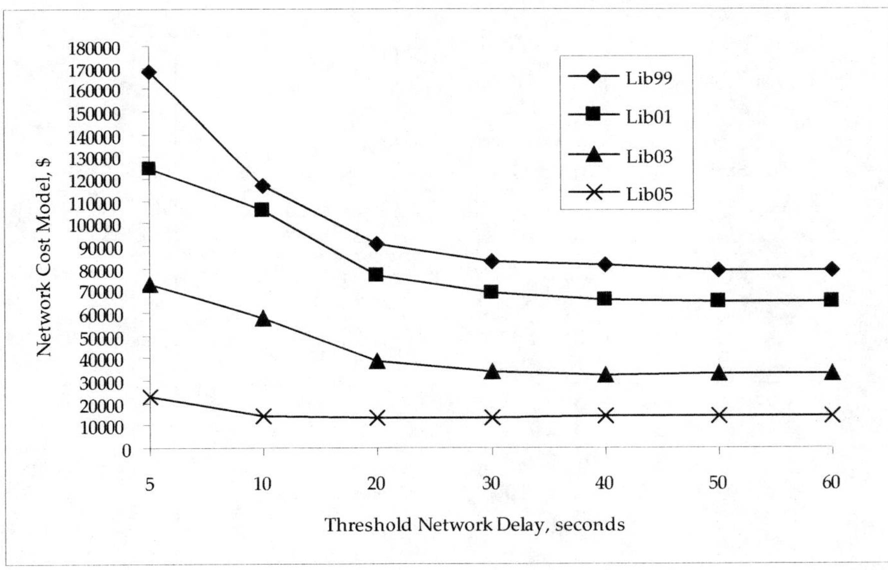

The network *homogeneity factor* (HF) is an output parameter of iCAD ranging (0.0, 1.0]; HF = 1.0 indicates that all the allocated network devices are selected from the same technology. Otherwise, HF indicates the ratio of the maximum number of allocated network devices of the same technology to the total allocated network devices in the three-level network. In the case using LIB99 and TND = 60.0 seconds, the best network design has an estimated average network delay equal to 37.53 seconds and its HF is equal to 0.54, which indicates that 54% of the allocated network devices are an ATM technology. In the case using LIB01 and TND = 60.0 seconds, the best network design has an estimated average network delay equal to 35.46 seconds and its HF is equal to 0.54, which indicates that 54% of the allocated network devices are also an ATM technology. In the case using LIB03 and TND = 60.0 seconds, the best network design has an estimated average network delay equal to 35.46 seconds and its HF is equal to 0.46, which indicates that 46% of the allocated network devices are also an ATM technology. The ATM technology disappeared in all the designs while using LIB05, and the Ethernet technology dominates 62% of all allocated network devices.

In the case using LIB99 and TND = 5.0 seconds, the best network design has an estimated average network delay equal to 4.91 seconds

Table 7. Three network designs produced by iCAD while connecting to LIB99, LIB01, LIB03, and LIB05 respectively (threshold network delay = 60.0 seconds)

Tasks	**Network Devices allocated from Using LIB99 by iCAD**	**Network Devices allocated from Using LIB01 by iCAD**	**Network Devices allocated from Using LIB03 by iCAD**	**Network Devices allocated from Using LIB05 by iCAD**
Backbone	IP Router, 100Kpps, 5 ports, $7,000	IP Router, 100Kpps, 5 ports, $7,000	IP Router, 100Kpps, 5 ports, $2,800	IP Router, 100Kpps, 5 ports, $300
Site 1	IP Router, 50Kpps, 5 ports, $3,000	IP Router, 50Kpps, 5 ports, $7,000	IP Router, 50Kpps, 5 ports, $1,200	IP Router, 50Kpps, 5 ports, $150
Group 1	ATM, 45Mbps, 15 ports, $9000	ATM, 45Mbps, 15 ports, $7000	ATM, 45Mbps, 20 ports, $4500	Ethernet, 100Mbps, 14 ports, $1500
Group 4	ATM, 75Mbps, 10 ports, $8,000	Ethernet, 100Mbps, 6 ports, $4,200	Ethernet, 100Mbps, 6 ports, $2,940	Ethernet, 100Mbps, 6 ports, $500
Site 2	IP Router, 50Kpps, 5 ports, $3000	IP Router, 50Kpps, 5 ports, $3000	IP Router, 50Kpps, 5 ports, $1200	IP Router, 150Kpps, 5 ports, $500
Group 2	Ethernet, 10Mbps, 6 ports, $200	Ethernet, 10Mbps, 16 ports, $440	Ethernet, 10Mbps, 12 ports, $168	Ethernet, 10Mbps, 6 ports, $25
Group 5	ATM, 45Mbps, 15 ports, $9,000	ATM, 45Mbps, 15 ports, $7,000	ATM, 45Mbps, 15 ports, $3,500	Ethernet, 100Mbps, 14 ports, $1500
Site 3	IP Router, 50Kpps, 5 ports, $3,000	ATM, 75Mbps, 5 ports, $4,000	IP Router, 50Kpps, 5 ports, $1,200	IP Router, 150Kpps, 5 ports, $500
Group 3	ATM, 45Mbps, 15 ports, $9,000	ATM, 45Mbps, 15 ports, $7,000	ATM, 45Mbps, 15 ports, $3,500	Ethernet, 100Mbps, 14 ports, $1500
Group 8	ATM, 45Mbps, 15 ports, $9,000	ATM, 45Mbps, 15 ports, $7,000	ATM, 45Mbps, 15 ports, $3,500	Ethernet, 100Mbps, 14 ports, $1500
Site 4	IP Router, 50Kpps, 5 ports, $3,000	IP Router, 50Kpps, 5 ports, $3,000	IP Router, 50Kpps, 5 ports, $1,200	IP Router, 50Kpps, 5 ports, $150
Group 6	ATM, 45Mbps, 10 ports, $6,000	ATM, 45Mbps, 10 ports, $4,000	ATM, 45Mbps, 10 ports, $2,000	Ethernet, 100Mbps, 10 ports, $1000
Group 7	ATM, 45Mbps, 10 ports, $6,000	ATM, 45Mbps, 10 ports, $4,000	ATM, 45Mbps, 10 ports, $2,000	Ethernet, 100Mbps, 10 ports, $1000
Design Summary	Total Cost = $78,960.13 Average Network Delay = 37.53 seconds Homogeneity Factor = 0.54	Total Cost = $64,400.10 Average Network Delay = 35.46 seconds Homogeneity Factor = 0.54	Total Cost = $33,468.10 Average Network Delay = 35.46 seconds Homogeneity Factor = 0.46	Total Cost = $13,8 85.10 Average Network Delay = 14.10 seconds Homogeneity Factor = 0.62

and has its HF is equal to 0.45, which indicates that 45% of the allocated network devices are an Ethernet technology. In the case using LIB01 and TND = 5.0 seconds, the best network design has an estimated average network delay equal to 4.84 seconds and its HF is equal to 0.77, which indicates that 77% of the allocated network devices are also an Ethernet technology. In the case using LIB03 and TND = 5.0 seconds, the best network design has an estimated average network delay equal to 4.94 seconds and its HF is equal to 0.38, which indicates that 38% of the allocated network devices are an IP router technology. The phenomena of the domination of Ethernet technology occurred during 2005 with HF equal to 62% of all allocated network devices. Tables 7 and 8 illustrate samples of the best eight network designs that are produced by iCAD using LIB99, LIB01, LIB03, and LIB05. Tables 7 and 8 illustrate all the design decisions, which are automatically made by the evolutionary approach to minimize the network-design cost.

Table 7 illustrates all design decisions made by iCAD to produce the best network design with TND set to 60.0 seconds. There are *three* new design decisions that are made by iCAD when it uses LIB01. There are also *three* new design decisions that are made by iCAD when it uses LIB03. During the period of 2001 to 2003, and with regards to midlevel performance, ATM technology is more cost-effective than Ethernet technology. And for the same period, an IP router is the ultimate choice for the backbone and site networks due to its capability to connect heterogeneous network components without interconnection penalty. During the network-evolution process, the network cost goes down and the performance goes up. Also, in this case the network homogeneity factor has evolved from 54% to 46%. Glancing at the last column in Table 7, we observed only two network technologies that are used mainly: Ethernet and IP router.

The high-performance network designs are illustrated in Table 8, where all the design decisions were made by iCAD to produce the best network design, with TND set to 5.0 seconds. There are *six* new design decisions that are made by iCAD when it uses LIB01. There are *nine* new design decisionsthat are made by iCAD when it uses LIB03. In this case, the network homogeneity factor has evolved from 45% to 77% to 38% where an IP router and ATM switch are more cost-effective than Fast Ethernet and GigaEthernet Hubs. Also, an IP router is the ultimate choice for the backbone network due to its capability to connect heterogeneous network components without interconnection penalty.

However, our empirical data for 2005 shows a new trend for high-speed networks where an ATM technology was not allocated as illustrated by all design decisions made by iCAD in the last column in Table 8. The Ethernet technology dominates all local area networks and the IP router technology takes over all site and backbone networks. We consider the Ethernet to be the comeback technology, which expects to continue for many years the domination at the subnetwork domains. Also, the IP router will continue be the choice to interconnect subnetworks together; therefore, it eliminates the needs for gateways to interconnect heterogeneous network technologies.

CONCLUSION

We introduced a research test-bed tool for automatically synthesizing application-specific networks and observing the network evolution. In this article, we described a call for a network metric to examine the network evolution overtime. We have developed a number of network-device libraries over the past seven years for many different network technologies, such as Ethernet, ATM, and IP router. And we attached these libraries into

Table 8. Three network designs produced by iCAD while connecting to LIB99, LIB01, LIB03, and LIB05 respectively (threshold network delay = 5.0 seconds)

Tasks	**Network Devices allocated from Using LIB99 by iCAD**	**Network Devices allocated from Using LIB01 by iCAD**	**Network Devices allocated from Using LIB03 by iCAD**	**Network Devices allocated from Using LIB05 by iCAD**
Backbone	IP Router, 150Kpps, 5 ports, $11,000	IP Router, 150Kpps, 5 ports, $11,000	IP Router, 200Kpps, 5 ports, $6,800	IP Router, 150Kpps, 5 ports, $500
Site 1	IP Router, 150Kpps, 5 ports, $11,000	Ethernet, 1Gbps, 4 ports, $6,000	IP Router, 100Kpps, 5 ports, $2,800	IP Router, 150Kpps, 10 ports, $950
Group 1	ATM, 100Mbps, 15 ports, $15,900	Ethernet, 100Mbps, 15 ports, $12,000	ATM, 100Mbps, 15 ports, $6,750	Ethernet, 100Mbps, 14 ports, $1,500
Group 4	Ethernet, 1Gbps, 6 ports, $13,000	Ethernet, 1Gbps, 6 ports, $8,000	Ethernet, 1Gbps, 6 ports, $5,600	Ethernet, 1Gbps, 6 ports, $5,000
Site 2	IP Router, 100Kpps, 5 ports, $7000	Ethernet, 100Mbps, 4 ports, $3,000	IP Router, 100Kpps, 5 ports, $2,800	IP Router, 150Kpps, 5 ports, $500
Group 2	ATM, 75Mbps, 10 ports, $8,000	Ethernet, 100Mbps, 6 ports, $4,200	Ethernet, 100Mbps, 6 ports, $2,940	Ethernet, 100Mbps, 10 ports, $1,000
Group 5	Ethernet, 100Mbps, 12 ports, $13,000	Ethernet, 1Gbps, 12 ports, $14,000	ATM, 100Mbps, 15 ports, $6,750	Ethernet, 100Mbps, 14 ports, $1,500
Site 3	IP Router, 150Kpps, 5 ports, $11,000	IP Router, 150Kpps, 5 ports, $11,000	IP Router, 150Kpps, 5 ports, $4,400	IP Router, 200Kpps, 5 ports, $500
Group 3	Ethernet, 100Mbps, 12 ports, $13,000	Ethernet, 100Mbps, 12 ports, $9,000	ATM, 100Mbps, 15 ports, $6,750	Ethernet, 100Mbps, 14 ports, $1,500
Group 8	Ethernet, 100Mbps, 12 ports, $13,000	Ethernet, 100Mbps, 12 ports, $9,000	ATM, 150Mbps, 15 ports, $8,000	Ethernet, 100Mbps, 14 ports, $1,500
Site 4	IP Router, 100Kpps, 5 ports, $7,000	IP Router, 100Kpps, 5 ports, $7000	IP Router, 150Kpps, 5 ports, $4,400	IP Router, 150Kpps, 5 ports, $500
Group 6	Ethernet, 100Mbps, 8 ports, $9,200	Ethernet, 100Mbps, 8 ports, $5,600	Ethernet, 100Mbps, 8 ports, $3,920	Ethernet, 100Mbps, 10 ports, $1,000
Group 7	Ethernet, 1Gbps, 10 ports, $25,000	Ethernet, 100Mbps, 10 ports, $7,500	ATM, 100Mbps, 10 ports, $4,250	Ethernet, 100Mbps, 10 ports, $1,000
Design Summary	Total Cost = $167,815.00 Average Network Delay = 4.91 seconds Homogeneity Factor = 0.45	Total Cost = $124,609.00 Average Network Delay = 4.83 seconds Homogeneity Factor = 0.77	Total Cost = $72,889.00 Average Network Delay = 4.94 seconds Homogeneity Factor = 0.38	Total Cost = $22,761.30 Average Network Delay = 4.94 seconds Homogeneity Factor = 0.62

an evolutionary design tool to select, integrate, and optimize a three-level enterprise network that can satisfy all design and performance constraints and at the same time minimizes the overall network-design cost.

As network technology changes, the design library is updated. By running the tool with the old and recent design libraries keeping the traffic flow constant, we plotted the cost and performance changes between the old and recent network technologies, which enabled us to predict the future change to a first order. The network evolution, which is represented by the gaps between the three cost-versus-performance curves, indicates that the network-design cost is decreasing at a doubling rate from 10.02-25.75% to 57-74% in every 24 months.

REFERENCES

Beigi, M., Calo, S., & Verma, D. (2004). Policy transformation techniques in policy-based systems management. In *Proceedings of the Fifth IEEE International Workshop on Policies for Distributed Systems and Networks*. Yorktown, NY.

Beigi, M., & Verma, V. (2001). Network prediction in a policy-based IP networks. In *Proceedings of the IEEE Globecom* (pp. 2522-2526). San Antonio, TX.

Bersekas, D., & Gallager, R. (1992). *Data networks* (2nd ed.). Englewood Cliffs, NJ: Prentice Hall Publishers.

Cosares, S., Deutsch, D., Saniee, I., & Wasem, O. (1995). Sonet toolkit: A decision support system for designing robust and cost-effective fiber-optic networks. *Interfaces, 25*, 20-40.

Doshi, B., Dravida, S., & Harshavardhana, P. (1995). Overview of INDT: A new tool for next generation network design. In *Proceedings of the IEEE Globecom*. Singapore.

Elbaum, R., & Sidi, M. (1996). Topological design of local area networks using genetic algorithms. *IEEE/ACM Transactions on Networking, 4*, 766-778.

Fahmy, H., & Douligeris, C. (1995). END: An expert network designer. *IEEE Network*, 18-27.

Frogner, B., & Cannara, A. (1999). Monitoring and Prediction of Network Performance. In *the Proceedings of the International Workshop on Advance Issues of E-Commerce and Web-Based Information Systems* (pp. 122-129). Santa Clara, CA.

Gerla, M., & Kleinrock, L. (1977). On the topological design of distributed computer networks. *IEEE Transactions on Communications, 25*, 48-60.

Gersht, A., & Weihmayer, R. (1990). Joint optimization of data network design and facility selection. *IEEE Journal on Selected Areas in Communications, 8*(9), 1667-1681.

Habib, S., & Parker, A. (2002). Computer-aided system integration tool for predicting enterprise network evolution. In *Proceedings of the Fifteen International Conference on Systems Engineering (ICSEng 2002)*, Las Vegas, NV.

Habib, S., Parker, A., & Lee, D. (2002). Automated design of hierarchical intranets. *Computer Communications, 25*(11-12), 1066-1075.

Michalewicz, Z. (1994) *Genetic algorithms + data structures = evolution programs*. Berlin, Germany: Springer-Verlag.

Mitra, D., Morrison, J., & Ramakrishnan, K. (1998). VPN Designer: A tool for design of multiservice virtual private networks. *Bell Labs Technical Journal, 3*(4), 15-31.

Moore, G. (1965). Cramming more components onto integrated circuits. *Electronics, 38*(8).

Nowell D. L., & Kleinberg, J. (2003). The link prediction problem for social networks. In *Proceedings of the Twelfth Annual ACM International Conference on Information and Knowledge Management (CIKM'03)* (pp. 556-559). New Orleans, LA.

Pierre S., & Legault, G. (1998). A genetic algorithm for design distributed computer network topologies, *IEEE Transactions on Systems, Man and Cybernetics, 28*, 249-258.

Saksena, V. (1989). Topological analysis of packet networks. *IEEE Journal on Selected Areas in Communications, 7*, 1243-1252.

Vekiarides, L., & Finkel, D. (1998). NETCAP: A tool for the capacity planning of ethernet LANs. In *Proceedings of the Sixth International Symposium on Modeling, Analysis and Simulation of Computer and Telecommunication Systems (MASCOTS 1998)* (pp. 198-203). Montreal, Canada.

Verma, D. (2002). Simplifying network administration using policy based management. *IEEE Network Magazine*.

This work was previously published in International Journal of Business Data Communications and Networking, Vol. 3, Issue 4, edited by Jairo Gutierrez, pp. 1-16, copyright 2007 by IGI Publishing, formerly known as Idea Group Publishing (an imprint of IGI Global).

Section IV
Utilization and Applications

Chapter XV
New Computer Network Paradigms and Virtual Organizations

Guillermo Agustín Ibáñez Fernández
Universidad Carlos III, Spain

INTRODUCTION

Origins of Computer Networks

A computer network consists of computers that communicate via any physical media through a network formed by links and nodes, the nodes being the computers. Computer networks have evolved along their short history. Computer networks have changed drastically in mission and implementation from the early projects supported by the Advanced Research Projects Agency (ARPA) and from other organizations, tracing back the origins to 1962. The ARPA network (ARPANET) consisted initially of a small set of nodes at research centres and universities, connected with links at 56 kbps across the United States. ARPANET was the core of the early Internet, a network for research centres and universities. Computer networks are based on the concept of packet switching within a shared communication medium, as opposite to circuit switching, the dominant paradigm for the precedent telegraph and telephone networks. In 1968 Paul Baran proposed a network system based on nodes that forward datagrams or packets from different users over a common line between computer systems from origin to destination. The packet switching paradigm provides resiliency of network against network node failures, the independent routing of datagrams per node makes possible that the datagrams reach their destination even in presence of multiple node failures.

Computer networks hardware and communication protocols evolved through time: the Network Control Protocol (NCP) evolved to the four layer (1978) TCP/IP protocol stack. TCP/IP became dominant against the complex seven layer Open Systems Interconnection (OSI) stack proposed (1977) by International Standard Organization (ISO), too complex for implementation and interoperability. A view of the Internet Engineering

Task Force (IETF) and Institute of Electrical and Electronic Engineers (IEEE) protocols stack is shown in *Figure 1.*

The role of IP protocol as simple interconnection protocol between networks using dissimilar LAN technologies (Token Ring, Token Bus, ATM LANE, Ethernet, Wi-Fi) has been essential to build the Internet. However, with the widespread deployment of high performance/cost Ethernet and its self compatibility (10/100/1 Gigabit/10 Gigabit), Ethernet is becoming more and more the interconnecting technology, and Metro Ethernet Services are offered by network service providers.

Computer networks are based on communication standards. The reference standardization organizations for computer networks, but not the only ones, are the Internet Engineering Task Force, which is the standard organization for Internet Protocols, and the Institute of Electrical and Electronics Engineers (IEEE) LAN MAN Group (IEEE 802) that elaborates and approves standards on the lower layers of protocol stack. The International Standards Organizations (ISO) and the International Telecommunications Union are also key organizations for computer networks, given the convergence of telecommunication and data networks.

The evolution of computer networks affects all protocol layers, including the application layer. The development of increasingly sophisticated computer applications that seamlessly integrate data and communications, creating a new virtual space where actors interact is the objective of "social" applications and collaborative software.

Wireless Networks Evolution

Besides the wired computer networks evolution, radio packet switched networks evolved in parallel. With the advances in radio and portability of computers, the Department of Defense sponsored the military project Survivable Adaptive Radio Networks (SURAN) program in the early 1980s to provide packet-switched networking in a battlefield without any fixed infrastructure. After significant work, the IEEE has issued since then with great success several 802.11 Wireless Local Area Network (WLAN) standards, evolving in characteristics and capacities: 802.11a/b/g/i. 802.11e currently focuses on controlled quality of service for services like real time voice, a very important service for Wi-Fi networks using Voice over IP applications like Skype. The evolution includes the Wireless Personal Area Networks

Figure 1. Protocol stack structure

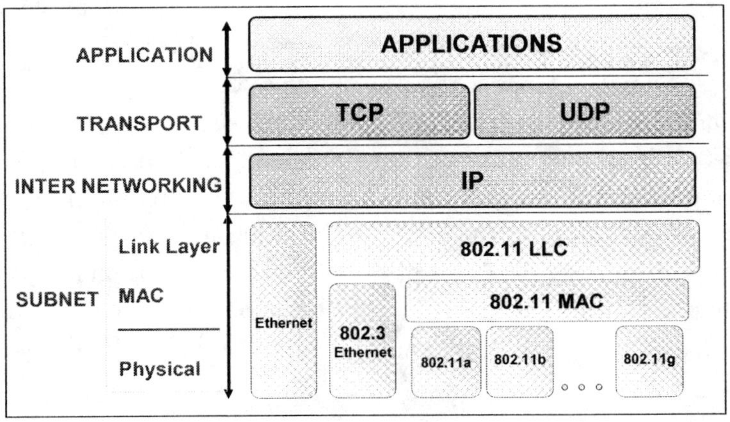

Table 1. IEEE 802 active working groups and study groups

IEEE 802 Active Working Groups and Study Groups
• 802.1 Higher Layer LAN Protocols Working Group
• 802.3 Ethernet Working Group
• 802.11 Wireless LAN Working Group
• 802.15 Wireless Personal Area Network (WPAN) Working Group
• 802.16 Broadband Wireless Access Working Group
• 802.17 Resilient Packet Ring Working Group
• 802.18 Radio Regulatory TAG
• 802.19 Coexistence TAG
• 802.20 Mobile Broadband Wireless Access (MBWA) Working Group
• 802.21 Media Independent Handoff Working Group
• 802.22 Wireless Regional Area Networks
• Link Security Executive Committee Study Group is now part of 802.1

(WPAN, IEEE 802.15) with the Bluetooth standard and Metropolitan Area Networks (WMAN, IEEE 802.16) also named as Wimax. *Table 1* shows the active groups at IEEE LAN/MAN Standards Committee (LMSC). Among them it is worth to mention the 802.21 Working Group, standardizing a Media Independent Handoff for seamless terminal operation across wireless and wired technologies.

VIRTUAL ORGANIZATIONS AND COMPUTER NETWORKS

Virtual organizations started as a way to enhance certain relationships inside a formal organization to focus on specific objectives, but they are currently a way of organizing teams on a per project basis. In a virtual organization persons interact differently: relationships are focused to the task, team member selection is based on specialized knowledge, relationships are more transient, responsiveness is high and confidence is required to overcome the distance barriers. The role of hierarchy for coordination is reduced and the organizations are flattened through information sharing. Virtual organizations operate differently to traditional hierarchies.

Virtual organizations have superiority in using communication technologies in that a formal organization does not restrict its use to fit to organization policies, allowing their full potential to be used adapting the tools and environment dynamically at each work phase's needs, instead of adapting to company standards, tools and rules. Virtual organizations, overcoming distance and time barriers, allow the most suited people to be assigned to the task. In VOs, communication and computer networks create the basic communication and information infrastructure and applications provide the services and interfaces to the users. Intermediate software (middleware) is used to integrate applications. The term "groupware" is used to describe the software applications for collaboration.

We can consider three "networking" dimensions of virtual organizations, that is, the virtuality based in telecommunications, the organizational teaming and the knowledge hyper linking. These three networking dimensions are implemented at the application layer of communication protocol stack as middleware that glues together different

Figure 2. Route changes due to node mobility in ad hoc networks

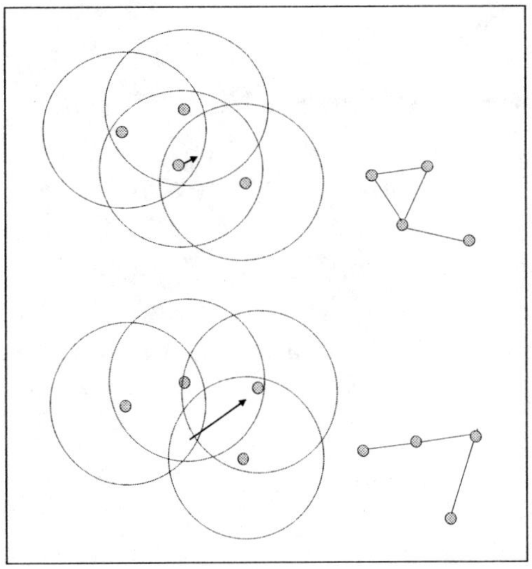

applications and communications, as shown in *Figure 3*. The role of computer network applications is towards getting a lower contact cost and an enhanced quality of the contact. Software applications leverage all the information available to enhance connectivity of individual in the virtual organization. Leading companies, like Cisco, are focusing on this model of networked virtual organization as the only possible model for best performing companies. It is worth noting that the virtualization paradigm is also used at computer networks themselves, where virtual networks are created at different layers as a mean to obtain *separate* networks inside a common infrastructure, for different uses (virtual private networks, virtual LANs, etc.). Virtual storage and virtual nodes (servers, routers) are also used to simplify and decouple the node functions from the real network infrastructure, as well as to provide alternative resource assignment and network reliability.

NEW COMPUTER NETWORKS PARADIGMS AFFECTING VIRTUAL ORGANIZATIONS

In recent years, several new paradigms have emerged in computer networks that are relevant to virtual organizations: the paradigms considered are peer-to-peer applications, wireless mesh networks and ad hoc networks.

Figure 3. Protocol stack and peer-to-peer (P2P)

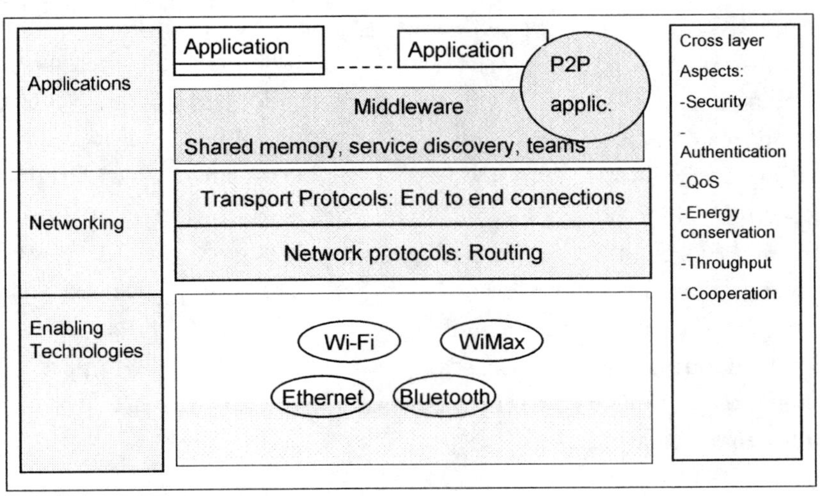

Ad Hoc Networks (MANETs)

Ad hoc networks are one of the main paradigm changes in the architecture of computer networks, although they have not yet obtained adequate standardization and implantation due to the complexities of routing in mobile environments. Mobile ad hoc networks (MANETs) are composed of wireless hosts, which may be mobile without using a pre-existing infrastructure. The routes between nodes may potentially contain multiple hops. Opposite to fixed networks, where there exist end nodes and routers, at MANETS each node acts as both terminal and router. Wi-Fi networks, standardized as 802.11, are not a kind of ad hoc network because they are single hop radio, no wireless routing exists. The main advantages of ad hoc networks are simple and fast deployment and saving of infrastructure. The drawbacks are reduced security, routing instabilities (routes change when the nodes move, see *Figure 2*) and limited throughput due to the sharing of radio media.

Routing in Ad Hoc Networks

Communication between non-neighbouring nodes in an ad hoc network requires the use of a routing protocol so that multi-hop paths may be discovered and made use of. A routing protocol used in an ad hoc network may be classified as one of two possible types: proactive or reactive. Proactive routing protocols (also known as table-driven protocols) are similar to the ones used in fixed networks. Each node calculates routing tables from the information received from other nodes of the network. Reactive routing protocols obtain the route when it is needed. So it is less affected by mobility but response is slow and overhead on the network is big because to obtain the route reply there is normally flooding applied. Research effort has been intense on routing protocols for ad hoc networks although no clear standard exists yet. AODV and dynamic source routing (DSR) are among the main protocols for ad hoc networks.

WIRELESS MESH NETWORKS

Wireless mesh networks (WMNs) (IEEE 802.11s) are composed of routers and mesh clients. Router nodes have minimum mobility forming a distribution backbone. Clients may be mesh (i.e., act as router) or conventional clients. Mesh clients may be fixed or mobile and can interconnect between them or with routers. Mesh networks integrate networks of different types like Wi-Fi (IEEE 802.11), Bluetooth, and so forth, using gateways and bridge functionality implemented at mesh routers. Mesh networks improve the restrictions in range, bandwidth, reliability and performance on any type of wireless networks like ad hoc networks, WLAN, WPAN and WMAN.

Wireless mesh networks can be seen as a subclass of ad hoc networks considering that mobility of some nodes is restricted. They can also be seen as a super class considering that wireless mesh networks can serve to integrate different types of networks and terminals, wireless and wired, mobile ad hoc and fixed.

Peer-to-Peer Applications

One of the most disruptive (Ocram, 2001) technologies is peer-to-peer applications (P2P).

P2P applications are disruptive because they break the specialized role of client and server, making all nodes active and cooperating. Peer-to-peer means that communication is between equals. Each peer is at the same time client to remote peers and server to remote peers. Nodes organize without the need of centralized servers. The information directory of attached nodes is distributed over all or part of the nodes. Well-known examples are E-mule, Skype, Napster and Chord.

A pure peer-to-peer network is an overlay, self-configuring, fully distributed and decentralized network formed by nodes acting both as servers and clients, sharing partially their resources and joining and leaving the network dynamically. But not all the so-called peer-to-peer are pure P2P applications. There are *pure* P2P and *hybrid* P2P architectures, depending on the auxiliary use or not of centralized servers for locating peers.

Napster, one of the better known P2P applications, is not pure peer-to-peer as it uses as central server indexing the contents at peers, and also controls communications between the peers. There are *hierarchic* P2P systems, such as the domain names service (DNS), where the nodes according to levels control and delegate to other nodes, and finally there are *pure* P2P systems, such as Gnutella or Freenet, where all peers work at the same level. Depending on the application, they can be *user centric* or *data centric* depending on the focus on the database of users versus a database of content.

Main Concepts of Peer-to-Peer Applications

The principle for data lookup in P2P is basically that a file is associated with a unique key (a globally unique identifier). This identifier is typically obtained by hashing the IP address. A range of identifiers is under the responsibility of a node (i.e., the identifiers in the range between B and A numbers are responsibility of B server). P2P applications route to identifiers instead to network addresses to create a closed virtual address space unrelated with IP network topology. A review of the most common concepts in use in peer-to-peer applications gives the following list:

- Self-organizing overlay network (bottom-up organization versus top-down, planned networks)
- Automatic host/user identifier generation. The most common form of identifier is obtained via hashing (e.g., SHA-1) of IP address or filename. This gives a constant length and uniformly distributed address space.
- Lookup of nodes is based on an identifier. Lookup is the main function of distributed hash tables.
- Logical proximity versus physical proximity in routing. Logical proximity allows fully distributed lookup, but physical proximity optimizes performance.
- Introspection. The network continuously optimizes its performance and adjusts internally for that.

Comparison with Client/Server Networks

Client/server networks have been predominant in the past on the Internet due to the success of the HTTP protocol. File transfer protocol (FTP) also uses the client/server model. Within a client/server network, a server provides all the necessary content or services. Servers are high capacity resources that provide service to a high number of clients so they have high processing power, communication bandwidth and storage capacity. The clients do not share resources so they do not need high capacity and only the communication bandwidth limits their speed. Client/server is a two-level centralized hierarchy with the advantages of centralized application control at the server and the disadvantages of inefficiency due to not sharing the resources of the clients. Being a centralized architecture, it is much less resilient than a decentralized P2P network, which provides service even with many nodes out of service. Distributed hash tables are one way to manage to distribute in pieces with redundancy the content across the nodes of a peer-to-peer network. Client/server networks are bandwidth inefficient at Internet, because the strict two-level hierarchy does not efficiently map to the decentralised architecture of the Internet.

P2P applications adapt very well to the needs of virtual organizations: lack of centralized infrastructure and directories, flexibility in joining and leaving the organization, distributed storage and shared resources. However, lookup of nodes is slower than in specialized client/server networks.

P2P and Ad Hoc Networks Conceptual Similarities

Peer-to-peer (P2P) as well as mobile ad hoc networks (MANETs) can create a network without the help of central entities. But while P2P applications operate at application layer, MANETs operate at network (IP) layer. There are strong similarities among the process of routing in ad hoc networks and the content discovery in P2P. One example is the use of routing to identifiers (**hashing** of the IP address to obtain an identifier). The problems of identity, security and anonymity in ad hoc and P2P applications are also similar.

Both ad hoc and P2P networking are considered disruptive technologies because they put into question the structure of the telecommunications industry (and even more the contents industry) through elimination of the content provider (P2P) and of the network operator role (ad hoc networks). An example is Skype for telephone calls.

APPLICATION TO VIRTUAL ORGANIZATIONS

Critical Issues

As it has been said, the three networking dimensions of virtual organizations are the virtuality based in telecommunications, the organizational teaming and the knowledge hyper linking. Thus the critical issues are all that affect these dimensions:

- Simplicity of configuration and flexibility
- **Security:** Authentication, identity spoofing, unauthorized access to networks, data and resources. Encryption and access control. The combination of security and simplicity to end-user is a difficult but essential goal. Although P2Ps exhibit lower security, their spread architecture makes it more difficult to attack and observe. Encryption is a way to obtain confidentiality.
- **Trust:** Chain of trust, the need for distributed trust mechanisms varies according to the sizes of virtual organizations.
- **Performance:** Latency of connection set up and information retrieval.
- Adequacy to create a technical and social infrastructure for collaboration and knowledge management.
- **Pervasive computing**: Beyond user friendliness, in some environments, pervasive computing may be important making the interaction natural and simple. And beyond pervasive computing, which tries to anticipate to user needs, we can think of *organizational intelligence*—adaptiveness to the user patterns to detect and also enforce organizational procedures and practices. An adaptive feature worth to mention from P2P is the adaptiveness of contents of Freenet, where a distributed storage algorithm works in such a way that contents tend to move gradually to be stored at nodes near the users of them; the network self optimizes continuously.
- **Virtual reality:** Virtual reality environments may enhance interaction making it more transparent to the user and effective, improving communication and trust (more realistic telepresence and virtual meetings).
- **Multilingualism:** Although automatic language translators are not yet a reality, mul-

tilingualism will have growing importance for virtual organizations, as computers will help eliminate language barriers between team members.

The priorities of these factors will vary according to the type of virtual organization: a virtual enterprise focuses on some goals, requirements—like security and responsiveness—and tools; a virtual university focuses on others, virtual reality may be essential in some environments. The criticality of the computer network appears only when the infrastructure becomes a key factor: an application requires specific network hardware and protocols to perform adequately. The same might apply to security; pervasive applications require specific network nodes to run smoothly; responsiveness requires protocols and networks with very low latency. Interoperability and support of seamless mobility are another two factors of growing importance.

FUTURE TRENDS

The trend from a computer centred organization, through business process centred, towards a team-centred and knowledge-centred organization will likely continue with the evolution from a manufacturing society to a knowledge-based society.

A possible trend might be the porting of traditional client/server applications like storage to peer-to-peer platforms for increased resiliency and flexibility. Although there is an important (especially for enterprises) issue of control to be carefully managed, P2P architectures are suited for distributed, flexible and resilient distributed storage, which makes them suitable for organizations with high degree of flexibility.

Trends towards increased security at all protocol layers are commonplace now since the strategic importance of networks and economic risk and damage is ever growing. Standards are under discussion at IEEE and IETF to enhance radically security at all protocol layers, focused on authentication and encryption.

There are two sources of innovation from P2P applications. One is that mechanisms for distributed networking used by P2P are conceptually new and applicable at network layer. New ideas for distributed addressing, lookup and routing currently used at peer-to-peer (P2P) networks might be applied to fixed and wireless IP networks to obtain more resilient, self-configuring and scalable networks. Addressing, configuration and routing in mobile, ad hoc and nomadic networks are not a fully solved problem. P2P applications such as Gnutella, Freenet, Plaxton and many others use novel approaches to addressing and routing. This is due to several new concepts that these networks apply: distributed hash tables, content addressable networks, and distributed object location and routing (DOLR). Peer-to-peer concepts may be applied to network layer functions: addressing, routing and node lookup.

The second source of innovation is that the combination of P2P and ad hoc (they act on upper and lower layers of protocol stack respectively) may create new services, providing mobility and dynamics to P2P applications and powerful content search to ad hoc networks. So we can think of Smart Mobs. Smart mobs appear when communication and computing technologies amplify human talents for cooperation, including revolutions (Rheingold, 2003).

Finally, the new science of networks (Barabasi, 2002) will also likely influence computer networks for virtual organizations as the fundamental laws of networking in many different fields are starting to be discovered (power laws).

CONCLUSION

Recent advances in P2P applications like Skype, high speed (10 Gb Ethernet) fixed and wireless mesh networks and the development of pervasive computing are changing the scenario for virtual organizations. When combined with wireless mobile mesh and ad hoc networks, they will make possible ubiquitous virtual networks.

The development of P2P has been however much faster than that of ad hoc regarding impact on existing networks. Mesh networks will likely balance the situation in the years to come. The rapid growth of social software and blogs is also creating new practices for relationships and communications. The requirements for performance for networks infrastructure will likely grow by orders of magnitude due to the explosion of multimedia traffic, virtual reality and 3D environments. Although there will be a variety of software and networks for virtual organizations according to their type, the predominant models will likely be new combinations of the online cooperative games with 3D environments, office and multimedia conferencing applications. We can think of a "virtual office" where the workers are remotely present. New paradigms for these kinds of organization and social desktops will likely emerge.

REFERENCES

Akyildiz, I., Xudong, W., & Wang W. (2005). Wireless mesh networks: A survey. *Computer Networks and ISDN Systems*, *47*(4), 445–487.

Barabasi, A. (2002) *Linked: The new science of networks*. Cambridge, MA: Perseus Books Group.

Filos, E. (2005, December) *Smart organizations in the digital age*. European Commission, Directorate-General Information Society and Media. Retrieved from http://www.ve-forum.org/

Foster, I., & Iamnitch, A. (2003). *On death, taxes, and the convergence of peer-to-peer and grid computing* (LNCS 2735, pp. 118-128). Berlin: Springer-Verlag.

Foster, I., Kesselman, C., & Tuecke, S. (2001). The anatomy of the grid: Enabling scalable virtual organizations. *International Journal of High Performance Computing Applications*, *15*(3), 200-222.

IEEE LAN/MAN Standards Committee. (n.d.). Retrieved from http://www.ieee802.org/

Internet Engineering Task Force. (n.d.). Retrieved from http://www.ietf.org

Mezgár, I. (Ed.). (2006). *Integration of information and communication technologies in smart organizations*. Hershey, PA: Idea Group Publishing.

Ocram, A. (2001). *Peer-to-peer—Harnessing the power of disruptive technologies*. O'Reilly & Associates.

Preston, S.M. (1999). *Virtual organization as process: Integrating cognitive and social structure across time and space*. Retrieved from http://www.msu.edu/~prestons/virtual.html

Perkins, C.E. (2001). *Ad hoc networking*. New Jersey: Addison-Wesley.

Rheingold, H. (2003). *Smart mobs: The next social revolution*. Perseus Publishing.

Second Life. (n.d.). Retrieved from www.secondlife.com

Stevens W.R. (1994). *TCP/IP illustrated, Volume 1: The protocols*. Addison Wesley Professional Computing Series. Addison-Wesley Professional.

Wikipedia. (2006). Retrieved from www.wikipedia.org

KEY TERMS

Ad Hoc Networks (MANETs): Self-organising, self healing, mobile wireless network that does not rely on infrastructure in which mobile nodes are responsible for discovery of each other and subsequent cooperation so that communication is possible.

Communications Protocols: The set of standard rules for data representation, signalling, authentication, and error detection required to send information over a communications channel (Wikipedia, 2006).

Grid: Grid problem is defined as flexible, secure, coordinated resource sharing among dynamic collections of individuals, institutions, and resources referred to as virtual organizations. In such settings, unique authentication, authorization, resource access, resource discovery, and other challenges are encountered

Grid Technologies: Grid technologies address It is this class of problem that is addressed by Grid technologies. Next, the authors present an extensible and open Grid architecture, in which protocols, services, application programming interfaces, and software development kits are categorized according to their roles in enabling resource sharing.

Middleware: Describes the software used to interconnect software applications. Middleware allows building complex distributed applications to support virtual organizations. Examples of middleware are application servers and Web servers.

Overlay Network: A computer network built using a subset of the nodes on top of another network. Links in the overlay network may be formed by several logical links in the network below. Many peer-to-peer applications are overlay networks.

Packet Switching: Each packet is routed independently at each node and sent to next node through communication links shared by other nodes. It is opposite to circuit switching, where resources of the network are reserved between origin and destination prior to the communication and released at the end.

Peer-to-Peer Applications (P2P): A P2P is a network established between autonomous and cooperating nodes that self organize on a peer basis sharing resources of nodes, rather than centralized servers operating on a client-server basis. The peer-to-peer model is the standard at Internet for many protocols and telephone communications; the best-known applications are Freenet, eMule, Skype.

Routing: Routing is the process of selection of path for the packets in a computer network. Normally it is performed using dynamic routing tables that change with network topology, or statically configured in small networks.

Routing Protocols: Protocols used by network nodes (routers) to perform routing. The information interchanged by the nodes to obtain the routing tables is identified as the control plane and the traffic packets forwarded is the user plane.

Wireless Mesh Networks: Wireless mesh networks (WMNs) consist of mesh routers and mesh clients, where mesh routers have minimal mobility and form the backbone of WMNs. They provide network access for both mesh and conventional clients. (Akyildiz, Wang, & Wang, 2005)

This work was previously published in Encyclopedia of Networked and Virtual Organizations, edited by G.D. Putnik and M.M. Cunha, pp. 1066-1073, copyright 2008 by Information Science Reference, formerly known as Idea Group Reference (an imprint of IGI Global).

Chapter XVI
Software Agent Technology for Supporting Ad Hoc Virtual Enterprises

Jarogniew Rykowski
The Poznan University of Economics, Poland

ABSTRACT

This chapter introduces a new idea of using software agents for supporting ad hoc virtual enterprises and similar forms of temporal business-to-business collaboration. It seems that current information and telecommunication technologies, based on information interchange and local data processing, are not flexible enough to deal with modern business requirements, especially dynamic and temporal business relations, heterogeneity of hardware, software and communication means, and data complexity. The proposed approach differs in the distribution of both data and programs for data treatment at-the-place and just-in-time. The distributed and remotely executed programs, software agents, are autonomous entities, targeted on obtaining preprogrammed goals, and working in the name and under the authority of their owners. The authors hope that the proposed techniques for agent preparation, distribution, and execution make the whole system safe and secure, providing an efficient environment for a wide spectrum of temporal and ad hoc business collaboration.

INTRODUCTION

The idea of a virtual enterprise (VE), a value-added business built upon different, distributed, autonomous units, resulted in the rapid evolution of traditional business models. As the technical possibilities grow, stable business units are becoming less and less geographically- and timely-restricted, and relations among these units are more and more dynamic and case-oriented. It looks like the VE evolution is going to change traditional enterprises into a set of autonomous business units, able to establish a virtual enterprise to achieve given business goals at a given place and time with minimum efforts and costs, and maximum profits.

Imagine such pool of autonomous enterprises, both real and virtual, ready to be in a business.

The enterprises are heterogeneous, taking into account both their internal organization, and computer/telecommunication (IT) infrastructure. Suddenly, a business opportunity appears, for example, a possibility to organize a jubilee of a famous pianist. The cooperation is scheduled for two to three weeks only, and the coordinated activities, for example, a philharmonic concert and an official reception in a palace, are occasional, as they will probably never happen in the future. Moreover, the situation is changing all the time; there are new limitations and requirements coming that cannot be identified from the beginning, for example, a need for transportation for some disabled guests. In the future, another jubilee, for example, for a Nobel prizewinner, would require quite different activities, such as the organization of a scientific congress. Even if at the first view the jubilees are similar, the business activities and relations are completely different, and the business partners to be involved cannot be determined in advance.

Current, fixed, and costly IT technologies are not well-suited to deal with such temporal and evolving VEs. Existing proposals for building a virtual enterprise, mainly distributed applications based on such technologies as CORBA (2002) and Voyager (SOA platform, 2005), multi-databases (Hurson, Bright, & Pakzad, 1993), Web services (2002), and Semantic Web (DAML, 2006), are built upon two basic assumptions: (1) The relations between different VE units are quite stable and long-lasting; and (2) the client-server model is a dominant way of interaction among VE units. As a consequence, each unit implements a set of well-defined services (interfaces, programs, etc.) to be used at request by other business parties. This approach seriously limits an efficient implementation of evolving relations among cooperating units for at least two reasons. First, the information flow must be initialized by the client party in the online mode (information polling). Due to this poll-only mode, automatic detecting of server-side information changes and "pushing" them to clients is hard to maintain. Second, it is quite difficult to adopt services and interfaces of a single unit to the specificity of another unit; the client must be ready to adopt itself to the server-side standards. Server-side parameterization is usually limited, due to both technical and economical reasons.

To solve the problem of supporting ad hoc VEs (ahVE), we propose autonomous software agents (Caglayan & Harrison, 1997; Franklin & Graesser, 1996; Nwana, 1996; Wooldridge & Jennings, 1995). The role of an agent is twofold. First, an agent is used as an information broker and wrapper, to adjust data format (both syntax and semantics) to the specificity of communicating units. The brokerage/wrapping algorithm may be programmed either by one or by both parties. Second, an agent may act as a monitor and asynchronous notifier about important data changes. What is "important" is programmed in agent code and variables. The agents may interact with local IT systems of the VE units, as well as with other agents, and with humans. The agents are executed in the scope of agent computing environment (ACE) framework, being a set of agent servers (Rykowski, 2003a). An agent server may be located at any host, including dedicated hosts belonging to unit's local area networks. The agents may be moved among agent servers (Rykowski, 2005a), and according to the situation each agent may be executed at client-side (i.e., in unit's local area network), at server-side (i.e., in a network of another unit), or at an external network host (i.e., on a separate host outside units' networks).

How do the proposed agents differ from (for example) Web services and other competitive technologies? There are three main advantages to using the agents in comparison with the classical approaches: substantially reduced amount of work needed to establish a connection between two cooperating VE units, unrestricted individualization of relations among units, and a possibility of off-line, server-side monitoring of critical information changes.

Current fixed servers and services operate with a global schema, common for all the cooperating units and possible business cases. Thus, by default, such services must be able to deal with any data flow related with the internal unit's information. A complexity of such a general service is quite high. In contrast, an agent is created just for a single business case. Thus, such an agent operates on a restricted amount of data, just a part (usually small) of the global schema. As a consequence, a complexity of the agent code is substantially reduced, and the agent may be prepared in a short period of time, usually hours rather than weeks, as for most of the present systems.

The agents are created by and for particular VE units. Thus, a behavior of these agents may be exactly adjusted to the requirements and expectations of both cooperating parties. Moreover, as the agents may operate at both server- and client-side, the adjustment covers not only the receiver, but also the source of information. Individualization of agents is not restricted to a parameterization of a generic access to servers and services; each agent is programmed (both code and variables) to deal with an individual business case. Indeed, once programmed, an agent may be used in future collaborations, with the same or other units. What is different in such reuse in comparison with current systems is that each agent instance is different. Thus, a new agent instance built upon a previous one may be in turn (slightly) adjusted to new requirements and specificity of a new business case.

In contrast to the current client-server approaches, the server-side, "remote" agents may be executed autonomously, with no online control of the agent owner. Such agents may detect "important" information changes at server-side (i.e., in a network belonging to the corresponding party) and asynchronously inform the agent owner (other agents and humans) about such changes. Mixing off-line monitoring, asynchronous notifying, and online requests greatly improves the possible ways of data interchange.

Similar to current proposals, the agents may use a set of predefined tools and utilities, implemented as server- and client-side agents, or executed in a given place of a network. Such usage covers as well existing systems and servers, including Web services, and Semantic Web utilities. However, as agents may be used as programmable brokers to external information sources, it is possible to individualize an access to each independent resource/service, as well as to monitor information changes at server-side.

Taking into account technical aspects of ad hoc collaboration by the use of software agents, one may notice that some important problems, being main barriers for wide usage of traditional approaches, are tempered. First, there is no need for maintaining a global schema and ontology, common for all the VE units. Instead, a set of distributed agent-wrappers is provided, each of them implementing a cooperation of two parties at a time only. Second, individualization is possible for each interaction, on both sides of the communication link. Third, the network is able to continuously monitor critical information changes, with no need for manual, cyclic inspection. The monitoring agents are executed near the information sources, thus reducing the growing needs for powerful communication links. Due to the above features, using software agents one obtains a powerful mechanism to establish ad hoc efficient cooperation between units of a temporal virtual enterprise.

The remainder of the chapter is organized as follows. In the section, "Current Approaches for Building VEs and AHVEs," the principal restrictions of the current approaches to build VEs and ad hoc VEs are presented and discussed. In "Agent Technology for Maintaining Ad Hoc VEs," a way of creating and maintaining ad hoc VEs is presented based on software agents and the agent computing environment (ACE) framework. Several architectural and implementation aspects are discussed, mainly agent distribution, agent programming, supporting connections with external information systems and (tele)communication channels, and the overall security policy. Basic architecture and functionality of the ACE agent-based framework is presented, and a typical scenario is discussed of using the framework for establishing and maintaining a sample ad hoc VE. The "Related Work" section contains a comparison with similar work based on the agent technology and related with virtual enterprises, while "Conclusion" points out some conclusions and directions for future work.

CURRENT APPROACHES FOR BUILDING VES AND AHVES

In this chapter, we discuss principal restrictions of current approaches to build a VE: (1) a necessity of providing global schema (view, ontology), common for all the cooperating business partners; (2) technical limitations related with the commonly-used client-server and similar models, (3) costs and efforts while providing cooperation and data interchange mechanisms; and (4) basic limitations related with the software agent technology.

In most of current implementations of a VE, a basic assumption is made that that a *global view* (schema, ontology) exists that is a super-view of all the views of business partners. Thus, to cooperate in the scope of a VE, a wrapper is needed to adjust private IT systems (both ontology and information exchange) to a common schema. Wrapper-based architecture is commonly used due to its numerous advantages: a clear and uniform implementation, well-documented information flow, well-defined interfaces to information sources, and so forth.

The term *global* does not have to be related with the centralized (i.e., single) place of data management, storage, exchange, and so forth. Thus, two global-view architecture subtypes are introduced: centralized information management, and common ontology. In the first case, a single software instance is provided to control information flow in the whole system. This architecture type was preferred in the early or prototype implementations of VEs, for example, Concordia mobile agent (2002). Simplicity of this architecture type usually contrasts with a lack of scalability and strong specialization for a specific business case. In the second case, a distributed approach is preferred; however, all business parties involved in the information exchange must a priori agree to a way of cooperation (data format, ontology, message passing, server interface, etc.). This architecture type is now more popular, mainly due to technical progress in the area of distributed systems and new technologies as software agents, Web services, Semantic Web, and so forth.

Regardless the architecture type and the implementation, the global-view approach is characterized by several disadvantages. First, the common view (ontology) is usually imposed upon business partners by the strongest partner, to minimize costs related with information and communication adjustment. Smaller business units have no choice and must obey the obligatory view unless they want to take part in a VE. In the case of relatively equal partners, a process of establishing a common view may take a lot of time, as everybody wants to cash in on a situation. The wider that a VE is, the lower are the chances to establish a single, consistent common view. Second, to minimize efforts related with future collaboration (with the same or other business partners), an owner of information source usually provides information in a general (universal) and extended (parameterized) way, even if some information/functionality is not needed at the moment. Thus, an interface/functionality of a local system of a business partner is usually quite complicated and, sometimes in a large part, unused for a long time. Moreover, an excess of parameters, modes, message types, and so forth, may substantially complicate an implementation of the information flow at the partners' side. And last but not least, changes in the view (ontology) may force re-implementing large system parts. Note also, that reengineering is usually not addressed by a community of business partners; an implementation of an information wrapper (from the global to a local, private view) is a private property of a party. Even if some parties slightly differ, their private information systems are kept disjoined.

On the contrary, in a non-centralized approach, only the business units directly interested in the specific cooperation agree on a common sub-view (ontology). Thus, individualized, small, specialized wrappers are used, related only with the information requested by the partners. So far, such architecture has not been widely discussed in the literature. However, in the reality many ad hoc implementations are based on such assumption, to minimize implementation costs and time. Note that it is hard to speak about global (data) consistency of a non-centralized system, due to a lack of a single management point. Note also that an ad hoc non-centralized system usually evolves to a centralized one, as one of the (at the beginning equal) business partners becomes more and more strong and eclipses the others, imposing new "common" sub-views, services, interfaces, and so forth. We think that our proposal may fill this gap, providing a reasonable trade-off between a centralized and non-centralized architecture, especially for "light" application areas and fuzzy, dynamically changed cooperation among different business partners in the scope of an ad hoc VE.

From a technical point of view of a cooperating business partner, a typical way of providing information is to provide a service to access the necessary information from the outside. Service providers, that is, owners of data sources, decide who is authorized to access the information and how the access is done. The software implementing an access to information is optimized from the information provider point of view, to achieve maximum system efficiency and security. On the other hand, the data source is used by several business partners, who are characterized by different requirements and possibilities, who are using different hardware, and who are connecting via different communication means. For the partners, a possibility of personalization of a way of accessing a data source and further using the collected information would be very useful. However, there is a trade-off between an abundance of different expectations of all potential partners, and software complexity. Due to hardware and software limitations, as well as costs of software development and use, software personalization is typically restricted to some simple built-in mechanisms including user (i.e., partner) authentication, stored profiles, preferences, dynamic cookies, sessions, and so forth. All these personalization mechanisms are internal parts of the server software. As such, they cannot be changed by the partners. As a consequence, from the business partner point of view, the level of individualization of the way of accessing a data source is often unsatisfactory. Therefore, unsatisfied partners must extra-process data at the client-side. In this case, data flow increases significantly though unnecessarily, because usually most of the transmitted data is filtered out during data processing at the client-side.

One may suppose that recently proposed n-tier architecture (CORBA, 2002), where several software layers are distinguished, responsible for storing, accessing, supplying, formatting, and presenting the information, may contribute to the solution of the personalization problem. However, if taking into account fixed functionality of the tiers and the fact that this functionality is common for all the business partners involved in the information exchange, one may see that the above-mentioned problem is not relaxed.

The second main limitation of most current data sources is a fact of forcing information polling rather than pushing. A data source passively waits for external demands, while the business partners must periodically poll for new information to compare with the previous one and eventually to detect information changes. This approach generates huge network traffic and consumes client and server resources. Thus, polling for information should be replaced by pushing information by the data source to the partners' systems, that is, making the source active. Some techniques have been proposed towards this goal, mainly in the domain of databases, namely, database triggers, and active databases. Database triggers are usually implemented as a set of procedures auto-executed by the database management system once a given event occurs. Triggers are widely used as a mechanism for online verification of database integrity (Oracle Triggers, 2006). Triggers may be used for personalization; however, as they were not designed to this goal, such personalization is restricted. First, triggers are common for all the database users. As such, triggers are usually made by system designers, while users cannot easily personalize them. Second, triggers cannot modify query results (including both the format and the contents of the information requested). Third, current trigger implementations are limited to "insert-update-delete" commands only; application of triggers to read-only queries (i.e., a casual access in the scope of VE application area) is not possible.

Active databases (Ceri & Franternali, 1997; Paton, 1999) are, in some sense, an extension of the triggering mechanism (in a sense of a concept; not as a historical successor). In active databases, an automatic system reaction is programmed by a set of mutually-connected event-condition-action (ECA) rules. The idea of active databases (ACT-NET Consortium, 1996) seems to be abandoned, mainly due to the impossibility of development of efficient methods of database/rules management. The prototypes never reached the commercial status, and a lot of basic functionality has never been implemented, including an efficient debugger for rules and actions triggered, query optimizer specialized for ECA rules, and multi-user engine. These facts, together with the technical difficulties of individualizing the set of rules for different users, makes the concept of active databases practically useless in the scope of individual efficient information access.

Using the above-mentioned "severe" software technologies as Web servers and Web services, database management systems, and so forth, together with the approach of providing a single generic service for all possible business cases, leads to a situation where the cost of providing a useful set of services is quite high. Thus, service owners tend to use the services for a long time, in a stable manner. A lack of dynamic adaptation of services to changes in both overall situation and individual business cases seriously limits the use of traditional technologies in the scope of ad hoc, thus highly dynamic and unpredictable VEs. As a result, short-term business relations are usually supported by a manual work of the human staff, with the minimal use of modern telecommunication and information technologies.

As follows from the above contradictions, a generic, "severe," "one place" centralized implementation is inadequate for an efficient implementation of a set of distributed, temporal, and ad hoc services of a VE. A better solution is to implement and use several individual instances of specialized sub-services, possibly distributed across the network, executed both off-line and online (i.e., on demand). Choosing a place, form, and time of execution of a given instance should be case- and user-dependent. In a typical case, the software related with the information sources (both passive and active) is to be executed preferably at the server-side, while the software related with the partner-specific information processing should be executed either at the client-side, or at a selected host in the network, if the end-user system/communication line is not powerful enough. However, the above assignment (as well as service complexity) should be adjusted to the specificity of a particular business case.

Nevertheless, a technology is needed to combine execution of distributed programs, owned by different business parties, and orchestrated to achieve case-specific business goals under different restrictions. It seems that the software agent technology may be suitable here. A software agent is a program, executed at a given place, characterized by: (1) autonomy—agents process their work independently without the need for human management; (2) communication—agents are able to communicate with one another, as well as with humans; and (3) learning—agents are able to learn as they react with their environment and other agents or humans. An agent may be programmed by its owner, thus allowing personalization of its behavior. Agents may be executed in different places, according to owners' needs and possibilities of the end-user hardware.

In the next sections, we discuss some architectural and implementation issues related with using a software agent technology to implement an efficient environment for maintaining an ad hoc VE. We also provide a comparison of several already proposed systems, both commercial and scientific prototypes.

AGENT TECHNOLOGY FOR MAINTAINING AD HOC VES

In this chapter, we discuss some basic architectural issues of an agent-based system to maintain ahVEs. We propose a generic framework capable of creating and controlling a set of ahVEs, implemented as distributed networks of cooperating software agents. The basic assumptions of the proposed framework are the following:

- Each ahVE has an individual set of the agents, and these agents are distributed across the network according to a specificity of the ahVE.
- The agents may be used as wrappers to existing software systems, and the information is owned by all the ahVE parties.
- The agents are executed in a safe manner, taking into account an overall security policy of the environment the agent is currently executed in, agent owner's access rights, and a specificity of a business case.
- The agents are under exclusive control of their owners, except for the situation where it is in a conflict with a security policy mentioned above. In particular, the agent owners are able to program agent behavior (i.e., program code and variables), as well as to choose a place and time of agent execution.
- The techniques to program agent behavior are adjusted to the needs and possibilities of the agent owners. In particular, there are some "light" techniques to program temporal and prototype agents, and some "severe" techniques to program agents offering stable, public, multi-user services. A stress is put on efficient reengineering of agents, both for parallel (i.e., multi-user) and future applications.

In the next subsections, a discussion is provided which is addressed to each of the above architectural issues.

Distribution of Agents

The key question related with mobile agents is how to distribute agents in order to perform the tasks in the optimum way. (Chess, Harrison, & Kershenbaum, 1995; Milojicic, 1999) Recently we observe an evolution in answering this question (Figure 1). Let us analyze this evolution taking into account applying software agents to classical client-server architecture (Figure 1A). In the first implementations, agents were executed at server-side, in a dedicated agent environment (Figure 1B). These agents, in a natural way, were prepared and controlled by the server owner. As a consequence, agent behavior and access policy were optimized from the server point of view (cf. section "Current Approaches for Building VEs and AHVEs"). From the client point of view, this architecture seriously limits individualization of the access to the server.

Figure 1. Distributing agents—possible extensions of the classical client-server model

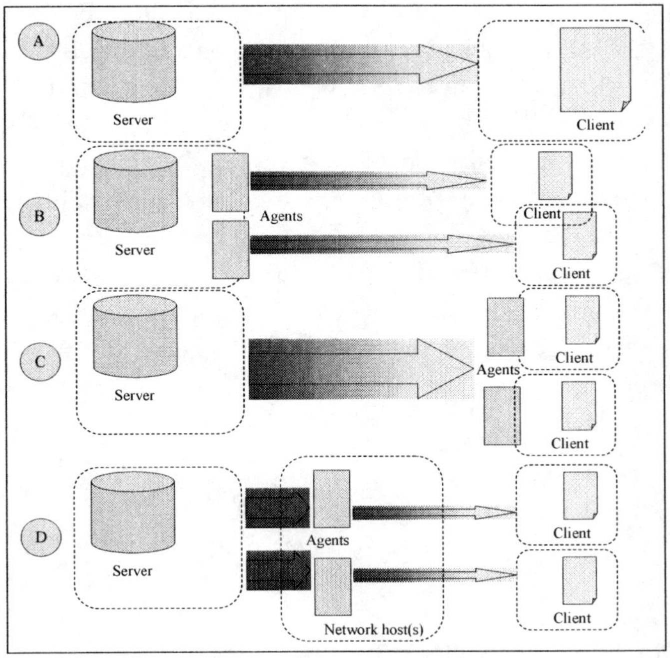

Together with a progress in communication and personal computers, some tasks were in a natural way shifted to the client-side, at the beginning, mainly the tasks related with data formatting and presentation. Thus, it is also quite natural to execute agents at client-side, in a safe (from a point of view of the client) environment (Figure 1C). However, some new problems arise related with the information access, both technical and psychological. The main technical problem is to provide an efficient communication link. As the transfer rate is substantially increased (in comparison with the server-side approach), the communication costs rise. Note that some data are filtered out just after the transmission by the agent (imagine an SQL query executed at the client-side with tuples collected from the server ...); from the client point of view, and taking into account the final results, high transmission costs are usually unjustified. Note also that client environment may be not powerful enough to efficiently perform all the agent tasks (in turn, imagine a mobile phone running the above SQL query ...). Thus, a new architecture, already mentioned n-tier architecture, has been proposed, where agents are executed in a specialized host located somewhere in the network. In a natural manner, the n-tier architecture may be adjusted to the agent technology. An agent from a network host is able to contact the server in an efficient and non-expensive manner, and at the same time this agent may be contacted by the clients, also using different communication means. The basic restrictions mentioned above for both the case B and C are relaxed, as the agents are exclusively controlled neither by the server owner, nor by the clients. We think that moving agents to the network-side provides a reasonable trade-off between "pushing" (server-side) and polling (client-side) data access (cf. the discussion in section "Current Approaches for Building VEs and AHVEs"). However, choosing an optimum place of agent execution is user- and business-case dependent, and thus this decision must be left to the individual business partners.

We restricted the above discussion to the classical client-server architecture. Similar analysis may be performed for the P2P and other architectures as well; however, the conclusions would be similar.

Note that a new strategy is needed to store, maintain, and execute individual, migrating agents. A new set of agent interfaces, dynamically optimized according an execution place, access rights, accessible communication means, and owners' preferences is needed as well. In the next subsections, we discuss security policy and implementation issues of such an agent-based distributed system.

Basic Techniques of Agent Programming

In order to choose an optimum way of programming agents' behavior, we have to analyze four basic issues: (1) uniformity of an agent interface, (2) security aspects, closely related with (3) programming technique and language, and (4) agent code re-engineering.

Key issues of the network-side heterogeneous environment, portability and mobility of agents (Chess, Harrison, & Kershenbaum, 1995; Kotz & Gray, 1999; Milojicic, 1999; Schiemann, Kovacs, & Rohrle, 1999), force the agent interface to be unified for the whole system. A natural approach is to use a traditional single-method interface, used successfully for many years in the area of operating systems. Each agent is equipped with a single method parameterized by a set of invocation parameters. We do not analyze here all the "administrational" methods, that is, for verifying access rights, agent migration, storing and searching agents, and so forth. Number and types of parameters are not determined in advance. In such cases, only the syntax is unified, while the verification of the semantics of the invocation parameters is left to the agent owners, to be dealt individually for each business case. This approach is similar to an execution of an application by an operating system, except that some invocation parameters may be set dynamically at run-time, independently of both the agent and its owner. This is in turn similar to the set of operating-system-defined variables, except that current values of variables are set up for each individual execution separately, and some parameters depend not only on the operating system, but also on current communication mean, access rights, date and time, and so forth.

To choose an optimum way of defining agent behavior, that is, agent code, we may in turn take a look at a typical operating system. For such a system, there are usually two kinds of executable programs: compiled binary applications, and textual shell scripts. The applications are typically quite complicated and complex software, to be used by many users, sometimes in parallel, executed or accessed remotely, and so forth. These programs are "installed" in the system under several restrictions (user rights, security policy, disk size, etc.), usually by administrators or system designers. Ordinary users cannot interfere in the application code; such users are able to parameterize an invocation (execution) of this code only. In contrast, shell scripts are light textual files, sometimes prepared in an ad hoc manner by the ordinary users. An amount of work necessary to define a shell script is usually small; however, script functionality is thus restricted. Typically, shell scripts are used mainly for file manipulation and the invocation of other scripts and installed applications. Thus, scripting languages are not computational-complete, in contrast to application-programming languages, as for example C/C++ or Java.

We propose to use a similar approach for the agent-based environment for maintaining ad hoc VEs. Users are able to define two kinds of agents: public service agents, being counterparts of operating system applications, and Private Agents, equivalent to the shell scripts. Public *service agents* (SA) are agents created by trusted users (i.e., those users who have appropriate access rights), to be applied by many partners at a mass scale. SAs are treated as trustful system elements. Their efficiency is of primary concern. Thus, SAs are programmed in Java. Java was chosen for its universality, portability, efficiency, big support for using with different Internet services (built-in support and libraries for HTTP, XML, (2005), SOAP (2003), KQML (2003), SQL, and other standards), and openness for combining with other software (Bigus & Bigus, 1998). Standard Java security checking is applied. If needed, some additional security mechanisms may be added by programmers, including user account and password checking, token verification, ciphering, and so forth. A way of usage of a given SA cannot be changed by an ordinary user; however, it may be parameterized during the invocation. SAs are used once there is a need for mass access by many users to given information, providing this information in a standardized form and with optimum effort (from the system point of view). The most frequently used SAs are the following: wrappers, cache utilities, brokers, and so forth.

The *private agents* (PA) are created and controlled by their owners. There is not a single user, including system designers and administrators, apart from the agent's owner, who is entitled to execute private agent code and access private agent variables. Unless directly ordered by its owner, the agent cannot be accessed by any other agent and service. Several PAs belonging to the same owner (directly or indirectly, via other agents) and SAs may be used to compose a complex private agent (CPA). All the agents composing a CPA are arbitrarily connected, forming a directed graph. A single PA is distinguished to be an entry point to the CPA. The CPA composition and internal links among their components may vary in time (even at run-time). CPAs belonging to a given user may be activated either on the owner's demand, or after detecting an information change. In the first case, it is the CPA owner who initiates the agent activation, asking for some data. Once the execution of all component agent activities is over, the CPA response is generated and sent to the owner. Then the CPA becomes inactive until the next demand. In the second case, it is the CPA which initiates the activation, usually after detecting an information change in one of the periodically-observed SAs and external services. Once a change is detected of any interest to the CPA owner, an appropriate message is generated and sent. Thus, the CPA may be used for continuous monitoring of changes of information provided by the other agents (and indirectly, by external data sources).

As mentioned above, agents are created and executed on behalf of their owners. An agent is controlled by its owner, that is, the code is defined by the owner or by a programmer on behalf of the agent owner, and the internal agent variables (agent's state) are defined and accessible for the owner only. Agent's behavior, programmed in its code and variables, is determined by the owner and used for his/her/its individual purposes. Although theoretically owners have full control over the behavior of their agents, in reality they are not obliged to set up all their agents individually. There are three predefined classes of users (agent owners), taking into account users' ability to program agents' behavior. The non-advanced, naïve users use predefined, standard, fixed CPAs. The advanced users use standard CPAs as well; however, the level of parameterization is much higher. The experts are able to develop their own PAs (and SAs, on the condition that they have enough access rights).

As it was previously stated, for the system agents, Java is a natural candidate for a basic programming language. As for the "scripting" agents, the answer is not so obvious. Popular scripting languages are specialized in either providing an access to files and applications (operating system shells) or page formatting (JavaScript, PHP, and similar languages). It is quite difficult to adopt an existing scripting language to the specificity of an agent environment. Moreover, as it is discussed below, according to the security aspects, standard scripting languages are not well-suited for unrestricted, user-defined, remotely-executed programs. For a trade-off between overall system efficiency, and agent privacy and efficiency, a specialized language is proposed to program the behavior of private agents, based on XML and imperative programming. The language is equipped with several non-standard mechanisms to improve efficiency and security (cf. next subsections). Although there are many other possibilities to store agent code and variables, XML looks to be the optimum solution for the following reasons:

- XML permits to use a single language for all agent-related data: code, variables, and messages (communication). From the system point of view, the interfaces and mechanisms for storing, searching, migrating, and executing agents across the system are uniformed, as they all use XML data with no semantic checking.
- There are plenty of analyzers for XML syntax checking, that may be easily extended by the semantic analyzer (even individualized for each agent).
- XML-based analysis is independent of local resources and parameters (operating system, back-end database support, communication means, etc.).
- XML is easy to translate to any well-defined format, for example, WML or HTML, and also to the plain text, by the use of the XSL transformations.
- XML is supported by many software tools, for example, back-end database support, above-mentioned XSL transformations, and so forth.

Sample definition of an agent code and variables is given in Figure 2. Note that the agent code is similar to the popular scripting languages, except for the specificity of the XML format (i.e., a way of pointing tags and their parameters). The XML-defined agent code is interpreted and carefully checked during the interpretation; however, compiled parts of system code (i.e., other agents) may be called from it in a transparent way. Thus, assuming the system code (provided as a set of system agents with appropriate access tokens given to private agents) is safe and efficient, the execution of the agent code is also safe and efficient, both from the system and the user points of view. Note that even if the agent code is interpreted, the overall system performance may be

Figure 2. Sample private agent (code and variables) defined in the XML dialect

```
<AGENT name="sending_sms">
<CALL name="send_sms">
  <PARAMETER name="recipient" value="48600123456"/>
  <PARAMETER name="message" value="Everything is OK..."/>
</CALL>
</AGENT>

<VARIABLE NAME="x1" VALUE="1"/>

<VARIABLE>
  <NAME>x2</NAME>
  <VALUE>1</VALUE>
</VARIABLE>

<VARIABLE>
  <NAME><CALL method="compute_name"/></NAME>
  <VALUE><CALL method="compute_value"/></VALUE>
</VARIABLE>
```

high due to the fact that most of the computations are performed inside the compiled Java-based system agents.

More details about the agent XML-based programming language are given in Rykowski & Cellary (2004), Rykowski (2003a), and Rykowski (2006).

Taking into account the above-mentioned programming techniques and languages, Java and XML, the question arises: Do we really need to program the agents in an imperative manner? Why not use any existing, declarative, agent-definition language? We choose the imperative approach for programming agents due to the following reasons. First, a single agent should have a possibility to combine several agents (and external systems and communication means, cf. the discussion below) of different input/output data, location, kind, and purpose. Moreover, such a combination has to follow frequent changes in the environment. Thus, it is not possible to use declarative or skeleton-based agents, as they are not flexible enough to deal with different agents and services, different user requirements, and dynamic changes of the environment. Instead, imperative programming should be used. Using imperative code, the agent's owner may program any behavior of the agent. In the declarative approach, code generation is limited by the declarations or skeletons defined by the system designers. Moreover, while collecting information coming from different sources, additional data processing is needed, wrapping, formatting, presenting, and so forth. Such processing must be defined in an imperative programming language, even if all the data sources are declarative-programmed.

Second, the data and knowledge interchange between agents and external data sources are not standardized (as in general they cannot be). There are some proposals for information- interchange standards (e.g., SOAP (2003) and FIPA (2006); (FIPA Personal Assistant, 2006); however, these proposals cover physical data transfer only, with limited support for semantic data processing. The knowledge representation and sharing standards (e.g., DARPA Agent Markup Language [2006], KQML and OWL [Web Ontology Language, 2004]) deal with data semantics and ontology. However, these standards do not concern data wrapping and formatting. As a consequence, while collecting data in different formats coming from different sources, additional data treatment is needed. This task must be performed in an imperative programming language, as one cannot foresee declarations for all possible standards, connections, communication means, and so forth.

Third, an environment for maintaining ad hoc VEs should be very flexible, due to the highly dynamic changes in the information flow. As one cannot foresee all the possible interactions among business partners, it is not possible to provide any base for creating and further using declarative agents. Moreover, for temporal and evolving cooperation, it is not justified to provide a set of agent declarations (skeletons, patterns, specialized ADLs, etc.) prepared in a traditional, "severe" manner. We think that the declarative approach would substantially reduce a possibility of individualization of agent functionality and behavior, forced by both the evolving user expectations and dynamic changes in the environment.

Fourth, as the cooperation is more and more stable, one may think about replacing "light" programming techniques with the "heavy" ones, for example, replacing a "scripting" prototype-agent with its Java-based compiled, much more effective counterpart. Thus, we propose to use XML-based private agents to start the cooperation, and Java-based system agents to provide a stable, efficient, long-lasting information exchange.

Combining two facts, imperative mode of agent programming, and individualization of agent behavior, we have found that reengineering of agent code is limited. As agents are independent entities, designed and used individually by their owners, it is very probable that a lot of programming work is repeated many times by different business parties. Many business tasks are similar, and thus there are many agents providing almost identical functionality. To avoid this problem, an *inheritance hierarchy* is proposed for the code and the variables of agents. This hierarchy links all the agents into a single directed graph. We propose to use late-binding for implementing agent hierarchy, based on identifiers of agent internal elements (e.g., a named part of a code, or a variable). A place of an agent in the hierarchy may be dynamically changed during its execution, even by the agent itself, as it is determined by one of its variables. Multiple inheritances are allowed, as defining an ordered set of "parent" agents from which the agent derives "missing" elements. Eventual cycles are detected at run-time, thus they do not block the execution.

Apart from effective reengineering, the inheritance mechanism may be used for three additional purposes: effectively storing the agent code, manipulating access rights, and auto-creating new agents. To the first goal, a set of agents is provided being "patterns" of other agents. These agents are equipped with standard code to serve typical business tasks; however, they are not directly executed by the users. Instead, the user's agents derive all the code from one or more "pattern" agents. As long as the code is not changed by the agent owner, private code of an agent is empty, and only the derived code is stored and further used (read, compiled, and executed). If, however, an agent owner modifies any named element, the modified value takes precedence over the inherited one and this modification is used, changing usual agent's behavior. Note that only parts of the code and some variables are usually changed, thus the space needed to store the changed parts is relatively small. Note also, that all the changes to the "shared" code and variables are immediately propagated to all the inheriting agents, allowing online and "hot-swap" actualizations.

To the second goal, the inherited access rights are used. Once an agent has a parent with a given access right, this right is inherited and acts as a private grant. Thus, it is enough to provide a few "empty", non-executable agents with certain access grants and add these agents to the inheritance hierarchy for the agent to grant or revoke access to given services for this agent. Note that it is similar to granting and revoking a single access right to a group of users at once, thus facilitating the administration of grants.

Finally, the inheritance mechanism may be also used to generate some "standard" agents automatically. A new "empty" agent is automatically created that derives code and variables (mainly access rights) from given "pattern" agents. As long as the newly-created agent is not modified by the agent owner, this agent shares all the system variables and the code with the parent(s), thus avoiding unnecessary use of disk space.

Gateways to External Software Systems and Telecommunication Channels

It is quite obvious that the agents must have a possibility of making contact with external software systems and with the humans (agent owners and users). To this goal, as well as for security and flexibility reasons, specialized system agents are provided, so-called gateways. There are two types of gateways: *gateways to external software systems*, and *telecommunication gateways*. These agents are able to provide bidirectional communication with external software systems, via a computer network, and with humans, via standard telecommunication facilities. As for the first case, gateways are able to connect with any network-accessible external software system, located at the same host, and remote ones, respectively (cf. section "Current Approaches for Building VEs and AHVEs"). To this goal, a gateway is programmed to fit to a specificity of a software system this agent is connected to, taking into account proprietary interfaces, communication standards, data formats, and so forth. Shifting this task to a system agent rather than leaving it for the system core improves overall system extensibility and flexibility.

Most of the gateways are used as brokers and wrappers between an external environment and the population of agents. A gateway is able to adjust the information collected as a result of cooperation with an external system to the specificity of the agents interested in using this information. There are three main advantages of such brokerage. First, the agents are not forced to deal with proprietary standards of external software systems. Gateways are able to provide a common, stable standard for accessing many different external software systems in the same, standardized way. As a consequence, the complexity of a private agent's code may be substantially reduced. Second, from a system point of view, an access to external systems may be supervised, if needed. Such supervision may cover specific external systems, communication means, agents and their users, and so forth. Third, the gateways are able to bill user agents (and, indirectly, agent owners) for each access to an external software system. The billing procedure, included in a gateway code, may be set up for a specific connection, communication channel, time, information content, user rights, and so forth.

Putting stress on pushing mode of executing agents and alerting their users, we introduced the second group of gateways, telecommunication utilities, connecting agents with selected public telecommunication systems, to enable contacts with humans via communication channels of different type and purpose. In general, two basic types of communication channels are available: textual and Web-based. A textual channel is able to exchange flat (unformatted) text messages, usually among humans and agents. Physically, textual channels may use such media as an e-mail SMPT/POP3 connection, short message system (SMS)/MMS connection with a telecommunication network, a voice gateway, and so forth. Once sent by a textual message, an agent acts as a chatterbot (Zillman, 2003), analyzing the message via keyword extraction and analysis (Jurafsky & Martin, 2000; Weizenbaum, 1976). The semi-natural access to an agent in a chatterbot manner (Rykowski, 2005b) is especially useful for non-advanced users, and for users who are temporarily handicapped due to limited hardware possibilities and communication costs. For example, an SMS message may be used to check the most important information during a journey, and a stationary PC is further used to get the complex information when the user is back at home.

Web-based channels are used to access an agent via a WWW/WAP page, and from specialized ACE applications. These channels use personal, semi-automatic formatting of both contexts and presentation of the data to be sent. To this goal, XSL-T technology (The Extensible Stylesheet Language Family, 2002) was adopted with XSL transformations defined in a personal manner and stored in private agent variables (Rykowski & Juszkiewicz, 2003). In a case of a conversation with a human, automatic detection of an end-user device may be used, thus restricting the communication. For example, a small textual message is sent to a mobile phone using WAP connection; a similar message with the same contents but with some additional formatting is sent to a PDA device; and a full text and graphic message is sent to a stationary PC.

The number and types of the telecommunication gateways used (including some specific parameters, as a phone number for an SMS center, an address for a SMTP/POP3 server, etc.) is local-administrator dependent. Note once again that these gateways are implemented as system agents; thus one may easily extend a given Agent Server by some specific communication means.

From a technical point of view, gateways may be used to collect information from any kind of an external software system, on the condition that there is a communication means available to the external system. In particular, information may be collected from a Web server, a Web service

(Web Services Activity, 2002) and its extensions (e.g., UDDI directories [UDDI Version 2, 2005]), Semantic Web system (DAML Semantic Web Services, 2006), a database, and so forth. In such a way, agents may be used for combining several data sources of different type and purpose into a single, consistent service (Rykowski & Cellary, 2004).

Security Policy

The most important problem related with mobile, user-defined, imperative agents' code is to achieve a reasonable level of global system safety. Note that executing "remote" agents means, from a local system point of view, executing alien code, unknown and potentially dangerous. Maybe in several cases this is more psychological than real menace; however, imperative migrating agents are usually used in closed, mutually-trusted environments. As a consequence, additional mechanisms must be introduced to ensure that the system is resistant/immune to malicious agents. However, additional security checking lessens efficiency of code execution. Thus, a reasonable trade-off must be found between the level of security checking and the overall system efficiency. We think that the division to compiled, Java-based system agents, and interpreted, XML-encoded private agents solves this problem. A system agent is created by a trusted system designer, to be massively used by other agents. Here, the agents' code is trusted, so security checking may be relaxed; system agents may be executed more efficiently. On the contrary, private agents are carefully inspected at run-time, in order to avoid "dangerous" cases. The interpretation and security checking slows down agent execution; however, as most of the massive computations are performed by the compiled system agents, overall system efficiency is high enough.

The above-mentioned run-time inspection covers: checking access rights, mainly to access other agents and (indirectly) external software systems and communication channels; limiting a maximum time of a single execution of an agent; and quoting memory space required by an agent for its code and variables.

Checking access rights applies to both system and private agents. There are two types of access rights: system- and agent-based. System-based access rights are checked before agent execution. The most frequently used system-based access rights are the following: a right to execute an agent in a given place, a right to execute compiled code of an agent, a right to migrate to any other place, and so forth. Agent-based access rights are verified at run-time (i.e., during agent execution), by other agents, involved in a communication with a given agent. It is each agent's responsibility to verify appropriate access rights and to react in a case of access violation. An access right may be related with any part of agent code and any agent variable. For example, there may be a grant for modifying a variable value, for accessing an external information source, for human communication by given communication means, and so forth. Introducing agent-based access rights extends security level to the point required by the agent owner. Note that, as for the agent code, agent-based access rights, as well as a process of their verification, may be set up individually for each agent.

Verifying maximum execution time and memory quota is related with the private agents only; system agents rely on standard Java security mechanisms. In addition, system agents are executed as separate parallel threads, and there are some mechanisms to verify whether a running thread is still active and behaves correctly from the JVM point of view. The maximum execution time is set up for each agent individually, depending on overall security policy, current system load, and (to some extent) individual access rights. During an interpretation of each XML node, current execution time is compared with the maximum. Once the time is over, an execution is unconditionally stopped with an error indication. Thus, a malicious (or simply badly programmed, with no intention for unwanted behavior) agent trying to execute "forever" cannot block the whole system. Note that such feature is not a standard mechanism for any programming language, and must be programmed manually. Even if theoretically the above-mentioned execution threads may be stopped once it is executed too long, it is quite easy to break this protection by modifying standard Java security mechanisms. Similar, a check is performed after each modification of any private agent variable. A total amount of memory required by the agent variables is computed (in the delta mode, i.e., only the affected variable is recomputed) and compared with the quota. Once the quota is exceeded, agent execution is stopped and the new value of a variable is not set.

Agent Computing Environment

The architecture described in the previous sections has been implemented in the scope of the *Agent Computing Environment* (ACE) project. ACE system consists of a distributed set of *agent servers*, characterized by different functionality of system agents, and different communication means. Taking into account a specialization of an agent server for certain tasks and localizations, we may distinguish three basic classes of agent servers: source-side servers, connected directly to data sources and located at the same host (local area network) as a given data source is connected; middle-side servers located in the network, not directly linked with any particular

Figure 3. Distributed set of cooperating agent servers

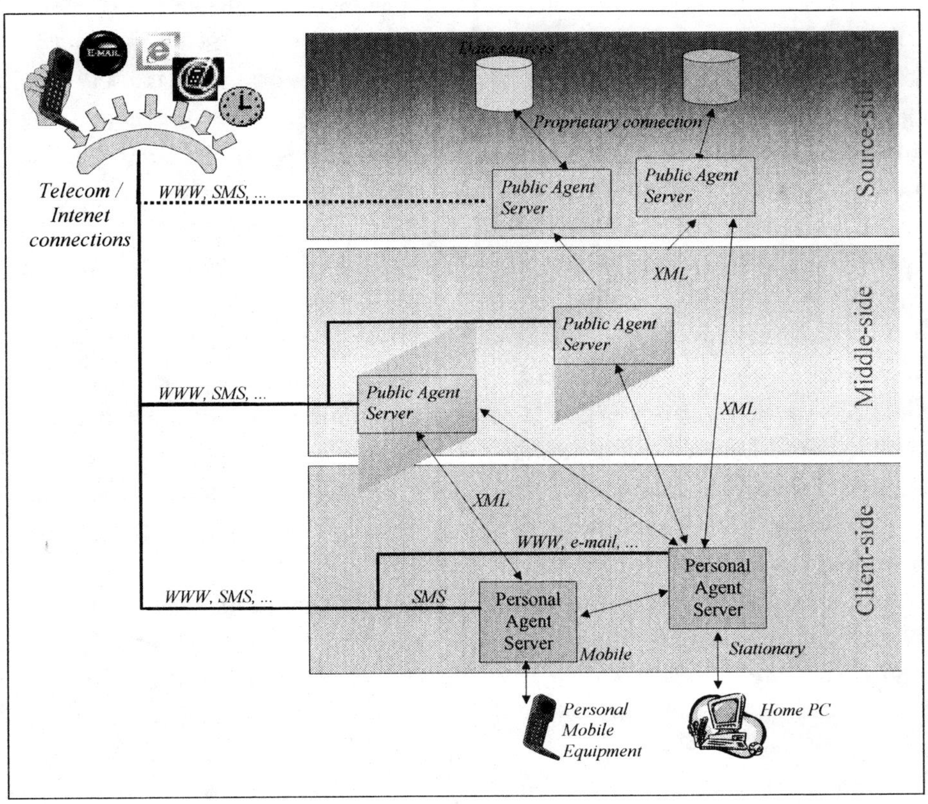

data source; and client-side servers for personal usage of a given agent owner (Figure 3). The servers are connected to each other, and they are able to transfer private agents among each other as well as remotely execute the agents previously transferred to a given place.

A functionality of a *source-side agent server* is optimized towards reliable and efficient access to selected data sources, from the point of view of the information owner (i.e., given business party). For security reasons, using private agents in such a server is substantially limited. The system is reduced to a set of gateways, able to standardize an access to the data source(s) connected, with limited support for public telecommunication facilities. Gateways are equipped with several mechanisms supporting efficient, parallel, multi-user access to the data sources as, for example, cache memories, proxies, synchronizers, semaphores, locks, query optimizers and serializers, and so forth.

Middle-side agent servers are located in the network, at a selected host. In contrast to the source-side agent servers, middle-side agent servers store and execute both system and private agents. A typical task list for system agents covers: brokering among source-side agent servers and user agents; wrapping and formatting messages exchanged by the population of agents; and providing access via different telecommunication means (WWW/WAP, SMS/MMS, e-mail, etc.). A stress is put on efficient access to the agents by many telecommunication channels and standards. User-defined agents from the private pool are usually devoted to the task related with comparing information and detecting changes that are "interesting" for users and other agents.

Architecture of a *client-side agent server* strongly depends on technical and communicational possibilities of an end-user hardware/software the server is running at. It is up to the user to locate his/her private agents either in a selected middle-side Agent Server, or in the private client-side server. In the first case, the network traffic may be substantially reduced; however, remotely-executed user agents are less secured (from the user point of view) and less efficient. In the second case, all the user agents are executed in a trusted (still, only from the user point of view) environment; however, a lot of information must be transferred among distributed agents.

For an agent server executed at a portable/mobile device, a stress is put on fast and user-friendly agent-to-human communication. The technical capabilities of the device strongly limit the possibilities of executing the agents (small memory, limited battery time, difficult management, etc.). Thus, usually only a few private agents are located in a mobile agent server capable of performing some simple tasks as, for example, final formatting of an alert message, filtering incoming messages, generating sound alerts, and so forth.

Typical Scenario of Creating and Maintaining an Ad Hoc VE

As already mentioned in the section, "Current Approaches for Building VEs and AHVEs," the system may be used by the beginners and advanced users at the same time, depending on individual preferences. Below, a strategy is proposed for personalization of the private agents according to the growing user level. The strategy assumes that at the beginning the user representing a business party is not able to program the agents. So, the VE representative (i.e., another business party) offers a set of predefined agents, with functionality common for all the users of the same kind. However, as time goes by, the user's expectations grow, and the system is ready to satisfy the user with new/extended functionality.

1. A client addresses a given ad hoc VE, determining some personal preferences: mobile and stationary devices used, identifiers (e.g., a phone subscriber number, IMEI device number, PKI signature, IP/DNS address, e-mail account, etc.), preferred communication means, and so forth.

2. The VE representative proposes a set of predefined SAs and PAs to be installed for the new VE member located in the local area network of the VE representative or in a selected network host. This set of agents is capable of serving the client's requests in a predefined way that is common for all the parties.
3. The VE representative proposes a set of PAs to be installed at client-side, to facilitate business contacts. It is up to the user to accept this set of agents or not. In the latter case, only the PAs executed remotely (from the point of view of the new member) are used. However, once a user accepts PAs agents in his/her private hardware/software environment, he/she may profit in using individual data treatment (cf. "Current Approaches for Building VEs and AHVEs"); better adjustment to end-user devices; more personalized formatting and presentation methods; and other functionality related with client-side information processing. Note that, as the new VE member may inform the VE representative about detailed characteristics of the devices used and preferable communication channels, the generated client-side PAs may be customized, even if the server-side PAs are quite standardized (at the beginning, however, cf. Point 5 below).
4. The set of accepted agents is installed and activated each time the change of information is detected related with VE. The detected changes are filtered and eventually sent to the client (usually, to an agent, and further via given communication channel directly to the end-user device, e.g., a mobile phone).
5. Once the client is not satisfied with a subscription or PA behavior, he/she has some rights to redefine both data and code of his/her agents. For advanced users, it is possible to fully personalize the behavior of the system, both at server-, and at the client-side. For non-advanced users, it is still a possibility to define preferable communication channel, ways and timings of sending alerts, formats and contents of messages generated by the agents (e.g., small SMS alerting vs. formatted HTML e-mails). In this way, system functionality grows as user requirements grow.

Note that even if the users change the behavior (i.e., the code) of their private agents executed at the server-side, the overall system security is not reduced due to continuous run-time inspection of "untrusted" (user-defined) agents (cf. section "Agent Computing Environment"). Note also that private agents are executed only in two "trusted" (from a user point of view) environments: a VE representative local area network, and private "network" of the new VE member, mobile phones included. Sending messaging among user agents may be encrypted (e.g., using PKI cryptography), increasing the overall confidence level of business contacts.

As an example of using the above-presented scenario, we may consider a warehouse selling and transporting large amounts of items to ad hoc (i.e., single-transaction) customers. Each customer is interested in obtaining real-time information about payments, loading to the trucks, further position of these trucks, and exact date of delivery. In addition, some alerts would be quite useful, starting from eventual problems with money transfer, and finishing with just-in-time delivery reports and delays. In the traditional approach, mobile phones are used to get the latest info about the transaction and delivery state. In such a case, these are the members of the customer's staff who would initiate the telephone calls ("polling" mode). However, as this is an ad hoc business activity, it is not possible to make a direct call to the warehouse by an ordinary employee of the company. Only the employee directly responsible for this ad hoc business activity (usually one of the company's management staff) has detailed knowledge about the cooperation, knowing warehouse's telephone numbers, staff names, responsibility areas, and so forth. Thus, this person is overloaded by "setting up" telephone connections, propagating some important information changes (i.e., about delivery delays), and performing more similar low-level business activities, rather than concentrating on high-level management.

To improve the quality of the business process, we introduce ACE agents as being brokers between the warehouse and the customer. This process is well known by the warehouse, as it was already performed several times, however, with different business partners. Thus, the warehouse is in charge for proposing a set of standard agents, operating in one of the warehouse-owned hosts, and responsible for the support of the business activities between the warehouse and the customer. The customer accepts this set of agents, providing some parameters, such as telephone numbers of the employees directly responsible for particular sub-activities (i.e., a storeman, a driver, an accountant, etc.). Such information, in turn, is well-known by the customer, however, not necessary by the warehouse. Once the business process is started, these agents and telephone numbers are used for propagating useful information about the current state of this process, as well as some alerts if something goes wrong (i.e., not in a usual/arranged way). Note that the information messages and alerts are individualized, taking into account the area of responsibility of the employee to whom the message is addressed. For example, an alert about a delay of the money transfer is sent to the accountant, while the exact date and time of the delivery is sent to the storeman.

If both business partners are satisfied with the transaction, probably the process will be repeated in the future. In such case, the ACE agents prepared for the first cooperation may be reused. Moreover, both business partners may use just-gained experience to improve the agent functionality. For example, the storeman's agent may play a significant role in planning the transport, by setting up possible departure/arrival dates, layout of the items on a truck, additional protection, and so forth. To this goal, the agent is partially moved to the personal communication device of the storeman (e.g., a palmtop with GSM modem). In case of a significant delivery delay, the storeman's agent is able to send an alert to the management staff. In turn, agents of the management staff are able to generate alerts and inquiries to the agents of the warehouse staff. Note, however, that the above-mentioned additional agent functionality must be designed by the customer rather than the warehouse, even if the agents are to be executed in one of the warehouse's hosts. Thus, the costs of individualization and improvement of the customer-side agents are left to the customer. Note also, that the decision for the automatization and/or automated support for a particular business sub-activity is left to the business partner for which the sub-activity is realized.

RELATED WORK

There were already some attempts to use agent technology in the scope of virtual enterprises. A good survey of using software agents, Web services with UDDI/WSDL, and other extension technologies to create VEs may be found in Petri and Bussler (2003). There are, as well, some scientific prototypes for an agent-based virtual enterprise; these are enumerated and discussed below.

Oprea proposed to use software agents for the coordination of different activities of business partners inside a VE (Oprea, 2003a). The concept resulted in a prototype of an agent-based environment for establishing and maintaining VEs (Oprea, 2003b). The basic assumption for the proposal is to use a stable set of predefined roles and software agents related with these roles, such as a coordinator, a member, a broker, and so forth. Agents are expected to coordinate the services and information exchange, to be able to follow complex negotiation protocols, and to perform other socially complex operations. Recently, the proposal resulted in a real-business application for establishing a set of VEs in the area of house-developers (Oprea, 2005).

The proposal concentrates on coordination of a set of autonomous agents; however, an evolution of agent roles is not taken into consideration. Due to the fixed set of roles and, in turn, agent types, the ad hoc and temporal activities are not modeled. Implementation of the prototype is based on JADE, a Java-based FIPA-compliant agent platform. The implementation allows for programming the agents in the imperative way; however, security aspects are not taken into consideration, and a priori assumption is taken that standard Java security mechanisms are enough for the distributed execution of the agents. In contrast, the ACE environment is concentrated on the security aspects for secure, remote execution of user-defined agents.

In the metaMorph II project (Shen & Norie, 2004), being an extension of MetaMorph I, ABCDE, and DIDE projects, agents are used as a mechanism for knowledge sharing in a distributed environment. The authors introduce a common vocabulary in order to provide efficient, global information sharing. The main project objective is to integrate the enterprise's activities such as design, planning, scheduling, simulation, and execution with those of its suppliers, customers, and partners into a distributed intelligent open environment. Two fixed classes of agents are introduced: resource agents, and mediator agents. As each mediator must contain detailed knowledge about the coordinated agents, the overall structure (network) of agents is fixed. The system behaves well in such stable applications as supply chain management. However, probably such a system would fail in a dynamically-composed and evolving structure of an ad hoc enterprise. Other drawbacks of the proposal in the scope of an ad hoc VE are related with fixed, generic way of representation of business knowledge (logic). As for an ad hoc VE, detailed, static representation of business knowledge of each VE business party would be quite difficult, because such knowledge cannot be collected in advance, and thus it is hard to represent in a formal way.

Petersen, Divitini, and Matskin (2001) extended a typical representation of business units to the uniform representation of different business partners, humans included. In the proposal, a VE is treated as a team of partners that have common goals and are committed to fulfilling these goals. A VE is a combination of entities: agents, human beings, and business enterprises. The classical VE life cycle is assumed with stable roles of business partners, fixed before any business cooperation. This assumption limits a possibility of establishing an ad hoc VE, because the final commitment of the whole ad hoc VE is hard to reach, and this evolving process takes some time to get to a stable state, involving several business-to-business activities among the partners.

The idea has been further extended to the AGORA proposal of an agent-based architecture for supporting VEs (Petersen, 2003; Petersen, Rao, & Matskin, 2003). In AGORA, several common predefined agent classes are introduced: initiator, partner, interested partner, and potential partner. Agents representing business partners are matched on the base of requirements (defined for a business goal) and skills and capabilities (defined for a business party). Fixed predefined algorithms allow for establishment of a VE as well as for contract negotiations inside the VE. The implementation is based on FIPA ACL and XProlog system, with a possibility of mapping Prolog clauses to FIPA messages. Similar to the MetaMorph system, AGORA proposal is not flexible enough to deal with ad hoc VEs.

Aerts, Szirbik, and Goosenaerts (2002) deal with the problem of modeling an ad hoc VE as a set of distributed agents. The authors put stress on the problem of high evolution of an ad hoc VE, in contrast to the stable and fixed supply chain management. The authors also discuss an interesting concept of an extended enterprise, stating that a company is made up not just of its employees, its board members, and executives, but also its business partners, its suppliers, and even its customers (Jagdev & Thoben, 2001). The project is concentrated on VEs that manufacture products, with an emphasis on fast and timely delivery, rather than on integration at the planning level. As a result, a group of enterprises is modeled that collaborate on the production of a set of related products, such as aircraft subsystems, whole complexity transcends the abilities of a single enterprise.

The main implementation issue of the proposal, docks and service bridges, seems to be less flexible than the agent servers of the ACE framework with specialized agent-gateways, as the docks form fixed system infrastructure. In such a way, a bridge serves as a single, multi-use wrapper to local software systems, with no possibilities of individualization. A common ontology is assumed, and a translation should be provided for each business partner. The bridges also serve as firewalls for incoming agents; however, nothing is said about security policies and inspection mechanisms to overcome security problems. For non-advanced and new-coming users, so-called agent providers are introduced, offering communication and application services to customers. We think that the derivation hierarchy of ACE agents is more flexible and easy-to-use. Moreover, ACE agents of a standard functionality may be generated automatically without direct human control, while the Agent Providers need a substantial amount of human management.

The DIP project (Data, Information and Process Integration with Semantic Web Services), started in 2004, has in mind to balance between ad hoc and global integration strategies (DIP, 2006). DIP's main objective is to develop and extend Semantic Web and Web service technologies by producing a new technology infrastructure in which different Web services can discover and cooperate with each other automatically. Although the implementation is based on Web services rather than agents, we think that the main goal of the DIP project is somehow similar to our evolving agent-supported cooperation in the scope of the ACE framework, being ad hoc and non-formal at the beginning and as time goes by, becoming more and more precise and stable. The idea of evolving ontology in the DIP methodology may be compared with the idea of evolving user-managed agents' code for the network of ACE agents. However, our proposal is concentrated on the ad hoc cooperation phase only. Thus, we put stress on the fact that the evolution of agents is managed by the agent owners as a result of evolving requirements and skills of the business parties, while the evolution of DIP ontology is more centralized and performed in a formal way.

CONCLUSION

Establishing and maintaining a VE requires a lot of efforts. The amount of work is substantially enlarged in a case of an ad hoc VE, with a highly dynamic environment and ongoing evolution of the requirements and expectations of the business partners involved. Due to the fact that the business activities related with an ad hoc VE will probably never happen in the future, using fixed, severe technologies as, for example, Web services is not economically justified. Moreover, as the individual goals of the business partners are frequently redefined on-the-fly, it is hard to provide a stable, commonly-shared knowledge about the business case. Such knowledge is a base necessary for most of the current attempts to provide an environment supporting creation and maintenance of VEs. Thus, the current proposals either assume a common, predefined ontology (and further common schema), to be set up by all the business partners before any cooperation takes place, or restrict the application area to such with well-defined generic business knowledge, for example, e-markets, supply chain management of a traditional enterprise, and so forth. In any case, all the business partners must agree with the common ontology and, as a consequence, the partners must either adjust their own way of doing business to the ontology specification, or to provide some mapping between "global" and "local" private ontologies.

In the case of an ad hoc VE, there is a limited possibility to establish a common ontology for the business partners prior to the start of the cooperation. First, the business case is usually not fixed, and the knowledge about this case is usually gathered on-the-fly, as the business evolves. Second, there are no generic rules and knowledge for similar business cases, as there were not present in the past, and they will probably never happen in the future. Third, the cooperation among business partners possibly involved in the cooperation in the scope of an ad hoc VE is usually restricted in time. Thus, it is not economically justifiable to invest in any severe technology dealing with generic knowledge (ontology) exchange and management. And fourth, if we take into consideration one-to-one relations between the business partners (a usual case for the ad hoc cooperation), only a part of the global knowledge is usually needed for a single relation. Then, instead of providing some complex tools for the ontology (schema) mapping and exchange, one need to precise only a part of the total knowledge with minimum efforts. Note, however, that this task must be individualized for both the business case and the business partners involved in this case. As a consequence, a need arises for a "light" technology to effectively and securely maintain such temporal and ad hoc business activities. We proposed to adopt a software agent technology, with a possibility of defining and controlling agents' behavior directly by the agent owners, that is, business partners involved in cooperation in the scope of an ad hoc VE.

Comparing with the other proposals related with using the agent technology to support VEs, the presented agent-based framework, based on agent computing environment (ACE) is much better suited for a specificity of ad hoc VEs. First, no assumption is made for creating common business knowledge prior to establish a VE. Instead, business partners develop individually their agents to use a part of this knowledge, necessary to establish one-to-one business contacts. As the control of agents' behavior is left to the agent owners, and there are no global restrictions related with a common schema and/or ontology, the agents may be quickly adopted to the changes in the environment and evolving expectations of the business partners. In such way, longer is the cooperation between business partners, better the contacts are supported by the partners' agents and their internal knowledge, and less work must be performed manually by the human staff.

Second, the ACE agents are imperatively programmed by their owners, being able to deal with any business case. Using the declarative technique of agent programming, which is a key for most of the current proposals, would strongly restrict overall system flexibility. The natural fear of remote execution of the imperative agents' code is tempered by introducing two classes of the agents: the public agents, designed by the trusted users and used as "black boxes" by the others, and the private agents, to be exclusively used by their owners.

Third, ACE agents make it possible to mix traditional, human-based cooperation and automated information exchange, by introducing several human-to-agent communication channels: from traditional (Web/WAP pages and e-mail) to modern mobile communication channels, as for example SMS/MMS, push-to-talk (PTT) and voice gateways for instant messaging. Even if the business activities are performed manually by the human staff, there is still a large application area for ACE agents being remote monitors and notifiers, sending alerts while "something interesting" happen. What is "interesting" is individually programmed by the agent owners, and may be dynamically changed as the overall situation evolves. For example, imagine an alert about a just-closed credit line, sent automatically by the use of a secret SMS message to the company's representative being in the train of business negotiations. Even if such information is not incorporated in the total knowledge about the VE, knowing this one fact may drastically change the style of cooperation between the business partners.

Fourth, the ACE framework does not control the internal functionality of the private agents provided by the business partners. The framework itself is a generic, safe environment for the establishment of an ad hoc VE of any kind and purpose. Thus, starting from a weak support for an ad hoc VE, the business partners may continuously enlarge agents' possibilities, either by providing gateways to existing and newly-proposed external software systems, or by incorporating the new ideas directly in the agents' code. Once enough knowledge is gathered about the business activities, some formal approaches may be adopted, replacing the manual work of a human staff by the automated data interchange according to well-defined business rules. The imperatively programmed ACE agents are well suited for both these cases. Note that the new functionality does not have to be provided for all of the business partners at once. Instead, at the beginning only a small group of business partners may be involved, and, as time goes by, more and more business partners incorporate new solutions in their business logic. For most of the current proposals, it is not possible to differentiate business partners in such a way.

It seems that the idea of a continuous evolution of the business relations, starting from manual cooperation of the humans, and finishing on automatized cooperation of standalone and individualized software agents, needs further research. The proposed ACE agents are, on one hand, a good starting point for beginning the cooperation by creating an ad hoc VE. On the other hand, several existing proposals may be incorporated dealing with stable and well-defined business relations, changing an ad hoc VE to "real," "severe" and long-lasting business relations. The ACE agents may be also used for providing a kind of a prototype of a VE. Once enough experience is gathered about business relations while using such a prototype, a targeted VE may be designed by the use of more sophisticated, but also more costly, software technologies.

REFERENCES

ACT-NET Consortium. (1996). The active database management system manifesto: A rule-base of ADBMS features. *ACM Sigmod Record, 25*(3), 40-49.

Aerts, A. T. M., Szirbik, N. B., & Goosenaerts, J. B. (2002). The flexible ICT architecture for virtual enterprises. *Computers in Industry, 49*(3), 311-327.

Bigus, J. P., & Bigus, J. (1998). *Constructing intelligent agents with Java. A programmer's guide to smart applications.* New York: Wiley Computer Publishing.

Caglayan, A., & Harrison, C. (1997). *Agent sourcebook: A complete guide to desktop, Internet, and intranet agents.* Toronto: John Wiley & Sons.

Ceri, S., & Fraternali, P. (1997). *Designing database applications with objects and rules.* Addison-Wesley.

Chess, D., Harrison, C., & Kershenbaum, A. (1995). *Mobile agents: Are they a good idea?* Research report. Yorktown Heights, NY: IBM, T. J. Watson Research Center.

Concordia Mobile Agent. (2002). Concordia: An infrastructure for collaborating mobile agents. *Mitsubishi Electric ITA, Horizon Systems Laboratory.* Retrieved from http://www.cis.upenn.edu/~bcpierce/courses/629/papers/Concordia-MobileAgentConf.html

CORBA. (2002). *3- and n-Tier Architectures*. Retrieved from http://www.softeam.com/technologies_architectures_distribuees.php, http://www.corba.ch/e/3tier.html#Why%203-tier?

DAML Semantic Web Services. (2006). Retrieved from http://www.daml.org/services

DARPA Agent Markup Language. (2006). Retrieved from http://www.daml.org/

DIP on-going project—public deliverables (2006). Retrieved from http://dip.semanticweb.org/

FIPA. (2006). Retrieved from http://www.fipa.org/

FIPA Personal Assistant Specification. (2006). Retrieved from http://www.fipa.org/specs/fipa00083/XC00083B.html

Franklin, S., & Graesser, A. (1996). Is it an agent, or just a program? A taxonomy for autonomous agents. In *Proceedings of the 3rd International Workshop on Agent Theories, Architectures, and Languages*. Springer-Verlag.

Hurson, A., Bright, M., & Pakzad, S. (1993). *Multidatabase systems: An advanced solution for global information sharing*. IEEE Press.

Jagdev, H. S., & Thoben, K. D. (2001). Anatomy of enterprise collaborations. *International Journal of Production Planning and Control, 12*(5), 437-451.

Jurafsky, D., & Martin, J. H. (2000). *Speech and language processing: An introduction to natural language processing, computational linguistics, and speech recognition*. Prentice-Hall.

Kotz, D., & Gray R. S. (1999). Mobile agents and the future of the Internet. *ACM Operating Systems Review, 33*(3), 7-13.

Milojicic, D. (Ed.). (1999). Trend wars—mobile agent applications. *IEEE Concurrency, 7-8*, 80-90.

Nwana, H. (1996). Software agents: An overview. *Knowledge Engineering. Review, 11*(3), 205-244.

Oprea, M. (2003a). Coordination in an agent-based virtual enterprise. *Studies in Informatics and Control, 12*(3), 215-225.

Oprea, M. (2003b). The agent-based virtual enterprise. *Journal of Economy Informatics, 3*(1), pp. 15-20.

Oprea, M. (2005). A case study of agent-based virtual enterprise modelling. *Lecture Notes in Artificial Intelligence, 3690*, 632-635.

Oracle Triggers (2006). Retrieved from http://www-rohan.sdsu.edu/doc/oracle/server803/A54643_01/ch15.htm

Paton, N. (Ed.). (1999). *Active rules in database systems*. Springer-Verlag.

Petersen, S. (2003). An agent-based evaluation framework for supporting virtual enterprise formation. In *Proceedings of the IEEE 12th International Workshop on Enabling Technologies, Infrastructures for Collaborative Enterprises (WET-ICE2003)* (pp.159-164) Linz, Austria.

Petersen, S., Rao, J., & Matskin, M. (2003, October 13-16). Virtual enterprise formation with agents: An approach to implementation. In *Proceedings of the 2003 IEEE/WIC International Conference on Intelligent Agent Technology (IAT-2003)* (pp. 527-530) Halifax, Canada. IEEE Press.

Petersen, S., Divitini, M., & Matskin, M. (2001). An agent-based approach to modelling virtual enterprises. *International Journal of Production Planning & Control, 12*(3), 224-233.

Petrie, C., & Bussler, C. (2003). Service agents and virtual enterprises: A survey. *IEEE Internet Computing, 7*(4), 68-78.

Rykowski, J. (2003a, September). Agent technology for secure personalized Web services. In *Proceedings of the 24th International Scientific School ISAT 2003,* Szklarska Poręba, Poland (pp. 185-193).

Rykowski, J. (2003b). Databases as repositories for software agents. In B. Thalheim & G. Fiedler (Eds.), *Emerging database research in East Europe. Proceedings of the Pre-conference Workshop joined with the 29th VLDB Conference*, Berlin, Germany (pp. 117-123).

Rykowski, J. (2005a). ACE agents—mass personalized software assistance. *Lecture Notes in Artificial Intelligence, 3690,* 587-591.

Rykowski, J. (2005b, April). Using software agents to personalize natural-language access to Internet services in a chatterbot manner. In *Proceedings of the 2nd International Conference Language and Technology L&T'05,* (pp. 269-273), Poznan, Poland.

Rykowski, J. (2006). Management of information changes by the use of software agents. *Cybernetics and Systems, 37*(2-3), 229-260.

Rykowski, J., & Cellary, W. (2004). Virtual Web services - Application of software agents to personalization of Web services. In *Proceedings of the 6th International Conference on Electronic Commerce ICEC 2004,* Delft, The Netherlands (pp. 409-418). ACM Publishers.

Rykowski, J., & Juszkiewicz, A. (2003). Personalization of information delivery by the use of agents. In *Proceedings of the IADIS International Conference WWW/Internet 2003,* Algarve, Portugal (pp. 1056-1059).

Schiemann, B., Kovacs, E., & Röhrle, K. (1999, October). Adaptive mobile access to context-aware services. In *Proceedings of the 3rd International Workshop on Mobile Agents* (pp. 190-201), Palm Springs, FL.

Shen, W., & Norie, D. H. (2004). An agent-based approach for information and knowledge sharing in manufacturing enterprise networks. *International Journal of Networking and Virtual Organizations, 2*(2), 173-190.

SOAP. (2003). *Version 1.2, Part 1: Messaging Framework, W3C Recommendation.* Retrieved from http://www.w3.org/TR/soap12-part1/

Specification of the KQML Agent-Communication Language. (2003). *The DARPA Knowledge Sharing Initiative, External Interfaces Working Group.* Retrieved from http://www.cs.umbc.edu/kqml/kqmlspec/spec.html

The Extensible Stylesheet Language Family. (2002). *XSL.* Retrieved from http://www.w3.org/Style/XSL/

The Voyager SOA Platform. (2005). *Recursion Software.* Retrieved from http://www.recursionsw.com/Voyager/2005-09-13-Voyager_SOA_Platform.pdf

UDDI Business Registry Version 2. (2005). Retrieved from https://uddi.ibm.com/ubr/registry.html

Web Services Activity. (2002). Retrieved from http://www.w3.org/2002/ws/

Web Ontology Language OWL. (2004). Retrieved from http://www.w3.org/2004/OWL/

Weizenbaum, J. (1976). *Computer power and human reason. From judgment to calculation.* S.Franc.

Wooldridge, M., & Jennings, N. R. (1995). Intelligent agents: Theory and practice. *Knowledge Engineering Review, 10*(2), 115-152.

Wooldridge, M., & Jennings, N. (1995). Agent theories, architectures, and languages: A survey. In M. Wooldridge, & N. Jennings (Eds.) *Intelligent Agents* (pp. 1-22). Berlin: Springer-Verlag.

XML Extensible Markup Language. (2005). Retrieved from http://www.w3.org/XML

Zillman, M. P. (2003). *Chatterbot resources and sites*. Retrieved from http://chatterbots.blogspot.com/

This work was previously published in Knowledge and Technology Management in Virtual Organizations: Issues, Trends, Opportunities and Solutions, edited by G. D. Putnik and M. M. Cunha, pp. 306-333, copyright 2007 by IGI Publishing, formerly known as Idea Group Publishing (an imprint of IGI Global).

Chapter XVII
Wireless Networks Based on WiFi and Related Technologies

Rajendra V. Boppana
University of Texas at San Antonio, USA

Suresh Chalasani
University of Wisconsin-Parkside, USA

ABSTRACT

Multihop wireless networks based on WiFi technology offer flexible and inexpensive networking possibilities. Applications of multihop wireless networks range from personal networks within consumer homes to citywide departmental networks to wide-area vehicular ad hoc networks. In this chapter, we focus on multihop ad hoc networks with communication among user devices and access points, where available, without the restriction that the user devices need to be within the radio range of access points. We first describe pure WiFi networks and their limitations. Next we discuss mixed networks based on WiFi and other wired and wireless technologies to provide robust city-scale networks. This chapter also explores security issues and vulnerabilities of wireless networks. An emerging application of WiFi ad hoc networks-RFID (radio frequency identification) networks based on the WiFi technology for warehouses and large retail stores-is presented. This chapter also presents another emerging application of WiFi-based networks: vehicular ad hoc networks for automobiles.

INTRODUCTION

Cellular and WiFi (wireless fidelity) are currently the most popular and actively pursued wireless technologies for consumer and business use. A cellular network consists of several base stations, each covering a small geographical region. Together the base stations cover a wide region such as a city. To be useful, the entire region of interest must be covered without gaps by these base stations. This requires billions of dollars of investment in network infrastructure consisting of wireless spectrum and base stations. An entirely different type of wireless network is made possible by the wireless fidelity (WiFi) technology. This technology (based on the IEEE 802.11 standard (IEEE Computer Society LAN/MAN Standards Committee, 1999)) enables wireless communication on an ad hoc basis. In the simplest configuration, a wireless access point (denoted, *hotspot*) can be used to share an Internet connection among several tens of users in a small area such as a conference room or a coffee shop. The network connectivity is limited to the radio range (about 50 meters) of the hotspots, and communication among users must go through the access point. Such networks, called wireless local area networks (WLANs), are already extensively used by businesses and academic campuses. In this chapter, we focus on multihop ad hoc networks with communication among user devices and access points, where available, without the restriction that the user devices need to be within the radio range of access points.

The WiFi technology is inexpensive due to two factors: (a) the use of free, unlicensed radio spectrum at 2.4 GHz and 5.8 GHz bands obviates heavy investment in private, dedicated radio spectrum, and (b) the widespread use of WiFi equipped PDAs, laptops, and even phones provides significant business opportunities and justification for deploying ad hoc WiFi networks. Since all WiFi devices must comply with the IEEE standard, WiFi products from multiple vendors can be mixed and matched for seamless operation. This has driven the cost of individual WiFi devices low, which in turn, made deployment of WiFi-based networks covering medium to large areas an attractive and, even necessary, business investment.

In this chapter, we describe various trends in the design and deployment of wireless networks based on WiFi and other technologies (Gast, 2005; Macker & Corson, 1998; Murthy & Manoj, 2004). The rest of the chapter is organized as follows. The section "Ad Hoc Wireless Networks" describes pure WiFi networks and their limitations. "Mixed Wireless Networks" describes mixed networks based on WiFi and other wired and wireless technologies to provide robust city-scale networks. "Security in Wireless Networks" describes security issues and vulnerabilities of wireless networks. "RFID Wireless Networks" describes an emerging application of WiFi ad hoc networks: RFID (radio frequency identification) networks based on the WiFi technology for warehouses and large retail stores. "Vehicular Ad Hoc Networks" describes another emerging application of WiFi-based networks: vehicular ad hoc networks for automobiles. The "Summary" section summarizes the chapter.

AD HOC WIRELESS NETWORKS

An ad hoc wireless network is an impromptu network formed by several wireless devices, such as PDAs, laptops, and phones, without relying on an existing network infrastructure (Perkins, 2000). These devices (denoted as nodes) may be mobile and use a common wireless technology such as WiFi. To facilitate communication among the nodes that are not directly in the radio range of one another, the other nodes act as intermediate routers, just like routers in the Internet. Such networks are useful in military combat situations, where a group of soldiers must be connected to exchange information, or in emergency rescue operations, where there is no network infrastructure or the existing infrastructure has been destroyed. Because of frequent topology changes due to node mobility and due to wireless interference, the existing networking software used for the Internet is not suitable for these ad hoc networks. Consequently, extensive research on routing protocols and transport protocols has been conducted to make ad hoc networks suitable for general-purpose use.

Routing protocols are broadly classified into proactive and reactive protocols. A proactive protocol keeps track of all possible routes within the ad hoc network, and disseminates routing information with others, periodically, or whenever network topology changes significantly. On the other hand, reactive or on-demand routing protocols learn and maintain active routes only. When a new route is needed for a new connection, source of the connection broadcasts, network-wide, a route request (RREQ). The intended destination responds by a route reply (RREP) containing the path information. This process is called route discovery. To minimize the number of transmissions and speedup route discovery, the intermediate nodes that have the requested route may respond to a RREQ.

The two most commonly used transport protocols are user datagram (UDP) and transmission control (TCP) protocols. For compatibility reasons, an ad hoc wireless network must support these protocols. However, TCP is tuned for use on wired networks and does not work well for multihop wireless networks.

Despite several years of research, the performance of current multihop wireless networks is unpredictable. To illustrate the performance issue, we present in Figure 1 the overall network throughput of an 8-node ad hoc network with 7 constant bit rate (CBR) connections, representative of voice over IP traffic, over UDP transport layer (Boppana, 2006). Even though nodes are stationary, the performance varies widely with time, owing to noise and interference caused by transmissions in the network.

Mobility makes it harder to sustain performance in an ad hoc wireless network. To illustrate the performance issues further, we simulated a 50-node mobile ad hoc network in a 1,000 m x 1,000 m field using the ns-2 network simulator (Fall & Varadhan, 1997). Each node has a transmission range of 250 m and nodes move in random directions with an average speed of 10 m/s (22.5 miles/hour) and a top speed of 20 m/s. We used a 2 Mbps channel rate for easier analysis. (The current WiFi technology provides various channel rates ranging from 1 Mbps to 54 Mbps, though in practice, the higher rates are used only when the communication nodes are close to each other.)

First, we illustrate the available bandwidth (BW) for a connection without taking any contention or interference for wireless channels (Dyer, 2002). The available BW is based on the number of hops required to reach from a specified source node to its destination node.

Next, we present performance of this network under CBR traffic load. We varied the load from

Figure 1. Throughput of a wireless ad hoc network with stationary nodes. The vertical bars indicate throughputs over 1-second intervals. The horizontal line indicates the throughput averaged thus far. The ad hoc network is built using off-the-shelf Linksys 54G routers reprogrammed with Linux operating system. Ad hoc on demand distance vector (AODV) routing protocol is used to discover and maintain routes.

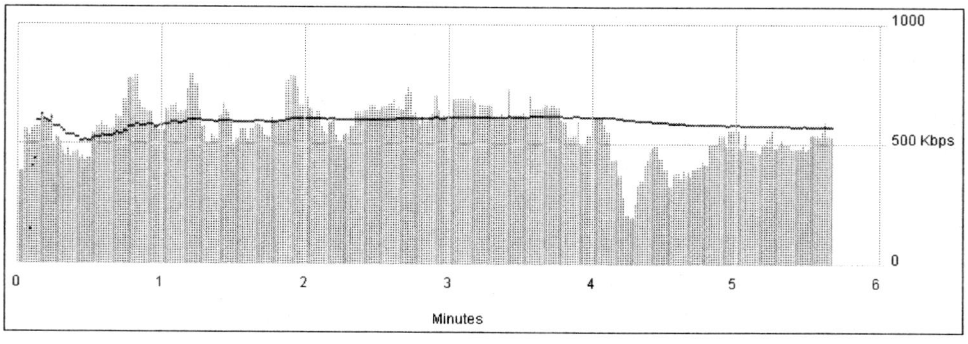

Figure 2. Capacity of a single transport-layer connection in a simulated ad hoc network. Owing to the nature of shared transmission space, the capacity of a connection varies inversely proportional to the number of hops from sender to receiver.

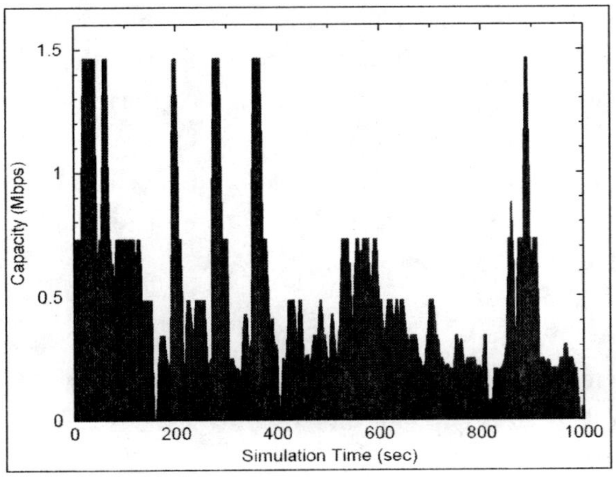

Figure 3. Packet delivery rates of various ad hoc network routing protocols

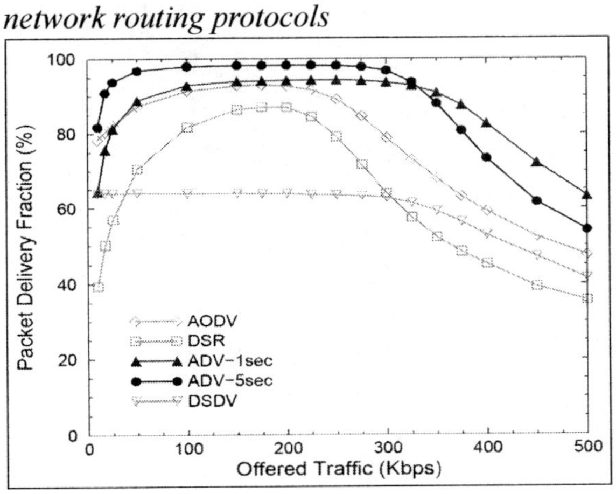

very low to very high, gradually, and measured the performance of the network. We simulated four recent routing protocols: destination sequenced distance vector (DSDV) (Perkins, 2000), adaptive distance vector (ADV) (Boppana & Konduru, 2001), ad hoc on demand distance vector (AODV) (Perkins, Belding-Royer, & Das, 2003), and dynamic source routing (DSR) (Johnson, Maltz, & Hu, 2003). The delivery rate (fraction of injected packets that are delivered to destinations) for this network with various routing protocols is indicated in Figure 3. DSDV and ADV are proactive routing protocols and AODV and DSR are on demand routing protocols. Two variants of ADV are shown based on the amount of time a packet is buffered within a node when there is no route. It is clear that the ability of a mobile ad hoc network to deliver depends greatly on the traffic load, type of routing protocol used, and choice of parameter values (such as buffer time).

Despite these performance issues, ad hoc networks are likely to be the dominant form of local area networks used in future for several reasons.

- Technology developments will make the basic WiFi protocol robust and improve nominal speeds further (Varshney, 2003). Recently, evolving WiFi technology based on MIMO antennas is shown to sustain higher data rates than a fast ethernet.
- Extensive ongoing research on networking software will result in better routing and transport protocols that will exhibit better performance characteristics (Boppana & Zheng, 2005; Desilva, 2004; Dyer, 2002).
- There are no alternatives to mobile ad hoc networks for military combat situations. In fact, Department of Defense (DoD) is one of the early and largest funding agencies for research in this area.
- Wireless networks streaming audio and video will be ubiquitous in consumer homes. Ad hoc networks are particularly attractive because they require no new wiring and satisfy location and space constraints easily (IEEE CCNC, 2006). Already, many consumers with high-speed broadband access have a WiFi-based network (in infrastructure mode using one access point or in multihop mode using additional WiFi extender devices) connecting multiple laptops wirelessly

within their homes. Apple's Airport Express is a commercial product designed to stream audio over WiFi channels. The newer WiFi technology based on IEEE 802.11n or the ultra wideband (UWB) wireless technology will likely be used for high-resolution video streaming due to higher BW offered by this technology. However, UWB will be used to complement WiFi networks rather than replace them.

- WiFi based ad hoc networks are the starting point to other types of networks, such as RFID networks and vehicular ad hoc networks (VANETs) (IEEE CCNC, 2005).

In the next two sections, we address twin deficiencies of ad hoc networks: predictable performance and security. First, we describe how to make the performance of WiFi networks robust, and then how to address some of the security issues that require attention in wireless networks.

MIXED WIRELESS NETWORKS

Given the weaknesses of ad hoc wireless networks, the area covered by them tends to be small. Instead, mixed networks consisting of fixed infrastructure nodes and mobile user nodes are suitable for a medium-range network spanning, for example, a metropolitan area (Boppana & Zheng, 2005). Point-to-point wired, cellular, or WiMAX (based on the IEEE 802.16 standard (IEEE 802.16, 2004) for metropolitan area wireless networks) wireless links among fixed nodes and wireless links for all nodes can be used for connectivity. These networks take advantage of reliability and high bandwidth of wired infrastructure backbone, and flexibility and low cost of wireless links using ad hoc networking concepts. Because these networks make use of ad hoc networking, there is no need for fixed nodes to cover all the desired area. If a fixed node is unavailable as a neighbor, a mobile node can send its data through other mobile nodes to the destination or to the nearest fixed node.

We illustrate this with an example network shown in Figure 4. This network has several mobile nodes that can communicate only via WiFi links and several relatively stationary nodes (denoted, infrastructure nodes) with point-to-point (p2p) links among them. A network of this type can provide multiple paths among user nodes. For example, node 8 in the upper left portion of the network can go through 12 and 13 or A and C to reach node 16. Ad hoc routing is used in cases when a user node is not near an infrastructure node. For example, node 10 can reach node 4 via node 6.

With the advent of new technologies, it is feasible to design such mixed networks. The WiFi is a popular short haul (for distances less

Figure 4. A mixed network with mobile user and fixed infrastructure nodes, denoted by circles and diamonds, respectively. The infrastructure nodes are interconnected by point-to-point links, denoted by dashed lines, for infrastructure support and to provide multiple paths. All nodes are capable of using a common wireless technology, such as WiFi. The radio range of infrastructure nodes is indicated by a circular shaded region.

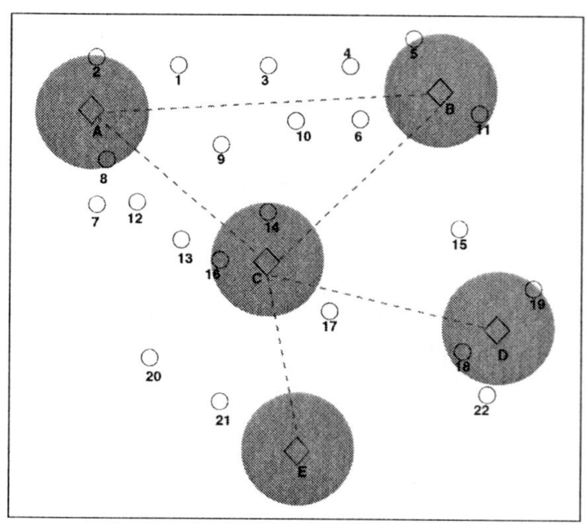

than 300 m) wireless link protocol. The fixed infrastructure nodes and p2p links among them are not difficult to set up. The p2p links can be wired links or long-haul wireless links. For example, the new IEEE 802.16 (IEEE 802.16, 2004) is an example of long-haul (for distances less than 10 Km) wireless link protocols. The infrastructure nodes can be already-existing fixed nodes connected via p2p links (for example, access points connected to the Internet), or semi-permanent nodes that remain stationary for a few hours and have p2p links implemented using a different wireless technology. More importantly, elaborate design and implementation to ensure complete geographical coverage by fixed nodes is not necessary, since gaps in the coverage can be managed using ad hoc networking, provided there is enough node density.

To see the performance benefits of mixed networks, we simulated a 60-node network in a 1,500 m X 1,500 m field. We used the Glomosim network simulator (Zeng, Bagrodia, & Gerla, 1998). There are 0, 4, or 9 fixed nodes, and the remaining nodes are mobile with speeds ranging from 1 to 29 m/s. The nominal WiFi link speed is 2 Mbps, and p2p links are full-duplex 2 Mbps. The fixed nodes are placed in a grid pattern, and only adjacent fixed nodes are connected to each other by a p2p link. We used ADV and AODV as the routing protocols for the pure ad hoc network (the network with 0 fixed nodes), and ADV static (ADVS), modified version of ADV to take advantage of p2p links where possible, for the other two networks. ADVS*n*F indicates the performance of mixed network with *n* fixed nodes. The delivery rates are given in Figure 5. (See Boppana & Zheng, 2005 for more information.)

Adding a few p2p links (12 links with 9 fixed nodes) improves the delivery rate and overall performance of the network significantly. It is even more illustrative to see the delivery rate, throughput, and packet latencies of a 1,000-node network in a 6 Km × 6 Km area. There are 0, 9, or 25 fixed nodes placed in a grid pattern with only adjacent fixed nodes connected by p2p links. The network with 0 fixed nodes denotes the pure ad hoc network.

These results clearly illustrate the performance benefits of using different link technologies in a mostly WiFi based ad hoc network. They also offer unique business opportunities (Markoff, 2006).

- Mixed networks are easy to set up since the number of stationary nodes required is small

Figure 5. Performance of 60-node mixed and pure ad hoc networks

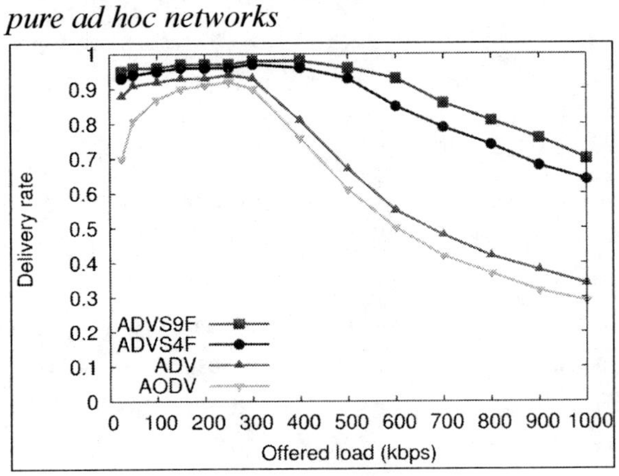

Figure 6. Delivery rates of 1,000-node mixed networks

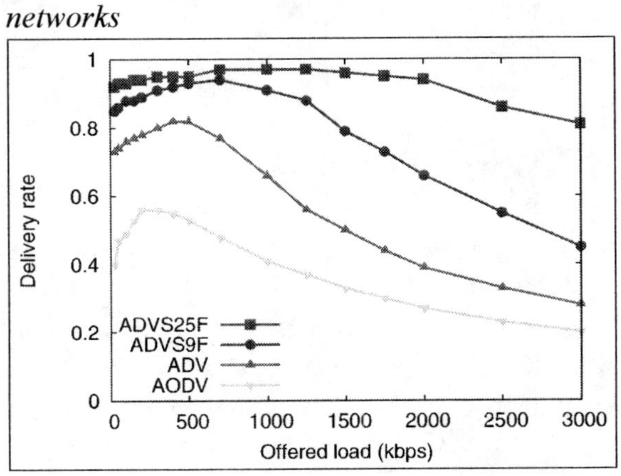

Figure 7. Throughput of 1,000-node mixed networks

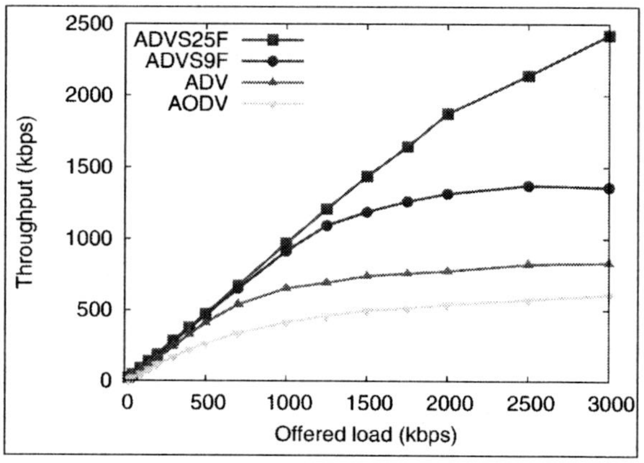

Figure 8. Packet latencies for 1,000-node mixed networks

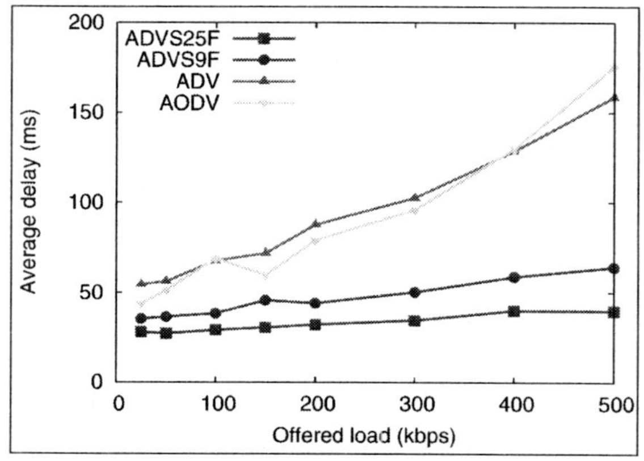

(2.5% in the 1,000-node network example). Owing to the use of ad hoc networking concepts, they are not likely to suffer the irksome gaps that are common in cellular networks. In fact, the existing cellular networks can be improved using these techniques. Several cellular networking companies are actively pursuing this type of networks to complement cellular networks.

- They lower the cost of setting up a metropolitan area network to the extent that citywide organizations, such as municipal government agencies or delivery service companies, can set up their own mixed network to provide wireless broadband access without having a telecom company as the carrier.

SECURITY IN WIRELESS NETWORKS

Besides performance, security is an important issue in wireless ad hoc networks. The traditional security issues on the Internet are keeping data confidential and unaltered. The most common solution is to encrypt the data by the source application and decrypt it by the destination application (Schneier, 1996). Intermediate nodes cannot examine the contents and cannot alter it without being detected by the destination. Since only some applications need it, this is implemented as an end-to-end solution (that is, the host computers or applications at both ends of a connection invoke and manage the security features). These techniques are also applicable to secure data communication on wireless networks.

In this section, we address a different type of security problem: crippling the network with false route information (Hu, Perrig, & Johnson, 2002; Marti, Giuli, Lai, & Baker, 2000; Zhou & Haas, 1999;). These attacks are on the control traffic rather than data traffic. Wireless networks are more susceptible to this type of attack than wired networks for two reasons: (a) physical access to a network port is not necessary with wireless networks, (b) peer dissemination of routing information and network topology leads to highly leveraged, hard to detect hacker attacks on wireless networks. The issue of unauthorized access to network can be addressed using wireless link-level encryption and decryption (wireless protected access or WPA) and server-based authentication (Varshney, 2003). We describe the second issue in detail.

The attacks on control traffic or routing protocol can be classified into two categories.

- Denial of service (DoS) or resource consuming attack
- Falsifying routes and dropping/delaying data packets

We first describe the impact of the DoS attack. In a routing protocol such as AODV, route discoveries depend on network-wide dissemination (called flooding) of RREQ control packets from a source node seeking route to its destination node. A RREQ broadcasted by a source is rebroadcasted by its neighbors to their neighbors. This is repeated until the destination receives a copy of this RREQ and responds with an RREP control packet that establishes the route between source and destination. A single RREQ broadcasted by a source node results in up to (n-1) additional broadcast transmissions in the wireless network, where n is the number of nodes in the network. This feature can be exploited by a malicious node to launch highly leveraged denial-of-service attacks in mobile ad hoc networks. These malicious nodes behave like the normal nodes in all aspects except that they initiate frequent control packet floods. This is hard to detect since any normal node with frequently broken routes could legitimately initiate frequent route discoveries.

Figure 9. Loss of throughput with bogus route discoveries by a malicious node in a 100-node mobile ad hoc network. The offered CBR load to network is kept constant at 300, 400, or 500 Kbps, and the throughput achieved is measured as a function of attack rate by the malicious node. The attack rate of zero RREQs/second denotes the normal network.

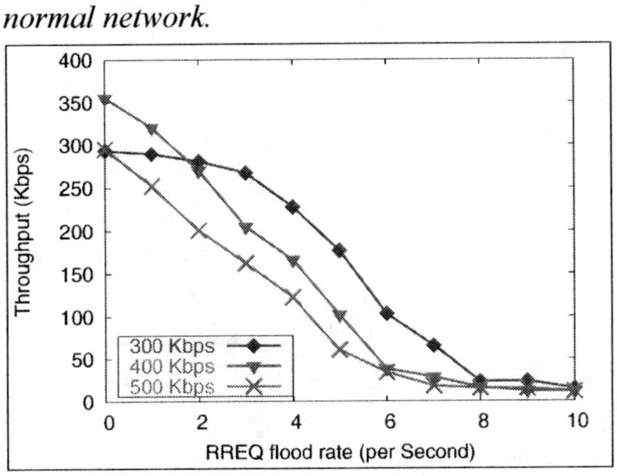

Figure 9 shows the loss of throughput in a 100-node mobile ad hoc network with AODV as the routing protocol and one malicious node initiating routing attacks. Even 1 RREQ/s by the malicious node causes measurable drop in throughput (Desilva & Boppana 2005).

Fortunately, a simple and inexpensive solution to this problem exists. Using statistical profiling of control activity of other nodes, each node can independently determine overactive nodes and effectively shut them off from causing permanent damage to network performance (Desilva & Boppana 2005). With this solution implemented, the performance of the network under attack is shown in Figure 10. Regardless of the attack rate, the normal network throughput is sustained.

Figure 10. Effectiveness of statistical profiling in the example ad hoc network under DoS attack

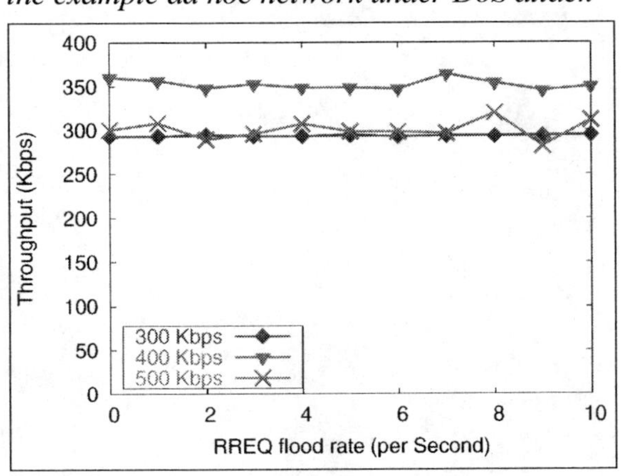

The other type of security attack on routing protocol is based on falsifying routes by the malicious node in order to place itself in the path of an active route. This often involves the malicious node claiming a better route than any other node to reach a destination. Data packets received on this route are dropped or delayed arbitrarily by the malicious node. This type of attack is called the blackhole attack. The impact of such attacks can be severe on network performance.

Figure 11 illustrates the impact of a blackhole attack by five malicious nodes in a 100-node network with AODV routing protocol. The malicious nodes send false RREPs in response to 1% of RREQs they hear. The detection of such an attack is expensive.

The proposed solutions to mitigate such attacks use hashing and symmetric cryptographic techniques (Hu et al. 2002; Zhou & Haas 1999). This makes the solution even more expensive than the attack itself since each control packet must be verified. Further research is needed to develop efficient solutions to these damaging, but low frequency attacks.

Figure 11. Impact of blackhole attack over time n a 100-node mobile ad hoc network. There are five malicious nodes sending false RREPs to 1% of RREQs they hear. The attack starts at 200 seconds and stops at 800 seconds. The traffic load is 200 Kbps. The two thick lines (at 200 and 800-second periods) indicate the start and end of the attack. Vertical bars indicate delivery rates for 5-second periods. The jagged horizontal line indicates the average network throughput from the most recent event—start of simulation, start of attack, or end of attack.

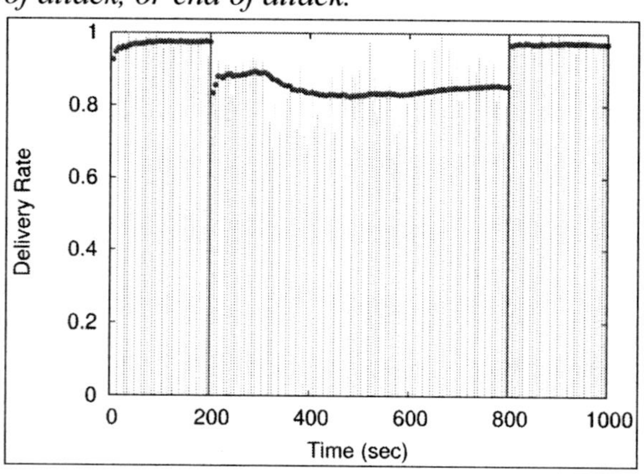

RFID WIRELESS NETWORKS

Several organizations, including Wal-Mart and Proctor & Gamble (P&G), are currently testing and deploying radio frequency identification (RFID) technology in their supply chains. In addition, the Department of Defense has mandated that its suppliers tag their products at the pallet level using RFID tags. The potential advantages of RFID technology in the supply chain are numerous. RFID technology has the ability to provide up-to-the-minute information on sales of items, and thus can give an accurate picture of the inventory levels. This accuracy may lead to reduction in inventory levels, thus causing a reduction in inventory costs. RFID technology at the pallet level has the potential to automate the distribution of goods in the supply chain between manufacturing plants, warehouses, and retail stores of different organizations, which in turn might reduce labor costs. RFID tags allow companies to identify all items, thus cutting down losses from lost or misplaced inventory.

For the purposes of this section, we assume the supply chain is comprised of the manufacturer, distributor, retailer, and the consumer. As an item with an RFID tag moves from one location to another location in the supply chain, it may be read at several different locations in the supply chain. We define an *RFID transaction* to be an event that corresponds to the reading of an RFID tag by an RFID reader. Each RFID transaction generates data including the RFID tag (EPC), the reader id, and other relevant pieces of information.

The transition of an item with an RFID tag from the manufacturer to the consumer is depicted in Figure 12. In this paper, we assume that the RFID tags are applied at the item, case, and pallet level. For some items, this hierarchy-items in cases and cases in pallets-may not be applicable, and for some items this hierarchy may need to be changed. However, the discussion in this chapter can be readily extended to other hierarchies. As an item is manufactured, an RFID tag is placed on the item, which generates the item creation RFID transaction at the manufacturing facility. Placing an item into a case, placing the case into a pallet, as well as loading a pallet into a delivery truck generate different RFID transactions at the manufacturing facility. At the distributor's warehouse, placing the pallet into a warehouse shelf, and loading the pallet onto a delivery truck (to be delivered to the retail store) generate RFID transactions. In a retail store, events such as shelf replenishment, movement of an item from one shelf to another (possibly because of item misplacement), and sale of an item generate RFID transactions. At the consumer's home, a futuristic model suggests that the consumer's refrigerator (or the storage area if the item does not need to be refrigerated) will be equipped with an RFID tag reader; this results in RFID transactions being generated when an item is placed in the refrigerator and when an item is taken out of the refrigerator, these events possibly triggering a refrigerator replenishment RFID transaction. For more information on RFID transactions and RFID tag reader designs, the reader is referred to Chalasani and Sounderpandian (2004) and Chalasani, Boppana, and Sounderpandian (2005). In each of the facilities (manufacturing facility, the distributor's warehouse, the retail store, and the consumer's home) the tag readers are interconnected, as shown in Figure 13. Given the current advances and adaptation trends, the Wi-Fi is likely to be the dominating and economical wireless technology for intertag reader communication. Though there are no commercial implementations, intertag reader network will alleviate communication and performance load on the back-office computer system. In this section, we explore two representative designs of RFID tag reader networks (or simply RFID networks) (Passmore, 2004; RFID.com, 2006).

One may use wired Ethernet (wired) links or wireless links to set up RFID networks. Instead, using only wired or wireless links, mixing both technologies may be advantageous. Combining

Figure 12. Transition of an item from the manufacturer to the consumer in the supply chain and the relevant RFID transactions

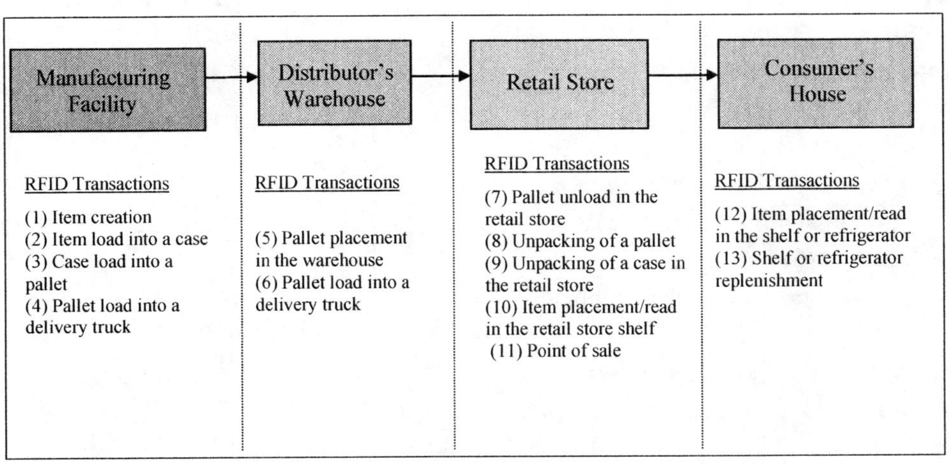

Figure 13. An RFID ad hoc tag reader network; ovals represent tag readers, while squares indicate the RFID tags on items, cases or pallets

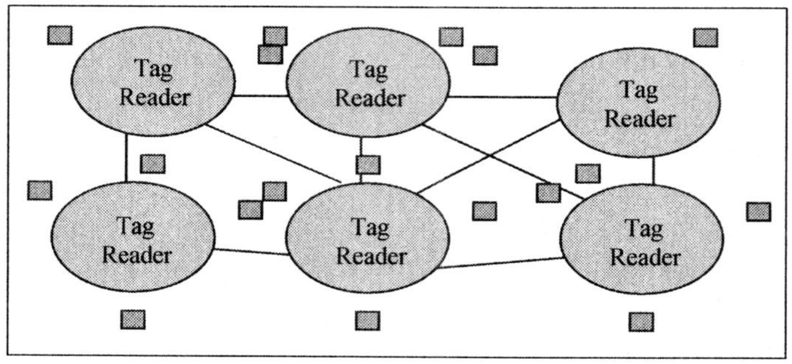

wired and wireless network technologies to interconnect tag readers (denoted, nodes) provides the benefits of wired network robustness and low-maintenance costs with the flexibility of wireless network for adapting to changing needs. In this type of network, all nodes have wireless capability to communicate among themselves. In addition, some of the nodes have Ethernet connections. There are two possible scenarios.

- Most of the nodes have both wireless and Ethernet connectivity, while the remaining nodes communicate via wireless links only. This scenario models the situation where most of the tag readers are stationary and interconnected by Ethernet. There are a few temporary tag readers that connect to the tag reader network via wireless links only.
- Most of the nodes have wireless capability only and the other nodes have both wireless and Ethernet connectivity. This scenario models the situation where most of the tag readers have only wireless capability and a few of the nodes are interconnected by a different technology and link type to improve connection reliability and performance.

We simulated both types of networks using the Glomosim simulator. We used 81 nodes arranged in a 9 × 9 grid in a 1-sq.-Km field, as shown in Figure 14. The distance between adjacent nodes was 100 meters, and wireless radio transmission range was 125 m. The traffic load consisted of 500-byte packets generated at constant bit rate (CBR). All nodes were stationary for the duration of the simulation (600 seconds). All wired connections were in a near neighbor mesh pattern, in which

Figure 14. Layout of RFID tag reader network. Adjacent nodes are separated by 100 m. The size of the field is 1 Km × 1 Km. Black (dark shade) nodes are the 9 wireless nodes in Scenario A and 9 nodes with wired connections among them in Scenario B. Light shade nodes are the 4 nodes with wired connections among them in the Scenario B.

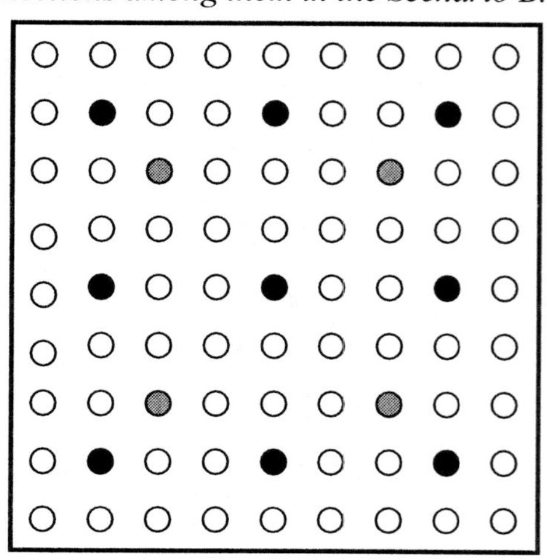

each node was connected to its adjacent nodes in the grid. The maximum link speed was 2 Mbps (million bits/second) for both wired and wireless links. Though these rates are not representative of what is currently available, the simulation results will be helpful in understanding performance degradation with wireless links in Scenario A and performance improvement with wired links in Scenario B. We used uniform traffic pattern in which each node sends data to all the other nodes with equal probability. We varied the offered load to the network by varying the time between consecutive packets generated by nodes.

For Scenario A, we compared the performance of two networks: one with all wired connections, and another with 9 of the nodes (indicated in black in Figure 14) spaced evenly in the grid pattern with only wireless connectivity. Static routes are used since most of the nodes are wired and mobility is not a practical option. Figure 15 gives the performance impact of wireless vs. wired connections. Up to certain loads, the 9 nodes with only wireless links do as well as they would with wired connections. Beyond that, the wireless transmission interference (interference range is 2-3 times that of valid reception range) limits achievable throughput. For the data we present, loads up to 4 packets/second/node (20 Kbps/node) can be handled by the nodes with wireless links.

For Scenario B, we simulated three different network configurations: no wired links (denoted, 0f), wired links among 4 nodes (denoted in blue in Figure 14) that form an equidistant 2x2 grid (4f), and wired links among 9 nodes (black nodes in Figure 14) that form a 3x3 grid (9f). Though the nodes are stationary during the simulation, the primary reason for using an all-wireless or mostly wireless network is the flexibility of relocating tag readers as needed; to adapt to such network topology changes, a dynamic ad hoc routing protocol must be used. In the two configurations with static links, 4 or 12 wired links with 2-Mb/s data rate are used. Figure 16 shows the throughputs of these three networks. While adding 4 wired links to an all wireless network provides about 20% improvement in peak throughput, 12 wired links provide as much as 67% improvement. Since most of the nodes are wireless only, the network is very flexible.

It is noteworthy that the results for Scenario B are not directly comparable to those for Scenario

Figure 15. Impact of wireless links in an all-wired network; throughputs of nine nodes placed in a 3 × 3 grid pattern are examined

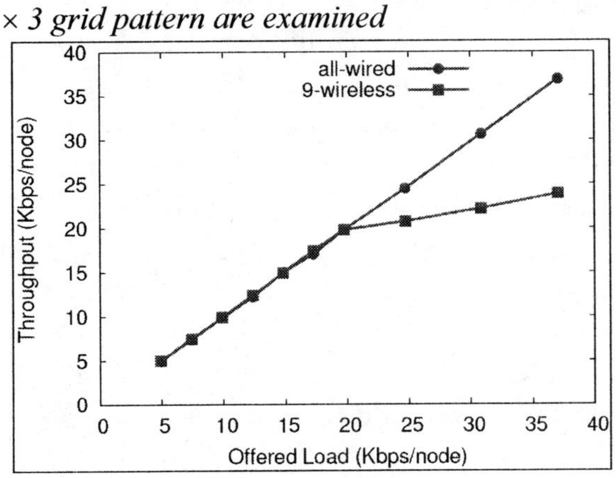

Figure 16. Impact of wired links in an all-wireless network

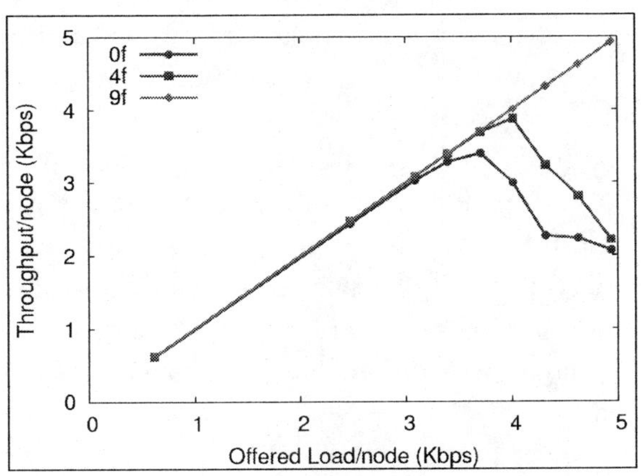

A since we used static routes in the latter case. For comparison purposes, we simulated the two, wired networks with no static routes and ADV as the dynamic routing protocol. While the all-wired network is not affected, the performance of 9-wireless nodes case is reduced to about 12 Kbps/node, about 140% higher than that of the 9f case in Scenario B. But this performance comes at a cost of wiring 72 nodes vs. only 9.

Using WiFi technology for warehouse networks is inevitable and beneficial. RFID networks represent good examples of using the technology, initially considered suitable for military and emergency civil applications, to improve business productivity.

VEHICULAR AD HOC NETWORKS

Vehicular ad hoc networks (VANETs) are exciting and rapidly growing examples of ad hoc networking in practice (ACM Sigmobile, 2005; IEEE CCNC, 2005). The US FCC allocated 75 MHz of spectrum in the 5.8-GHz band for dedicated short-range communication (DSRC) in VANETs (FCC, 2003). A new wireless standard, wireless access for vehicular environments (WAVE), denoted IEEE 802.11p and based on WiFi, is being developed. Compared to the ad hoc networks described earlier, there are significant differences in the use and demands placed on VANETs.

Two types of networking concepts will be used to network moving vehicles: (a) One-hop wireless connectivity from a vehicle to a roadside wireless access point (V2R) and (b) multihop connectivity among vehicles (V2V). V2R connectivity is similar to Internet connectivity using WiFi hotspots with mobility added. V2V connectivity will use ad hoc networking concepts similar to mixed networks described earlier. Since vehicular movement is predictable, VANETs can be designed to perform reliably and efficiently under network overload.

VANETs will be used to improve driver safety in addition to providing Internet connectivity to users. Consequently, VANETs will be required to support two classes of applications: safety and nonsafety applications. Emergency road conditions, medical facility location, and passage of emergency vehicles, such as ambulances, fire trucks, and police cars, are examples of safety applications. Access to Internet, streaming multimedia to consumers in vehicles, road congestion advisory, and information of nearby service facilities, such as restaurants and stores, are examples of nonsafety applications. To accommodate both types of applications, multiple channels, with some of them dedicated for safety applications and the others for applications in the order of priority, will be used. In contrast, WiFi is designed to use one channel for connectivity; multiple radios need to be used for multichannel operation in MANETs and hotspot networks.

VANETs differ from MANETs in terms of energy constraints and security. Since the radios will be powered by the batteries and engines in vehicles, there will be few energy constraints. On the other hand, security and disruption to network connectivity will be more problematic compared to MANETs. Security is more problematic since hackers with unlimited power and without the need to physical access to a network port can launch the type of attacks currently used on the Internet more easily. Also, since VANETs use wireless links, which have less BW than the wired links used for the Internet, denial-of-service attacks can be devastating. Also, lack of energy constraints mean radios can use high power levels for transmissions (intentionally or inadvertently), which will cause excessive radio interference and reduce V2V network performance.

VANETs provide significant new business opportunities for high-technology companies, service providers, and local businesses. There are significant impediments as well. Involvement of the automobile industry will slow down the process of developing and deploying new technologies for VANETs. A good example is the audio system offered in new automobiles. While the audio players in the consumer market advanced considerably, most manufacturers continue to offer 20-year-old technology as standard equipment, and do not provide any option to connect user equipment. Owing to the intended use of VANETs, there will be significant regulatory constraints by federal, state, and local government agencies.

SUMMARY

Multihop wireless networks based on WiFi technology offer flexible and inexpensive networking possibilities for various purposes ranging from personal networks within consumer homes to citywide departmental networks to wide area vehicular ad hoc networks. While the business significance of citywide ad hoc networks may not be clear in developed countries, these networks will play a crucial role in reaching consumers in rural areas in developing countries. RFID networks improve the efficiency and productivity in the areas of manufacturing and distribution. The biggest and most profitable area for business opportunities may very well be vehicular ad hoc networks. Though the potential of wireless networks seems limitless, the technology available today does not provide dependable network performance of reliability. Extensive research is being conducted on improving the network software and hardware. The future wireless standards, such as IEEE 802.11n, which uses advanced MIMO antennas and new routing and transport protocols, will facilitate designing ad hoc networks that will be suitable for consumer use.

ACKNOWLEDGMENT

Rajendra Boppana's research was partially supported by NSF grants EIA-0117255 and AIA grant F30602-02-1-0001. Suresh Chalasani's research was supported in part by summer research grants awarded by the University of Wisconsin system.

REFERENCES

ACM Sigmobile. (2005). *ACM International Workshop on Vehicular Ad Hoc Networks*. New York: ACM Press.

Boppana, R. V. (2006). *On setting up a WiFi ad hoc network testbed*. Retrieved December 2006, from http://www.cs.utsa.edu/faculty/boppana/projects

Boppana, R. V., & Konduru, S. P. (2001). An adaptive distance vector routing algorithm for mobile, ad hoc networks. In *Proceedings of IEEE Infocom*. IEEE.

Boppana, R. V., & Zheng, Z. (2005). *Designing ad hoc networks with limited infrastructure support*. Presented at the IEEE Consumer Communication and Networking Conference (CCNC). IEEE.

Passmore, D. (2004). RFID: Network implications, *Business Communications Review*. Retrieved December 2006, from http://www.bcr.com/bcr-mag/2004/11/p16.php

Chalasani, S., Boppana, R. V., & Sounderpandian, J. (2005). RFID tag reader designs for retail store applications. In *Proceedings of the 11th Americas Conference on Information Systems (AMCIS)*. Association for Information Systems.

Chalasani, S. & Sounderpandian, J. (2004). RFID for retail store information systems. In *Proceedings of the 10th Americas Conference on Information Systems (AMCIS 2004).* Association for Information Systems.

Desliva, S. A. (2004). *Techniques to mitigate traffic overload and protocol inefficiencies in mobile ad hoc networks.* PhD dissertation, CS Department, University of Texas at San Antonio.

Desilva, S. & Boppana, R. V. (2005, March). *Mitigating malicious control packet floods in ad hoc networks.* Presented at the IEEE Wireless Communications and Networking Conference.

Dyer, T. D. (2002). *Design and analysis of adaptive routing and transport protocols for mobile ad hoc networks.* PhD dissertation, CS Department, University of Texas at San Antonio.

Fall, F. & Varadhan, K. (1997, November). *NS notes and documentation.* The VINT Project, UC Berkeley, LBL, USC/ISI, and Xerox PARC. Retrieved from http://www-mash.cs.berkeley.edu/ns

Federal Communications Commission. (2003, December 17). *Amendment of the Commission's Rules Regarding Dedicated Short-Range Communication Services in the 5.850-5.925 GHz Band.* FCC 03-324.

Gast, M, (2005). *802.11 wireless networks: The definitive guide* (2nd ed.). O'Reilly Media, Inc.

Hu, Y.-C., Perrig, A., & Johnson, D. B. (2002). Ariadne: A secure on-demand routing protocol for ad hoc networks. In *Proceedings of the 8th ACM International Conference on Mobile Computing and Networking.* New York: ACM Press.

IEEE CCNC. (2005). *The automobile as a network interface.* Presented at the Technology Application Panel, Session 3, IEEE Consumer Communications and Networking Conference.

IEEE CCNC. (2006). *Proceedings of the IEEE Consumer Communication and Networking Conference 2004-2006.* IEEE.

IEEE Computer Society LAN/MAN Standards Committee. (1999). *Part 11: Wireless LAN, medium access control (MAC) and physical layer (PHY) specifications* (Standard ANSI/IEEE 802.11).

IEEE 802.16 Working Group on Broadband Wireless Access Standards. (2004). *IEEE 802.16 standard: WirelessMAN standard for broadband wireless metropolitan area networks* (Standard ANSI/IEEE 802.16).

Johnson, D., Maltz, D., & Hu, Y. (2003). *The dynamic source routing protocol for mobile ad hoc networks.* IETF MANET Working Group, Internet Draft 2003.

Macker, J. P., & Scott Corson, M. (1998). Mobile ad hoc networking and the IETF. *ACM Mobile Computing and Communications Review, 2*(1). New York: ACM Press.

Markoff, J. (2006, February 6). Venture for sharing Wi-Fi draws big-name backers. *New York Times.* Retrieved December 2006 from http://www.nytimes.com/2006/02/06/technology/06mesh.html

Marti, S., Giuli, T., Lai, K., & Baker, M. (2000). Mitigating routing misbehavior in mobile ad hoc networks. In *Proceedings of ACM/IEEE International Conference on Mobile Computing and Networking* (pp. 255-265). New York: ACM Press.

Perkins, C. E. (2000). *Ad hoc networking.* Boston: Addison Wesley.

Perkins, C. E., Belding-Royer, E. M., & Das, S. R. (2003, July). *Ad hoc on demand distance vector (AODV) routing* (RFC 3561). IETF.

RFID-101.com. (2006). *Online guide to RFID technology and products.* Retrieved from http://www.rfid-101.com

Schneier, B. (1996). *Applied cryptography* (2nd ed.). New York: John Wiley & Sons.

Siva Rama Murthy, C., & Manoj, B. S. (2004). *Ad hoc wireless networks: Architectures and protocols*. Prentice Hall.

Varshney, U. (2003). The status and future of 802.11-based WLANs. *IEEE Computer, 36*(6), 102-105.

Zeng, X., Bagrodia, R., & Gerla, M. (1998). Glomosim: A library for parallel simulation of large-scale wireless networks. In *Workshop on parallel and distributed simulation* (pp. 154–161). IEEE.

Zhou, L., & Haas, Z. J.. (1999). Securing ad hoc networks. *IEEE Network Magazine, 13*(6).

This work was previously published in E-Business Process Management: Technologies and Solutions, edited by J. Sounderpandan and T. Sinha, pp. 155-174, copyright 2007 by IGI Publishing, formerly known as Idea Group Publishing (an imprint of IGI Global).

Chapter XVIII
An Immune Systems Approach for Classifying Mobile Phone Usage

Hanny Yulius Limanto
Nanyang Technological University, Singapore

Tay Joc Cing
Nanyang Technological University, Singapore

Andrew Watkins
Mississippi State University, USA

ABSTRACT

With the recent introduction of third generation (3G) technology in the field of mobile communications, mobile phone service providers will have to find an effective strategy to market this new technology. One approach is to analyze the current profile of existing 3G subscribers to discover common patterns in their usage of mobile phones. With these usage patterns, the service provider can effectively target certain classes of customers who are more likely to purchase their subscription plans. To discover these patterns, we use a novel algorithm called Artificial Immune Recognition System (AIRS) that is based on the specificity of the human immune system. In our experiment, the algorithm performs well, achieving an accuracy rate in the range of 80% to 90%, depending on the set of parameter values used.

INTRODUCTION

A sample dataset of 24,000 mobile phone subscribers is used for this study, in which 20,000 records are known to be second generation (2G) network customers and the remaining are 3G network customers. The target field is the customer type (2G/3G). About 75% of the dataset have the target field available and are used for training purposes; while the remaining quarter of the dataset has a missing target field and are meant for prediction.

The objective of the classification is to correctly predict as many 3G customers as possible from the prediction set and obtain insights on the characteristics of the existing 3G customers to be used as reference for their marketing strategy.

To perform this task, we use a novel algorithm called the Artificial Immune Recognition System (AIRS) proposed by Watkins, Timmis, and Boggess (2004), which is based on the processes in biological immune systems.

OVERVIEW OF AIRS

The Human Immune System

The function of the human immune system is to identify and destroy foreign invaders (antigens) which are possibly harmful to the body. It does this through an innate and nonspecific response (mediated by macrophages) and also with an adaptive and specific response (mediated by lymphocytes). An innate response is not directed towards any specific antigens, but against any invaders that enter the body. The adaptive response is mediated mainly by two types of lymphocytes, B-cells and T-cells. The AIRS approach is modeled based on the behavior of B-cells, hence only the behavior of B-cells will be described here. On the surface of each B-cell are receptors that are capable of recognizing proteins of a specific antigen. Through costimulation and suppression of each other, similar B-cells form networks that can recognize similar antigens.

When antibodies on a B-cell bind with an antigen, the B-cell becomes activated and begins to proliferate. Thus, it means that only B-cells which are able to recognize the invading antigen will proliferate and produce clones (a process known as clonal selection). New B-cell clones are produced which are exact copies of the selected B-cells, but then undergo somatic hyper-mutation to generate a wider range of antibodies, so as to be able to remove the antigens from the body. A small quantity of B-cells remains in the system after the invading antigens have been removed. These B-cells act as an immunological memory to allow the immune system to produce a faster response to similar antigens that might re-infect the body in the future.

The AIRS Algorithm

Processes in biological immune systems have inspired the design of AIRS. An artificial recognition ball (ARB) is used to represent a set of *identical* B-cells. The ARBs in the system will compete for B-cells in order to survive (in the evolutionary sense); therefore an ARB with no B-cell will be removed from the system. A fixed number of B-cells are allowed in the system, so the least stimulated ARB will not be able to get any B-cells, and will therefore be removed from the system. We will at times refer to a B-cell as an ARB, only because an ARB is simply a representation of many B-cells of the same specification. When a new training data record (antigen) is presented to the system, each B-cell is cloned in proportion to how well it has matched the antigen according to the principle of clonal selection. The mutation rate used in the cloning process is *inversely* proportional to how well it matches the antigen. During the mutation process, new clones undergo a process of somatic hyper-mutation, where each attribute of the clones is varied slightly to provide a wider range of response to the training data record. Eventually, the clone with the best fit to the presented antigen will be retained as a memory cell. The memory cells are retained in the system to provide faster response should the system become re-infected with similar antigens.

AIRS relies heavily on finding the similarity (or difference) between a pair of customer records, therefore a proper distance measure needs to be defined. Hamaker and Boggess (2004) conducted a survey of distance measures that can be used in conjunction with AIRS. Based on this survey, we use the *heterogeneous value difference metric*

(HVDM) as our distance measure. The HVDM measure uses the *Euclidean distance* measure for numerical data fields and a *value difference metric* (VDM) for categorical data fields. Consider x and y as two customer records that are going to be compared, each having G data fields and having one of C possible classifications (where $C = 2$, either being 2G or 3G). All numerical fields are normalized to the range of [0, 1]. The function $HVDM(x, y)$ returns the HVDM distance of x and y according to the formulas:

$$HVDM(x,y) = \frac{1}{\sqrt{G}}\sqrt{\sum_{g=1}^{G} hvdm(x_g, y_g)^2}$$

$$hvdm(x_g, y_g) = \begin{cases} \sqrt{vdm(x_g, y_g)} \\ |x_g - y_g| \end{cases}$$

g is categorical field
g is numerical field

$$vdm(x_g, y_g) = \sum_{c=1}^{C}(P(c \mid x_g) - P(c \mid y_g))^2$$

where $P(c|x_g)$ denotes the probability that a customer record has class c given the value x_g.

Based on our judgment, VDM is the most effective measure for categorical fields compared to other distance measures as presented in Hamaker and Boggess (2004) because it relies on the statistics observed in the training data, which more likely reflects the real-world situation. We will now present the algorithm (full details may be found inWatkins et al., 2004). The symbols that we will be using to define the algorithm are explained as follows:

- There are n antigens, and there are G data fields and *one categorical* class in each antigen. The class of the antigen may take the value of $\{1, 2, \ldots, nc\}$, where nc is the number of possible values in the target class.
- MC denotes the set of memory cells. mc represents an individual member of MC and $mc.c$ represents the class of the memory cell, while $mc.f_i$ represents the value of the ith feature (data field) in the memory cell.
- AB denotes the set of ARBs and ab represents a single ARB. $ab.c$ denotes the class and the ith feature of an ab respectively.
- ag represents an antigen (training data record), and $ag.c$ and $ag.f_i$ represent the class and value of the ith feature (training data field) in the antigen, respectively.
- $MC_c \subseteq MC$ denotes the set of memory cells with the class c.
- $ab.stim$ denotes the stimulation level of an ARB ab.
- $ab.resources$ denotes the number of resources (B-cells) currently held by an ARB ab.

The user-defined parameter values we used in our experiments to control the behavior of the algorithm are presented as follows:

- *numEpochs*: Number of passes through the training data (= **3**).
- *numResources*: number of B-cells (resources) allowed in the system (= **200**).
- *clonalRate*: Number of mutated clones that a given ARB is allowed to produce. An ARB *ab* is allowed to produce at most (*clonalRate* * *ab.stim*) new clones. This product is also used to control how many B-cells will be allocated to an ARB (= **10**).
- *mutationRate*: A value between [0, 1] that determines the likelihood that a given feature or class of an ARB will be mutated (= **0.1**).
- *hypermutationRate*: number of mutated clones that a given memory cell is allowed to inject into the system. A memory cell *mc* injects at least (*hypermutationRate* * *clonalRate* * *mc.stim*) mutated clones (= **10**).

- *distance*: The distance metric used (= ***Heterogeneous Value Difference Metric***).
- *ATS*: affinity threshold scalar, a value between [0, 1] that will be multiplied with the affinity threshold (*AT*) to produce the cutoff value for memory cell replacement (= **0.2**).
- *stimThreshold*: a value between [0, 1] used as a stopping criterion for training on specific antigen (= **0.95**).
- *k*: The parameters that specify how many nearest neighbors participate in the classification process (using *k*-nearest neighbor) (= **15**).

These parameter values were determined by performing multiple tests on the training data (using 10-fold cross validation) to verify their effectiveness. The training process of the algorithm is shown in Figure 1. Each phase will be discussed separately followed by the classification process.

The Initialization Phase

During this phase, all data preprocessing that is needed will be performed. Details of the data preprocessing involved will be elaborated in a later section. After the preprocessing, the affinity threshold (AT) of the system is calculated by averaging the distance between all training data. It should be noted that the affinity between two antigens should always be between 0 and 1, therefore:

Figure 1. Process flow of the AIRS training process

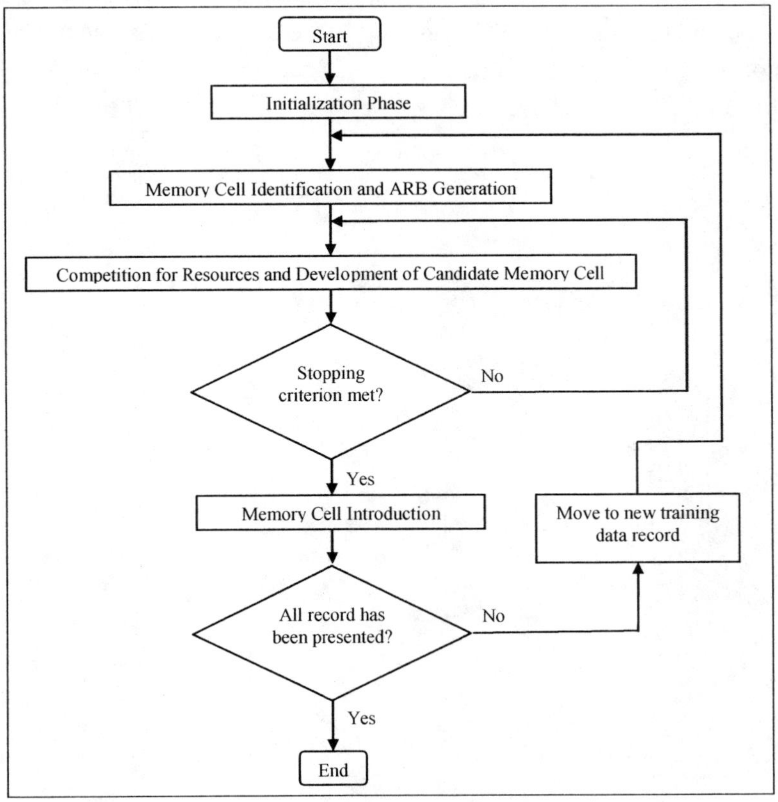

$$AT = \frac{\sum_{i=1}^{n}\sum_{j=i+1}^{n} affinity(ag_i, ag_j)}{{}^{n}C_2}$$

where *affinity*(*x*, *y*) will return the distance of the two antigens *x* and *y*, depending on which distance metric is used (HVDM for this specific case). Seeding of memory cells, when desirable, will also be performed in this phase.

Memory Cell Identification and ARB Generation

In this phase, the best matched memory cell is identified and ARBs are generated from the best matched memory cell that is identified. The first step is to identify the memory cell mc_{match} from $MC_{ag.c}$ (the set of memory cells with the same class label as the presented antigen *ag*) which has the highest stimulation. The level of stimulation between two records (antigens) *x* and *y* is defined as *stimulation*(*x*, *y*) = 1 – *affinity*(*x*, *y*). If $MC_{ag.c} = \{\}$ then $mc_{match} = ag$ and *ag* is added to $MC_{ag.c}$.

After mc_{match} has been identified, *NumClones* number of clones of mc_{match} will be produced as ARBs. The number of clones is *NumClones* = (*hypermutationRate* * *clonalRate* * mc_{match}.*stim*) where *stim* denotes the current stimulation level. It should be noted that *NumClones* is proportional to the stimulation value of mc_{match} to the antigen. Each numerical feature in mc_{match} can only mutate in a constrained range centered at the original value. The width of the range is inversely proportional to the stimulation value *stim*. This implies that a heavily stimulated mc_{match} will produce clones that are closer to it because it has a limited range for mutation, which is reasonable since the original mc_{match} already matches well with the antigen; hence it is very likely that best-fitted B-cell can be found in the neighborhood of mc_{match}. A categorical feature does not have this constraint and can be mutated freely. All of the produced clones are added into the system as ARBs and will have the same class label as the original mc_{match}.

The pseudo code for this phase is shown in Figure 2. maxStim($MC_{ag.c}$, *ag*) returns a memory cell in the set of $MC_{ag.c}$ with the maximum stimulation to the antigen *ag*. mutate(mc_{clone}) mutates the clone of a memory cell according to the rule described previously. *AB* represents the set of ARBs (or set of representative B-cells) in the system, and |*AB*| denotes the size of the set *AB*.

Figure 2. Memory cell identification and ARB generation

```
if (MC_ag.c = {})
        mc_match = ag
        MC_ag.c = {mcmatch}
else
        mc_match = maxStim(MC_ag.c, ag)
endif
mc_match.stim = stimulation(ag, mc_match)
NumClones = hypermutationRate * clonalRate * mc_match.stim
AB = {}
while ( |AB| < NumClones) do
        mc_clone = mc_match
        mc_clone = mutate(mc_clone)
        AB = AB ∪ mc_clone
endwhile
```

Competition for Resources and Development of a Candidate Memory Cell

The objective of this phase is to find the ARB in the system (which is produced from the previous phase) which is the most stimulated when presented with an antigen. This ARB may then be retained as a memory cell. Recall that an ARB is a set of *identical* B-cells, therefore its characteristics can also be represented by a single cell. To find the most stimulated ARB, we allocate resources (B-cells) to all ARBs in the system based on their stimulation level to the current antigen. An ARB *ab* is given (*ab.stim* * *clonalRate*) B-cells.

However, as B-cells allowed in the system are limited (set by the value of *numResources*), if the number of allocated B-cells exceeds the allowed number of B-cells; some B-cells need to be removed from the system. B-cells will be

Figure 3. Competition for resources and development of a candidate memory cell

```
resAlloc = 0
foreach (ab ∈ AB) do
        ab.resources = ab.stim * clonalRate
        resAlloc = resAlloc + ab.resources
endfor
while (resAlloc > numResources) do
        numRemove = resAlloc – numResources
        ab_remove = minStim(AB)
        if (ab_remove.resources ≤ numRemove)
                AB = AB – {ab_remove}
                resAlloc = resAlloc – ab_remove.resources
        else
                ab_remove.resources = ab_remove.resources – numRemove
                resAlloc = resAlloc – numRemove
        endif
endwhile
s = avgStim(AB)
newAB = AB
foreach (ab ∈ AB) do
        NumClones = clonalRate * ab.stim
        MU = {}
        while ( |MU| < NumClones) do
                ab_clone = ab
                ab_clone = mutate(ab_clone)
                MU = MU ∪ ab_clone
        endwhile
        newAB = newAB ∪ MU
endfor
AB = newAB
if (s < stimThreshold)
        // repeat this phase
else
        //continue to next phase
endif
```

removed starting from the ARBs with the weakest stimulation, until the number of B-cells in the system equals to or are less than *numResources*. In this way, ARBs which are weakly stimulated by the antigen will be removed from the system. The surviving ARBs are examined by the algorithm to test whether they are stimulated enough by the antigen to cease further training. The average stimulation level s is computed by averaging the stimulation level of all surviving ARBs. If $s \geq$ *stimThreshold* then the stopping criteria is met.

Regardless of whether the stopping criterion is met or not, the surviving ARBs are then allowed to produce (*stim* * *clonalRate*) clones which undergo somatic hyper-mutation. The mutation rule is the same as the rule described in the previous section where numerical fields have constrained ranges for mutation. If the stopping criterion is met, the most stimulated ARB in the system is chosen as the candidate memory cell $mc_{candidate}$. The algorithm then proceeds to the next training phase, otherwise this phase is repeated.

The pseudo code for this training process is shown in Figure 3. minStim(*AB*) returns an ARB with minimum stimulation from the set of ARBs *AB*. avgStim(*AB*) returns the average stimulation value of ARBs in *AB*. *newAB* and *MU* are temporary storage locations for newly created ARBs.

Memory Cell Introduction

In this phase, we will decide whether $mc_{candidate}$, identified in the previous phase as the ARB with highest stimulation, should be retained as a memory cell and whether $mc_{candidate}$ should replace mc_{match}. $mc_{candidate}$ is retained as memory cell if it is more stimulated compared to mc_{match} with respect to the current antigen. Furthermore, if the affinity between $mc_{candidate}$ and mc_{match} is less than the threshold (*AT* * *ATS*), mc_{match} is going to be replaced by $mc_{candidate}$. It means that if $mc_{candidate}$ and mc_{match} have strong affinities (the distance between the two is sufficiently large), one memory cell cannot be used as a substitute for the other. If the affinity is weak, one memory cell can be regarded as a substitute of the other, and since $mc_{candidate}$ is more stimulated than mc_{match}, $mc_{candidate}$ is used as the memory cell.

Once the process is finished, the algorithm completes training for one antigen. If there are any other antigens (other data records) that need to be trained, the algorithm returns to train the new antigen. If all antigens have been trained, the training phase is finished and the produced memory cells can be used for classification.

Produced Classification Model

The algorithm produces a set of memory cells as a result of the training. In a clustering algorithm, each of the memory cells may be visualized as a cluster center. Using the parameter values specified before, we observed that the number of memory cells produced is only around 10% of the number of training data records.

Classification is performed using a weighted k-nearest neighbor approach; where k most stimulated memory cells have the right to vote for the presented antigen class. Since we have an imbalanced training data (3,000 3G customers against 15,000 2G customers), we can expect that there will be more memory cells representing the 2G customers compared to 3G customers; hence unweighted voting puts the 3G customers at a high probability of being incorrectly classified as 2G customers. To compensate for this, we set the ratio for a vote of 3G:2G = 5:1 (proportional to the ratio of training data available). Testing has shown that a weighted k-nearest neighbor approach performs more effectively compared to the unweighted approach.

As for the k value used to determine how many neighbors get to participate in the voting process, we used $k = 15$. By experimentation, we vary the value of k from $k = 1$ to 15 and observe that the solution quality increases. When k is higher than 15, there are some cases where the solution quality is reduced.

EXPERIMENTAL DESIGN

We are provided with 24,000 customer records, from which 18,000 records are to be used for training purposes, while the remaining 6,000 are for prediction. There is one categorical target field, which is customer type (2G/3G). Of the 18,000 training data records, 15,000 are 2G customers, and 3,000 are 3G customers. Each record in the dataset consists of 250 data fields. The data field can be categorized into two types; *numerical* and *categorical*.

Data Preprocessing

Before we supply the data into the AIRS algorithm, we need to preprocess the data. The data preprocessing that is done are normalizing numerical value, handling missing values, and converting categorical variables from literal strings to integer indices for easier processing.

For each numerical data field, we perform min-max normalization. First, we scan the data to find the minimum and maximum value of the field, then we normalize each value in this field to the range of [0,1] by converting it to

$$\frac{val - \min}{\max - \min},$$

where *val* is the value to be converted, *min* is the minimum value, and *max* is the maximum value.

For missing values that occur in the dataset, we simply replace them with a global value "MISSING," however, other strategies such as statistical regression might be more effective in handling the missing value.

For categorical data fields, we give an index to each possible value that might appear in the data and change the values from literal strings to indices. This has the advantage of more efficient processing because the algorithm only needs to deal with numbers instead of strings.

In addition, we also need to compute the conditional probability value $P(c|x_g)$ which is needed if we use HVDM distance measure. $P(c|x_g)$ denotes the probability that a customer record has class c given the value x_g.

We also note that in the dataset, there are some data fields in which only *one* value would appear in all customer records for the entire training data (e.g., **HS_CHANGE**, **TOT_PAS_DEMAND** for a numerical data field, and **VAS_SN_STATUS** for a categorical data field). We would not get any useful information from these fields since there is only *one* value that appears in the data and no other value to compare it to. Therefore, for these data fields, they can be safely removed from the data.

Training, Validation, and Parameter Tuning

To decide on which parameter values would yield the best results, we try several sets of parameter values. Validation is performed using 10-fold cross validation. We divide the training data randomly into 10 equal parts. For every run of the algorithm, nine parts will be used for training, and the remaining part will be used for validation purposes. We train the system with the training data, and we analyze the accuracy of the prediction on the validation set. By performing the validation methods repeatedly with different sets of parameter values, we can pick the set of parameter values which yield the best results. This is further illustrated in the next section.

RESULTS AND DISCUSSIONS

Obtained Results and Analysis

To perform the experiment, we perform a 10-fold cross validation method 10 times for each set of

Table 1. Accuracy rate of AIRS with varying sets of parameters

Number of training passes	*k* value for k-nearest neighbor			
	1	7	11	15
1	85.53%	86.54%	86.98%	87.22%
2	86.37%	87.33%	87.49%	87.82%
3	86.71%	87.47%	87.81%	88.08%
4	86.97%	87.69%	88.04%	88.26%
5	86.90%	87.81%	88.19%	88.32%

parameter values. The result shown in Table 1 is the average of the obtained values. Due to the time constraint, we only modify the number of training passes (*numEpochs*) and the value of *k* for classification using *k*-nearest neighbor.

As we can see from Table 1, as *k* increases, the accuracy rate generally increases, however, we also observe that sometimes, for *k* value higher than 15, the accuracy rate decreases. Therefore, we set the *k* value to 15. The same can be observed in the number of training passes, when the number of training passes is greater than 3, sometimes, the accuracy decreases. In addition, too many training passes increases the time involved in training the algorithm. Hence, we set the number of training passes to 3. Other values, such as the clonal rate, hyper-mutation rate, and number of allowed B-cells in the system are the default values for the algorithm which are initially used in the original source code by Hamaker and Watkins (2003).

Dataset Analysis

The discussion in this section will be based only on the training data, since we will need to know whether a record is incorrectly classified as false positive (classified as 3G although the actual class is 2G). Since our algorithm does not produce the characteristics of 3G customers, therefore the false positives obtained in our experiment might be a useful insight to decide the characteristics of customers that are likely to change their 2G subscription to 3G. If a false positive is found, it is possible that this record is closer to the characteristics of 3G customers rather than 2G customers, therefore, by analyzing the characteristics of all the false positives that are found, we can find some common characteristics that differentiates the 3G customers from the 2G customers.

To perform this experiment, we randomly divide the training data into 90% training set, and 10% testing set. After training the algorithm with the training set, we classify the testing set and obtain the false positives produced by the algorithm. This experimentation is repeated several times, and the statistics of the false positives obtained for each experiment is compared to the statistics of the entire training data whether there are a constant deviation between the statistics of the false positives and the entire data. For numerical data field, we compared the average and standard deviation, while for categorical data field, we compared the frequency count.

We find some almost constant, strong deviation between the statistics of false positives and the statistics of training data in the following fields. The field name uses the original name that is found in the dataset.

- **MARITAL_STATUS**: *More singles* can be found in the false positives compared to the entire data; this suggests that singles are more likely to purchase a 3G subscription.

- **OCCUP_CD**: *Less* customers with *other (OTH)* occupation is found in the false positives; therefore, customers with occupation {EXEC, POL, STUD, MGR, HWF, ENG, CLRC, SELF, GOVT, TCHR, SHOP, FAC, AGT, MED} (this occupation code are other codes that can be found in this data field) might be more likely to change to 3G.
- **HIGHEND_PROGRAM_FLAG**: *More* customers with *high end* programs are found in the false positives. This is expected since a 3G plan can also be considered a high-end program.
- **TOP1_INT_CD, TOP2_INT_CD, TOP3_INT_CD**: *Considerably more* customers in the false positives with values for this parameters set to *other than NONE*. It is probable that customers who make international calls are more attracted and able to afford a 3G subscription.
- **VAS_GPRS_FLAG**: *More* customers who own a *GPRS Data Plan* (code 1) can be found in the false positives. This is expected, since customers without GPRS are less likely to appreciate the improvement provided by 3G, and therefore, might not find a 3G subscription attractive.
- **LOYALTY_POINTS_USAGE**: Customers in the false positives set have *considerably higher average of loyalty points*. Loyal customers might be more easily persuaded to switch to a 3G subscription.
- **TOT_TOS_DAYS**: Customers in the false positives set have *a lot less average total temporarily on suspended* days compared to the entire data. Customers who are routinely using their mobile phone are more likely to purchase a 3G subscription.
- **AVG_CALL_FRW_RATIO**: *Less call forwarding* is utilized by customers in the false positive set. Customers who are carrying their mobile phone everywhere, therefore they do not need call forwarding services, are more likely to purchase a 3G subscription which would enable them to do more with their mobile set.
- **AVG_MIN_OBPK, AVG_MIN_OBOP**: *More minutes are spent in outbound calls in peak and off-peak period* by customers in the false positives set. Customers who make a lot of calls might be interested in the video phone feature provided by a 3G subscription.
- **AVG_MIN_FIX**: More minutes are spent by customers to call fixed line numbers in false positives set. Customers who are using their mobile phone actively are more likely to purchase a 3G subscription.
- **AVG_MIN_T1:** More top 1 minutes spent by customers in false positives set. Customers who are using their mobile phone actively are more likely to purchase a 3G subscription.
- **AVG_CALL_1900**: *Considerably more than 1,900 calls* by customers in the false positives set. Customers who are using their mobile phone for social use as well as for communication are more likely to purchase a 3G subscription.
- **AVG_REVPAY_AMT, REVPAY_FREQ**: *Slightly more reverse payment* in average for both the amount and the frequency can be detected in the false positives set.
- **CONTRACT_FLAG**: *Slightly more* customers *with contract* are found in the false positives. This may be caused by the price of 3G mobile phones that are still expensive; therefore, an attractive contract plan might attract more customers to switch to 3G.
- **AVG_VAS_QTUNE**: Customers in the false positives set *downloaded more quick tunes* than the average of the entire data. However, other services such as quick games, text, or pix do not exhibit any statistical deviation from the data.

- **AVG_VAS_GBSMS**: *More e-mail SMSs* are sent by customers in the false positives set. These customers might be interested in the functionality provided by 3G.

CONCLUSION

In this article, we described our approach in mining interesting patterns from the dataset of mobile phone customers to identify whether the customer is currently subscribed to a 2G or 3G subscription plan. To solve this problem we use the AIRS algorithm which operates using the principles of the human immune system. We experimented with different sets of parameter values to find the optimal values that are suitable for our purpose. Our experimentation shows that this algorithm is very effective; with proper parameter values it can achieve an accuracy rate near 90%. After a classification model was produced, we analyze the training data for patterns found in the classification model. These patterns that we found can hopefully be used to identify the common characteristics of 3G and 2G customers. We also discuss how these patterns can assist a mobile phone service provider in marketing their 3G subscription plan.

REFERENCES

Goodman, D., Boggess, L., & Watkins, A. (2003, July). An investigation into the source of power for AIRS, An artificial immune classification system. In *Proceedings of the 2003 International Joint Conference on Neural Networks.*

Hamaker, J., & Boggess, L. (2004, June). Non-Euclidean distance measures in AIRS, an artificial immune classification system. In *Proceedings of the 2004 IEEE Congress on Evolutionary Computation* (pp. 1067-1073).

Hamaker, J., & Watkins, A. (2003). Artificial immune recognition system (AIRS) Java Source Code.

Marwah, G., & Boggess, L. (2002, September). Artificial immune system for classification: Some issues. In *Proceedings of 1st International Conference on Artificial Immune Systems (ICARIS)* (pp. 149-153).

Watkins, A. (2001, November). *AIRS: A resource limited artificial immune classifier.* Unpublished master's thesis.

Watkins, A., & Boggess, L. (2002, May). A new classifier based on resource limited artificial immune system. In *Proceedings of IEEE Congress on Evolutionary Computation* (pp. 1546-1551).

Watkins, A., & Timmis, J. (2002, September). Artificial immune recognition system (AIRS): Revisions and refinements. In *Proceedings of 1st International Conference on Artificial Immune Systems (ICARIS)* (pp. 173-181).

Watkins, A., Timmis, J., & Boggess, L. (2004, September). Artificial immune recognition system (AIRS): An immune-inspired supervised machine learning algorithm. *Genetic Programming and Evolvable Machines, 5*, 291-317.

This work was previously published in International Journal of Data Warehousing and Mining, Vol. 3, Issue 2, edited byD. Taniar, pp.54-66, copyright 2007 by IGI Publishing, formerly known as Idea Group Publishing (an imprint of IGI Global).

Chapter XIX
Collecting Consumer Behavior Data with WLAN

Patrik Skogster
Rovaniemi University of Applied Sciences, Finland

Varpu Uotila
Turku School of Economics, Finland

ABSTRACT

Knowing consumers' shopping paths is an essential part of successful retailing. Good space management requires accurate data about consumer behavior. Traditionally, these data have been collected through, for example, panel interviews, camera tracking, and in-store observation. Their nature is more or less subjective. Modern technology makes it possible to use more objective methods, such as wireless local area network (WLAN) and radio frequency identification (RFID). In this article we examine the possibilities WLAN provides information systems studies. The empirical data is collected from a large DIY (do-it-yourself) store. The results show that WLAN has great potential for accurate and objective data collection processes and modeling data in retailing.

INTRODUCTION

Most of the customer's in-store behavior is made in an unconscious state. Because of this, the customers are afterward unable to explain their purchase decisions in more detail. Most of the purchase decisions are made inside a store. The challenge of the retail business is to create an environment where the customer has a bilateral relation with the store, to optimize the customer's use of time, and to offer a buying experience that the consumer wants to renew later (Soars, 2003). The pleasure produced by the store environment is a significant reason for the extra time used by the consumer inside the store. In fact, in these occasions consumers spend more money than intended (Donovan, Rossiter, Marcoolyn, & Nesdale, 1994).

The purchasing behavior of a customer inside the store has been studied for several dozens of years. Various studies have been made on unplanned buying, store and product types, the demographic features of the consumer profiles, and the effect of the internal campaigns of the store. Also, the data-processing and decision-making processes, which take place in the store, have been researched. For example, Park, Iyer, and Smith (1989) studied how the time customer spends and his/her earlier knowledge of the store can affect the making of unplanned purchases. These can appear as failure in the making of planned purchases and changes in other purchasing behavior.

Consumers' buying behavior has been studied with in-store videos and interviews (Underhill, 1999). However, according to Larson, Bradlow, and Fader (2005), the results of these studies are limited to general recommendations, which only increase the convenience of the customers. Only a few studies, which contain really large data sets, have been conducted on the customers' complete shopping paths so far. This has not been possible earlier with traditional data collecting methods (Larson et al.; Sorensen, 2003).

From the retailers' point of view, space management is a significant factor in a successful store (Soars, 2003). A good store is one where the largest possible amount of products is in the sight of the largest possible amount of customers as long as possible. Also, the placement of the products must be realistic concerning logistic issues. As Underhill (1999) states, a good store is the kind where the products are placed along the customers' routes and sight in a way that makes the customer consider buying. Still, how can the retailer know where their customers really walk and spend their time?

The purpose of this study is to explore how information about customer traffic can be collected from a large DIY (do-it-yourself) store with the help of modern technology and how the collected data can be analyzed. The main focus is on collecting customer traffic data with the help of WLAN (wireless local area network) and analyzing them with different models using geographical information systems (GIS). A similar study has not been done before within the DIY context and Scandinavian service industry. Therefore, this study contributes to different practical solutions but also to theoretical discussion.

The empirical data of the study have been collected in a large, modern DIY unit located in Scandinavia. The store belongs to a large DIY chain. The study has been limited to concern the store in question and its customers in August 2006.

LITERATURE REVIEW

The consumer buying behavior process inside the store has been studied for several dozens of years. There have been attempts to research it in many different ways. For example, Köhne, Totz, and Wehmeyer (2005) use conjoint analysis, a classical technique, to identify consumer preferences in multi-attribute decision making for designing a new context in sensitive services. They evaluate consumer behavior by using a fictitious example of location-based services in a touristic setting.

In this article, four different data collecting methods are briefly discussed: WLAN, radio frequency identification (RFID), camera tracking, and in-store observation. These methods can also be combined. Another possible technique for collecting data on consumer behavior in the store is interviewing the customers as they leave. This kind of memory-based technique can lead to high inaccuracy in the results. However, this technique makes it possible to collect a very inclusive picture of the customers' shopping experience (Phillips & Bradshaw, 1991). This collecting method is not studied in this article as it clearly deviates from the other methods. Also, Bluetooth technology can be mentioned as a method for collecting consumer behavior data.

People and machines can be tracked with wireless local area networks without any additional devices and by using the already existing wireless network. That is why the installing of this tracking system in the desired outdoor or indoor space can be very fast. The WLAN tracking system consists of the network and the tags, which are being tracked. This technology has been used among others in health care centers and both mine and tunnel construction (Martikainen, 2006).

RFID is a common name for technologies that use radio waves to identify individual products. RFID consists of a reader and tags. The tag contains a microchip and an antenna (Jones, Clarke-Hill, Shears, Comfort, & Hillier, 2004). RFID is thought of as the successor of the bar code because with RFID technology, individual products can be identified instead of product categories (Spekman & Sweeney, 2006). In addition to retail warehouses, RFID technology is used in the health care sector, libraries, and the travel sector to follow luggage (Juban & Wyld, 2004). RFID has also been used to track the shopping paths of customers in a store. Sorensen (2003) researched over 200,000 shopping paths in a supermarket. According to him, the number of the studied shopping paths in earlier studies has been a few hundred at its largest. He estimates being the first who has studied tens of thousands shopping paths in detail. In Sorensen's research, the customers' shopping trolley or basket was tracked with a small tag placed at the bottom of it. Also, Larson et al. (2005) used the RFID technology in their study. They studied the customer routes in an American grocery market. An RFID tag had been placed at the bottom of every shopping trolley and was located at intervals of 5 seconds. RFID readers had been placed throughout the store. In their study, 27,000 shopping paths were collected and every third route was analyzed. In the study, attention was not paid to the customer's actual purchases or to different sales tactics of the store. Larson et al. did valuable work, especially on changing the routes of different lengths to a comparable format. They analyzed the routes by a grouping algorithm developed by them. It made the handling of the large amount of spatial information possible quantitatively. The customer routes varied between 2 minutes and 2 hours. One customer could be located as many as 1,500 times.

By tracking the customer with a camera, the shopping paths can be studied with fair easiness (Dodd, Clarke, and Kirkup, 1998). With this information, the product categories and the customer rotation can be planned so that they are as optimal as possible for the consumer. At the same time, it can also be found out how long the customer stays in the store and which offers or other messages in the store he or she responds to. Phillips and Bradshaw (1991) wrote how in-store customer behavior has been mainly studied using manual recording techniques. With this technique, the customer can be tracked around a store. Also, product handling, verbal contact with the store staff, and display viewing can be added to the study. It is also almost impossible to get an inclusive picture of customer behavior throughout the whole store. However, the labor intensiveness makes the technique very time consuming and difficult to obtain large sample sizes.

In a standard camera tracking survey, the store is first inspected and the camera locations are selected. Trial photographs are taken to check each location. The optimum number of cameras is a compromise between having the smallest possible amount of cameras to minimize the amount of time spent analyzing the film and having enough cameras for the adequate coverage of the store. Each camera takes one photograph every minute. Analyzing the film is the most labor-intensive part in the camera tracking study (e.g., Kirkup & Carrigan, 2000; Newman, Yu, & Oulton, 2002).

It is possible to observe and to follow the customers also from inside the store while they are shopping. It is very uncommon that the customer notices that he or she is being watched. A template of the store layout, on which all entrance and exit areas, tills, gondolas, and special product displays have been marked, can be used in the observation and tracking. In addition, it is possible to add the specific product categories that are being researched. When the shopper enters the store, his or her shopping path is marked on an individual template to show where the customer stopped to look, to browse, or to purchase and so on. Also, the day of the week, time of the day, and length of the time spent in the store are all recorded with observable characteristics of the customer. In addition, time spent at a specific location or display in the store can be recorded. The individual tracking forms can be processed for evaluations by computer (M. Johnson, 1999).

The effects on the quality of the data set when using different collecting methods have been collected in Table 1. In Tables 1 and 2, WLAN and RFID are viewed as one collecting method because they have very similar features but clearly deviate from camera tracking and in-store observation.

When a local area network or RFID is used, the customer can behave abnormally because he or she knows that his or her route is being tracked. However, the collecting of the data set is performed by a technical device, which makes the collecting process itself objective. When camera surveillance is used, the customer does not know that he or she is being followed, but when the collected data are recorded from filmstrip to the computer, there is a possibility for researcher interpretation. For in-store observation, the data are most vulnerable. The customer can notice that he or she is being followed. This can lead to changes in behavior. Also, the observations

Table 1. Effects on the quality of the data when using different collecting methods

	subject of data collection	way of data collection
WLAN/RFID	can affect	does not affect
camera tracking	does not affect	can affect
in-store observation	does not affect unless the consumer notices the observer	can affect

Table 2. Features of different data collecting methods

	advantages	disadvantages
WLAN/RFID	– possibility to track the customer, shopping cart, or basket very accurately – possibility for a very large data set – collecting data requires very little or no workforce	– memory limitations when the collected data set is very large – possible technical disturbance and problems before and during the data collection – possible labor intensity when setting up the system – setting up the system can be expensive but the actual collecting of the data is not
camera tracking	– possible to study also other customer behavior features besides the shopping paths – an easy way to get a mediocre picture of the customer traffic	– labor intensive – large data sets almost impossible to collect and to analyze – accurate shopping paths almost impossible to find out especially in large stores
in-store observation	– possibility to study consumer behavior in detail	– labor intensive – large data sets almost impossible to collect and to analyze

made by the person conducting the collection are subjective.

The different collecting methods are best suitable in different situations. Determining the best collecting method depends on the objectives of the study in question. Features of the four different data collecting methods have been collected in Table 2.

Camera tracking and observation are best suitable when studying detailed consumer behavior. With these methods it is, for example, possible to find out which products the customer considers, between which choices he or she chooses from, which product he or she finally takes, and if he or she has interaction with the store staff. The disadvantage of both methods is the labor intensiveness, which causes large data sets quite impossible to collect. Also accurate shopping paths and customer traffic information is hard to obtain. Camera tracking and in-store observation are more suitable for tracking customer traffic in a small store and observing consumer shopping behavior in detail.

The sample size depends on the objectives of the study: The more specific the information that is wanted, the larger the sample size should be (Uusitalo, 1991). For a large sample size, the best data collecting method is either RFID or WLAN. Because of the technology, both of them make the collecting of large data sets possible. However, it is not possible to collect data concerning other behavior features besides the shopping paths, and technical disturbance and problems can occur.

When using WLAN and RFID, the tag can be attached to the shopping cart or basket or be given to the customer to carry. When the tag is being given to the customer, a very large sample size is more difficult to reach because of the labor intensiveness. When the tag is attached to the cart or basket, no extra staff is needed and then very large sample sizes are possible. The placement of the tag affects what is measured in the study. When the tag is carried in the cart or basket, the whole path of the customer is not tracked. Only a quite good picture is obtained because the customer can leave the cart or basket at times when getting a product. When the customer carries the tag, a very precise picture of the customer's shopping path is obtained. Then the only limitation to the accuracy of the path is the location accuracy. The DIY store customer often moves in the store without a shopping cart or basket. If the tag would be attached to the carts and baskets, it might limit the number of the potential participants in the study.

WLAN and RFID are very similar as data collecting methods but quite dissimilar in their technical qualities. RFID is an identifying technology and not a location technology (Schwartz, 2004). WLAN as a data collecting system is easy to set up because the already existing wireless network can be used. When using RFID, readers must be put in the store as densely as possible for the tag to be recognized. For example, if the tag is to be recognized within a 2-meter radius, then the readers should be placed every two meters. This is why RFID as a location technology is difficult to move from one place to another. It is more suitable to a store, where customers' shopping paths are monitored on a long-term. Then the system can be set up there permanently or at least for a long time.

In the literature there has also been discussion on the limitations of these consumer follow-ups. Virtual reality or videotape methods (e.g., Hui & Bateson, 1991) using conjoint choice experiments could manipulate environmental variables to determine the magnitude of the effects that changing tenants and environments have on consumer responses. Also, as Wakefield and Baker (1998) conclude, including two or more stores with different environmental aspects and/or a variety of characteristics would overcome the concern of limited site research.

The use of various spatiotemporal data (data with time and location elements) and information usually greatly improves decision making in many different fields (Christakos, Bogaert, &

Serre, 2003). Examples can be found in Meeks and Dasgupta (2004). However, when using spatial and temporal information to improve decision making, attention must be paid to uncertainty and sensitivity issues (Crosetto & Tarantola, 2001).

Because the spatial data fusion process is by its origins a process that produces data assimilations, the challenges it is facing are largely related to the data handling process. Data integration processes, synchronous sampling, and common measurement standards are developed to optimize the consumer behavior data-fusion performance. This includes increasing both the data management process and data collection efficiency (e.g., Fischer & Ostwald, 2001; Rafanelli, 2003).

The collection of data and their availability can also be seen as a strategic matter. Roberts, Jones, and Frohling (2005) highlight the importance of making sense of networks that comprise many nodes and are animated by flows of resources and knowledge. The transfer of managerial practices and knowledge is essential to the functionality of these networks and resources. A survey made by Vanderhaegen and Muro (2005) reveals that almost all of the organizations (90%) making use of spatial data "experience problems with the availability, quality and use of spatial data." In general, the organizations using the widest range of data types experienced the greatest difficulties in using the data.

The quality of the spatial data is still only one of the many factors that must be taken into consideration within spatial data fusion. Clearly, the results of any spatial data fusion are only as good as the data on which it is based (R. G. Johnson, 2001). One approach to improve data quality is the imposition of constraints upon data entered into the database (Cockcroft, 1997). The proposal is that "better decisions can be made by accounting risks due to errors in spatial data" (Van Oort & Bregt, 2005).

METHODOLOGY

In this study, the data were collected with the help of WLAN. The features of WLAN as a data collection method are the tracking system's easy installation in a large store, the possibility of collecting large data sets, the objectivity of the collecting method, the possibility to track customers with less than 1-meter accuracy, and the portability of the tracking system to other stores in the chain. There is also the possibility of disturbance and technical problems. WLAN is probably the newest technology used in tracking customers. It was chosen because it has some major advantages compared to other tracking methods and provides a valuable and useful tool for stores in retail.

The data set was collected by locating a tag carried by the shopping customer with the help of the wireless local area network. The data was collected in a modern DIY store in Scandinavia. The store of this study has more than 20 departments. There are four service points and an information desk in the store. There is also a small cafeteria located in connection with the main entrance. The store has a total of 5,100 square meters of selling space.

There were 30 tags in use in August 2006 during a time period of 2 weeks. Data were collected from 4 to 6 hours per day. The customer routes were located every day with variance between mornings and evenings so that after the collecting period, tracked shopping paths could be found from nearly all opening hours from different days of the week. The data were collected neither during the first opening hour of the morning nor during the last closing hour of the evening. The amount of customers during these times was so small that the results would not have been reliable. The sample size of the study was 866 (Table 3). An average of 72 shopping paths was located

Table 3. Located routes

Time		Week 1						Week 2						
		Th	Fr	Sa	Mo	Tu	We	Th	Fr	Sa	Mo	Tu	We	Σ
a.m.	8–9	0	10	0	0	5	0	11	0	0	10	0	5	41
	9–10	0	11	13	0	10	0	13	0	15	14	0	12	88
	10–11	0	11	16	0	15	0	17	0	17	18	0	22	116
	11–12	0	7	16	0	19	0	9	0	20	11	0	16	98
p.m.	0–1	0	11	18	0	12	0	17	0	9	8	0	15	90
	1–2	0	10	14	0	7	4	7	11	14	8	0	0	75
	2–3	0	0	5	10	0	14	0	16	1	0	14	0	60
	3–4	6	0	0	14	0	18	0	23	0	0	18	0	79
	4–5	20	0	0	15	0	9	0	15	0	0	14	0	73
	5–6	18	0	0	15	0	13	0	10	0	0	18	0	74
	6–7	25	0	0	10	0	11	0	18	0	0	8	0	72
	Σ	69	60	82	64	68	69	74	93	76	69	72	70	866

during 1 day. The store was open on weekdays from 7:00 in the morning to 8:00 in the evening. On Saturdays, the store was open from 9:00 a.m. to 3:00 p.m. The number of the tracked shopping paths per day varied between 60 and 93.

A scanning circuit located the tag at intervals of 10 seconds. The customer got the tag immediately in the entrance area of the shop. The customer was given instructions to keep the tag with him or her during the whole shopping trip and to give it to the nearest cash desk when leaving the store. If there were more people in the group, the same person had to carry the tag all the time. When the customer was given the tag, the identification number of the tag, the time, the gender and age of the customer, and the size of the group were recorded.

The customer was not asked directly any information; the figures are estimates of the recorder. The age was approximated in one of the four different groups. These groups were 18 to 30 years, 30 to 45 years, 45 to 60 years, and over 60 years. Seventy percent of the customers were men and thirty percent were women. Over half of the customers shopped alone, one third with one other person, and the rest in a bigger group. Forty percent of the customers were approximated being between 30 and 45 years old. About one third was between 45 and 60 years, and the rest were over 60 years or between 18 and 30 years old. The proportions of different types of customers varied a little depending on the time of the day and the day of the week.

The majority of the customers who participated in the study were men. In the total sample there were slightly over 30% of women and nearly 70% of men. The customers' age and gender share can be seen in Figure 1. The percentages in the different classes of pillars represent a relative share of the whole sample of the study.

The majority of the customers were between 30 and 45 years old. There were 40% of them. Less than one third of the customers were 45 to 60 years old. There were almost as much customers in the 18- to 30-year group than in the over-60-year group. Both represent approximately one eighth of the total population. The male customers' relative share is the biggest in the over-60-

Figure 1. Gender and age of the participants

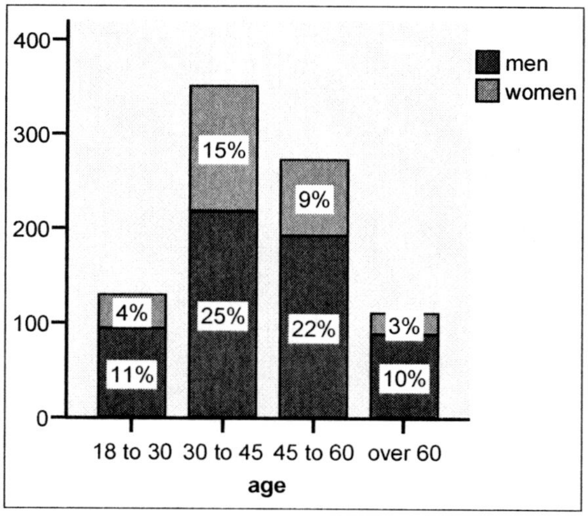

Figure 2. The size of groups

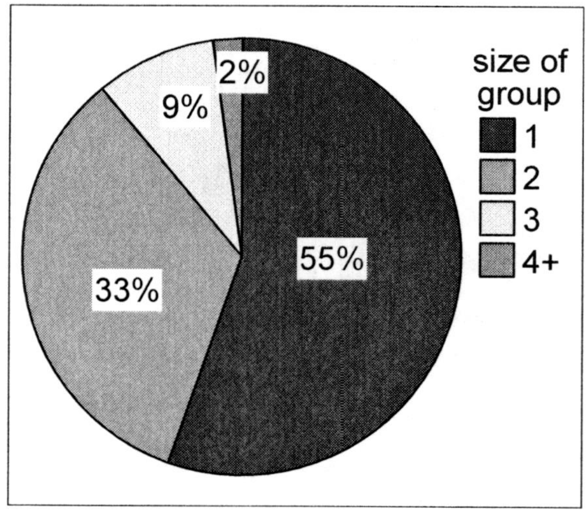

year-old customer group. However, there were a majority of male customers in all the age groups. The customers' ages are subjective estimates of the recorder. The majority of the customers who participated in the study were shopping alone. There were more than half of them. The shares of groups of different sizes of the total population can be seen in Figure 2.

About one third of the customers were shopping with one other person and 11% of the customers were shopping in groups of three or more people. There were no major variations in the group sizes during different times. Lone shoppers formed the majority at all times. The largest shares of lone shoppers were found on the weekdays in the mornings (between 8 a.m. and noon). A typical customer in this study was 30 to 45 years old, male, and shopping alone.

Taking the tag was totally voluntary. By taking the tag, the customer gave permission

Table 4. Proportion of refusals

Time	Week 1						Week 2						
	Th	Fr	Sa	Mo	Tu	We	Th	Fr	Sa	Mo	Tu	We	Σ
Were offered the detector	79	79	107	80	85	90	99	110	100	88	82	83	1 082
Refused	10	19	25	16	17	21	25	17	24	19	10	13	216
Proportion of refusals (%)	12,7	24,1	23,4	20,0	20,0	23,3	25,3	15,5	24,0	21,6	12,2	15,7	20,0

Table 5. Proportion of cash transactions

Time	Week 1						Week 2					
	Th	Fr	Sa	Mo	Tu	We	Th	Fr	Sa	Mo	Tu	We
Proportion (%)	33	19	15	18	22	23	27	24	16	26	25	31

to track his or her shopping paths in the store. The majority of the customers to whom the tag was offered took it. On average, every fifth customer denied. The number of customers to whom the tag was offered, the number of customers who refused, and their share can be seen in Table 4.

The number of the collected routes was compared with the amount of the cash transactions of the time in question (Table 5). The cash transactions describe the number of the groups better than the total amount of customers, but the tags were also distributed to groups instead of individual customers. It was approximated that on average, a fifth of the customers' routes were located. During the busiest days, more than 30% of the shopping paths were tracked, and about 15% were tracked during the quietest days.

The numbers are only suggestive because they have been calculated from the same time period that the customers were given tags. In other words, the customers still had tags when the cash transactions were not being counted anymore. These numbers do not include the groups, which did not make purchases.

ANALYSIS

The whole sample of the study was analyzed by modeling with the help of spreadsheets (Microsoft Excel), statistical software (SPSS), and GIS software (MapInfo and ArcInfo). The GIS software was used especially in the illustrating process. The models represented here concentrate on studying the customer traffic as a whole instead of explaining individual shopping paths.

In the first model, all the locations are marked on the base plan of the store as pixels. This way, detailed data can be obtained about which parts of the store the customers visited. The location accuracy is at its best within a half a meter. The weakness of this pixel model is the fact that from the map it cannot be seen if there have been more locations in one pixel. The amount of the locations in smaller aisles especially cannot be easily verified from this kind of figure.

In the second model, the large square model, the base plan of the store is divided into 21 squares in the horizontal direction and into 11 squares in the vertical direction. Consequently, 219 different square locations were formed on the base plan; 182 of these squares included tracked customer locations. The rest of the 37 squares are located in areas that were inaccessible to the customers, in the cafeteria, or near the entrance doors. The area of one square is about 25 square meters in reality. The disadvantage of this model is the large size of the squares. This is why the same base plan has been added with the location points from the pixel model. Combining the large squares and the exact customer traffic locations, the whole figure gives a better overview of the customer traffic. The clear advantage of this model is that from the base plan, it is easy to see where the customers spend their time in the store.

The third and fourth models are similar to the large square model. In these models, the base plan is also divided into squares but the squares are smaller than in the second model. In the medium square model, the base plan of the store is divided into 42 squares in the horizontal direction and 23 squares in the vertical direction. Altogether, 906 squares are formed on the base plan from which 520 squares include tracked locations. The amount of squares without tracked locations is fairly large because, for example, large gondolas and other store furniture take up a whole square, making it inaccessible to the customers. The area of one square of the third model is approximately 6 square meters in reality. In the fourth model, the small square model, the used squares are the smallest. In the model, the base plan of the store is divided into 105 squares in the horizontal direction and 56 squares in the vertical direction, forming over 5,000 squares on the base plan. One square equals about 1 square meter in reality. Although the information provided by the picture is more

Figure 3. Pixel model

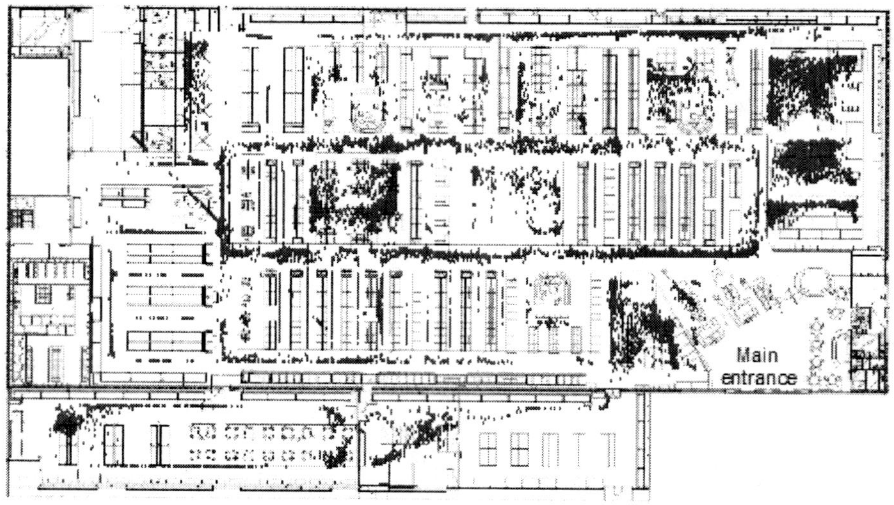

Figure 4. Large square model

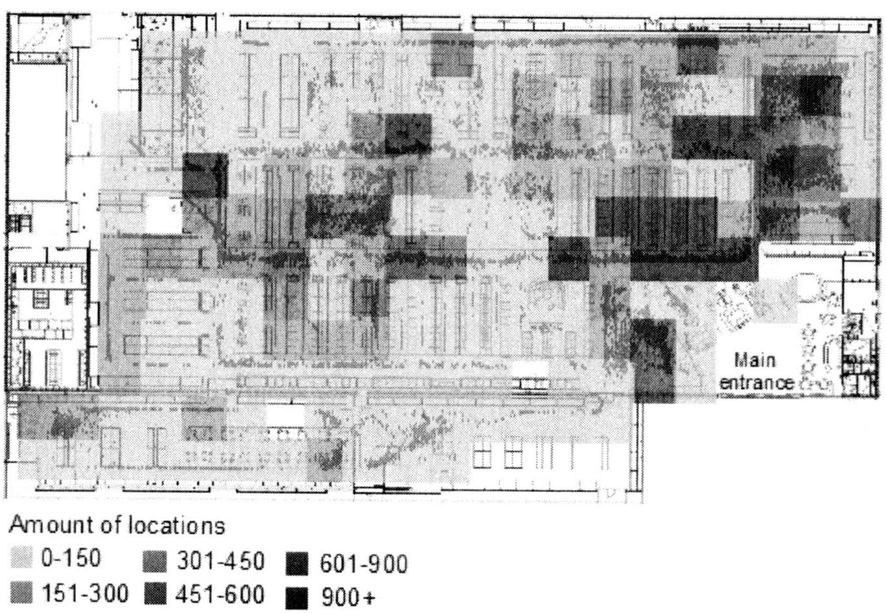

accurate than in the second or third model, the picture seems to be, in a way, chaotic.

In Figures 7, 8, and 9, it can be seen how at the different times of day and on different weekdays the consumer behavior is not significantly different.

CONCLUSION AND DISCUSSION

The retailing industry needs objective information about consumer behavior. The cornerstone of the strategic planning processes are to give the right services and products in the right place at the right time. An excellent picture of customer traffic can be obtained when collecting data with wireless local area networks and analyzing them with spreadsheets and geographical information systems. As seen in this study, it is possible to analyze the customer traffic data sets in several different ways.

The purpose of this article was to examine how information about customer traffic can be collected from a large DIY store with the help of modern technology and how the collected data can be analyzed. Consumer behavior research methods and processes must be capable of accepting wide ranges of data types, accommodating natural-language extracted text reports, historical data, and various types of spatial information (maps, charts, images). Therefore, the processes must have learning abilities. These processes must develop adaptive and learning properties, using the operating process to add to the knowledge base while exhibiting sensitivity to unexpected or too-good-to-be-true situations that may indicate countermeasures or deceptive activities. The performance of the processes needs to be improved exponentially. When the amount of processed data increases and analysis becomes more complicated, efficient and linked data structures are required to handle the wide variety of data. Data volumes for global and even regional spatial databases will be measured in millions of terabits, and short access times are demanded for even broad searches. In this article, WLAN technology provided the technological platform. As a result, this technology provided an enormous amount of valuable data. The continued handling of this data was done mainly by GIS tools. The results can therefore be shown in several visual formats and more specific analysis can be done. Actually, only the researcher's imagination can limit further analysis. The data and tools do not create barriers.

When different consumer data collecting methods are used, the quality of the data can weaken in two different ways. When collecting the data, the customers may alter their behavior if they know that they are being tracked. The second possible weakness is caused by the data collector. Personal beliefs and attitudes can affect the observations. In other words, the observations are subjective.

The subjectivity can appear when observations are collected or when they are interpreted. Different models are best suited in different situations since they have their own advantages and disadvantages. By using only the pixel model, wrong conclusions can be drawn from the customer traffic. On the other hand, the weakness of the large square model is the fact that the used squares have a rather big size. Together, the pixel and the large square models give a good overview of the customer traffic. The medium square model and the small square model give more exact information about the customer traffic, but for an overview they are somewhat vague. They are more suitable when analyzing the different departments or sections of the store.

The results of this study can be applied to practical issues in many ways. First, it helps to allocate the right products near the routes that consumers prefer. Second, personnel workload can be rationalized with the help of the systems described. When the actual consumer route information is accompanied with data about the time of the day, a retailer can divide the available personnel resources to the most appropriate service desks during the day and the week. The practical implications can be applied directly to consumer behavior analysis, store management, and, for example, in the utilization of consumer hot spots—places that gather consumers' interest.

Theoretical implications are concentrated on the technology and methodology comparison. This article gives implications on the consumer behavior theories. The methodology tests theories in practice with a large sample size.

As with any single study, there are still a number of limitations in this article. First, caution must be used in extending the results of this research to other settings. While the research area was a single DIY store that was similar to many large DIY stores across Scandinavia in the store assortment, it must be remembered that all locations and stores are unique. There are no two identical surroundings and consumer profiles.

There may also have been factors influencing the tenant values that were not captured in the data, such as good or bad relationships with neighboring shops. Third, there might have been a "grass is greener" effect, meaning that shopping is more enjoyable outside the consumers' customary stores. When an individual consumer within a store context is away from home and enjoying the novelty of another store, it may make the consumer act out of the ordinary. In this study, the locations of the customers' permanent residences were not fully studied. In the area are numerous summer cottages and also a major highway passes the store. Therefore, it can be supposed that nonlocals performed a significant amount of the examined shopping paths.

It is important to note the limitations inherent in a limited site study. Because the differences in shopping location alternatives (i.e., by including surveys and/or follow-ups in several stores in the study) were not directly assessed, it is tenuous to conclude that consumer perceptions were not generalized. Given this limitation, future research examining the effect of changes on affective and behavioral responses should more precisely incorporate consumer behavior differences. For example, before and after studies (e.g., Downs & Haynes, 1984) could measure the impact of renovations and tenant additions.

Because the research area analyzing the consumer's behavior with modern technology is a very new field of research, there is only a small amount of previous research done and the relevant amount of source material is very limited. For example, a similar study has not been done before within the DIY context. Source material about processing the customer traffic data was especially almost impossible to find. Studies that have been made in the field of retail business have been clearly emphasized in the perishables business, but there is only a limited amount of material concerning the DIY trade. Therefore, this study gives new methodological viewpoints. This is also the main limitation of this study. More empirical data and analyses are definitely needed in order to achieve more reliability and validity.

Because in a DIY store a similar study has not been carried out earlier, very many technical problems had to be solved. The sample size of the study is so big that sufficient comprehensiveness has been reached to guarantee the validity and reliability of the study. Validity can be defined as the ability of a study to reflect the true state of affairs. In this study the shopping paths do not have large variance on different days of the week and times of the day. Therefore, the internal validity of this study is acceptable. Since the data were collected from only one DIY store, it is quite clear that the external validity needs to be enhanced through a comparative study in other representative DIY units.

The research design (quantitative, descriptive) and methodology (empirical follow-up, sampling, and literature review) allowed for neither comparisons between various research groups under controlled conditions, nor a longitudinal study. The study was conducted only during a 2-week period. The outcomes are almost entirely based on hard evidence rather than the perception of participants, for example, norm- or criterion-referenced tests.

Outcomes reported here, for example, changes in behavior, depend on narrow information about consumer masses in a certain geographical area (visitors in a certain DIY store) rather than on a representative sample providing deep knowledge of the behavior of certain consumers in the store. However, the amount of collected data gives justification to the results.

The technology used can also give an impact on the reliability and validity of the results. Arumugam, Doufexi, Nix, and Fletcher (2003) report the impact of Bluetooth interference on the WLAN system and vice versa. Modern mobile phones usually have Bluetooth capability, and their impact on the results of this study has not been measured.

In this study, attention has been paid to reliability during the whole research process. In the store, there are only a few areas in which there are no locations. These areas are the cash desk area and the outer corner on the side of the main entrance in the outer shop store. The missing locations can possibly be explained by the fact that the local area network has not covered the area in question. The second alternative to the missing locations is the fact that there have continuously been many active tags simultaneously side by side in the area. This has possibly caused disturbance to the location system. Therefore, attention should be paid also to technological issues. Larson et al. (2005) have developed a research frame and performed a study very similar to this one. The results of their study were generalized. Also in this study external validity comes true. In this study the sample size is big; this marks off the random results. Thanks to random sampling, the data of the study present the target population well. The data collecting method can be considered objective and the location exactness of the data collecting method sufficient for reliable information. Thus, it can be supposed that the results of the study are similar when repeated. In other words, the study can be considered as rather reliable.

REFERENCES

Arumugam, A. K., Doufexi, A., Nix, A. R., & Fletcher, P. N. (2003). An investigation of the coexistence of 802.11g WLAN and high data rate Bluetooth enabled consumer electronic devices in indoor home and office environments. *IEEE Transactions on Consumer Electronics, 49*(3), 587-596.

Christakos, G., Bogaert, P., & Serre, M. (2002). *Temporal GIS: Advanced functions for field-based applications.* Berlin, Germany: Springer.

Cockcroft, S. (1997). A taxonomy of spatial data integrity Constraints. *GeoInformatica, 1*(4), 327-343.

Crosetto, M., & Tarantola, S. (2001). Uncertainty and sensitivity analysis: Tools for GIS-based model implementation. *International Journal of Geographical Information Science, 15*(5), 415-437.

Dodd, C. A., Clarke, I., & Kirkup, M. H. (1998). Camera observations of customer behavior in fashion retailing: Methodological propositions. *International Journal of Retail & Distribution Management, 26*(8), 311-317.

Donovan, R. J., Rossiter, J. R., Marcoolyn, G., & Nesdale, A. (1994). Store atmosphere and purchasing behavior. *Journal of Retailing, 70*(3), 283-294.

Downs, P. E., & Haynes, J. B. (1984). Examining retail image before and after a repositioning strategy. *Journal of the Academy of Marketing Science, 12*, 1-24.

Fischer, G., & Ostwald, J. (2001). Knowledge management: Problems, promises, realities, and challenges. *IEEE Intelligent Systems, 16*(1), 60-72.

Ganesh, J. (2004). Managing customer preferences in a multi-channel environment using Web services. *International Journal of Retail & Distribution Management, 32*(2/3), 140-146.

Hui, M. K., & Bateson, J. E. (1991). Perceived control and consumer choice on the service experience. *Journal of Consumer Research, 18*, 174-185.

Johnson, M. (1999). From understanding consumer behavior to testing category strategies. *Journal of the Market Research Society, 41*(3), 259-288.

Johnson, R. G. (2001). *United States Imagery and Geospatial Information Service geospatial transition plan.* Bethesda, MD: National Imagery and Mapping Agency.

Jones, P., Clarke-Hill, C., Shears, P., Comfort, D., & Hillier, D. (2004). Radio frequency identification in the UK: Opportunities and challenges. *International Journal of Retail & Distribution Management, 32*(3), 164-171.

Juban, R. L., & Wyld, D. C. (2004). Would you like chips with that? Consumer perspectives of RFID. *Management Research News, 27*(11/12), 29-44.

Kirkup, M., & Carrigan, M. (2000). Video surveillance research in retailing: Ethical issues. *International Journal of Retail & Distribution Management, 28*(11), 470-480.

Köhne, F., Totz, C., & Wehmeyer, K. (2005). Consumer preferences for location-based service attributes: A conjoint analysis. *International Journal of Management and Decision Making, 6*(1), 16-32.

Larson, J. S., Bradlow, E. T., & Fader, P. S. (2005). An exploratory look at supermarket shopping paths. *International Journal of Research in Marketing, 22*(4), 395-414.

Martikainen, O. E. (2006). Complementarities creating substitutes: Possible paths towards 3G, WLAN/WiMAX and ad hoc networks. *Info, 8*(4), 21-32.

Meeks, W. L., & Dasgupta, S. (2004). Geospatial information utility: An estimation of the relevance of geospatial information to users. *Decision Support Systems, 38*(1), 47-63.

Newman, A. J., Yu, D. K. C., & Oulton, D. P. (2002). New insights into retail space and format planning from customer-tracking data. *Journal of Retailing and Consumer Services, 9*, 253-258.

Park, C. W., Iyer, E. S., & Smith, D. C. (1989). The effects of situational factors on in-store grocery shopping behavior: The role of store environment and time available for shopping. *The Journal of Consumer Research, 15*(4), 422-433.

Phillips, H. C., & Bradlow, R. P. (1991). Camera tracking: A new tool for market research and retail management. *Management Research News, 14*(4/5), 20-22.

Rafanelli, M. (2003). *Multidimensional databases: Problems and solutions.* London: Idea Group Publishing.

Roberts, S. M., Jones, J. P., & Frohling, O. (2005). NGOs and the globalization of managerialism: A research framework. *33*(11), 1845-1864.

Soars, B. (2003). What every retailer should know about the way into the shopper's head. *International Journal of Retail and Distribution Management, 31*(12), 628-637.

Sorensen, H. (2003). The science of shopping. *Marketing Research, 15*(3), 30-35.

Spekman, R. E., & Sweeney, P. J., II. (2006). RFID: From concept to implementation. *International Journal of Physical Distribution & Logistics Management, 36*(10), 736-754.

Underhill, P. (1999). *Why we buy: The science of shopping.* New York: Simon & Schuster.

Vanderhaegen, M., & Muro, E. (2005). Contribution of a European spatial data infrastructure to the effectiveness of EIA and SEA studies. *Environmental Impact Assessment Review, 25*(2), 123-142.

Van Oort, P. A. J., & Bregt, A. K. (2005). Do users ignore spatial data quality? A decision-theoretic perspective. *Risk Analysis, 25*(6), 1599-1610.

Wakefield, K. L., & Baker, J. (1998). Excitement at the mall: Determinants and effects on shopping response. *Journal of Retailing, 74*(4), 515-539.

This work was previously published in International Journal of Information Systems and Supply Chain Management, Vol. 1, Issue 2, edited by John Wang, pp. 57-75, copyright 2008 by IGI Publishing, formerly known as Idea Group Publishing (an imprint of IGI Global).

Section V
Critical Issues

Chapter XX
Distributed Resources Management in Wireless LANs

Ghassan Kbar
American University in Dubai, UAE

Wathiq Mansoor
Zayed University, UAE

ABSTRACT

This paper introduces a new radio resource management technique based on distributed dynamic channel assignment, and sharing load among Access Points (AP). Deploying wireless LANs (WLAN) on a large scale is mainly affected by reliability, availability and performance. These parameters will be a concern for most managers who want to deploy WLANs. In order to address these concerns, a new radio resource management technique can be used in a new generation of wireless LAN equipment. This technique would include distributed dynamic channel assignment, and load sharing among Access Points (AP), which improves the network availability and reliability compared to centralized management techniques. In addition, it will help to increase network capacities and improve performance, especially in large-scale WLANs. Analysis results using normal and binomial distribution have been included which indicate an improvement of performance resulting from network balancing when implementing distributed resources management at WLANs.

INTRODUCTION

WLAN technology is rapidly becoming a crucial component of computer networks that has been widely used in the past few years. It provides mobility as well as essential network services where wire-line installation has proved impractical. Wireless LAN technology evolved gradually during the 1990s, and the IEEE 802.11 standard was adopted in 1997 (Crow, 1997; IEEE 802.11, 1997). The inclusion of the newer IEEE 802.11g versions of the standard offers a firm basis for high performance wireless LANs. Companies and organizations are investing in wireless networks at a higher rate to take advantage of mobile, real-time access to information. While first generation IEEE 802.11 technology is adequate for residential and small office/home office (SOHO) customers, the same is not always true for enterprise customers. In fact, some chief information officers (CIOs) and information technology managers are reluctant to deploy wireless LANs. Among their concerns are security, reliability, availability, performance under heavy load, deployment, mobility and network management. While security is often mentioned as manager's greatest worry about wireless, some of their other concerns, such as reliability, availability, performance and deployment, can be addressed through radio resource management techniques. The use of such techniques would encourage the rapid deployment of wireless infrastructure with much greater flexibility than has previously been available. The current wireless network products do not scale as well as they might in large-scale enterprise networks. IEEE 802.11 wireless networks have become increasingly popular and more widely deployed. This puts pressure to expand the functionality of wireless LAN equipment to become suitable for large scale operations. Although IEEE 802.11 task groups and study groups are working to improve the standard, there is a need for much improvement to suit the future functionality that will be added to wireless equipment.

Enterprise managers want to deploy wireless networks with several important qualities. These include: high security, highly reliable and available WLANs with very little downtime and high performance (i.e., capability of high throughput and low latency). The ideal wireless network has reliability, availability and performance criteria similar to wired enterprise networks. In addition, it should be possible to deploy wireless networks very quickly and without the need for extensive and time-consuming site surveys. Furthermore, the networks should have the flexibility needed to support load balance and changes in the radio environment. Radio resource management (RRM) forms the basis of quality of service (QoS) provisioning for wireless networks (Kayiazakos, 2004). It is an intense research area due to the wireless medium's inherent limitations and the increasing demand for better and cheaper services. There are many benefits of RRM: timely guarantees of resources for key applications, enhanced network planning and management and efficient bandwidth utilization. Improving the mobility management has been addressed in Ush-Shamszaman (2005) based on dividing the location management into two levels: intra- and inter- mobility. This will reduce the amount of signaling traffic, but still doesn't address the problem of reliability and availability. Supporting security, reliability and QoS in a dynamic environment has been discussed in DaSilva (2004), using modified routing protocol OSPF-MCDS over WLANs. It manages the bandwidth allocation using a decentralized policy-based network management scheme. In this approach, load balancing would be improved but at the expense of generating more traffic between the APs required to exchange signaling needed for the protocol OSPF-MCDS. In addition, there would be an increase of latency for terminal association moving across the boundary of multiple APs.

WLAN performance is dependent on the radio propagation environment in which the wireless LAN operates. The radio propagation environment may change from time to time, affecting connection speeds and error rates. In a manufacturing environment, for example, where the multipath environment changes as equipment is moved about, it is quite possible for a link to fail completely even if the mobile is stationary. Network management personnel in information technology departments are often unable to manage the network all the way to the user's mobile computer, and may be dependent on users to notify them of some types of problems. On the other hand, many CIOs see network management solutions extending all the way to users' laptops and other handheld devices. In WLANs, when AP is called on to serve a high number of users, it is likely to become overloaded, and the resulting congestion may significantly degrade the service received by users dependent on that AP. These issues in network management that are affected by unpredictable network load lead to problems of reliability and availability. The ideal is for WLANs to provide similar reliability and availability, and, to the extent possible, wireless LAN designers strive to provide the kind of service to which wired network users have become accustomed. Wireless users are likely to experience slower service, which is the result of the often lower transmission speeds and higher error rates on wireless links. Raw data rates on wireless LANs are typically lower than those on Ethernet networks. For example, at one time most Ethernet networks provided 10 Mb/s service to users. At that time, wireless LAN technology provided no more than 1 or 2 Mb/s raw data rate to the user. Today the numbers have increased considerably, but there is still a gap. Ethernet networks can provide 100 Mb/s or even 1000 Mb/s service to users. While IEEE 802.11n promises higher speeds, wireless LANs currently in use operate at speeds up to 11 Mb/s in the case of IEEE 802.11b, and up to 54 Mb/s in the case of IEEE 802.11a/g. But these numbers make wireless LANs sound much faster than they really are. So, for example, an IEEE 802.11b link may step down from the data rate of 11 Mb/s to 5.5, 2, or even 1 Mb/s according to network congestion. Overhead bits dramatically reduce the effective data rate available as described in IEEE 802.11, 1997; IEEE 802.11a, 1999; O'Hara (1999); and Van Nee (1999). Congestion further reduces the throughput experienced by a user because AP and the mobile computers it serves share a single radio channel. As with Ethernet, when the traffic level or number of active computers is high, congestion occurs and poor performance is the result. With switched Ethernet, however, each station can have a segment of its own. On the other hand, all stations using an IEEE 802.11 AP share the same bandwidth resource, and congestion is likely to be particularly severe in areas of high user density (Hills, 2004; O'Hara, 1999). It is highly desirable for wireless LAN equipment to include provisions to mitigate this problem and try to eliminate or reduce the effect of congestion.

Deploying WLANs seems easy, but in fact, the deployment of a well designed, large-scale wireless LAN requires a careful site survey and design, both of which can be difficult and time consuming. This is due to selection of AP locations and assignment of radio channels to the APs. The design of WLAN is usually based on signal strength measurements and on consideration of radio propagation issues. This is challenging because the building is a three-dimensional space, and an AP located on one floor of the building may provide signal coverage to adjacent floors of the same building and perhaps to other buildings. First, selection of AP location should be done to provide complete coverage of the target space without undue coverage overlap. Consideration of the characteristics of the radio propagation environment in which the wireless LAN is being deployed can be difficult but is important in a WLAN design (Hills, 2001). In a coverage-oriented design, one would like to space the APs as far apart as possible while still providing complete coverage of the target space. This will minimize equipment and installation costs, and it will also allow the minimization of coverage overlap between APs operating on the same radio channel. Such "co-channel overlap" degrades performance (Hills, 2001). Second, channel assignment is normally done in a way that minimizes co-channel overlap. This is because, with carrier sense multiple access with collision avoidance (CSMA/CA), the IEEE 802.11 multiple access scheme, co-channel overlap causes interaction between stations in different cells, degrading performance (Hills, 2001). A good site survey and design for a large-scale enterprise WLAN requires radio expertise. Since most data communications personnel lack this kind of expertise, enterprise organizations often provide personnel with the necessary training or hire an outside company to handle deployment. Careful site survey and design are time consuming but are important to the successful deployment of first-generation wireless LAN networks. Although many customers attempt to shorten the process, they may experience performance problems resulting from a less than adequate design.

Centralized management architecture has addressed the above issues to improve reliability, availability, performance and deployment effectiveness in enterprise and other large-scale wireless LANs (Hills, 2004). These improvements arise from the radio resource management algorithms contained in the software running on the intelligent switches that control APs. In this approach, the software controlling the APs attempts to optimize performance without having any direct control over client behavior, and this limits the effectiveness of the approach. The above article discussed how radio resource management is beginning to be used to mitigate some of the problems in enterprise wireless LANs. However, centralized management will still suffer from availability problems if the centralized node fails. A solution to this problem has been addressed in this paper using distributed network management that has the same benefit described in Hills (2004): reliability, availability, performance and deployment effectiveness in enterprise and other large-scale WLANs, but at higher network availability and fault tolerance as would be described in the following sections.

DISTRIBUTED WLAN NETWORK MANAGEMENT ARCHITECTURE

In a first-generation IEEE 802.11 wireless LAN, the network's intelligence is distributed among the APs. However, managing the association between mobile terminal and AP is controlled by the mobile terminal according to signal strength. This management strategy wouldn't involve the AP

Figure 1. Distributed APs management techniques

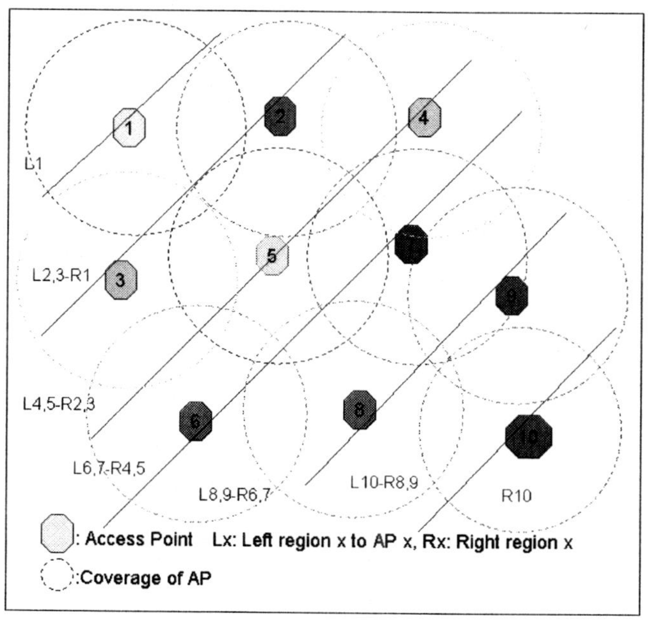

in making the decision for the association which might result in overloading some APs while other APs might be under-loaded, consequently causing congestion. In addition, channel frequency allocation for APs is done at the design and deployment stage. Radio resource management techniques that apply centralized management (Hills, 2004), require access to information that must be gathered across a number of APs, and the techniques involve control decisions that apply to a number of APs, not just one. The centralized management would reduce the congestion but it suffers from low network availability if the central point fails.

In order to reduce the congestion at the AP, and at the same time maintain high network availability, distributed dynamic network management across multiple APs has been suggested in this paper as described in Figure 1, as shown in , the dotted circle indicating the coverage of a particular AP. This coverage might overlap with an adjacent AP, where interference is caused if both APs are running at the same channel radio frequency. At the same time, mobile terminals falling within the overlapped region would have the choice to associate with either AP. To solve the problem of interference, adjacent APs should be assigned different channel radio frequencies. This can be done statistically through the design and deployment phases, or through dynamic channel assignment. The distributed dynamic channel assignment described in the next subsection has the advantage of dynamically allocating different channel frequencies to different APs. This would reduce the interference and provide high network availability at the same time.

The coverage region of every AP is divided in two regions, left and right hemispheres. This region division would be used to distribute load among adjacent APs that have coverage overlap, where a mobile terminal would choose to associate with one AP according to its location in the left or right hemisphere of the AP as described in the subsection, "Load Balancing and Channel Association." This will reduce congestion and provide load balancing between different APs.

Distributed Dynamic Channel Assignment

The performance of a network depends, in part, on the assignment of radio channels to APs. This assignment is often done using a manual process in which the designer attempts to assign the channels in a way that minimizes co-channel overlap. The coverage areas, and therefore the channel assignments, are dependent on, among other things, the radio propagation environment. Since the radio propagation environment changes, one cannot be sure that the channel assignments valid at the time the network was designed will continue to be so. However, distributing channel assignments between APs according to channel frequencies that minimize the overlap between these frequencies would reduce the interference between these channels as described in Table 1. In this table, the assignment of channel frequencies to adjacent APs, is spread from minimum to maximum frequency, which causes less interference between these APs. For example, adjacent APs 2, 2 and 3 are assigned channel numbers 1, 6 and 11 consecutively, and APs 4, 5 and 6 are assigned channel numbers 2, 7 and 10 consecutively. Distributed Dynamic Channel Assignment would add extra flexibility to assign a proper channel number to different Aps. This is done dynamically, according to AP index number, where channel frequency is assigned according to the AP index, as mapped in Table 1. This dynamic assignment would help in replacing failed APs with others using the same index number. Each AP requires an index number that can be set at the initialization stage.

Distributed dynamic channel assignment would be done according to AP index belonging to the same network, where all APs are assigned a unique index, starting from one at the left hand side and increment by one for the next adjacent

Table 1. AP frequency allocation and indexing

Channel Number	Channel Frequency (GHz)	AP Index associated with Channel number for minimum interference
1	2.412	1
2	2.417	4
3	2.422	7
4	2.427	9
5	2.432	11
6	2.437	2
7	2.442	5
8	2.447	8
9	2.452	10
10	2.457	6
11	2.462	3

AP. Then the AP would be assigned a channel frequency according to the Table 1 assignments. Since there are 11 to 14 channel frequencies (Crow, 1997; IEEE 802.11, 1997; Van Nee, 1999) available for use according to different standards (North America, most of Europe, France and Japan), different frequencies would be assigned for the first 11 APs, then these frequencies can be re-assigned for the remaining APs that exceed 11, where interference is unlikely since the APs assigned the same frequency are not adjacent and therefore would not cause signal interference. In addition, that interference between the first 11 channel frequencies would be at a minimum since adjacent AP assigned frequencies are distributed to have non adjacent frequencies as described in Table 1.

Load Balancing and Terminal Association

Since an AP and its associated clients share a limited bandwidth resource, APs can become overloaded, leading to congestion and poor network performance. On the other hand, a client may be able to communicate quite successfully with two or more APs, but at the expense of possibly causing congestion at an AP. Thus, one would like to have a wireless LAN that is capable of distributing client associations among APs more or less uniformly so that no one AP is overloaded. WLAN equipment with this capability can enhance network performance considerably. Association between a client and an AP begins with an association request that is initiated by the client. This association request is normally preceded by the client's transmitting one or more probe requests on channels it selects. In each of these probe requests, the client asks for a response from all APs operating on that channel and that are able to receive the client request. This tells the client which APs are within radio range, and the signal strengths received from the APs give an indication of which APs will be able to provide higher-quality service. Before sending an association request, a client should also have previously sent an authentication request that has been granted. The method by which a client decides with which AP to request association is not specified in the IEEE 802.11 standard. In addition, client cards produced by different manufacturers, use different algorithms for requesting an association.

The association between mobile terminals in distributed dynamic management can, however, be controlled, not only according to signal strength, which causes network congestion at some APs, *but also according to its location in the right or left hemisphere of the AP*. To determine the location of a mobile terminal within the network map, the same procedure described in Kbar (2005) can be used for wireless terminals and compared to the coordinate of the AP that detected its strong signal. This management technique works by distributing the load between adjacent APs where mobile terminals fall within the overlapped region and would be associated only with one AP located to its right hemisphere. If a mobile terminal is located within the overlapped region, the signal strength would not be used to determine which AP it will associate with. In fact, the mobile terminal location would be used to distribute load to a different AP, where it will associate with an AP that is located at the right hemisphere of itself. As shown in Figure 1, if a mobile terminal located anywhere within the region of AP-1 without overlapping with another AP, it always associates with this AP-1. However, if it locates within the overlapped region of AP-1 and AP-2, it will associate with AP-2 when the terminal is located at the left hemisphere of AP-2. The golden rule here is to associate all mobile terminals that fall within two APs (overlap region) with an AP that locates at the right of the terminal, or the left hemisphere of the AP. In other words, the AP in the right of the terminal (left hemisphere of AP region) would be the preference to associate with mobile terminals located in the overlap region with other APs. These actions are expected to significantly improve the performance of the WLAN. They also will make the site survey and design process easier because these techniques can, to some degree, compensate for errors in designing the network.

ANALYSIS RESULTS USING NORMAL DISTRIBUTION

For proving that our method is decreasing the load in heavily congested areas where there are a lot of APs in the regions, we focus on AP-5 in Figure 2, since it is in the center where there are many terminals that will heavily decrease the bandwidth available to them. Let us assume that there are 10

Figure 2. Load balancing

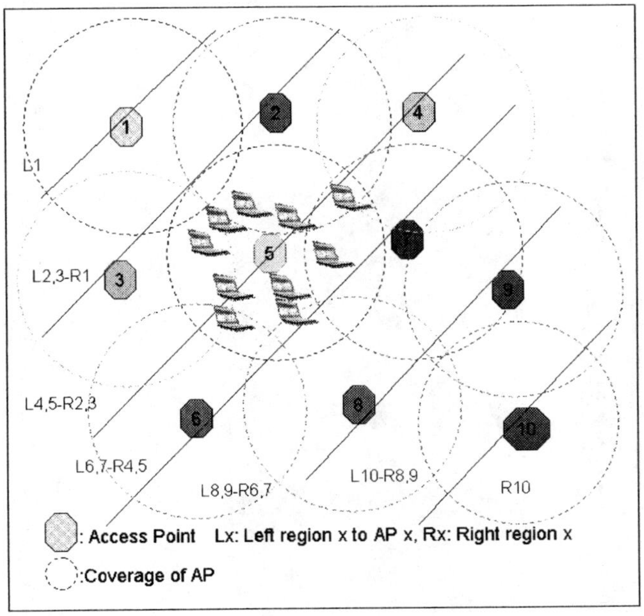

terminals as shown in Figure 2. They are usually associated with AP-5 in the normal association method since they all are closest to AP-5.

While using our method, only the five terminals on the left hemisphere of AP-5 will be associated with is plus another on the right hemisphere since it is within the coverage of AP-5 only. The other four will be associated to AP4, AP6, AP7 and AP8, with one each since they are located on the right hemisphere of AP-5 and on the left hemisphere of the other APs. This example shows that 60% of the terminals associate with AP-5 and 40% associated with other APs. Considering the same assumption that 40% of the terminals might associate with other APs according to uniform distribution when there are 20, 30, 40 and 50 terminals, we can draw the distribution of terminal association as shown in Figure 3. The graph in Figure 3 shows the relationship between number of terminals and the number of associated terminals to AP-5 for both the distributed method and the strength-based method.

Figure 3. Association distributions

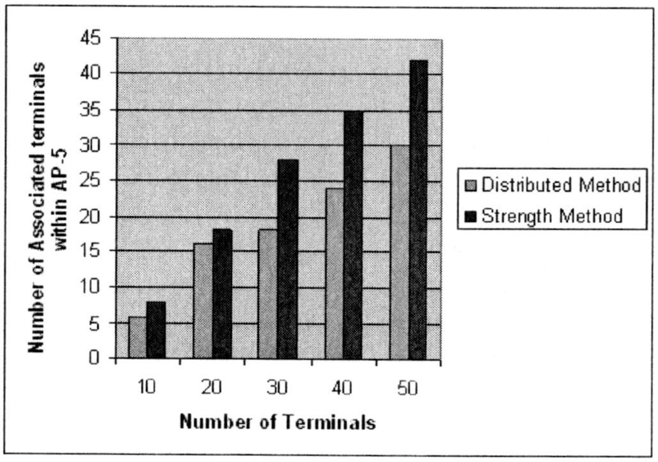

Figure 4. Average bandwidth allocation

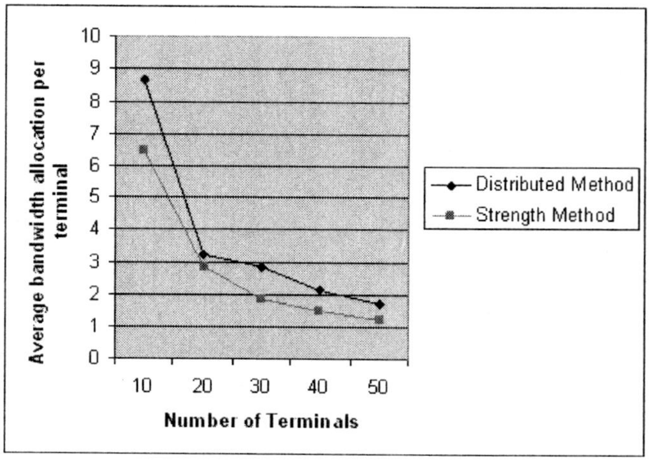

It is clearly shown in Figure 3 that around 60% of the terminals are associated with AP-5 in the case of using a distributed method, while around 75% to 100% of the terminals are associated with AP-5 when using strength-based method. We assume a uniform distribution of the terminals within the region since the random distribution is not a reliable condition for evaluation. The result also proves that by using the distributed method there will be better utilization of the bandwidth over the strength method as shown in Figure 4. Under the assumption that the maximum bandwidth of an AP is 54 Mb/sec, according to IEEE-11g standard as shown in Figure 4, the average bandwidth allocation per terminal is 9 Mb/sec when using the distribution method for a total number of 10 terminals; the average bandwidth drops to 6.5 Mb/sec using the strength method under the same condition for 10 terminals. The difference in bandwidth allocation using the two methods approaches each other when the number of terminals increases to 50. As shown in Figure 4, the average bandwidth allocation per terminal is 1.5 Mb/sec when using the distribution method for a total number of 50 terminals, while the average bandwidth drops to 1.2 Mb/sec using the strength method under the same condition of total 50 terminals.

The last AP (AP-10) would accept all associations since terminals falling within its right hemisphere don't have a choice to associate with another AP. This will increase the load for the last AP to the same level as the strength method.

ANALYSIS USING BINOMIAL DISTRIBUTION

The binomial distribution gives the discrete probability distribution $P_p(n/N)$ of obtaining exactly n successes (terminal falling only within a particular AP range, e.g., AP-5) out of N Bernoulli trials or terminals falling within the coverage of a particular AP and its neighboring APs (where the result of each Bernoulli trial is true with probability p and false with probability $q = 1 - p$). The binomial distribution is therefore given by:

$$P_p(n|N) = \binom{N}{n} p^n q^{N-n} \quad (1)$$

$$\frac{N!}{n!(N-n)!} p^n (1-p)^{N-n} \quad (2)$$

where $\binom{N}{n}$ is a binomial coefficient.

Assume that the desired probability of terminals falling within one Access Point (e.g., AP-5) only is $p = 0.8$. Therefore the probability of failure q = $1 - p = 0.2$. The total number of terminals (trials) is $N = 50$. According to binomial distribution of formula 2, the probability of at least up to i terminals is falling within only one AP (e.g., AP-5) is:

$$p(x \le i) = \sum_{k=0}^{i} \frac{n!}{(n-k!)k!} p^k (1-p)^{n-k} \quad (3)$$

The probability that more than i terminals are falling within the same AP is given by:

$$p(x > i) = 1 - p(x \le i)$$
$$= 1 - \sum_{k=0}^{i} \frac{n!}{(n-k!)k!} p^k (1-p)^{n-k} \quad (4)$$

Applying formula 4 (http://www.stat.sc.edu/~west/applets/bionmialdemo.html), we can obtain the probability for more than i terminals falling within AP-5 according to different values of i {i=20 to 48} as shown in Figure 5. As shown in this figure, the probability would reach one when the desired probability is 0.8 if the number of terminals is 32. The number of terminals would increase to 42 if the desired probability increases to 0.95.

According to distributed management and binomial distribution, the total number of terminals that fall within AP-5 can be obtained by the following formula as shown in Figure 6:

Figure 5. Probability distribution

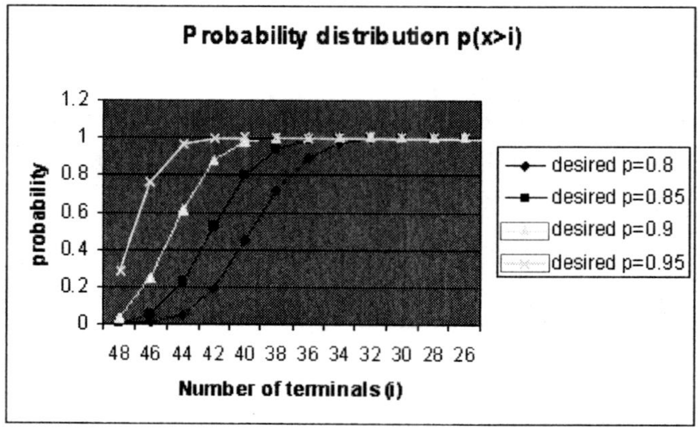

Figure 6. Total number of terminals at different desired probabilities

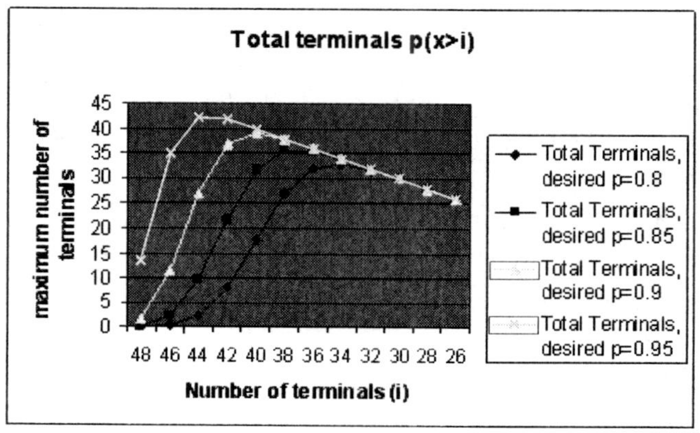

$$T = i \times p(x > i) \tag{5}$$

As shown in Figure 6, the maximum total number of terminals depends on the desired probability. When the desired probability is 0.95, the maximum number is 43 terminals. This number drops to 32 when the desired probability is reduced to 0.8.

Using signal strength terminal management, the total number of terminals that fall within AP-5 is N. Hence, the advantage of using distributed management is the reduction of the number of terminals associated with a particular AP (e.g., AP-5) by:

$$N - T = N - i \times p(x > i) \tag{6}$$

As shown in Figure 7, the total number of terminals falling within the same access points using signal strength is N (50 terminals) independent of the desired probability. The total number of terminals using distributed management depends on the desired probability, where the total number would be 32 at desired probability of 0.8, and increasing

Figure 7. Number of terminals associated with same AP

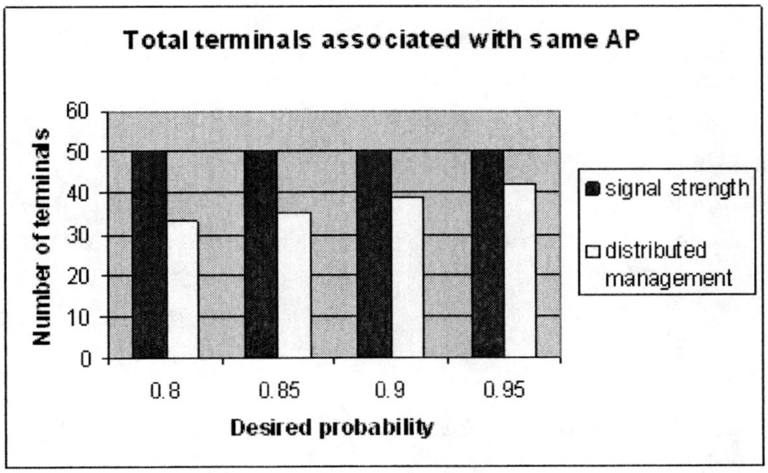

to 43 when the desired probability increases to 0.95. Hence, the total number of terminals falling within the same AP using distributed management drops by 1-32/50 (35%). This would allow terminals to have better bandwidth allocation using the distributed management scheme.

CONCLUSION

The architecture that has been described has the potential to improve performance and deployment effectiveness in enterprise and other large-scale wireless LANs, and at the same time maintain a high network availability and reliability. These improvements arise from the distributed dynamic resource management deployed in mobile terminals and APs.

The suggested technique assists the APs to optimize their performance by dynamically allocating different frequencies to adjacent APs in order to reduce the signal interference. In addition, the associations between mobile terminals and APs are distributed to different APs according to their location in the left or right hemisphere of an AP. This will reduce the congestion in APs and consequently improve the WLAN large-scale network performance.

REFERENCES

Crow, B.P. Widjaja, I. Kim, L.G. Sakai, P.T. (1997). IEEE 802.11 wireless local area networks. *IEEE Communication Magazine, 35*(9), 116–26.

DaSilva, L. A., Midkiff, S. F., Park, J. S., Hadjichristofi, G. C., & Davis, N. J. (2004). Network mobility and protocol interoperability in ad hoc networks. *IEEE Communications Magazine.*

Hills, A. (2001). Large-scale wireless LAN design. *IEEE Communication Magazine,39*(11), 98–104.

Hills, A, A. (2004). Radio resource management in wireless LANs. *IEEE Radio Communications.* Retrieved from http://www.stat.sc.edu/~west/applets/binomialdemo.html

IEEE 802.11 (1997). IEEE Ssandard. *wireless LAN medium access control (MAC) and physical layer (PHY) specifications.*

IEEE 802.11a (1999). *IEEE standard. wireless LAN medium access control (MAC) and physical layer (PHY) specifications: High speed layer in the 5 GHz band.*

Kbar, G., & Mansoor, W. (2005). Mobile station location based on hybrid of signal strength and time of arrival. *International Conference on Mobile Business ICMB2205*, Sydney. Australia.

Kyriazakos, S., & Karetsos G. (2004). Practical resource management in wireless systems. [Book reviews] *IEEE Communications Magazine*.

O'Hara, B., & Petrick, A. (1999). *The IEEE 802.11 handbook: A designer's companion*. IEEE Press.

Ush-Shamszaman, Z., & Abdur Razzaque, M. (2005). A mobility management scheme in all-ip integrated network. In *Proceedings of the 23rd IASTED International Multi-Conference Parallel and Distributed Computing And Networks* (pp. 15-17).

Van Nee, R., Awater, G., Morikura, M., Takanashi, H., Ma Webster, M., & Halford, K. (1999). New high-rate wireless LAN standards. *IEEE Communication Magazine*, 37(12), 82–88.

This work was previously published in International Journal of Business Data Communications and Networking, Vol. 2, Issue 4, edited by J. Gutierrez, pp. 46-58, copyright 2006 by IGI Publishing, formerly known as Idea Group Publishing (an imprint of IGI Global).

Chapter XXI
An Empirical Examination of the Impact of Wireless Local Area Networks on Organization of Users

Lei-da Chen
Creighton University, USA

Ravi Nath
Creighton University, USA

EXECUTIVE SUMMARY

In recent years, the concept of nomadic computing has received considerable attention from the business community. As an early form of nomadic information environment (NIE), wireless local area network (WLAN) has gained tremendous popularity with organizations. Using mostly anecdotal evidences, WLAN equipment manufacturers and practitioners claimed that WLAN brought dramatic improvements in the forms of productivity gains and attainment of convenience, flexibility, mobility, and time saving to organizations and their employees. However, very little academic research has been conducted to verify these claims and further our understanding of this new phenomenon. By surveying end users and managers, this study investigates the impact of WLAN on users and their work. Finally, recommendations to researchers, managers, WLAN technology providers, and equipment manufacturers also are provided.

INTRODUCTION

Improvements in wireless communication technologies have spawned interest in nomadic computing. Nomadic computing refers to an environment in which nomad users have access to computing resources; communication capabilities; and services that are transparent, integrated, convenient, and adaptive (Kleinrock, 2001). Such an environment offers users unprecedented capabilities to access and to distribute information, when they are on the move. While consumers are embracing some of the new wireless connectivity technologies, such as WiFi, for their homes, the biggest demand for these capabilities is likely to originate from the business community. Nomadic computing promises to enhance the level of mobility in computing and communication for employees both within and beyond organizational boundaries. Business organizations are quick to take notice of the value created by these capabilities. Key benefits of this unfettered computing include improved employee productivity, quick response to inquiries and requests, and enhanced customer services.

Growth in nomadic computing is driven by incessant advances in wireless and mobile technologies and the business need for mobility. Industry experts are predicting a growing trend toward a new way of doing business based on wireless and mobile technologies—mobile commerce (m-commerce). According to Balasubramanian, Peterson, and Jarvenpaa (2002), m-commerce refers to communicating and conducting business transactions using mobile devices. While many wireless technologies promise to revolutionize the conduct of business, organizations often fail to make a business case for investing in these technologies (Goldman, 2001; "Use Tech as a Tool," 2002). In addition, the rapid technological innovations in this field have left many IT managers still trying to sort out the different technology platforms and the type of business applications that these technologies would support effectively. Many questions remain to be answered in the area of nomadic computing providing researchers with ample research opportunities, as outlined by Lyytinen and Yoo (2002).

It is imperative to understand the business value of nomadic computing. Balasubramanian, et al. (2002) suggested that mobile technologies relax spatial and/or temporal constraints of activities. For example, with mobile technologies, a worker in the field can check e-mail at any time. Without the technologies, this activity would be limited by both spatial and temporal constraints (i.e., one can only check e-mail when one is at a location where a computer and a network connection are present). In the same vein, Chen and Nath (2004) proposed a model that helps managers to determine the value of mobile and wireless applications. The model stipulates that the value of mobile and wireless applications is a function of the user's immediacy of information needs and user mobility.

Even though a national nomadic information environment (NIE) is somewhat possible by using services provided by various national wireless providers, many small pockets of NIEs have emerged and continue to grow. Most of these NIEs with limited geographic reach serve employees within the organization's physical boundaries. Open standards, such as wireless fidelity (Wi-Fi) and bluetooth, allow organizations to develop these NIEs with relative ease and low costs. Data and information can be shared seamlessly among different devices and networks within a limited geographic area. Such NIEs are often referred to as wireless local area networks (WLANs). Many organizations have adopted the IEEE802.11b and IEEE802.11g technology, two of the Wi-Fi standards, to provide wireless access to users within a local geographical area (e.g., building, campus, airport, coffee shop, hotel). According to the 2001 NOP World-Technology (2001) study, the market penetration of WLANs in the U.S. reached 10% in 2000, and users credited WLAN with attainment of convenience, flexibility, mobility, time saving, and productivity gains. A more recent study conducted as part of the PEW Internet and American Life Project showed that 17% of Internet users have logged on to the Internet using a wireless device such as Internet-connected mobile phones and WiFi-enabled laptops (Rainie, 2004). While IEEE802.11b and IEEE802.11g are the most popular WLAN solutions among U.S. businesses today, newer technology solutions (e.g., IEEE802.11a and IEEE802.11i) are promising higher data speed and enhanced security in the near future (Funk, 2005). There is also a significant potential for WiFi technology in the consumer market. By 2009, according to Datacomm Research Company, WLAN equipment sales will triple in terms of number of units sold, primarily due to the growing use of WLAN in home entertainment applications (LAN Product News, 2005).

In the case of WLAN, while user mobility is limited to a small geographical area (e.g., an office or a building), it supports a wide range of immediacy of information needs. Therefore, while WLAN delivers important business value to organizations, the value proposition can be amplified, if user mobility is extended beyond organizational boundaries; that is, if the spatial constraints are relaxed. A new development to overcome the range limitation of WiFi is also under way. Another promising technology, WiMax, supported by the WiMax Forum and consisting of more than 200 industry members, is designed to have a bandwidth of 40 Mbps over a five-mile range (Wright, 2005).

Most research efforts in WLAN have been devoted to technical issues. For example, Xu and Walke's (2001) study examined the design issues of self-organizing broadband wireless networks. Hsu, Qian, and Ilyas (2003) proposed and analyzed two algorithms for establishing ad hoc WLANs. In another study, Lehr and McKnight (2003) compared the technical standards of 3G and WiFi, two competing technology standards for broadband wireless Internet access, and recommended a complementary strategy that allowed both standards to coexist. At the same time, there is a dearth of empirical research dealing with the impact of WLAN on users and their work environments. Most of the evidence in support of WLAN is anecdotal and conceptual (Keane, 2002; Malladi & Agrawal, 2002; McGarvey, 2002). Therefore, there is a need for a systematic, scientific, and empirical investigation that evaluates the user and work benefits of WLAN. Specifically, the proposed research seeks to investigate the impact (benefits and risks) of WLAN on users from both the end user's and the IT manager's perspectives. These findings will provide an unbiased assessment of WLAN benefits and risks and, perhaps, corroborate the claims made by practitioners and equipment manufacturers. Furthermore, understanding users' perceptions of WLAN is crucial to technology providers and equipment manufacturers, as this knowledge can guide and help them to design user-centric products and services. Also, the findings of this study should provide insights to academicians and practitioners in this evolving area of nomadic computing.

This research attempts to answer the following research questions:

- **Research Question 1:** What are the underlying constructs explaining the impact of WLAN on users?
- **Research Question 2:** What are the perceived benefits and risks of WLAN to users?
- **Research Question 3:** What factors are related to user satisfaction with WLAN?

WIRELESS LAN

The last few years have witnessed the emergence of a wide range of portable devices, including portable PCs, PDAs, handheld computers, and handsets. Many of these devices have wireless capabilities that are either embedded or attached. At the same time, the workforce is becoming increasingly mobile (Daniel, 2000). To increase the agility of an organization, employees need to have access to core corporate information from anywhere and at anytime. As a result, organizations often find themselves needing to extend their existing network to places that cannot be reached economically with cables (Singer, 2001). WLANs offer organizations a flexible and cost-effective alternative. They are especially useful in situations in which real-time information access is desired and physical network wiring is infeasible (Malladi & Agrawal, 2002). Today, WLANs are gaining popularity for the following reasons. First, WLAN technologies, especially WiFi, have reached maturity. The technical standard has received broad industry support, and hence, a high level of interoperability can be expected from WLAN products from different vendors. Furthermore, the speed of WLANs as defined by 802.11x wireless Ethernet standards have improved to be comparable to the bandwidth of wired LAN with relative reliability. Finally, the costs of WLAN equipment have decreased dramatically to make it financially feasible for many organizations to adopt them (Conover, 2000). As a result, significant growth in the WLAN market was witnessed in the last few years. Besides home and office wireless networks, many retail and service businesses have created WiFi hot spots to provide wireless Internet access to their customers (OECD, 2003).

As mentioned earlier, WLANs offer several benefits to organizations. First, WLAN can help to improve employee productivity, which leads to cost savings for organizations. For example, a healthcare organization is achieving savings of $1,000 a day per pharmacist by delivering timely information to the pharmacists' handheld devices via an in-house WLAN (Keane, 2002). Furthermore, by providing employees with vital information from anywhere and at anytime, employees can be more proactive and react more quickly to business problems and opportunities. At Carlson hotels, managers use WLAN to access all the information they need in order to manage the properties in real time on Pocket PCs. As a result, managers quickly can spot any problems or opportunities at their hotel and react to them (McGarvey, 2002). Also, a number of high-end hotels are improving customer service by experimenting with undesked staff who can check guests in at the lobby, parking lot, or conference room (McGarvey, 2002). Several major stadiums are bringing wireless point of sale (POS) to customers' seats in order to generate more sales (Baldwin, 2002). When combined with voice-over IP (VoIP) technology, WLANs can transmit voice conversations, giving WiFi users wireless access to the corporate telephone system with reduced costs (Wang, Liew, & Li, 2005). In summary, anecdotal evidences suggest that WLANs provide users with convenience, mobility, and flexibility at work. They also help to reduce costs and enhance the level of service for organizations.

In spite of many benefits that WLAN applications can provide, they do have their weaknesses. First, WLAN lacks the security, reliability, range, and bandwidth to which most users are accustomed with wired LAN (Stanley, 2002). An investigation by *Business Week* (Green, Rosenbush, Crockett, & Holmes, 2003) identified a number of challenges faced by WLAN technology, including unclear standards, spotty security, limited range, hidden costs, and the lack of interoperability. One of the most widely cited problems with WLAN is security. The nature of the radio transmission technology used in WLANs inevitably opens the possibility of eavesdropping, masquerading, and unauthorized use by hackers (Elliott & Phillips, 2004; Regan, 2003). IT professionals also are facing many complications during the implementation of WLAN due to their lack of experience in this new technical area. This problem is worsened further by the emergence of a wide range of wireless devices (Shapland, Gavurlin, & Chartoff, 2002). The conflict between the capacity of WLAN and the growing number of organizational applications makes the management of WLANs more difficult. The problem especially is pronounced when a firm uses VoIP over its WLANs. Network managers are finding out that only a few VoIP stream sessions can be supported simultaneously without affecting the performance of WLANs (Wang et al., 2005). In addition to the technical difficulties, it is still too early for organizations to determine the real return on investment (ROI) of WLAN, and many organizations are seeing little usage of WLAN by users after its implementation (OECD, 2003). Also, the uncertainty about the direction of WLAN technologies and government regulations governing the use of WLAN is leading to hesitation on the part of organizations to make sizable investments in WLAN projects (Markoff, 2002).

THEORETICAL DEVELOPMENT

The body of MIS literature over the last few decades offers many studies on the adoption of new media, such as personal computer networks (Lucas & Spitler, 2000; Nath, 1988; Remenyi & Money, 1991), the Internet (Alba et al., 1997; Chen, Gillenson, & Sherrell, 2004; Jarvenpaa & Todd, 1997), and intranets (Phelps & Mok, 1999; Lai, 2001). While each has its unique characteristics, the media in these studies, notably LANs, the Internet, and intranets, bear remarkable resemblance to WLAN.

A common theme found in literature on the adoption of new media is user satisfaction. Many research projects were dedicated to the evaluation of user satisfaction with media and the search for the determinants of user satisfaction with media. User satisfaction was identified as one of the measures for IS success by DeLone and McLean (1992). It has been used widely as the single measure of IS success due to (1) its high degree of face validity, (2) availability of reliable instruments for measure, and (3) conceptual weakness and unavailability of other measures. Much research has contributed to the development of an instrument for measuring user satisfaction (Bailey & Pearson, 1983; Doll & Torkzadeh, 1988; Ives, Olson, & Baroudi, 1983; Palvia, 1996; Phelps & Mok, 1999; Woodroof & Burg, 2003).

Another common theme found in this body of literature was risk. While each new medium had obvious benefits to users and organizations, the benefits often were coupled with risks that were disruptive. Identifying risks of new media is crucial, as it prompts organizations to reevaluate their decisions to adopt the new medium and to proactively develop ways to address these risks. As Phelps and Mok (1999) found with intranets, these risks often were not limited to technical risks. Business risks and organizational risks were found to be more complex and persistent. Based on the previous observations, this study intends to investigate the benefits and risks associated with WLAN and their relationship with user satisfaction.

The nascence of this field also is reflected by the lack of theories specifically designed to address issues in ubiquitous computing environments such as WLANs. Classic frameworks, such as the Technology Acceptance Model (TAM) and the Innovation Diffusion Theory (IDT), often are applied in order to study the adoption of new technologies. Venkatesh and Davis (2000) later proposed and validated a theoretical extension of TAM—TAM2—by including constructs related to social influence processes (i.e., subjective norm, voluntariness, and image) and cognitive instrumental processes (i.e., job relevance, output quality, result demonstrability, and perceived ease of use). Nevertheless, these theories tend to be too general in nature and fail to address the issues uniquely relevant to the technology under study (Legris, Ingham, & Collerette, 2003). Studies employing these theories often have to extend the models in order to include constructs unique to the technology under study (Chen et al., 2004; Devaraj, Fan, & Rajiv, 2002). In addition, these theories require researchers to utilize a confirmatory research design. In the case of WLAN, little empirical research on its impact on users exists; therefore, an exploratory approach is employed by this study.

There is limited literature that addresses some of the issues under study here. Lyytinen and Yoo's (2002) framework provides a broad view of NIE. This framework was used to understand the research implications of NIE for IS researchers. The framework for NIE stipulates that the three key drivers that influence and enable both nomadic computing infrastructure and services are mobility, digital convergence, and mass scale. As users move away from a stationary computing environment to an NIE, the need for mobility has enormous impact on the design of devices, content, and networks. Second, digital convergence refers to open standards that allow heterogeneous devices and networks to share information seamlessly. Finally, mass scale is reflected by the wide availability of NIE at a global level and the high usage level demonstrated by users, and is essential to the attainment of true mobility and digital convergence. These three interweaving factors influence the design and deployment of NIE infrastructure and services. Based on this framework, Lyytinen and Yoo (2002) generated the following eight research themes for NIE: individual-level services, infrastructure for individual level, team-level services, infrastructure for team level, organizational-level services, infrastructure for organizational level, interorganizational-level services, and infrastructure for interorganizational level. The eight research themes by and large are applicable to WLAN. This study attempts to address one of the eight themes—individual-level services—especially the question of what impact WLAN has on users through the three aforementioned research questions. A more recent study by Lyytinen, Varshney, et al. (2004) laid out more detailed research directions for researchers interested in studying NIE. The study suggested that researchers explore the application, network, data, security, and privacy issues relevant to NIE in order to understand the impact of NIE. Based on this recommendation, this study will focus on WLAN's potential impact on operational efficiency, decision-making effectiveness, collaboration and communication, network service quality, privacy, and data security from the perspectives of both users and network managers.

Jessup and Robey (2002) underscored the importance of social issues when studying ubiquitous computing environments. They claimed that new technologies, such as wireless technology, that enable mobile work environments inevitably would cause social consequences. Researchers should conduct three levels of social analysis: individual, team, and organization. This study again focuses on one aspect of the individual level of social analysis by addressing the impact of WLANs on users' qualities of work life.

The existing literature mentioned previously offers this study some general direction. While understanding the magnitude of value delivered by WLAN is important, it is even more critical to identify the areas from which the value comes. Equally important is the identification of factors that are preventing WLAN from realizing its full potential. This study will expand existing research by delving into these questions. An exploratory research design is utilized for this study.

RESEARCH METHODOLOGY

Questionnaire Development

In order to develop questionnaire items that measure the impact of benefits and risks, a multi-stage approach was undertaken. In stage one, the existing literature on WLAN was synthesized. Table 1 summarizes the WLAN impact areas and the supporting literature. These areas encompass issues such as the quality of the user's work life, operational efficiency, decision-making effectiveness, collaboration and communication, network service quality, privacy, and data security. From this information, an initial list of statements indicating the potential impact of WLAN on users was created. Each item was phrased so that it can be rated using a five-point Likert scale (1 = strongly disagree; 5 = strongly agree). In the next stage, this list of items was shown to several WLAN users and three IT managers responsible for WLAN in their organizations. They were asked to comment on the completeness of these items and to suggest additions, deletions, and modifications. Feedback

Table 1. Underlying areas, description and sources

Underlying Area	Description	Sources
Operational Efficiency	WLAN has been accredited for improving operational efficiency by reducing the time and costs for data entry, information search, and execution.	NOP World-Technology, 2001; Keane, 2002; Malladi and Agrawal, 2002
Decision-Making Effectiveness	WLAN has been accredited for reducing information float so that the right information can be accessed by the right person at the right time to make decisions.	Sweeney, 2000; McGarvey, 2002; Green, et al., 2003
Collaboration and Communication	WLAN has been accredited for improving connectivity between workers for collaboration and communication.	NOP World-Technology, 2001; Wang, et al., 2005
Network Service Quality	WLAN is considered slower, less reliable, and harder to use than the wired LAN.	Sweeney, 2000; Malladi and Agrawal, 2002; Shapland, et al., 2002; Cantwell, 2003; Wang, et al., 2005
Privacy and Security	WLAN is considered less secure than the wired LAN, leading to users' privacy and security concerns.	Stanley, 2002; Green, et al., 2003; Regan, 2003; Elliott & Phillips, 2004
Quality of Work Life	WLAN has been accredited for improving user mobility and flexibility and making working anywhere anytime possible.	Jessup & Robey, 2002; Cantwell, 2003

from these two groups (users and managers) resulted in significant revisions to the initial list of items. In order to refine the list further, two faculty members in IT independently reviewed the list and provided feedback. As a result of this process, 23 items were retained in order to gauge the impact of WLAN on end users. These items are listed in Table 2.

User satisfaction is a key construct that has been used widely in IS literature as a surrogate measure for system success (DeLone & McLean, 1992). This research intends to identify factors that are related to user satisfaction with WLAN. In order to ascertain user satisfaction with WLAN, five items developed by Phelps and Mok (1999) within the context of intranet user satisfaction were adapted for the WLAN environment. This set of statements was chosen over other user satisfaction instruments, since it is designed specifically for a network environment. Other measures are most suitable to end-user computing satisfaction (Bailey & Pearson, 1983; Doll & Torkzadeh, 1988; Ives et al., 1983). The five items are:

- The extent to which the user intends to use WLAN for work.
- Whether the user would recommend co-workers to use WLAN.
- The user's perception of whether implementing WLAN represented a positive move for the organization.
- The user's perception of whether he or she was better off or worse off using WLAN.
- The user's overall satisfaction with WLAN.

Table 2. WLAN user impact items

Compared to wired LAN
I1. WLAN is more convenient for the user.
I2. WLAN gives the user more mobility.
I3. WLAN allows the user to be more flexible at work.
I4. WLAN has made the user's life easier at work.
I5. WLAN improves collaboration among users.
I6. WLAN improves organizational knowledge sharing.
I7. WLAN improves organizational communication.
I8. WLAN improves the user's productivity.
I9. WLAN allows the user to reduce errors at work.
I10. WLAN allows the user to connect other handheld devices to the network more easily.
I11. WLAN saves the user time at work.
I12. WLAN improves the user's operational efficiency.
I13. WLAN reduces paperwork for the user.
I14. WLAN improves the user's decision-making capabilities.
I15. WLAN allows the user to better serve customers.
I16. WLAN allows the user to access up-to-date information.
I17. WLAN allows the user to be more proactive.
I18. WLAN is slower than wired LAN.
I19. WLAN is NOT as reliable as wired LAN.
I20. WLAN is NOT as easy to use as wired LAN.
I21. WLAN is NOT as secured as wired LAN.
I22. I am more concerned about my privacy on WLAN.
I23. I am concerned about someone intercepting information if I use WLAN at work.

Each of the five items was measured using a five-point Likert scale (1 = strongly disagree; 5 = strongly agree).

It is well established that there is usually a discrepancy in the perceived benefits and risks between IT professionals who design and implement the system and end users who use the system. It is worthwhile to investigate the magnitude of this discrepancy within the context of WLAN. Therefore, the same set of questions measuring the perceived benefits and risks of WLAN was administered to both IT professionals and end users. The two sets of responses will be compared in order to measure the degree of discrepancy.

Two separate questionnaires—one for WLAN users and one for network administrators—were developed. In addition to the relevant items discussed previously, each questionnaire had a section that elicited demographic information regarding the respondent and the WLAN in the respondent's organization.

Data Collection Procedure

Data collection posed a greater challenge than the authors expected. Besides the fact that only a small percentage of firms had implemented WLAN, many organizations remained secretive about their WLAN use due to security reasons. The data collection process began with identifying a number of organizations that had adopted WLAN. This information was obtained through direct contact with key persons of organizations in a major Midwestern metropolitan area. WLAN

Table 3. User and WLAN profile

User Profile

Gender	**n**	**%**
Male	51	77.3
Female	15	22.7

Age	**n**	**%**
25 – 34	25	37.9
35 – 44	25	37.9
45 – 54	16	24.2

Frequency of WLAN use at Work	**n**	**%**
1 (Not at All)	2	3.0
2	18	27.3
3	22	33.3
4	12	18.2
5 (Very Frequently)	12	18.2

Applications Using WLAN	**n***	**%**
E-mail	52	78.8
Corporate data access	44	66.7
Internet and WWW	49	74.2
Data input	39	59.1
Data processing	25	37.9
File sharing	30	45.5
Calendaring/Scheduling	36	54.5
Other	9	13.6
* Multiple responses were provided.		

WLAN Profile

WLAN Coverage	**n**	**%**
Corporate-Wide	3	21.4
Building-Wide	4	28.6
Department-Wide	1	7.1
Workgroup-Wide	2	14.3
Not Reported	4	28.6

continued on following page

Table 3. continued

Areas Using WLAN	n**	%
Information Technology	10	71.4
Sales and Marketing	6	42.9
Human Resources	4	28.6
Accounting	5	35.7
Finance	6	42.9
Manufacturing	4	28.6
Customer Services	5	35.7
Supply Chain Management	4	28.6
Legal	3	21.4
Research and Development	5	35.7
Other	2	14.3
** A WLAN can cover multiple functional areas.		

# of WLAN Users	n	%
Less than 25	3	21.4
25 – 99	2	14.3
100 – 499	0	0
500 – 999	4	28.6
1000 or more	1	7.1
Not Reported	4	28.6

Length of Time Using WLAN	n	%
Less than 6 months	0	0
6 months – 1 year	3	21.4
1 year – 2 years	3	21.4
2 years – 3 years	2	14.3
Over 3 years	2	14.3
Not Reported	4	28.6

users in 14 organizations with WLAN were identified. The IS or business managers of these organizations who worked directly with WLAN were contacted. They were asked whether their organizations would be willing to participate in this study. Fourteen organizations agreed to participate in this study. The 14 organizations included two higher-education institutions, two computer hardware manufacturers, two communications companies, two transportation companies, a consulting firm, a computer software company, a consumer product manufacturer, a healthcare organization, a law firm, and a wireless service firm. Upon receiving their permission, questionnaires were sent to the contact person of each organization. The contact person was responsible for distributing the questionnaires to WLAN users and to network managers at his or her organiza-

Table 4. Eigen values and percentage variation explained

Factor	Eigenvalue	% Variance
1	7.65	33.3
2	2.48	10.8
3	2.10	9.1
4	1.88	8.2
5	1.38	6.0

tion. Completed questionnaires were returned directly to the authors by the respondents using postage-paid envelopes provided by the researchers. Respondents were assured that the information collected would be kept confidential. Two hundred user surveys and 50 network manager surveys were distributed. Note that in many organizations, there were multiple network managers in charge of WLANs. Sixty-six user responses and 10 network manager responses were completed and, thus, usable for this study.

ANALYSIS AND RESULTS

User Profile

Table 3 shows the profiles of end users and WLANs in their organizations. Note that the studied organizations represent a broad spectrum of industries with WLAN usage in many functional areas within an organization. Not surprisingly, in 10 of the 14 organizations, the IT group was using WLAN, while other areas such as sales and marketing and finance were not too far behind. Also, given the newness of the technology, a majority of the WLANs had been implemented during the past three years. With respect to WLAN users, they used WLAN for a myriad of applications, such as e-mail, Internet, corporate data access, data input and processing, and calendaring and scheduling.

Factor Analysis

The 23 WLAN impact items shown in Table 1 were factor analyzed in order to identify the underlying constructs (factors). One important issue in factor analysis is the case-to-variable ratio. Stevens (1986) recommends a ratio of 5:1 to guarantee a robust and reliable factor analysis procedure. However, researchers such as Fuller and Swanson (1992) have utilized ratios as low as 2:1. In our analysis, the case-to-variable ratio is roughly 3:1 (66 users and 23 variables). This ratio is adequate, given the suggested ratio guidelines, but one needs to be cautious in interpreting the results. Principal components analysis was used in order to extract the initial factors, and the number of factors was determined by using the eigenvalue-greater-than-one criterion (Kaiser, 1974). Table 4 shows the initial statistics for the five factors with eigenvalues greater than one.

Note that the five-factor solution explains nearly 67% of the variation. Next, in order to obtain the most meaningful configuration of the five factors, several orthogonal rotation methods were tried. Varimax rotation yielded the most interpretable and meaningful factors. Table 5 shows the matrix of final factor loadings. A cutoff value of 0.40 was chosen to bond an item to a factor. Numbers exceeding 0.40 in absolute value are bolded in Table 5. An examination of the factor loadings matrix revealed that while most items bonded with only one factor, four items (I1, I6,

Table 5. Matrix of factor loadings

Item	Factor 1	Factor 2	Factor 3	Factor 4	Factor 5
I9	**0.682**	-0.113	0.338	-0.099	0.016
I11	**0.747**	0.343	0.260	-0.005	0.012
I12	**0.770**	0.378	0.214	-0.038	-0.084
I13	**0.816**	0.081	0.217	0.077	0.019
I14	**0.782**	0.032	0.277	-0.095	0.134
I16	**0.762**	0.220	-0.106	0.108	-0.024
I17	**0.714**	0.355	0.067	0.131	0.141
I1	0.168	**0.566**	0.023	-0.028	**0.419**
I2	0.117	**0.756**	-0.008	0.032	-0.017
I3	0.078	**0.717**	0.353	0.094	0.073
I4	0.367	**0.630**	0.364	0.195	0.103
I8	**0.478**	**0.556**	0.284	0.076	0.085
I15	0.343	**0.649**	0.106	0.176	0.077
I5	0.164	0.226	**0.779**	0.030	0.011
I6	**0.437**	0.034	**0.774**	0.120	-0.035
I7	0.350	0.294	**0.649**	-0.034	0.101
I21	-0.098	-0.041	0.094	**0.820**	-0.094
I22	0.102	0.083	0.016	**0.891**	0.044
I23	0.059	0.206	-0.015	**0.802**	0.099
I10	0.260	0.230	-0.236	-0.141	**0.399**
I18	0.225	**-0.464**	0.132	0.125	**0.671**
I19	-0.088	0.018	0.085	0.068	**0.869**
I20	-0.056	0.318	-0.014	-0.021	**0.734**

I8, and I18) bonded with two factors, since their factor loadings exceeded 0.40. In these instances, it was decided to assign the item to the factor with the highest loading. A thorough and careful examination of the items associated with each of the five factors (see Table 5) leads us to describe the five factors as follows:

- **Factor 1:** The seven items in this factor deal primarily with enhancing users' efficiencies, reducing errors and paperwork, saving time, improving decision making, and allowing the user to be more proactive. In light of this, the factor is named Efficiency and Effectiveness (EE).
- **Factor 2:** Items bonding with this factor reflect how WLAN enhances users' work lives by providing mobility, flexibility, and convenience, and by making the user more productive. Thus, this factor is named Quality of Work (QW) Life.
- **Factor 3:** All three items in this factor focus on improved communication, collaboration, and knowledge sharing among co-workers. Therefore, this factor is called Collaboration and Communication (CC).

- **Factor 4:** Items in this factor pertain to information security and privacy issues. This factor is named Security and Privacy (SP).
- **Factor 5:** Items in this factor address issues of WLAN reliability, speed, and ease Network Quality (NQ).

The five user impact factors, the associated items, and their mean and standard deviations are reported in Table 6. The score for each factor is computed by averaging values of items comprising the factor across all cases. Note that QW life has the highest mean value (4.03), followed by CC (3.34) and EE (3.24). The remaining two factors, SP and wireless NQ have the lowest scores of all the factors (2.78 and 2.61, respectively). This means that users do not find the quality of the WLAN to be better than that of the wired network. Also, the security and privacy issue concerns are nearly the same as they would be in a non-WLAN environment. The main benefits of wireless networks lie in how untethered network access enhances users' qualities of work vis-à-vis convenience, mobility, and the ability to access information anywhere. Furthermore, WLANs provide a platform that is highly conducive to communication, collaboration, and information sharing. In sum, WLAN tends to make users more effective in what they do and to improve their operational efficiency.

Satisfaction with WLAN

Five indicators of users' affinities with WLAN adapted from Phelps and Mok (1999) are considered. Table 7 shows the mean and standard deviation of the five items. Note that the mean for each measure is more than 3.00, reflecting a reasonable degree of agreement with each item. Next, in order to ascertain the degree of relationship among the five WLAN impact factors and the five satisfaction measures, correlation coefficients between the factors and measures were calculated, as shown in the second panel of Table 6. As the correlation table shows, Efficiency and Effectiveness (EE), Quality of Work (QW) Life, and Collaboration and Communication (CC) correlate significantly with all user satisfaction items. Wireless Network Quality (NQ) correlates significantly with two items: the extent to which the user intends to use WLAN for work and the user's overall satisfaction with WLAN. Security and Privacy (SP), however, only correlates significantly with one user satisfaction item: the user's perception of whether implementing WLAN represents a positive move for the organization. These findings suggest that IT managers need to focus on improving the benefits related to EE, QW, and CC that are brought by WLAN in order to enhance overall end-user experience. Nevertheless, while the correlations between the SP and NQ factors and since the majority of the user satisfaction items were not statistically significant, we should not underestimate or ignore WLAN security, privacy, and quality issues. User perceptions are reflections of users' actual experiences, and thus, their satisfaction with WLAN will be affected negatively only when episodes of security breach, invasion of privacy, and/or deteriorated wireless network quality occur.

IT Manager Perspectives

Scores for the five WLAN impact factors were calculated for IT managers. Figure 1 shows a comparison of the mean scores of these factors for IT managers and users. Note that IT managers' scores are higher than those of WLAN users for each of the five factors. For Wireless Network Quality (NQ), the mean for IT managers is statistically significantly higher than that for users ($t = 2.96$; $p = 0.004$) at the 0.01 level of significance. For the other four factors, the two groups (managers and users) do not differ significantly.

Table 6. Factors and associated items

		Mean	SD
Factor 1: Efficiency and Effectiveness (EE)		**3.24**	**0.82**
I9	WLAN allows me to reduce errors at work.	2.82	0.82
I11	WLAN saves me time at work.	3.36	1.12
I12	WLAN improves my operational efficiency.	3.55	1.04
I13	WLAN reduces paperwork.	3.08	1.09
I14	**WLAN improves my decision-making capability.**	**2.92**	**1.01**
I16	WLAN allows me to access up-to-date information.	3.62	1.05
I17	WLAN allows me to be more proactive.	3.35	0.97
Factor 2: Quality of Work Life (QW)		**4.03**	**0.63**
I1	WLAN is more convenient.	4.08	0.90
I2	WLAN gives me more mobility.	4.74	0.44
I3	WLAN allows me to be more flexible at work.	4.17	0.87
I4	WLAN has made my life easier at work.	3.74	0.97
I8	WLAN improves my productivity.	3.76	0.86
I15	**WLAN allows me to better serve my customers.**	**3.67**	**0.95**
Factor 3: Collaboration and Communication (CC)		**3.34**	**0.78**
I5	**WLAN improves collaboration among co-workers.**	**3.47**	**0.92**
I6	WLAN improves organizational knowledge sharing.	3.24	0.91
I7	WLAN improves organizational communication.	3.32	0.88
Factor 4: Security and Privacy (SP)		**2.78**	**0.93**
I21*	**WLAN is as secured as wired LAN.**	**2.59**	**1.14**
I22*	I am not more concerned about my privacy on WLAN than on wired LAN.	2.82	1.10
I23*	I am not concerned about someone intercepting information if I use WLAN at work.	2.92	1.06
Factor 5: Wireless Network Quality (NQ)		**2.61**	**0.73**
I10	WLAN allows me to connect other handheld devices to the network more easily.	3.09	1.00
I18*	WLAN is as fast as wired LAN.	1.95	1.00
I19*	WLAN is as reliable as wired LAN.	2.41	0.99
II20*	WLAN is as easy to use as wired LAN.	3.00	1.23

*. The item is worded negatively in the questionnaire.

Table 7. User satisfaction with WLAN

User Satisfaction Items	Mean	SD
US1: I intend to use WLAN for my work.	3.88	0.92
US2: I would recommend other staff to use WLAN.	3.85	0.97
US3: The introduction of WLAN has been a positive move for my organization.	3.94	0.89
US4: I am better off with WLAN.	3.70	0.94
GUS: Overall, I am satisfied with the WLAN in my organization.	3.62	0.97

Correlations between Impact Factors and User Satisfaction Measures

	User Satisfaction Measures				
Impact Factor	US1	US2	US3	US4	GUS
Efficiency and Effectiveness (EE)	0.524**	0.598**	0.481**	0.617**	0.443**
Quality of Work (QW) Life	0.695**	0.731**	0.736**	0.556**	0.567**
Collaboration and Communication (CC)	0.440**	0.564**	0.357**	0.319**	0.474**
Security and Privacy (SP)	0.196	0.225	0.292*	-0.002	0.211
Wireless Network Quality (NQ)	0.249*	0.210	0.182	0.067	0.391**

* significant at the 0.05 level.
** significant at the 0.01 level.

Figure 1. WLAN impact scores for users and IT managers

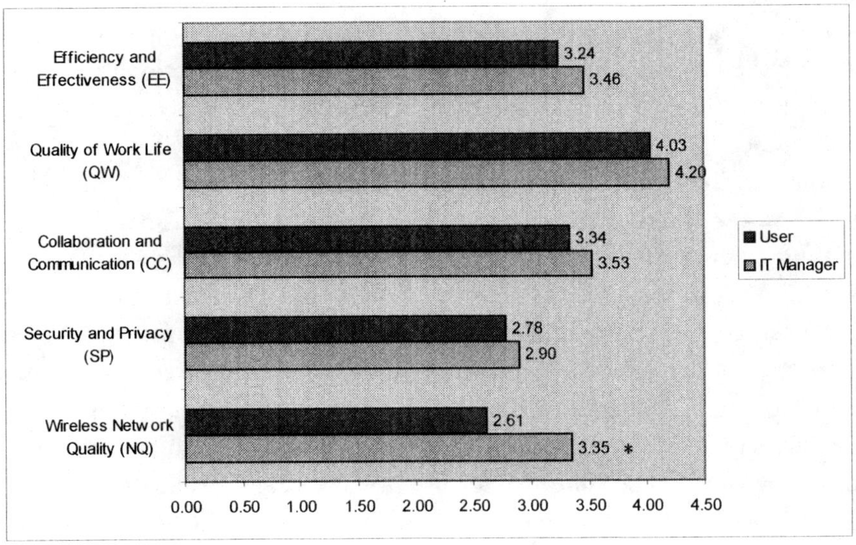

*significant at the .01 level.

SUMMARY AND CONCLUSION

This research identified five constructs underlying the impact of WLAN on users. The five factors were efficiency and effectiveness, quality of work life, collaboration and communication, security and privacy, and wireless network quality. In addition, users felt that WLAN improved their quality of work life, enhanced work efficiency and effectiveness, and led to better collaboration and communication. With respect to the issues of security and privacy and the quality of the wireless network environment, users were somewhat neutral.

The analysis further indicated that users were fairly positive with their experiences with WLAN, as measured by their intention to use it, whether they will recommend it to others, whether WLAN is a positive move for the organization, whether they are better off with WLAN, and the overall satisfaction with the WLAN. Also, three of the five impact factors (efficiency and effectiveness, quality of work life, and collaboration and communication) had statistically significant relationships with each of the five satisfaction indicators. The other two factors (security and privacy, and wireless network quality) did not correlate significantly across all satisfaction measures, indicating that there is something to be desired when it comes to network quality, security, and privacy aspects of a wireless environment.

For organizations, the findings of this study can help to design new business processes that account for the benefits of untethered computing. For IT managers, these findings can serve as a guide in educating WLAN users on the benefits and perils of a wireless environment. For wireless technology providers and equipment manufacturers, the findings of this study should help to understand better the needs of their customers in order for them to develop customer-centric products and services in the future.

There are several implications for practice of these research findings. First, even though the overall experiences of the WLAN users are positive, network managers must be proactive in educating users regarding the business benefits of the wireless environment, which include improved user efficiency, decision making, quality of work life, and communication and collaboration. Hopefully, more informed users would use the wireless environment more readily and with less hesitation. Second, IT managers need to address concerns of wireless security and user privacy. As wireless technology becomes more pervasive, security and privacy concerns will continue to be at the top of the list of issues needing attention. Once again, wireless computing with appropriate security measures that meet and exceed the needs of the nomadic users is warranted. Third, users expect reliable and efficient wireless networks, and IT managers must make every effort to deliver such an infrastructure.

IMPLICATIONS FOR FUTURE RESEARCH

This study represents an important step toward a better understanding of the impact of WLAN on users and organizational work. For IS researchers, this study provides insights into the role of WLAN within the realm of nomadic computing. These findings can serve as a guide for future research studies in this evolving and dynamic field of wireless computing. The five underlying factors of WLAN's impact on users can be considered as key variables in future investigations. Additional studies also are recommended to examine the relationship between organizational computing needs and nomadic computing environments. Besides considering the impact of network facilities, as this study has done, future studies also should take into account the impact of devices and applications. This will help to answer questions such as which mobile solutions are appropriate for corporate decision-making tasks. Thus, in sum, this research sets forth important initial directives for future research in nomadic computing.

LIMITATIONS

One limitation of this study is the relatively low case-to-variable ratio. The case-to-variable ratio for this study is roughly three. While prior exploratory studies with a similar case-to-variable ratio have proven to produce valid results, it is recommended that the findings of this study be interpreted with caution. Another limitation of this study is that it focuses only on users' perceptions of the benefits and risks of WLAN. WLAN's impact on organizations is not included in this study. In order to obtain a clearer picture of the roles of nomadic computing environments such as WLANs in today's organization, it is recommended that future studies examine the impact of WLANs on organizations by including constructs such as costs, management issues, and security measures. This will offer network managers valuable insights on how to utilize nomadic computing technologies in order to maximize their benefits to organizations.

REFERENCES

Alba, J., Lynch, J., Weitz, B., Janiszewski, C., Lutz, R., Sawyer, A., et al. (1997). Interactive home shopping: Consumer, retailer, and manufacturer incentives to participate in electronic marketplaces. *Journal of Marketing, 61*, 38-53.

Bailey, J. E., & Pearson, S. W. (1983). Development of a tool for measuring and analyzing computer user satisfaction. *Management Science*, 530-545.

Balasubramanian, S., Peterson, R. A., & Jarvenpaa, S. L. (2002). Exploring the implications of m-commerce for markets and marketing. *Journal of the Academy of Marketing Science, 30*(4), 348-361.

Baldwin, H. (2002). Sell where you want, when you want. *Mbusiness*, 29-30.

Cantwell, E. (2003). In-building wireless: How to keep a signal when you're indoors. *Wireless Business & Technology, 3*(3), 12-14.

Chen, L., Gillenson, M. L, & Sherrell, D. L. (2004). Consumer acceptance of virtual stores: A theoretical model and critical success factors for virtual stores. *Data Base*, *32*(2), 8-31.

Chen, L., & Nath, R. (2004). A framework for mobile business applications. *International Journal of Mobile Communications* (forthcoming).

Conover, J. (2000, August 7). Anatomy of IEEE 802.11b wireless. *Network Computing*, 96-100.

Daniel, D. (2000, November 24). Wired on wirelessness: No strings. *Computing Canada*, 18-19.

DeLone, W. H., & McLean, E. R. (1992). Information systems success: The quest for the dependent variable. *Information Systems Research, 3*(1), 60-95.

Devaraj, S., Fan, M., & Rajiv, K. (2002). Antecedents of B2C channel satisfaction and preference: Validating e-commerce metrics. *Information Systems Research, 13*(3), 316-335.

Doll, W. J., & Torkzadeh, G. (1988). The measurement of end-user computing satisfaction, *MIS Quarterly, 12*(2), 259-274.

Elliott, G., & Phillips, N. (2004). *Mobile commerce and wireless computing systems* (1st ed.). Essex, UK: Pearson Education Limited.

Fuller, A. L., & Swanson, E. G. (1992). Information centers as organizational innovations. *Journal of Management Information Systems, 9*(1), 47-68.

Funk, P. (2005). 802.11i secures wireless LANs. *Network World, 22*(12), 39.

Goldman, C. (2001). Data on aisle six! *Overland Park, 18*(17), 11A-15A.

Green, H., Rosenbush, S., Crockett, R. O., & Holmes, S. (2003). Wi-.i means business, *Businessweek*, 86-92.

Hair, J. F., Anderson, R. E., Tatham, R. L., & Grablowsky, B. J. (1984). *Multivariate data analysis*. New York: Macmillan Publishing.

Hsu, S., Qian, L, & Ilyas, M. (2003). An analytic study of two probabilistic models for establishing ad hoc WLANs. *Information Technology and Management, 4*(1), 55-67.

Ives, B., Olson, M., & Baroudi, J. (1983). The measurement of user information satisfaction. *Communications of the ACM, 26*(10), 785-793.

Jarvenpaa, S. L., & Todd, P. A. (1997). Consumer reactions to electronic shopping on the World Wide Web. *International Journal of Electronic Commerce, 1*(2), 59-88.

Jessup, L. M., & Robey, D. (2002). The relevance of social issues in ubiquitous computing environments. *Communications of the ACM, 45*(12), 88-91.

Keane, B. (2002). Lowering health care costs out-of-the-box. *Wireless Business & Technology, 2*(2), 36-38.

Kaiser, H. F. (1974). An index of factorial simplicity. *Psychometrika, 39*, 31-36.

Kleinrock, L. (2001). Breaking loose. *Communications of the ACM, 44*(9), 41-45.

LAN Product News. (2005). Wireless LAN equipment shipment to triple within 5 years. *LAN Product News*. Retrieved from http://cuhsl.creighton.edu/login?url=http://search.epnet.com.cuhsl.creighton.edu/login.aspx? direct=true&db=buh&an=16475413

Legris, P., Ingham, J., & Collerette, P. (2003). Why do people use information technology? A critical review of the technology acceptance model. *Information & Management, 40*(3), 191-206.

Lehr, W., & McKnight, L. W. (2003). Wireless Internet access: 3G vs. WiFi? *Telecommunications Policy, 27*, 351-370.

Lai, V. S. (2001). Intraorganizational communication with intranets. *Communications of the ACM, 44*(7), 95-100.

Lucas, H. C., & Spitler, V. (2000). Implementation in a world of workstations and networks. *Information & Management, 38*, 119-128.

Lyytinen, K., & Yoo, Y. (2002). Research commentary: The next wave of nomadic computing. *Information Systems Research, 13*(4), 377-388.

Lyytinen, K., Varshney, U., Ackerman, M. S., Davis, G., Avital, M., Robey, D., et al. (2004). Surfing the next wave: Design and implementation challenges of ubiquitous computing environments. *Communications of AIS, 13*, 697-716.

Malladi, R., & Agrawal, D. P. (2002). Current and future applications of mobile and wireless networks. *Communications of the ACM, 45*(10), 144-146.

Markoff, J. (2002). Military seeks to restrict wireless. *San Francisco Chronicle*, B1, B4.

McAdam, R. (2000). The implementation of reengineering in SMEs: a grounded study. *International Small Business Journal, 18*(4), 29-45.

McGarvey, R. (2002). Hospitality checks out wireless. *Mbusiness*, 18-23.

Nath, R. (1988). Local area networks: The network managers' perspective. *Information & Management, 14*(4), 175-181.

NOP World-Technology. (2001). Wireless LAN benefit study. CISCO thought leadership series.

Retrieved from http://newsroom.cisco.com/dlls/tln/WLAN_study.pdf

OECD. (2003). Overview of wireless LANs. *OECD Papers*, *3*(9), 7-11.

Palvia, P. C. (1996). A model and instrument for measuring small business user satisfaction with information technology. *Information & Management, 31*(3), 151-163.

Phelps, R., & Mok, M. (1999). Managing the risks of intranet implementation: An empirical study of user satisfaction. *Journal of Information Technology, 14*, 39-52.

Rainie, L. (2004). The rise of wireless connectivity and our latest findings: A PIP Data Memo. *Pew Internet and American Life Project*. Retrieved from http://www.usabilityviews.com/uv007099.html

Regan, K. (2003). Wireless LAN security: Things you should know about WLAN security. *Network Security*, 7-9.

Remenyi, D., & Money, A. (1991). A user-satisfaction approach to IS effectiveness measurement. *Journal of Information Technology*, *6*, 162-175.

Shapland, E., Gavurlin, S., & Chartoff, M. (2002). Making wireless LANs work for you. *Business Communications Review*, 30-34.

Singer, T. (2001). Wireless LANs come of age. *Plant Engineering, 44*, 46, 48.

Stanley, R. A. (2002). *Wireless LAN risks and vulnerabilities. Information Systems Audit and Control* [white paper].

Stevens, J. (1986). *Applied multivariate statistics for the social sciences*. Hillsdale, NJ: Lawrence Erlbaum.

Sweeney, T. (2000, November 13). Wireless LANs almost ready for widescale adoption. *Informationweek.com*, 286-292.

Use tech as a tool. (2002). *Men's Health*, 64.

Wang, W., Liew, S.C., & Li, V.O.K. (2005). Solutions to performance problems in VoIP over a 802.11 wireless LAN. *IEEE Transactions on Vehicular Technology*, *54*(1), 366-384.

Woodroof, J., & Burg, W. (2003). Satisfaction/dissatisfaction: Are users predisposed? *Information & Management*, *40*(4), 317-324.

Wright, M. (2005). WiMax wireless broadband. *EDN*, *50*(7), 44-50.

Xu, B., & Walke, B. (2001). Design issues of self-organizing broadband wireless networks. *Computer Networks*, *37*(1), 73-81.

This work was previously published in Journal of Electronic Commerce in Organizations, Vol. 4, Issue 2, edited by M. Khosrow-Pour, pp. 62-81, copyright 2006 by IGI Publishing, formerly known as Idea Group Publishing (an imprint of IGI Global).

Chapter XXII
Supporting Real-Time Service in Packet-Switched Wireless Networks

Maode Ma
Nanyang Technological University, Singapore

Zheng Xiang
Nanyang Technological University, Singapore

ABSTRACT

The requirement of providing real-time data service by wireless networks is a critical issue in the design of current wireless networks. Distributed Queueing Request Update Multiple Access (DQRUMA) has been proposed as a demand assignment medium access control (MAC) protocol to efficiently control the medium sharing by multiple mobile users. With the help of a packet transmission policy, DQRUMA has a potential to provide QoS service for the time-constrained burst data traffic. In this article, we study the problem of providing real-time service to fixed-size packets with or without time constraints in wireless networks. In particular, we propose a novel scheduling scheme for the DQRUMA protocol to control the packet transmission in packet-switched wireless networks. We have conducted extensive simulation experiments to evaluate the performance of the proposed algorithm and to compare its real-time performance with those of other packet transmission policies. This study proves that the new algorithm is an efficient transmission policy for the DQRUMA to support real-time service in wireless networks.

INTRODUCTION

With the recent development in wireless technology, wireless networks have been widely employed to support data and multimedia applications. One good example of the successful development of wireless networks is the deployment of wireless LAN systems. The wireless LAN has become a very popular wireless access network, which can be found in operation on university campuses, airports, company offices, homes, and even in cafés like McDonald's in recent years. Within the coverage of a wireless LAN system, anyone can enjoy the great mobility offered by the wireless technology in order to access huge amounts of information through the Internet with a notebook or a PDA. Wireless LANs have more functions, especially with mobility, to substitute the wired local area networks. Another example is the wireless access network or WiMAX, which is a promising technology for future broadband services due to its capacity to provide different kinds of communication services in various application environments and, at the same time, guarantee an agreed quality of service (QoS). Theoretically, there are two types of system structures of the wireless data networks. One is the centralized network with a base station to control the entire network. The other is the ad hoc network, in which every mobile works in peer-to-peer fashion. The wireless air interface consists of the wireless physical layer and data link control layer, which contains a medium access control (MAC) and a logical link control sublayers.

Based on the network structure, a medium access control (MAC) protocol is needed to schedule packet transmission over limited wireless channels. In particular, the design of a medium access control protocol that makes efficient use of the limited resources while satisfying the constraints on the transmitted packets and the network is highly expected. The collision-free MAC protocols proposed for the centralized structure of the wireless network can be divided into two categories: polling-based techniques and reservation-based techniques (Gummalla, 2000). The polling-based techniques (Tang, 1998; Zhang, 1991) assign transmission rights to different nodes by polling each node in the network ahead of transmission. In the reservation-based protocols (Bolla, 1997; Mikkonen, 1998; Wong, 1993), channel access requests are transmitted to inform the base station of the required timeslots and constraints on the transmission by competing the transmission channel or another channel; time slots on the transmission channel will be assigned for the individual mobile node, based on the channel access requests from them. Reservation-based techniques are more dynamic in nature to meet different QoS requirements of packet transmissions. The objective to design the reservation-based protocol is to efficiently schedule packet transmission while effectively meeting QoS requirements of the transmission and the constraints on the data and the network.

The DQRUMA is an efficient demand assignment MAC protocol proposed for centralized wireless networks (Karol, 1995). It is designed for fixed-length packets transmission in a TDM system. Its operation can be divided into two phases: a request access (RA) phase and a packet transmission phase. The DQRUMA uses two physical transmission frequencies that serve as up-link and down-link channels, respectively. The up-link channel is divided into a series of mini-slots used for requesting access (RA channel), each one followed by a slot for packet transmission (transmission channel). When a packet arrives in an empty buffer at a mobile, it will send in contention with other mobiles an access request (transmission request) to the Base Station (BS) on the up-link RA channel. When the BS successfully receives a transmission request from a mobile, it sets the corresponding entry in its Request Table (RT) to indicate that a mobile has a packet to transmit. The BS informs the reception of the transmission request by broadcasting the Access ID over the down link. When there is no acknowledgment received, the mobile will try to retransmit after a back-off interval, for which the truncated binary exponential back-off algorithm could be used. Slotted Aloha or some other random access MAC protocols could be adopted to access the RA channel. According to a desired scheduling policy, the BS chooses one of the mobiles that has a non-empty entry in the RT and broadcasts the Access ID over the down-link channel to allow the desired mobile to start its transmission. With cell transmission, it also includes Piggybacking request, without contention, to indicate it has more packets in its buffer.

Many scheduling policies have been proposed for the DQRUMA algorithm to schedule the transmission of the packets represented in the RT. Typical scheduling schemes include First Come First Serve (FCFS), Round-Robin (RR), and Guaranteed Bandwidth Minimum Delay (GBMD) (Pancha, 1995; Verikoukis, 2000). GBMD is proved to be an efficient scheduling scheme to provide guaranteed bandwidth for each traffic stream while attempting to minimize the system delay. However, all of the previous scheduling polices have not considered providing real-time service to the packets with time constraints. One of the important issues of wireless packet-switched networks is to provide real-time communication service for applications with time constraints. The most important aspect of the time-constrained applications is that a packet generated at a source must be received at its destination within a given amount of time. This time is referred to as *packet deadline*. If the delay of a packet in the network exceeds its time constraint, the packet will be considered lost. The lost packets will be dropped before transmission. To provide real-time service to time-constrained packets is to schedule and manage the transmission of the packets to meet their time constraints as much as possible.

There are numerous research results on scheduling algorithms to provide real-time service to the packets with time constraints in various network environments (Maode, 1999). A few reports on medium access control protocols, which can provide real-time service to time-constrained packets in wireless packet switched networks, have been found. Choi (2000) has considered the connection-based real-time service to real-time traffic flows. To schedule the transmission of real-time packets, it has taken the Earliest-Deadline-First principle. While scheduling the transmission of non-real-time packets, it has taken the simple Round-Robin way. Cho (2000) has provided real-time service based on the static priority and the traffic load of each traffic flow. The packets in real-time traffic flows will have higher priorities. The transmission of packets will be scheduled based on the priority of a packet and on the packet arrival rate of the traffic flow from which the packet comes. In this article, we explore the ideas in the previous results to develop a scheduling algorithm named Batch-Based Minimum Laxity First (BBMLF) in order for DQRUMA protocol to support real-time service in the wireless packet-switched networks. The beauty of the proposed algorithm is that it schedules the transmission of the packets by a dynamic priority scheme, making real-time burst traffic to be served within its deadline as much as possible.

The remainder of this article is organized as follows. The second section specifies our system model and the service provided by the network. The third section presents our algorithm and the analysis of its complexity. The fourth section shows our results from simulation experiments. Finally, the fifth section concludes the article with a summary.

SYSTEM MODEL

In this article, we consider packet transmission in a centralized wireless network. There are N mobile stations communicating with a base station through two radio frequencies in the network. One of them is the down link (from the BS to mobiles). The other serves as an up link (from mobiles to the BS). The down-link channel is used to broadcast control information to all the mobiles and to transmit specified data to the desired mobile from the BS. The up-link channel will be shared by all of the mobiles to transmit control information and data to the BS. The DQRUMA algorithm is used as the MAC protocol to control the shared medium. The mobiles are assumed to generate fixed length packets with or without time constraints. The up-link and down-link channels are time slotted. The basic time interval on the channels is the transmission time of one packet. A queue of packets to be transmitted is assumed to exist at each mobile node. According to the operation of the DQRUMA protocol, the packet transmission procedure in the specified wireless network will be as follows. Once a mobile has a packet to transmit, it first has to send a request in contention with others to the BS. The BS keeps a record of the request in its RT and informs the mobile of its reception, while BS continues scheduling the requests in its RT by a certain scheduling policy. Once a request is permitted, it will be informed to the mobile. Then the mobile can start transmission of this packet. The mobiles in the wireless network can be modeled as sources to generate packets with or without time constraints with a queue keeping each packet flow. The BS can be modeled as a server to provide transmission service to all of the traffic flows. This transmission procedure could be modeled as a service process in which one server provides transmission service to multiple traffic flows according to a certain service policy. We present our model of the packet transmission procedure controlled by the DQRUMA protocol in the centralized wireless network in Figure 1.

The service provided by the centralized wireless network to the packets, as described previously, is to transmit the real-time packets to meet their time constraints as much as possible; at the same time, the transmission of non-real-time packets also must be ensured. The real-time performance of the service of the wireless network can be described by mean packet loss rate for the packets with time constraints. The performance of the service of the network to the packets without time constraint can be measured by mean packet delay and system throughput.

Figure 1. System model of packet transmission

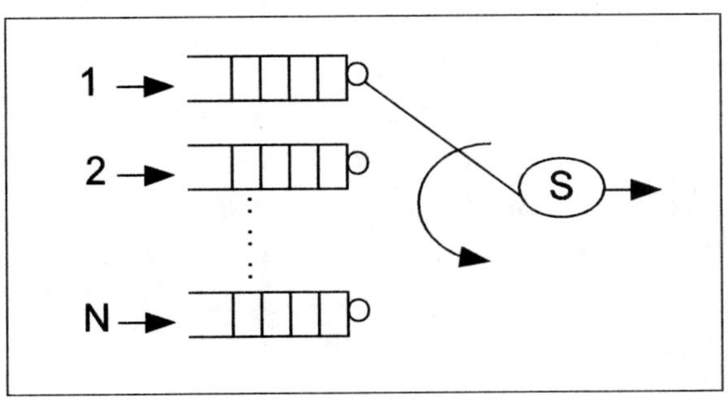

SCHEDULING ALGORITHM

In general, a scheduling algorithm of the MAC protocol to schedule packet transmission should consider two issues: the resources assignment and the packet transmission ordering. Since, in the system model of the centralized wireless network, there is only one server to provide the transmission service, there is no resource assignment issue to be considered. The function of the scheduling algorithm for the server is only to determine the packet transmission sequence from different traffic flows. To determine the packet transmission sequence, various scheduling algorithms have been proposed for the DQRUMA protocol. The FCFS scheduling determines the packet transmission sequence, based on the sequence of the packet's coming. However, it has been shown that there is no fairness to different traffic flows. The RR scheduling determines the sequence with the consideration of the fairness to different flows. It periodically scans every flow to pick up the first packet to transmit in each flow. It has distributed evenly the transmission right among all the flows. However, some packets that come into the flows that have just been scanned may experience a longer delay resulting in a larger delay deviation for the whole network. The GBMD scheduling can ensure the minimum delay and guarantee bandwidth for each flow. Its operation has two phases. One of them is to use a Guaranteed Service Queue (GSQ) in order to meet the bandwidth requirement of each flow so that fairness can be achieved. The other is to schedule packets at the head of each flow according to FCFS principle. Its operations rely on the operation of a counter for each flow with decremental value *r*. The implementation of the GBMD is complex. The value of *r* is difficult to set. If it is too large to 1, the GBMD is approaching the RR; vice versa, it goes to the FCFS. Based on the observations of these scheduling algorithms and the motivation to provide real-time service to the time-constrained packets in different traffic flows, we have an idea to develop a dynamic priority scheme as a scheduling policy for the server. We name it Batch-Based Minimum Laxity First Scheduling (BBMLF) algorithm.

We consider that the desired dynamic priority scheme should work, based on a batch to avoid starvation problem. The packet at the head of each queue in a certain time period will be included in a batch. We then assign a different priority to each packet in one batch according to characteristics of each packet and to the designed priority scheme. We consider the time constraint to each packet, which is represented by laxity of a packet, and the importance (or static priority) of a packet to be the factors of the priority assignment. We design the dynamic priority assignment scheme presented by the following formula to assign the priority to the packets in one batch:

$$P_i = T_{ni} + \alpha SP_i$$

where P_i, the dynamic priority assigned to packet *i*, T_{ni}, is the normalized relative laxity of packet *i*, which presents the urgency of the packet, SP_i is the static priority of packet *i*, which shows the different importance of different kinds of packets (the packet with the least value of SP_i has the highest priority), *a* is the normalized parameter for the static priority. The priority assigned to a packet is a dynamic one in the meaning that changes with time and network situation. It is also an integrated scheme in the meaning that it has combined different priority assignment schemes. By this priority scheme, the packet with the least value of P_i will be regarded as the packet with the highest priority, which will be transmitted first.

We describe the BBMLF algorithm formally as follows: We assume that there are *N* traffic flows and one server in the network as our system model. In each flow, the arrival of packets is a Poisson process. The packets may have time constraints, represented as Laxity that follows an Exponential distribution. The packets are indicated with different importance based on the traffic flows that exist. There is a virtual queue, named BBMLF queue, to temporarily hold the packets in one batch. The packets can be served (transmitted) in the following procedure:

BBMLF Algorithm:

A: Test whether BBMLF queue empty;
No, go to B.
Let the packet at the head of each flow enter the queue.
Sequence the packets in the order of the priority scheme.
Assign priority to each entered packet;

B: Serve one packet according to its priority;
Go to A.

The complexity of the BBMLF algorithm can be evaluated according to its operation. It can be found that the algorithm has only one sequencing procedure to schedule each packet. If the number of traffic flows (mobile nodes) is *N*, then the complexity of the algorithm to schedule the packets in one batch will be $O(N\log_2 N)$, in the worst case, by a popular sequencing algorithm.

EXPERIMENT RESULTS

In this section, we present the results of a set of performance comparison experiments. We study the performances of the network supported by DQRUMA MAC protocol with four scheduling algorithms, when an integrated traffic (including the packets with and without time constraints) applied to the network. They are the FCFS algorithm, the RR algorithm, GBMD algorithm, and the proposed BBMLF algorithm. The simulation model essentially has been built up by C programming language with the aid of SSS library, which has been described in Pollatschek (1995).

In the first group of simulation experiments, we take the following parameters. The number of traffic flows is 10. Packet arrival is a Poisson process. The mean packet arrival rate to each of all of the flows ranges from 0.01 to 0.1 packets per unit time. The packet time constraint is expressed as relative laxity, which is a random variable following an Exponential distribution with mean value as 5 time units. The length of a packet is fixed and assumed that the transmission time of the packet is one time slot. Each type of packets (real-time or non-real-time) occupies 50% of the total population.

Figure 2 presents the real-time performance of the system using four algorithms during the transmission of the different kinds of packets. The real-time performance is measured by the packet loss rate, which is the percentage of the number of real-time packets dropped compared with the total number of packets that have been transmitted. The performance of the BBMLF algorithm is shown to be the best among those of four algorithms in comparison. The BBMLF algorithm has much improved the real-time performance in that the packet loss rate of the BBMLF algorithm is only around than 40% of that of the RR algorithm and 50% of that of the GBMD algorithm when the traffic becomes heavy. It also improves the performance of the FCFS algorithm around 8%. The reason for this improvement is that when the order of the packet transmission is considered, only the BBMLF algorithm manages the packet transmission, based on the time constraints of the transmitted packets, while others have not counted it to be a decision factor. We can get a conclusion by the result that only with consideration of the time constraints of the packets can the packet transmission deserve very low packet loss rate.

Figure 3 presents the relationship between the average packet delay and the total traffic load in the network for all four algorithms. The figure shows that the FCFS algorithm performs best among all algorithms. However, the BBMLF algorithm has improved the network performance

Figure 2. Loss rate vs. traffic load

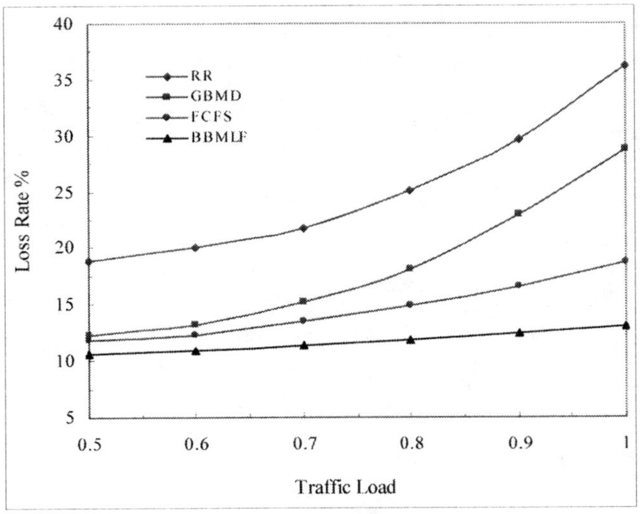

in a sense that the average packet delay of the BBMLF algorithm is only 20% of that of the RR algorithm and around 50% of that of the GBMD algorithm, when traffic intensity increases. The reason for this improvement is obvious: the order of the packet transmission has less flexibility by either the RR or the GBMD algorithm, resulting in relatively high packet delay, while the BBMLF algorithm is more free to change the order of the packet transmission according to dynamic information of the packets, resulting in receiving not only low packet loss rate but also low packet delay. Although the average delay of the BBMLF algorithm has no improvement to that of the FCFS algorithm, it has been shown to approach it very closely.

In the second group of simulation experiments, we take the following parameters. The number of traffic flows is the same as 10. Packet arrival is a Poisson process. The mean packet arrival rate to each of all the flows has been fixed, making the total traffic load in the entire network 0.9. The time constraint of packet, laxity, follows an Exponential distribution. The mean value of the laxity changes from 5 to 50 time units. The length of a packet is the same as designed before. Each type of packet (real-time or non-real-time) still occupies 50% of the total population.

Figure 4 presents the relationship between the packet loss rate and the packet laxity when the total traffic intensity is 0.9 for all four scheduling algorithms. It shows that the packet loss rates decrease for all four algorithms when the time constraints of the packets have been relaxed. The decreases of the loss rate go sharply, until the mean value of laxity is 25. Then, they change smoothly and go to convergence. This shows that, after 25, further relax on the time constraints will not have much effect on real-time performance. Also, the improvement achieved by the BBMLF algorithm will be decreased to a small extent, although the loss rate for the BBMLF algorithm still keeps to the lowest value. The reason for this fact is that with time constraints relaxed, more packets will be considered and treated as non-real-time packets, resulting in time-constraint-based scheduling losing its effectiveness.

Figure 5 shows the relationship between the average packet delay and the packet laxity when the total traffic intensity is 0.9 for all four schedul-

Figure 3. Average delay vs. traffic load

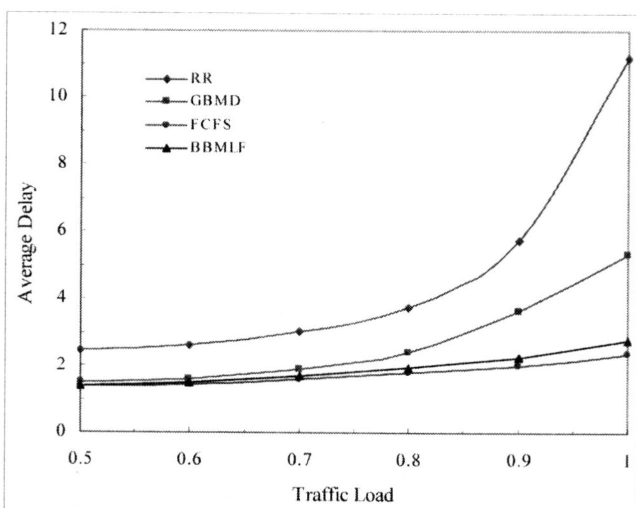

Figure 4. Loss rate vs. laxity

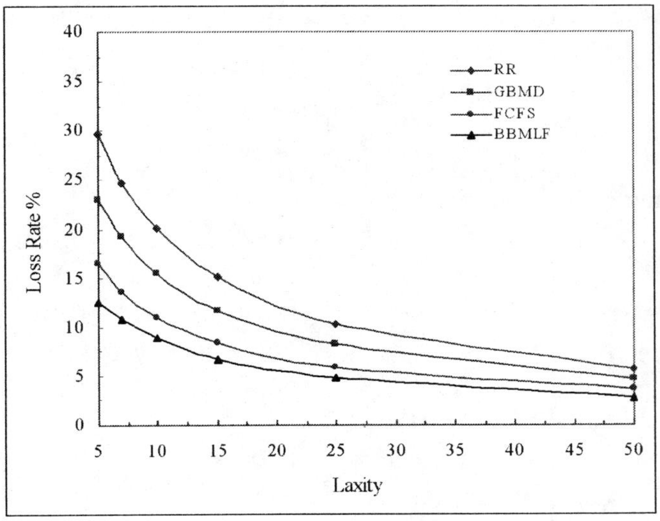

Figure 5. Average delay vs. laxity

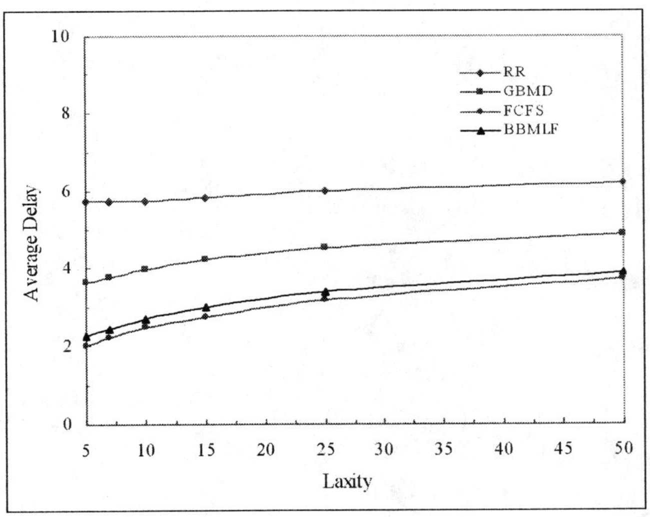

ing algorithms. It shows that the average packet delays remain almost unchanged or increase a little for all four algorithms, when the time constraints of the packets have been relaxed. It reveals that relax on the time constraints will not have much effect on the average packet delay, although there are small increases for the GBMD algorithm, the FCFS algorithm, and the BBMLF algorithm.

CONCLUSION

In this article, we have proposed a new packet transmission policy for the DQRUMA protocol to support real-time services in the packet-switched wireless networks. The proposed algorithm considers the order of the packet transmission, based on the individual time constraint to each packet and the static priority of each traffic flow. It has been shown that the algorithm largely can reduce not only the packet loss rate but also the average packet delay, when an integrated traffic is applied to the packet-switched wireless network so that the transmission of packets with or without time constraints could benefit in the network topology.

REFERENCES

Bolla, R., Davoli, F., & Nobile, C. (1997). A RRA-ISA multiple access protocol with and without simple priority schemes for real time and data traffic in wireless cellular systems. *Mobile Networks and Applications, 2*(1), 45-53.

Chandra, A., Gummalla, V., & Limb, J. O. (2000). Wireless medium access control protocols. *IEEE Communications Surveys & Tutorials, 3*(2), 1-14.

Cho, S. H., Yang, S. J., Heu, S., & Park, S. H. (2000). QoS oriented bandwidth management scheme for stable multimedia services on the IMT-2000 networks. In *Proceedings of the International Conference on Information Technology: Coding and Computing* (pp. 289-294).

Choi, S., & Shin, K. G. (2000). A unified wireless LAN architecture for real-time and non-real-time communication services. *IEEE Transaction of Networking, 8*(1), 44-59.

Karol, M. J., Liu, Z., & Eng, K. Y. (1995). An efficient demand-assignment multiple access protocol for wireless packet (ATM) networks. *Wireless Networks, 1*(3), 267-279.

Ma, M., Hamidzadeh, B., & Hamdi, M. (1999). Efficient scheduling algorithms for real-time service on WDM optical networks. *Photonic Network Communications, 1*(2), 161-178.

Mikkonen, J., Aldis, J., Awater, G., Lunn, A., & Hutchison, D. (1998). The MAGIC WAND: Functional overview. *IEEE Journal of Selected Areas in Communications, 16*(6), 953-972.

Pancha, P., & Karol, M. (1995). Guaranteeing bandwidth and minimizing delay in packet-switched (ATM) networks. In *Proceedings of the IEEE Global Telecommunication Conference, 2* (pp. 1064-1070).

Pollatschek, M. A. (1995). *Programming discrete simulations.* New York: Prentice Hall.

Tang, R., & Niu, Z. (1998). A polling-based scheduling scheme for guaranteed quality-of-service wireless multiple access to ATM networks. In *Proceedings of the International Conference of Communication Technology* (pp. 10.1-10.5).

Verikoukis, C. V., & Olmos, J. J. (2000). On the performance of the DQRUMA with different scheduling algorithms for multimedia traffic in an indoor environment for W-ATM networks. In *Proceedings of the Symposium on Communications and Vehicular Technology* (pp. 103-108).

Wong, W. C., & Goodman, D. J. (1993). Integrated data and speech transmission using packet reservation multiple access. In *Proceedings of the IEEE International Conference of Communications* (pp. 172-176).

Zhang, Z., & Acampora, A. S. (1991). Performance of a modified polling strategy for broadband wireless LANs in a harsh fading environment. In *Proceedings of the IEEE Global Telecommunication Conference* (pp. 1141-1146).

This work was previously published in International Journal of Business Data Communications and Networking, Vol. 2, Issue 1, edited by J. Gutierrez, pp. 32-43, copyright 2006 by IGI Publishing, formerly known as Idea Group Publishing (an imprint of IGI Global).

Chapter XXIII
Effect of Wireless Channels on the Performance of Ad Hoc Networks

Q. Nasir
University of Sharjah, UAE

M. Al-Dubai
University of Sharjah, UAE

S. Harous
University of Sharjah, UAE

ABSTRACT

Mobile ad hoc (MANET) network is a collection of wireless mobile nodes dynamically forming a temporary network without the use of any existing network infrastructure or centralized administration. To accomplish forwarding a packet to its destination, a routing protocol is used to discover routes between these nodes. This article presents a variety of results for packet-level simulations for the popular protocol—dynamic source routing (DSR)—when different channel models are used. Different radio propagation models representing the wireless channel have been proposed over the years, each one being suitable for a certain situation. The simplest model that represents wireless propagation is the freespace model. Other propagation models are the tworay ground reflection model and the shadowing model. Simulation results show that the performance metrics are highly affected by the channel model used, even the energy left or the number of nodes left alive are also different.

INTRODUCTION

A mobile ad hoc network (MANET) is a collection of wireless mobile nodes that can dynamically form a temporary network to exchange information without using any pre-infrastructure networks (Jayaputera & Taniar, 2005; Woesner, Ebert,

Schlager, & Wolisz, 1998). This may be done either because it may not be economically practical or physically possible to provide the necessary infrastructure or because the situation does not permit its installation. Some classic examples would be situations where friends or business associates would run into each other in an airport terminal and wish to exchange business cards, or in case of an emergency, a group of rescue workers may need to be quickly deployed. In such situations, a collection of mobile hosts with wireless network interfaces may form a temporary network without the aid of any established infrastructure or centralized administration. This type of wireless network is known as a mobile ad hoc network (MANET).

In the case where only two hosts, within the transmission range, are involved in the ad hoc network, no real routing protocol or routing decisions are necessary. But in many practical ad hoc networks, two hosts that wish to correspond may not close enough to be within wireless transmission range of each other. These hosts could communicate if other hosts between them also participating in the ad hoc network are willing to forward packets for them (Waluyo, Goh, Taniar, & Srinivasan, 2005).

In MANET, all nodes behave as routers and take part in discovery and maintenance of routes to other nodes in the network. Route construction should be done with a minimum of overhead and bandwidth consumption. These mobile devices are battery operated so extending the battery lifetime has become an important objective. Many researchers have recently started to consider power-aware design of network protocols for the ad hoc networking environment. As each mobile node in a MANET performs, the routing function for establishing communication among different nodes the "death" of even a few nodes due to energy exhaustion might cause disruption of service in the entire network (Papadimitriou & Tsaoussidis, 2005; Woesner et al., 1998).

Routing protocols in ad hoc networks are divided into two types, proactive (*table driven*), and reactive (*on-demand*) routing. In proactive routing protocols, each node maintains a routing table containing the information for every other node in the network. This leads to a significant overhead on network traffic to keep this information up-to-date. Examples of proactive protocols include destination sequence distance vector (DSDV) (Perkins & Bhagwat, 1994) and the fisheye state routing (FSR) (Pei, Gerla, & Chen, 2000). In reactive routing protocols, a discovery process determines the path to the destination only when the node has a packet to forward, that is, it reacts to a request to send data to a host. These types of routing algorithms are also referred to as on-demand routing protocols. Two prominent examples are dynamic source routing (DSR) (Johnson & Maltz, 1996) (Johnson, Maltz, & Broch, 2001) and ad hoc on-demand distance vector (AODV) routing algorithm (Woesner et al., 1998). In Akkaya and Younis (2004) and Broch, Maltz, Johnson, Hu, and Jetcheva (1998), the performance of some ad hoc routing protocols has been studied and compared using Ns-2 simulator. Simulation result show that DSR performs better than other routing protocols at different mobility rate and movement speed, that's why the performance of DSR has been studied further in (Iyer, Kanodia, & Mahsa, 2005; Nasir, Al-Dubai, & Harous, 2005).

The DSR protocol was developed by researchers at Carnegie Mellon University CMU (Fall & Varadhan, 2003) and is one of the protocols being considered for standardization by the Internet engineering task force (IETF). In DSR, a routing header is prefixed to each packet entering the network that specifies the complete sequence of nodes on which the packet should be forwarded. This route is obtained through route discovery. When a node has a packet to send for which it does not have a route, it initiates route discovery by broadcasting a route request packet. In a request packet is stored the sequence of nodes that the request visits as it is propagated through the network. If a node, say x, receives a request and has a route to the destination, then it sends a route reply packet to the requester with the new route formed by concatenating the node sequence in the request to the route stored by x. To limit how far a request is propagated, a time-to-live (TTL) field is attached to every request along with a unique request identifier. A node drops any request that it has either seen before or has lived beyond its TTL. To reduce the number of route discoveries, each node maintains a cache of routes that it has learned. A node may learn of a route through route discovery, or through other means such as snooping routes in route replies and data packets, or eavesdrop- ping on local broadcasts. This cache is updated through route error messages. A route error message is sent by a node when it discovers that a packet's source route is no longer valid (Johnson et al., 1996).

The route discovery protocol, as implemented by CMU, has two phases: a local broadcast (a ring-0 search) followed by a propagating search. The ring-0 search is initiated in the hope that a route can quickly be found in a neighbour's cache. If a route is not found within a small amount of time, a propagating search is attempted. If this fails, the protocol backs-off and tries again; eventually giving up if a route is not found. This procedure repeats until all of the packets queued for that particular destination are dropped from the queue, or a route is found. A packet may be dropped from the queue if a route has not been found within a pre-specified amount of time (the send buffer timeout interval), or if the queue is full and newly arriving packets force it out. Route discoveries for the same destination are limited by the back-off and retry procedure, which is initiated per destination (vs. per packet). Thus, regardless of the number of packets that need a route to the same destination, only one route discovery procedure is initiated.

Once a route is found and a packet is sent, there is the possibility that the route becomes "stale" while the packet is in fight, because of node mobility (a route is "stale" if some links on the route are broken). In such an instance, DSR uses a mechanism called packet salvaging to re-route the packet. When a node x detects that the next link in a packets route is broken, it first sends a route error message to the node that generated the packet's route to prevent it from sending more packets on the broken route. Then, node x attempts to salvage the packet by checking its cache to see whether another route exists to the packet's destination. If so, node x inserts the new source route into the packet and forwards it on that route; if not, the packet is dropped.

Ns-2 (Fall et al., 2003) support three types of wireless channel models, (a) *freespace* model, (b) *tworay* ground reflection model, and (c) *shadowing* model. In this article, we will study the performance of DSR under these channel models and what is the effect of these models on the energy of the network using different environment settings as well as other metrics that are discussed in the third section.

The rest of this article is organized as follow: the second section will explain the wireless channel models. The third section will present the simulation environments. Simulation results will be discussed in the fourth section, which will also conclude the article.

WIRELESS CHANNEL MODELS

Noisy, error-prone, and subject to frequent outages, conditions on a wireless link can be, on average, a hundred times worse than on a wired link. As a signal from a radio propagates through the environment, it is dramatically attenuated and distorted by physical phenomenon such as path loss, shadowing, and fading. The degree and nature of that depends on many factors such as the operating frequency, the characteristics of the terrain, and the mobility of the participating radios. Such phenomena are commonly categorized based on the timescale of their fluctuations, where path loss and shadowing are large-scale effects, and multipath fading is a small-scale effect. When we couple these effects with the inherent noise from the receiver's electronics and interference from competing transmissions, the original signal is often difficult to recover. The result is a potentially unreliable communication channel. In this section, we give an overview of these phenomena and their impact on three wireless channel communications, *freespace*, *tworay* ground, and *shadowing* channel models (Aguiar, & Gross, 2003; Rappaport, 1996; Waluyo, Srinivasan, & Taniar, 2005; Xia, 1993).

Path loss is defined as the difference in received signal power to transmitted signal power. The predominant cause is the dispersion of the radio wave as it propagates omni-directionally away from the transmitter. As the distance between radios increases, the received power falls because the energy density at a point on the wave decreases as the wave expands. Theoretically, the received power is inversely proportional to the square of the distance between the radios. This is the basis for the well-known Friis *freespace* path loss model (Aguiar et al., 2003; Nishimura, Hayashibara, Enokido, & Takizawa, 2005; Rappaport, 1996).

The *freespace* model, Propagation/FreeSpace, assumes a single, clear line-of-sight path between the transmitting and receiving nodes. The received signal power at a distance from the transmitter is given by:

$$P_r(d) = \frac{P_t G_t G_r \lambda^2}{(4\pi)^2 d^2 L}$$

Where P_t is the transmitted signal power, G_t and G_r are the transmitter and receiver antenna gains respectively, L is the system loss factor and it will be assumed to be 1 in the simulation, and λ is the transmitted carrier wavelength.

However, this model is not accurate for many practical environments because it assumes perfect *freespace* propagation. Instead, measurements show that signal power does decrease logarithmically but not always with an exponent of two. This is expressed in the widely used log-distance path loss model

$$P_l(d) = P(d_0) + 10n\log(\frac{d}{d_0})$$

Where $P_l(d)$ is the path loss (in dB), n is the path loss exponent, and $P(d_0)$ is the path loss (in dB) at some close-in reference distance d_0. Often, $P(d_0)$ is found by measurement. Appropriate values of n have been proposed for a number of different environments. They tend to fall in the range of 2-5, where higher values represent environments with a large number of obstructions, such as city streets and office buildings Nasir et al., (2005).

For large environments with distances greater than a hundred meters, the *tworay ground* propagation model is favoured over *freespace*. This model considers the aggregate effects of radio waves converging by two different paths on the receiver: the direct line-of-sight path and a second path reflecting off the ground. It is based on the observation that as the distance between the transmitter and receiver increases the power of the signal along the line-of sight (ray one) decreases to a point where the signal's refection off of the ground (ray two) becomes a significant part of the received signal. Unfortunately, the second ray arrives out of phase with the first so the two interfere destructively at the receiver, causing a more rapid decline in receive power. The *tworay* ground model is given in Watts by Xia (1993):

$$P_r(d) = \frac{P_t G_t G_r h_t^2 h_r^2 \lambda^2}{(4\pi)^2 d^4 L}$$

Where h_t and h_r are the antenna heights at the transmitter and receiver, respectively, and are usually set to $h_t=h_r= 1{:}5$ m; the remaining parameters hold the same meanings as in *freespace*. *Tworay* ground has been shown to yield better accuracy than *freespace* for long distances (Xia, 1993).

A hybrid propagation model, combining both *freespace* and *tworay* ground, has been implemented in Ns-2 by the Monarch group: a cross-over distance is determined by $d_c = (4\pi\ h_t\ h_r/\lambda)$, which represents the distance at which both models result in the same signal strength. For distances $r < d_c$, *freespace* is employed; for distances $r > d_c$, *tworay* ground is employed. For radio wave frequencies of 900 MHz and 2.4 GHz, the threshold distance d_c is equal to 86.14 m and 227.33 m, respectively (Xia, 1993).

Shadowing is defined as the attenuation in power that occurs when a signal passes through an obstruction in its path such as a building or a hill. The degree to which the signal attenuates depends on the composition of the material, its thickness, and its physical characteristics such as its reflection, refraction, and absorption coefficients in the signal's frequency band. Extensive measurements have been made for many common materials and building structures. Results show that path loss P_{ls} due to *shadowing* is log-normally distributed about the distance-dependent mean given by Aguiar et al. (2003) and Rappaport (1996).

$$P_{ls}(d) = P_l(d) + X_\sigma$$

Where X_σ is a zero-mean normal random variable with standard deviation σ. Attenuation due to environmental obstructions like rain, smog, and foliage can also be represented using a similar model. The fading is modelled as a Gaussian distribution with average 0, and standard deviation σ. Rather than focusing on the impact of n and σ on protocol behaviour, we pick representative values and use them to evaluate the impact of fading on MANET protocols.

Fading is a term often used interchangeably with attenuation. Here, it is used to mean small-scale multipath fading, which is the dramatic variation in signal power that occurs when transmissions from the same radio reflect off surfaces in the environment and arrive at the receiver at various times and with different phases, causing interference. The *tworay* model we described earlier is based on multipath fading, but considers only the large-scale effects on receiver power. In a multipath environment with no moving objects and a fixed transmitter, there are regions where the multipath rays combine constructively, resulting in a gain in power or destructively causing a fade in power. A receiver travelling through this environment will see dramatic variations in power as it moves from region to region, the rate of which is proportional to its speed. However, another speed-dependent effect arises which is related to its direction with respect to the arrival angle of each multipath ray. If the direction of the receiver is toward or away from a multipath ray, then a shift in frequency occurs, called the Doppler shift. The combination of the many Doppler-shifted multipath rays creates the rapid time-varying signal that is characteristic of multipath fading. In this the Doppler-shifted multipath is not taken into consideration (Rappaport, 1996).

Radio propagation models are used for calculating the received signal power of packets in wireless simulations. Based on the received signal power and *thresholds*, different actions may be taken for a packet. Either the signal is considered too weak, the packet is marked as containing errors and thrown away by the medium access control (MAC) layer, or it is received correctly. Currently, there are three radio propagation models implemented in Ns-2; the *freespace* model, the *tworay* ground reflection model, and the *shadowing* model (Rappaport, 1996).

Figure 1 shows the received signal power as the distance between the transmitter and receiver increases using different channels. It can be noticed

Figure 1. The received signal power as the distance between the transmitter and receiver

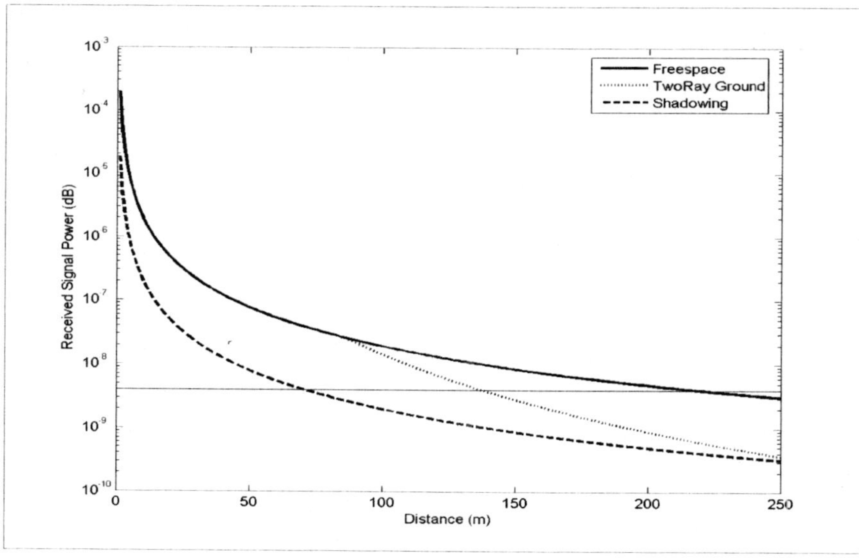

Figure 2. The received signal power measured continuously at distance 120 meter from the transmitter

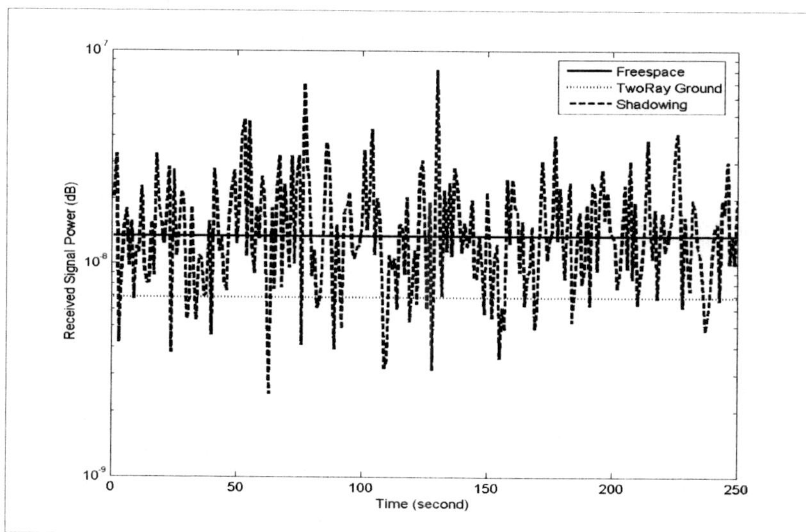

that as the distance increases, then the received signal falls and if the receiver sensitivity is 4.0e10-9, then the communication range of *shadowing* channel is only 75 meters, while it will be around 140 meters for *tworay* channel and the in the case of *freespace* channel, the communication range is more than 225 meters.

Figure 2 shows the received signal power measured continuously at distance 120 meter from the transmitter. It shows that the received power is not constant across time when *shadowing* channel model is used while it does not change when other channels are used.

SIMULATION ENVIRONMENTS

Ns-2 (Fall et al., 2003) and BonnMotion (2005) are discrete event simulators used specially for simulating network for research purpose. Ns-2 provides substantial support for simulation of transmission control protocol/user datagram protocol TCP/UDP, routing protocols such as DSR and AODV, and multicast protocols over wired and wireless networks. An extension developed by UCB Daedelus, CMU Monarch projects, and Sun Microsystems included wireless network communication in Ns-2. Features such as a complete wireless (radio) propagation models (*freespace*, *tworay* ground, and *shadowing*); data link and MAC layer models are added.

Ns-2 simulator has been used, which has support for simulating an ad-hoc wireless environment. Most existing network simulators employ random waypoint mobility to model how nodes move on a terrain (Johnson et al., 1996). Nodes in the random waypoint regime move according to the following rules: (1) each node picks a destination randomly within the simulation area and also picks a speed v that is uniformly chosen between vmin and vmax. Each node then moves toward the destination over a straight line with speed v.; (2) upon reaching the destination, a node pauses for some pause-time; (3) the node then picks the next destination and the process re-starts. Typically, the values of vmin, vmax, and pause-time are parameters of the simulation and are selected according to the requirements and operating environment of the application at hand. To generate random waypoint mobility mode, BonnMotion has been used (BonnMotion, 2005). BonnMotion is a Java software package, which creates mobility scenarios. The scenarios can also be exported to the network simulator Ns-2 and GlomoSim/QualNet.

Random traffic connections of constant bit rate (CBR) can be setup between mobile nodes using a traffic-scenario generator script. This traffic generator script is available under ~ns/indep-utils/cmu-scen-gen and is called *cbrgen.tcl*. It can be used to create CBR traffics connections between wireless mobile nodes. In order to create a traffic-connection file, we need to define the type of traffic connection CBR, the number of nodes and maximum number of connections to be setup between them, a random seed and in case of CBR connections, a rate whose inverse value is used to compute the interval time between the CBR packets. So the command line looks like the following (BonnMotion, 2005):

ns cbrgen.tcl [-type cbr] [-nn nodes] [-seed seed] [-mc connections][-rate rate].

The start times for the TCP/CBR connections are randomly generated with a maximum value set at 180.0s.

Four important performance metrics are evaluated:

- **Packet delivery fraction (PDF):** The ratio of the data packets delivered to the destinations to those generated by the CBR sources. It gives an idea about the extent of packet loss for a given algorithm. It is achieved as total received packets/total sent packets.
- **Normalize routing load (NRL):** The number of routing packets transmitted per data packet delivered at the destination. Each hop-wise transmission of a routing packet is counted as one transmission.
- **Average end-to-end delay of data packets:** This includes all possible delays caused by buffering during route discovery latency, queuing at the interface queue, retransmission delays at the MAC, and propagation and transfer times. It is measured in millisecond (ms).
- **Energy efficiency:** Total number of data delivered per Joule. It is measured in Kilobytes/Joule (KB/Joule).

The objective of our article is to measure the ability of the routing protocols to react to network topology changes while continuing to successfully deliver data packets to their destinations under different wireless channel models. To measure this ability, the basic methodology was to measure the aforementioned metrics for a variety of movement patterns for the mobile nodes, a variety of network load. The number of connections (sources/destination) is varied between 5 and 20 in steps of 5 connections to study the effect of the network load. Doubling the number of connections means doubling the network load, which obviously reduce the network performance because of high collision.

The DSR protocol evaluations are based on the simulation of 50 wireless nodes forming an ad hoc network, moving over a rectangular (1500m * 300m) flat space for 500 seconds of simulated time with different levels of speed, walker (low) speed (1m/s), bicycle speed (10m/s), and car (high) speed (20m/s). Additional levels of speed have been considered (5m/s and 15m/s) to achieve more accurate results. To have a fair comparison of the algorithms, identical movement patterns and workloads are used for all types of channels under study. Each run of the simulator accepts as input a scenario file that describes the exact motion of each node and another which details the traffic flows in the network during the simulation time.

The simulation generated trace files that contained a list of all major events such as packet transmission, receipt, and drops during the simulation. The movement scenario files of random waypoint mobility model generated by BonnMotion were characterized by a pause time. Each node begins the simulation by remaining motionless for pause time seconds. It then selects a random destination in the 1500m * 300m space and proceeds to that destination at a predefined speed. Each simulation ran for 500 seconds of simulated time. We ran our simulations with movement patterns generated for six different pause times: 0, 30, 60, 120, 300, 500 seconds. A pause time of 0 second corresponds to continuous motion, and a pause time of 500 corresponds to almost no motion. A set of five movement scenario files for each value of pause time are used to improve the accuracy of the results and to smooth out spikes due to extremely favorable or unfavorable movement scenarios.

The traffic sources being used are of constant bit rate (CBR) sources. During the course of the simulation, a different number of random (source/destination) connections were setup between the nodes in the network. We used a sending rate of four packets of 512 bytes each per second. TCP was not used as a source because TCP implements timeouts and retransmissions and also adjusts its sending rate according to the perceived network. As a result, both the time, at which each data packet is originated by its sender and the position of the node when sending the packet might differ between the protocols, preventing a direct comparison between them.

IEEE 802.11 MAC layer is used in the simulation study, which assumes that the power consumption is 0.2 Watts in receiving mode and 0.6 Watts in transmitting mode. In all simulation results, the energy consumption during idle period was ignored, and promiscuous receive mode was assumed to be enabled (Johnson et al., 1996).

SIMULATION RESULTS

Effect of Mobility

To study the effect of mobility on the network performance, the network load is considered to be constant such as 20 connections.

Figure 3 shows the effect of speed changes on performance metrics when pause time is zero (continues movement). The packet delivery fraction for freespace, tworay, and shadowing channel models are almost stable when increasing the speeds. Freespace channel model demonstrates

Figure 3. Routing packet effects using high network load, no pause, and different speeds 1, 5, 10, 15, and 20.

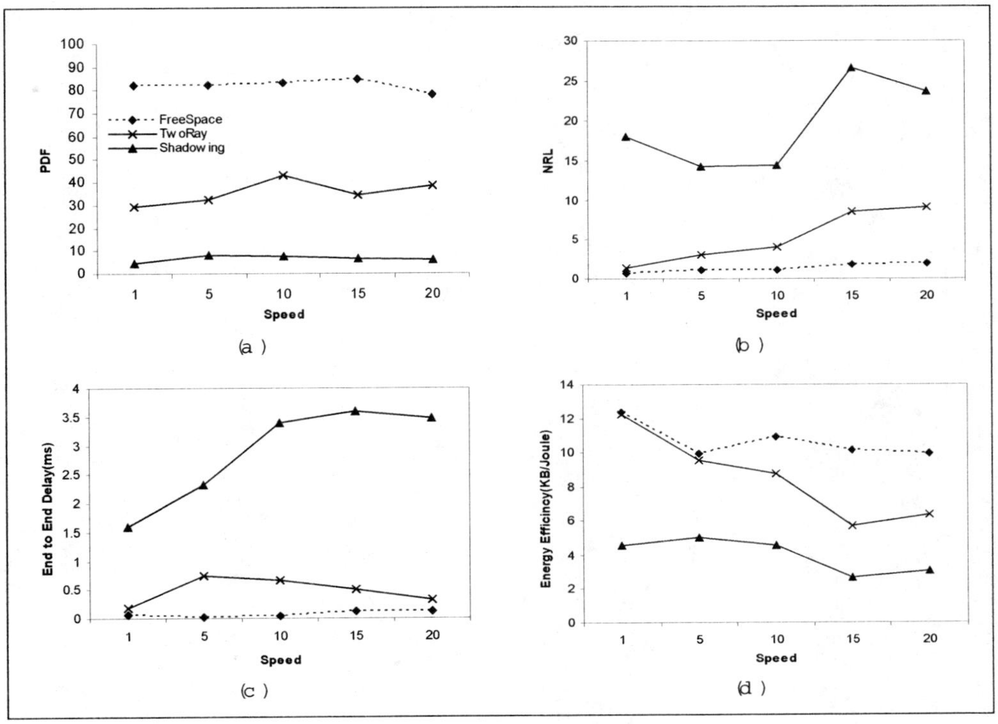

significantly higher packet delivery fraction than tworay and shadowing channel models by about 50 and 80 percent respectively, as can be seen in Figure 3a. Figure (3b) shows that as the speed increases, the NRL, (routing overhead), increases too. It also shows that shadowing channel model has higher effect on NRL when the speed is increased compared with other channel models. As it can be seen from Figure 3c, the shadowing channel model has lower energy efficiency compared with other channels. Also it shows that as the speed of mobile nodes increases, the energy efficiency decreases for all wireless channel modes.

In all cases, the performance of DSR protocol with shadowing channel model demonstrates significantly lower performance than other channel models. While with freespace, DSR protocol performs better as it is clear from Figure 3.

Figure 4 shows the effect of pause time changes on performance of wireless channel models when the speed is fixed at 20 m/s. The packet delivery fraction for freespace and shadowing channel models are almost stable when increasing the speeds, while it has a few changes for tworay channel. The PDF of tworay channel model decreases almost 10% when the speed increases. Freespace channel model demonstrates significantly higher packet delivery fraction than tworay and shadowing channel models by about 40 and 70% respectively, as it can be seen in Figure (4a). It is clear from Figure (4b) that as the pause time increases, the NRL of shadowing and tworay channel models decreases. It also shows that freespace channel has almost stable effect on NRL for all values of pause time. As the pause time of nodes during simulation increases, the routing overhead of all

Figure 4. Routing packet effect when speed is high(20m/s)), and different pause times 0, 30, 60, 120, and 300

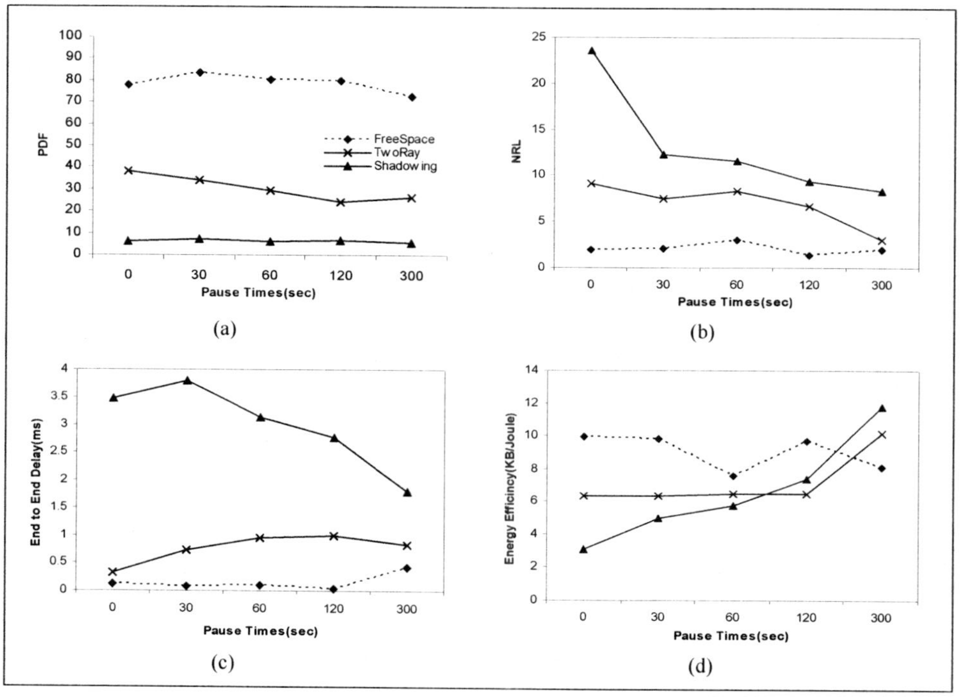

channel models decreases. It is clear from Figure (4c) that shadowing channel model has a longer delay than other channel models. The delay in the shadowing model decreases very fast as the pause time increases. The energy efficiency of shadowing channel model is low but it increases as the pause time increases. Thus, shadowing channel model can deliver traffic data per Joule efficiently better than other channel models when the pause time is high (300s) as it is shown in Figure (4d).

In all cases, see Figure 4, the performance of DSR protocol with shadowing channel model demonstrates significantly lower performance than other channel models. But the performance of DSR with shadowing performs better when the pause time becomes 300 seconds.

Impact of Network Load

To study the effect of network load on the network performance, the speed and the pause time are considered to be constant, walker speed (1m/s) and pause time 0 m/s (continuous movement) respectively. We vary the number of connections (5, 10, 15, and 20 connections) to vary the network load. Doubling the number of connections means doubling the network load, which obviously reduce the network performance because of high collision.

Figure 5 shows the effect of network load changes on performance metrics. The packet delivery ratio of all channel models performs inefficiently when the network load increases due to high network collision. The freespace channel

Figure 5. Routing packet effect when speed is (1m/s), no pause, and different connections 1, 5, 10, 15, and 20

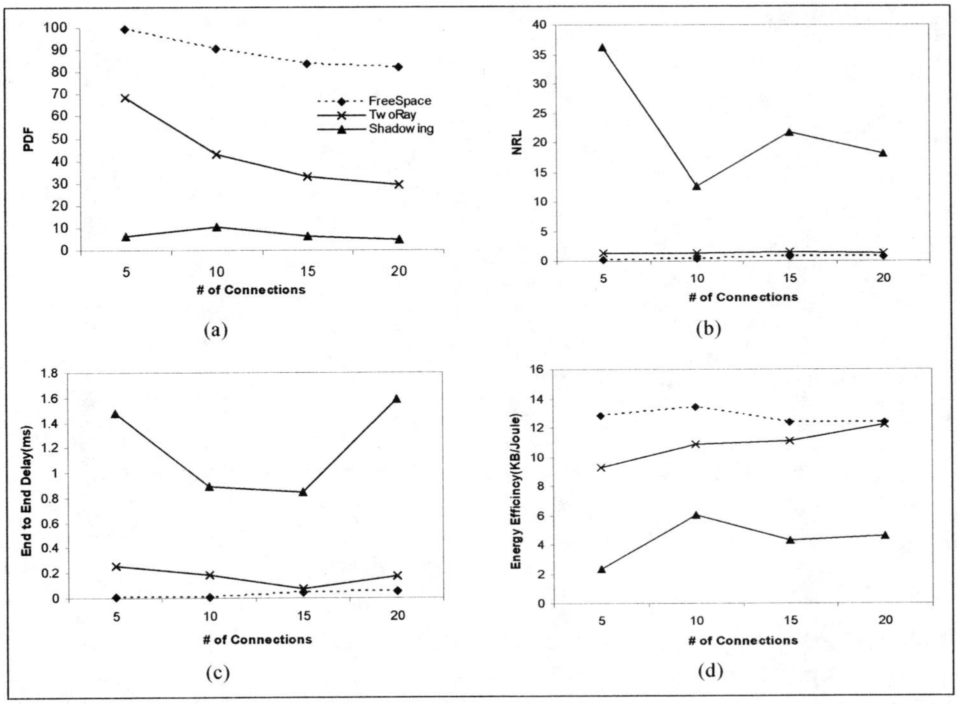

model demonstrates significantly higher packet delivery ratio than tworay and shadowing channel models by about 50 and 90 percent respectively, as it can be seen in Figure (5a). NRL of freespace and tworay channel models is very low compared to shadowing channel model, see Figure (5b). In Figure (5c), shadowing channel model shows a longer delay than other channel models. Energy efficiency of freespace channel model is almost stable for all network load levels, and outperforms tworay and shadowing channel models by (0~45) and (160~500) percent respectively, as it is shown in Figure (5d). In all cases, the performance of DSR protocol with shadowing channel model demonstrates significantly lower performance than other channel models. While with freespace, DSR protocol performs better as it is clear from Figure 5.

CONCLUSION

This article presents the effect of different wireless channel models on the dynamic source routing performance in MANET. In all cases, routing overhead is high when shadowing model is used, while more packets are delivered when either freespace or tworay models are used. The packet delivery fraction in freespace channel outperforms tworay and shadowing channels by an average percentage of about (35~50) and (75~85) respectively. The end-to-end delay is relatively short when a freespace or tworay channel models are used. The wireless channel model has strong effect on the energy efficiency (data traffic delivered per Joule) of the network. The energy efficiency in the network is high when freespace channel model is used, while it is low when shadowing model is used especially with high mobility. As the mobility increases, the energy efficiency of shadowing model decreases.

REFERENCES

Aguiar, A., & Gross, J. (2003). *Wireless channel models.* Technical Report, Telecommunication Networks Group, Technical University Berlin.

Akkaya, K., & Younis, M. (2004). A survey on routing protocols for wireless sensor networks. Ad Hoc Networks.

BonnMotion. (2005). *A mobility scenario generation and analysis tool.* Retrieved from http://web.informatik.uni-bonn.de/IV/Mitarbeiter /dewaal/BonnMotion

Broch, J., Maltz, D. A., Johnson, D. B., Hu, Y., & Jetcheva, J. (1998). A performance comparison of multi-hop wireless ad hoc network routing protocols. In *Proceedings of the 4th Annual ACM/IEEE International Conference on Mobile Computing and Networking,* Dallas, Texas, United States, October 25-30, 1998.

Fall, K., & Varadhan, K. (2003). ns Notes and Documentation. The VINT Project, UC Berkeley, LBL, USC/ISI, and Xerox PARC.

Iyer, S., Kanodia V., & Mahsa, M. (2005). *Performance analysis of DSR under statistical fading models.* Course project report. Retrieved from http://www2.ece.rice.edu/~kanodia/publications/html

Jayaputera, J., & Taniar, D. (2005). Data retrieval for location-dependent query in a multi-cell wireless environment. *Mobile Information Systems: An International Journal, IOS Press, 1*(2), 91-108.

Johnson, D. B., & Maltz, D. A. (1996). Dynamic source routing in ad hoc wireless networks. In T. Imielinski & H. F. Korth (Eds.), *Mobile computing.* Kluwer Academic Publishers.

Johnson, D. B., Maltz, D. A., & Broch, J. (2001). DSR the dynamic source routing protocol for multihop wireless ad hoc networks. In C. E. Perkins (Eds.), *Ad hoc networking* (pp. 139-172). Addison Wesley.

Nasir, Q., Al-Dubai, M., & Harous, S. (2005). A performance study of DSR under different wireless channel models. The *3rd International Conference on Advances in Mobile Multimedia (MoMM2005)* (pp. 283-289), September 19-21.

Nishimura, Y., Hayashibara, N., Enokido T., & Takizawa, M. (2005). Design of a hierarchical group to realize a scalable group. *Journal of Mobile Multimedia, 1*(3), 180-197, Rinton Press.

Papadimitriou, P., & Tsaoussidis V. (2005). On transport layer mechanisms for real-time QoS. *Journal of Mobile Multimedia, 1*(4), 342-363, Rinton Press.

Pei, G., Gerla, M., & Chen, T. W. (2000). Fisheye state routing: A routing scheme for ad hoc wireless networks. *Proceedings of ICC 2000,* New Orleans. IEEE.

Perkins, C., & Bhagwat, P. (1994). Highly dynamic destination-sequenced distance-vector routing (dsdv) for mobile computers. In *Proceedings of the Conference on Communications Architectures, Protocols, and Applications* (pp. 234-244). London England, ACM.

Rappaport, T. S. (1996). *Wireless communications: Principles and practice.* Upper Saddle River, NJ: Prentice Hall.

Waluyo, A. B., Goh, G., Taniar, D., & Srinivasan, B. (2005). On building a data broadcast system in a wireless environment. *International Journal of Business Data Communications and Networking, 1*(4), 14-36.

Waluyo, A. B., Srinivasan, B., & Taniar, D. (2005). Efficient broadcast indexing scheme for location-dependent queries in multi channels wireless environment. *Journal of Interconnection Networks (JOIN)*, Special Issue on Advanced Information Networking: P2P Systems and Distributed Applications, *6*(3), 303-321.

Woesner, H., Ebert, J., Schlager, M., & Wolisz, A. (1998). Power-saving mechanisms in emerging standards for wireless LANs: The MAC level perspective. *IEEE Personal Communications*, *5*(3), 40-48.

Xia, H. H. (1993). Radio propagation characteristics for line-of-sight microcellular and personal communication. *IEEE Transactions on Antennas and Propagation*, *41*(10), pp. 1439-1447.

This work was previously published in International Journal of Business Data Communications and Networking, Vol. 3, Issue 2, edited by J. Gutierrez, pp. 22-35, copyright 2007 by IGI Publishing, formerly known as Idea Group Publishing (an imprint of IGI Global).

Chapter XXIV
Key Issues in Mobile Marketing:
Permission and Acceptance

Stuart J. Barnes
University of East Anglia, UK

Eusebio Scornavacca
Victoria University of Wellington, New Zealand

ABSTRACT

The growth and convergence of wireless telecommunications and ubiquitous networks has created a tremendous potential platform for providing business services. In consumer markets, mobile marketing is likely to be a key growth area. The immediacy, interactivity, and mobility of wireless devices provide a novel platform for marketing. The personal and ubiquitous nature of devices means that interactivity can, ideally, be provided anytime and anywhere. However, as experience has shown, it is important to keep the consumer in mind. Mobile marketing permission and acceptance are core issues that marketers have yet to fully explain or resolve. This chapter provides direction in this area. After briefly discussing some background on mobile marketing, the chapter conceptualises key characteristics for mobile marketing permission and acceptance. The chapter concludes with predictions on the future of mobile marketing and some core areas of further research.

INTRODUCTION

The proliferation of mobile Internet devices is creating an extraordinary opportunity for e-commerce to leverage the benefits of mobility (Chen, 2000; Clarke, 2001; de Haan, 2000; Durlacher Research, 2002; Evans & Wurster, 1997; Kalakota & Robinson, 2002; Siau & Shen, 2003; Yuan & Zhang, 2003). Mobile e-commerce, commonly known as m-commerce, is allowing e-commerce businesses to expand beyond the traditional limitations of the fixed-line personal computer (Barnes, 2002a; Bayne, 2002; Clarke, 2001; Lau, 2003; Siau & Shen, 2003; Sigurdson & Ericsson, 2003). According to a study by Telecom Trends

International (2003), global revenues from m-commerce could grow from $6.8 billion in 2003 to over $554 billion in 2008.

Mobile commerce has a unique value proposition of providing easily personalized, local goods and services, ideally, at anytime and anywhere (Durlacher Research, 2002; Newell & Lemon, 2001). Due to current technological limitations, some problems, such as uniform standards, ease of operation, security for transactions, minimum screen size, display type, and the relatively impoverished web sites, are yet to be overcome (Barnes, 2002b; Clarke, 2001).

As each mobile device is typically used by a sole individual, it provides a suitable platform for delivering individual-based target marketing. This potential can improve the development of a range of customer relationship management (CRM) tools and techniques (Seita, Yamamoto, & Ohta, 2002). It is believed that in the near future marketing through the mobile phone will be as common a medium as the newspaper or TV. However, mobile marketing is unlikely to flourish if the industry attempts to apply only basic online marketing paradigms to its use; the medium has some special characteristics that provide quite a different environment for ad delivery, including time sensitivity, interactivity, and advanced personalization. Moreover, a key tenet is likely to be that consumers receive only information and promotions about products and services that they want or need; one of the most important aspects to consider is that wireless users demand packets of hyperpersonalized information, not scaled-down versions of generic information (Barnes, 2002c). Sending millions of messages to unknown users (known as spam) or banner ads condensed to fit small screens (Forrester Research, 2001) are doubtless unlikely to prove ideal modes of ad delivery to a captive mobile audience.

This chapter aims to explore the peculiarities of mobile-oriented marketing, focusing on issues of permission and acceptance, and some of the possible business models. The following two sections provide a basic review of the technological platform for mobile marketing and an introduction to marketing on the mobile Internet (focusing on advertising), respectively. The fourth section presents a conceptual definition and model for permission on mobile marketing applications, while section five provides a model for mobile marketing acceptance and examines a number of possible scenarios for mobile marketing, based on the previous analysis. Finally, the chapter rounds off with some conclusions, and further research questions, and provides some predictions on the future of wireless marketing.

THE TECHNOLOGICAL PLATFORM FOR MOBILE MARKETING

Kalakota and Robinson (2002) define mobile marketing as the distribution of any kind of message or promotion delivered via a mobile handset that adds value to the customer while enhancing revenue for the firm. It is a comprehensive process that supports each phase of the customer life cycle: acquisition, relationship enhancement, and retention. A variety of technological platforms are available to support mobile marketing. Here we describe briefly some of the principal components. (For a more detailed discussion, see Barnes [2002b, 2002c].) The m-commerce value chain involves three key aspects of technology infrastructure:

- **Mobile transport.** Current networks have limited speeds for data transmission and are largely based on second-generation (2G) technology. These "circuit-switched" networks require the user to dial up for a data connection. The current wave of network investment will see faster, "packet-switched" networks, such as General Packet Radio Service (GPRS), which deliver data directly to handsets, and are, in essence, always connected. In the near future, third-generation

(3G) networks promise yet higher transmission speeds and high-quality multimedia.

- **Mobile services and delivery support.** For marketing purposes, SMS (a text-messaging service) and WAP (a proprietary format for Web pages on small devices) are considered the key platforms in Europe and the United States, with iMode (based on compact hypertext markup language or cHTML) and iAppli (a more sophisticated version of iMode based on Java) taking precedence in Japan (WindWire, 2000). For PDAs, "Webclipping" is often used to format Web output for Palm or Pocket PC devices.
- **Mobile interface and applications.** At the level of the handset and interface, the brand and model of the phone or PDA are the most important part of the purchase decision, with "image" and "personality" being particularly important to young customers (Hart, 2000).

The next section explores the possibilities and experiences of using wireless marketing on these technology platforms.

MARKETING ON THE WIRELESS MEDIUM

The wireless Internet presents an entirely new marketing medium that must address traditional marketing challenges in an unprecedented way (WindWire, 2000). Key industry players in the value chain providing wireless marketing to the consumer are agencies, advertisers, wireless service providers (WSPs), and wireless publishers. For agencies and advertisers, the wireless medium offers advanced targeting and tailoring of messages for more effective one-to-one marketing. For the WSP, the gateway to the wireless Internet (e.g., British Telecom, AT&T, and TeliaSonera), wireless marketing presents new revenue streams and the possibility of subsidizing access. Similarly, wireless publishers (e.g., the *Financial Times, New York Times*, and CBS Sportsline), as a natural extension of their wired presence, have the opportunity for additional revenue and subsidizing access to content. At the end of the value chain, there is potential for consumers to experience convenient access and content value, sponsored by advertising (Kalakota & Robinson, 2002; WindWire, 2000).

Like the wired medium, marketing on the wireless medium can be categorized into two basic types: push and pull, which are illustrated in Figure 1. *Push* marketing involves sending or "pushing" advertising messages to consumers, usually via an alert or SMS (short message service) text message. It is currently the biggest market for wireless advertising, driven by the phenomenal usage of SMS—in December 2001, 30 billion SMS messages were sent worldwide (Xu, Teo, & Wang, 2003). An analysis of SMS usage has shown unrivalled access to the 15 to 24 age group—a group that has proved extremely difficult to reach with other media (Puca, 2001).

Pull marketing involves placing advertisements on browsed wireless content, usually promoting free content. Any wireless platform with the capacity for browsing content can be used for

Figure 1. Categorization of wireless marketing—with examples (Barnes, 2002c)

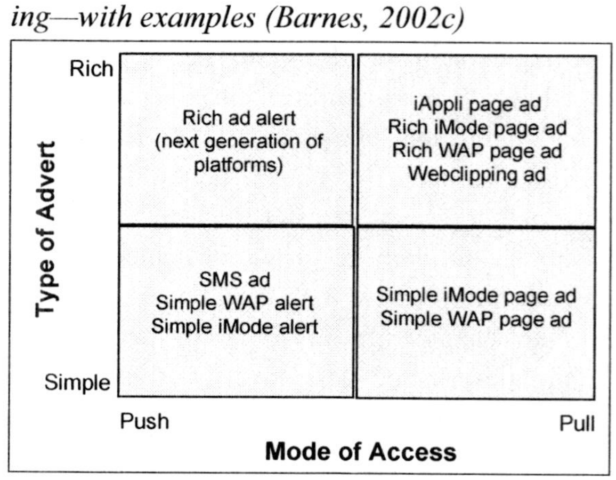

pull advertising. WAP and HTML-type platforms are the most widely used. Japan has experienced positive responses to wireless pull marketing, using iMode. Interestingly, wireless marketing in Japan has more consumer appeal than marketing on the conventional Internet. Click-through rates for mobile banner ads during the summer of 2000 averaged 3.6%, whilst those for wireless e-mail on iMode averaged 24.3%. Click-through rates for online banner ads on desktop PCs in Japan often average no more than 0.5 or 0.6% (Nakada, 2001).

Overall, current push services are very much in the lower left-hand quadrant of Figure 1. Until the availability of better hardware, software, and network infrastructure, services will remain basic. With faster, packet-based networks and more sophisticated devices, protocols and software, richer push-based marketing is likely to emerge, pushing the possibilities into the top left-hand quadrant.

PERMISSION ISSUES FOR MOBILE MARKETING APPLICATIONS

The discussion above has provided some insights about mobile marketing, particularly in terms of the wireless technological platform and basic applications of the medium. However, as yet, we have provided little conceptual discussion. The purpose of this section is to discuss the key variables of mobile marketing and present a conceptual model of permission for applications on this field.

In order for mobile marketing to reach its full potential of personalized information available anytime, anyplace, and on any device, it is necessary to understand the key characteristics of the mobile medium involved. We believe that any mobile marketing application should contemplate the following aspects:

- **Time and Location.** Although two different aspects, we consider them strongly related. An individual's behavior and receptiveness to advertisement is likely to be influenced by their location, time of day, day of week, week of year, and so on. Individuals may have a routine that takes them to certain places at certain times, which may be pertinent for mobile marketing. If so, marketers can pinpoint location and attempt to provide content at the right time and point of need, which may, for example, influence impulse purchases (Kannan, Chang, & Whinston, 2001). Feedback at the point of usage or purchase is also likely to be valuable in building a picture of time-space consumer behavior.
- **Information**. In particular, data given a context by the user. By itself, data do not contain an intrinsic meaning. It must be manipulated appropriately to become useful. Therefore, information can be defined as the result of data processing, which possesses a meaning for its receiver. Murdick and Munson (1988) point out that quantity of data does not necessarily result in quality of the information. The most important thing is what people and organizations do with the information obtained and its ability of extraction, selection, and presentation of information pertinent to the decision-making process should be considered as a decisive factor.
- **Personalization.** One of the most important aspects to consider is that wireless users demand packets of hyperpersonalized information, not scaled-down versions of generic information (Barnes, 2002c). The nature of the user, in terms of a plethora of personal characteristics such as age, education, socio-economic group, cultural background and so on is likely to be an important influence on how ads are processed. These aspects have already proven to be important influences on Internet use (OECD, 2001), and as indicative evidence has shown above, elements such

as user age are proving an important influence on mobile phone usage. The wireless medium has a number of useful means for building customer relationships. Ubiquitous interactivity can give the customer ever more control over what they see, read, and hear. Personalization of content is possible by tracking personal identity and capturing customer data; the ultimate goal is for the user to feel understood and simulating a one-to-one personal relationship. Through relational links of personal preferences, habits, mobile usage, and geographic positioning data the process of tailoring messages to individual consumers can become practical and cost effective.

The combination of the variables mentioned above allows us to understand one of the most important issues in mobile marketing: permission. Godin and Peppers (1999) refer to the traditional way of delivering marketing to customers as "interruption marketing." The authors suggest that instead of interrupting and annoying people with undesired information, companies should develop long-term relationships with customers and create trust through "permission marketing." The concept of permission marketing is based on approaching customers to ask for their permission to receive different types of communication in a personal and intimate way. It is well known among marketers that asking for a customer's permission is better and easier than asking for forgiveness (Bayne, 2002). In the wireless world, there is evidence to suggest that customers do not want to be interrupted—unless they ask to be interrupted (Newell & Lemon, 2001).

A mobile phone is a more personal environment than a mailbox or an e-mail inbox, and an undesired message has a very negative impact on the consumer (Enpocket, 2003; Godin & Peppers, 1999; Newell & Lemon, 2001). As mobile marketing has a more invasive nature than any other media, much attention must be given to permission issues in order to make the mobile marketing experience pleasant to the users. The information received must be of high value to gain the user's permission. It must produce a win–win situation between user and advertiser.

We understand permission as the dynamic boundary produced by the combination of one's personal preferences, that is, personalization, of time, location, and information. The user should be able to indicate when, where, and what information he/she would like to receive. Here are a couple examples of how mobile marketing can help consumers and businesses:

- You are getting ready to go to the airport and you receive a sponsored message saying that your flight is delayed for 4 hours. Because of this information, instead of spending 4 long and boring hours waiting at an airport lounge, you manage to have an enjoyable dinner with your friends.
- You let your wireless service provider know that you would like to receive during weekdays, from 12 p.m. to 1 p.m., information about the menu specials of all Italian restaurants costing less than $20 and within a 1-mile radius of where you are located.

Now, let us consider the situation if this information was not customer relevant, or time and location sensitive. For example, imagine the following scenario. You are on a business trip, it is 3:30 p.m. and you had to forgo lunch due to an important meeting. Next, your cell phone beeps and you receive an offer of a menu special of an unknown restaurant in your hometown. The value to the recipient of this information is zero; moreover, it is more likely to have a negative impact. Figure 2 helps us visualize the concept of permission on mobile marketing.

The idea of a message being sent directly to an individual's phone is not without legislative concerns. Indeed, all over the world, privacy and consumer rights issues lead to the promotion of

Figure 2. Concept of permission for mobile marketing

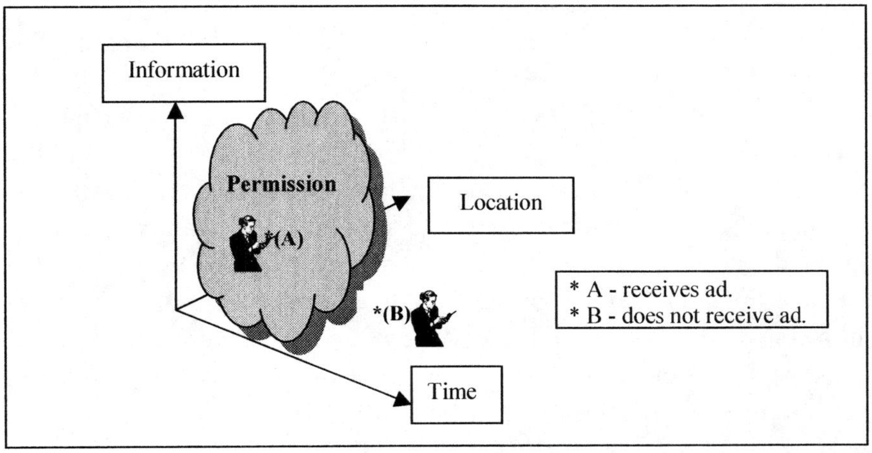

"opt-in" schemes. In essence, "opt-in" involves the user agreeing to receive marketing before anything is sent, with the opportunity to change preferences or stop messages at any time. Several current initiatives and industry groups, such as the Mobile Data Association, are helping to build standards of best practice for the mobile data industry (MDA, 2003).

As permission for mobile marketing applications should be dynamic, it is important to be able to identify customer responses to events. Stemming from the technological capabilities of mobile Internet-enabled devices, the measurement of reaction marketing is facilitated. As a consequence, the planning and justification of marketing expenditure becomes more precise. It also will help the identification of which mobile marketing strategies work and which do not. The constant feedback permits marketing strategies to be dynamically adjusted to produce better results for marketers.

ACCEPTANCE OF MOBILE MARKETING

Now we have discussed the technological and conceptual factors surrounding mobile marketing, let us examine the variables that influence customer acceptance. Specifically, this section aims to explore the few studies already accomplished on mobile marketing acceptance and provide a model that summarizes the main variables concerning this issue.

There is no doubt that mobile marketing is still at an embryonic stage. However, several recent studies help us to understand some key factors contributing to the penetration and acceptance of mobile marketing among consumers (Enpocket, 2002a, 2002b; Ericsson, 2000; Godin & Peppers, 1999; Quios, 2000). The study by Ericsson (2000) had a sample of approximately 5,000 users and 100,000 SMS ad impressions in Sweden; the Quios study (2000) examined 35,000 users and

2.5 million SMS ad impressions in the UK; and the Enpocket study (Enpocket, 2002a, 2002b, 2003) researched over 200 SMS campaigns in the UK, surveying over 5,200 consumers—after they had been exposed to some of the SMS campaigns—from October 2001 to January 2003. The results of the three studies tend to converge, each pointing out that more than 60% of users liked receiving wireless marketing. The reasons cited for the favorable attitudes to mobile marketing include content value, immersive content, ad pertinence, surprise factor, and personal context.

The Enpocket study (2002a, 2002b, 2003) found that consumers read 94% of marketing messages sent to their mobile phones. It is important to point out that all these customers had given permission to receive third-party marketing. Moreover, the viral marketing capability of mobile marketing was identified by the fact that 23% of the customers surveyed by Enpocket showed or forwarded a marketing message to a friend. Another interesting finding is that the average response rate for SMS campaigns (15%) was almost three times higher than regular e-mail campaigns (6.1%). If delivered by a trusted source such as a wireless service provider (WSP) or major m-portal, acceptance of SMS marketing (63%) was considered comparable to that of TV (68%) or radio (65%). Notwithstanding, SMS marketing delivered by another source was far less acceptable—at just 35% of respondents. Similarly, the rejection level of SMS marketing from a WSP or portal was just 9%, while SMS from other sources was rejected by 31% of those surveyed. Telesales was rejected by 81% of respondents.

The indicative evidence about customer trust was further strengthened by other findings from the surveys. For example, 74% of customers indicated that WSPs were the most trusted organisation to control and deliver SMS marketing to their mobile devices. Major brands such as Coca-Cola and McDonald's were preferred by only by 20% of respondents (Enpocket, 2002a). As a result of the close relationship with the user, SMS marketing typically helps to build stronger brand awareness than other medias (Enpocket, 2002b).

It is important to highlight that the statistics presented above are being materialized in the form of profits mainly by mobile marketing and content sponsorship. Some marketers are using the sponsorship revenue model by conveying brand values through association with mobile content that fits the company's product or corporate image (Kalakota & Robinson, 2002). Features such as mobile barcode coupons are allowing a better measurement and understanding of return on investment (ROI) for mobile marketing (12Snap, 2003).

The indicative evidence and discussion above provide strong hints towards three main variables that influence a consumer's acceptance of mobile marketing: user's permission, WSP control, and brand trust. Figure 3 presents a conceptual model for mobile marketing acceptance based on these factors. Note that user permission is weighted in the model (see below).

The model allows us to forecast eight scenarios for mobile marketing acceptance. Table 1 summarizes the different scenarios. An example for scenario 1 would be if a trusted brand such as Coca-Cola sent a marketing message through

Figure 3. Model for mobile marketing acceptance

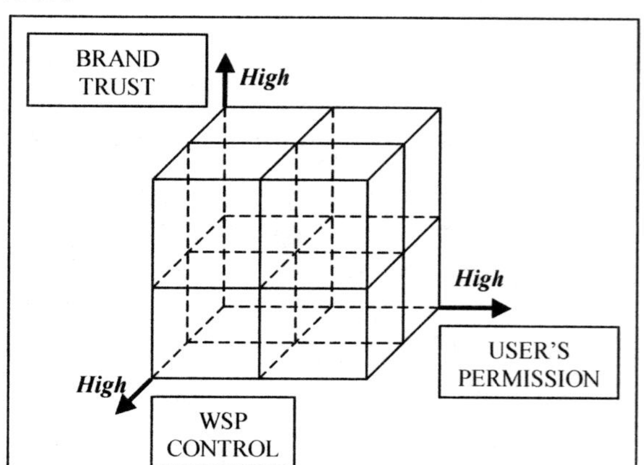

Table 1. Scenarios for mobile marketing acceptance

Scenario	Brand Trust	WSP Control	User's Permission	Acceptance
1	High	High	High	High acceptance
2	Low	High	High	Acceptable
3	High	Low	High	Acceptable
4	High	High	Low	Low acceptance
5	Low	Low	High	Acceptable
6	Low	High	Low	Low acceptance
7	High	Low	Low	Low acceptance
8	Low	Low	Low	Unacceptable

the user's WSP (e.g., Vodafone) with his/her permission. In this situation, all the variables have a high level and the message should be highly acceptable to the customer. At the opposite end of the spectrum, in scenario 8, an unknown company (brand) sends a message without WSP control and without the user's permission. Here, the probability of rejection is very high.

Scenarios 4 and 5 point out an element that requires further detailed investigation. We believe that the most important variable in this model is "user permission." For example, if Coca-Cola sends a message via an operator to a user who has not granted permission (scenario 4), it should have a lower acceptance than a brand with low trust that sends a message without WSP control to a customer who granted permission. This assumption is supported by the fact that the great majority of the consumers interviewed by Enpocket (2002a) are fearful that SMS marketing will become comparable to e-mail marketing with high levels of unsolicited messages.

The scenarios presented above are based on literature and on secondary data from the three studies previously approached. It would be interesting in the near future to substantiate this conceptual grid with primary data.

WSP control can directly affect how mobile marketing business models are configured. Based on the findings from the above analysis, we present two basic business models in which WSP control is the main differentiator (Figure 4).

Figure 4. Possible business models for mobile marketing

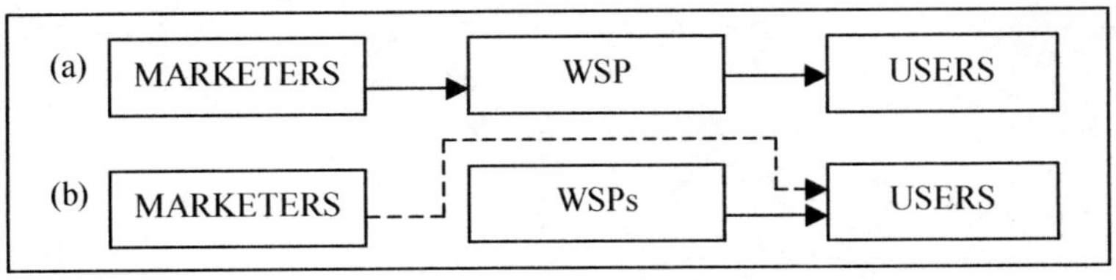

Figure 4a presents a model where the WSP has full control of the marketing delivery. On the other hand, Figure 4b shows a model where marketers can send messages directly to users without the control of the WSP.

The results of the studies presented by Ericsson (2000), Quios (2000), and Enpocket (2002a, 2002b, 2003) allow us to presume that the model presented by Figure 4a should be more successful than the one in Figure 4b. This assumption can also be supported by the fact that a WSP is usually more highly trusted by the consumers and possesses the technological capabilities to limit the delivery of messages. In addition, consumers interviewed by Enpocket (2002a) expressed a strong preference for the WSPs to become the definitive media owners and permission holders—possibly as a consequence of bad experiences with Internet marketing using nontargeted spam mail. Another issue to be taken into consideration is how WSP control can affect the revenue model for mobile marketing. In Figure 4a, the WSP can easily charge marketers for using its services, but in Figure 4b, this becomes a difficult task.

CONCLUSION

The immediacy, interactivity, and mobility of wireless devices provide a novel platform for marketing. The personal and ubiquitous nature of devices means that interactivity can be provided anytime and anywhere. Marketing efforts are potentially more measurable and traceable. Furthermore, technologies that are aware of the circumstances of the user can provide services in a productive, context-relevant way, deepening customer relationships. The convergence between marketing, CRM, and m-commerce represents a potentially powerful platform for wireless marketing.

Notwithstanding, it is important to keep the consumer in mind; the key to success is the management of and delivery upon user expectations. A key aspect of mobile marketing is likely to be obtaining permission from the users to send information to their mobile devices. Already, the wireless Internet has demonstrated the need for temperance; the wireless Internet is not an emulator of or replacement for the wired Internet, it is merely an additional, complementary channel for services. Further, aside from initial pilot investigations, it is not abundantly clear how consumers will respond to the idea of mobile marketing. Clearly, the issues concerning mobile marketing acceptance need to be further investigated. Alongside, a deeper investigation into business and revenue models is needed; for example, how can companies, marketers, WSPs, and consumers create a win–win environment? In addition, although it is expected that consumers will not tolerate receiving messages without permission, more work is still needed to explain how consumers give permission to receive mobile marketing.

Currently, wireless marketing is embryonic and experimental—the majority of wireless marketing is SMS based (simple push services—lower left-hand quadrant of Figure 1). The next generation of devices and networks will be important in the evolution of wireless marketing; higher bandwidth will allow rich and integrated video, audio and text. In addition, considerable effort is needed in building consumer acceptance, legislation for privacy and data protection, standardizing wireless ads, and creating pricing structures. If these conditions hold, wireless could provide the unprecedented platform for marketing that has been promised. Clearly, it is too early to tell, but future research aimed at examining these fundamental issues will help to further understand the implications of permission-based mobile marketing.

REFERENCES

12Snap. (2003). Mobile barcode coupons—The marketing revolution for marketeers. Retrieved May 18, 2003, from *www.12snap.com/uk/help/couponsshort.pdf*

Barnes, S.J. (2002a). Under the skin: Short-range embedded wireless technology. *International Journal of Information Management, 22*(3), 165–179.

Barnes, S.J. (2002b). The mobile commerce value chain: Analysis and future developments. *International Journal of Information Management, 22(2)*, 91–108.

Barnes, S.J. (2002c). Wireless digital advertising: Nature and implications. *International Journal of Advertising, 21*(3), 399–420.

Bayne, K.M. (2002). *Marketing without wires: Targeting promotions and advertising to mobile device users.* London: John Wiley & Sons.

Chen, P. (2000). Broadvision delivers new frontier for e-commerce. *M-commerce, October*, 25.

Clarke, I. (2001). Emerging value propositions for m-commerce. *Journal of Business Strategies, 18*(2), 133–148.

de Haan, A. (2000). The Internet goes wireless. *EAI Journal, April*, 62–63.

Durlacher Research. (2002). *Mobile commerce report.* Retrieved July 10, 2002, from *www.durlacher.com*

Enpocket. (2002a). Consumer preferences for SMS marketing in the UK. Retrieved March 13, 2003, from *www.enpocket.co.uk*

Enpocket. (2002b). The branding performance in SMS advertising. Retrieved March 13, 2003, from *www.enpocket.co.uk*

Enpocket. (2003). The response performance of SMS advertising. Retrieved March 13, 2003, from *www.enpocket.co.uk*

Ericsson. (2000). *Wireless advertising.* Stockholm: Ericsson Ltd.

Evans, P.B., & Wurster, T.S. (1997). Strategy and the new economics of information. *Harvard Business Review, 75*(5), 70–82.

Forrester Research. (2001). Making marketing measurable. Retrieved February 10, 2002, from *www.forrester.com*

Godin, S., & Peppers, D. (1999). *Permission marketing: Turning strangers into friends, and friends into customers.* New York: Simon & Schuster.

Hart, Peter D. (2000). *The wireless marketplace in 2000.* Washington, DC: Peter D. Hart Research Associates.

Kalakota, R., & Robinson, M. (2002). *M-business: The race to mobility.* New York: McGraw-Hill.

Kannan, P., Chang, A., & Whinston, A. (2001,). Wireless commerce: Marketing issues and possibilities. In *Proceedings of the 34th Hawaii International Conference on System Sciences*, Maui, HI.

Lau, A.S.M. (2003). A study on direction of development of business to customer m-commerce. *International Journal of Mobile Communications, 1*(1/2), 167–179.

Mobile Data Association (MDA). (2003). *Mobile Data Association.* Retrieved May 1, 2003, from www.mda-mobiledata.org/

Murdick, R.G., & Munson, J.C. (1988). *Sistemas de Informacion Administrativa.* Mexico: Prentice-Hall Hispano Americana.

Nakada, G. (2001). *I-Mode romps.* Retrieved March 5, 2001, from *www2.marketwatch.com/news/*

Newell, F., & Lemon, K.N. (2001). *Wireless rules: New marketing strategies for customer relationship management anytime, anywhere*. New York: McGraw-Hill.

NTT DoCoMo. (2003). Sehin Rain-Apu. Retrieved March 13, from *http://foma.nttdocomo.co.jp/term/index.html* (in Japanese)

Organisation for Economic Co-operation and Development (OEC). (2001). *Understanding the digital divide*. Paris: OECD Publications.

Puca. (2001). Booty call: How marketers can cross into wireless space. Retrieved May 28 2001, from *www.puca.ie/puc_0305.html*

Quios. (2000). *The efficacy of wireless advertising: Industry overview and case study*. London: Quios Inc./Engage Inc.

Sadeh, M.N. (2002). *M commerce: Technologies, services, and business models*. London: John Wiley & Sons.

Seita, Y., Yamamoto, H., & Ohta, T. (2002). Mobairu wo Riyoushitari Aiaru Taimu Maaketingu ni Kansuru Kenkyu. In *Proceedings of the 8th Symposium of Information Systems for Society*, Tokyo, Japan.

Siau, K., & Shen, Z. (2003). Mobile communications and mobile services. *International Journal of Mobile Communications, 1*(1/2), 3–14.

Sigurdson, J., & Ericsson, P. (2003). New services in 3G—new business models for strumming and video. *International Journal of Mobile Communications, 1*(1/2), 15–34.

Telecom Trends International. (2003). M-commerce poised for rapid growth, says Telecom Trends International. Retrieved October 27, 2003, from *www.telecomtrends.net/pages/932188/index.htm*

WindWire. (2000). *First-to-wireless: Capabilities and benefits of wireless marketing and advertising based on the first national mobile marketing trial*. Morrisville, NC; WindWire Inc.

Xu, H., Teo, H.H., & Wang, H. (2003,). Foundations of SMS commerce success: Lessons from SMS messaging and co-opetition. In *Proceedings of the 36th Hawaii International Conference on Systems Sciences*, Big Island, HI.

Yuan, Y., & J.J. Zhang (2003). Towards an appropriate business model for m-commerce. *International Journal of Mobile Communications, 1*(1/2), 35–56.

NOTE

An earlier and shorter version of this paper appeared as Barnes, S. J., & Scornavacca, E. (2004). Mobile marketing: The role of permission and acceptance. *International Journal of Mobile Communications, 2*(2), 128–139.

This work was previously published in Unwired Business: Cases in Mobile Business, edited by S. J. Barnes and E. Scornavacca, pp. 96-108, copyright 2006 by IRM Press (an imprint of IGI Global).

Chapter XXV
Consumer Perceptions and Attitudes Towards Mobile Marketing

Amy Carroll
Victoria University of Wellington, New Zealand

Stuart J. Barnes
University of East Anglia, UK

Eusebio Scornavacca
Victoria University of Wellington, New Zealand

ABSTRACT

Mobile marketing is an area of m-commerce expected to experience tremendous growth in the next 5 years. This chapter explores consumers' perceptions and attitudes towards mobile marketing via SMS through a sequential, mixed-methods investigation. Four factors were identified and proven as all having a significant impact on mobile marketing acceptance—permission, content, wireless service provider (WSP) control, and the delivery of the message, which guided the development of a revised and empirically tested model of m-marketing consumer acceptance. The findings also suggest that marketers should be optimistic about choosing to deploy mobile marketing, but exercise caution around the factors that will determine consumer acceptance. The chapter concludes with a discussion about directions for future research.

INTRODUCTION

One area of m-commerce that is expected to experience tremendous growth is global wireless advertising. It has been predicted that the mobile marketing industry will grow from $4 billion to $16 billion from 2003 to 2005 (Ververidis & Polyzos, 2002). Mobile marketing provides new revenue streams and the opportunities for subsidized access, along with the potential for customers to experience more convenient and relevant content value, sponsored by advertising (Barnes & Scornavacca, 2004). It is expected that 33% of cellular service provider's revenue will be coming from advertising and from payments and commissions from mobile commerce activities (Ververidis & Polyzos, 2002).

Wireless marketing allows effective targeting and tailoring of messages to customers to enhance the customer-business relationship (Barnes & Scornavacca, 2004). Studies on this new advertising medium indicate that mobile advertising campaigns can generate responses, which are as high as 40% compared with a 3% response rate through direct mail and 1% with Internet banner ads (Jelassi & Enders, 2004). Despite this phenomenal marketing potential, there has been very little research on mobile marketing and particularly through its most successful application, short message service (SMS) (Barnes & Scornavacca, 2004). According to GSM Association, cell phone users send more than 10 billion SMS messages each month, making SMS the most popular data service (Dickinger, Haghirian, Murphy, & Scharl, 2004). Conceptual frameworks and models identified in the literature provide insight into the critical success factors of m-commerce marketing; however, very few of these studies have empirically tested or generated models from a consumer's perspective (Barnes & Scornavacca, 2004; Dickinger et al., 2004; Scornavacca & Barnes, 2004).

The aim of this chapter is to explore consumers' perceptions and attitudes towards mobile marketing via SMS, and to empirically test Barnes and Scornavacca's (2004) m-marketing acceptance model. The following section provides a background to mobile marketing and identifies some of the prominent models in the m-business literature. It also examines the factors believed to influence consumer acceptance of mobile marketing. The third section discusses the methodology, while the fourth and fifth sections provide the results of the study and a revised model for mobile marketing acceptance. The chapter concludes with a discussion about the future for SMS mobile marketing, and directions for further research.

BACKGROUND ON MOBILE MARKETING

Mobile marketing can be defined as "Using interactive wireless media to provide customers with time and location sensitive, personalized information that promotes goods, services and ideas, thereby generating value for all stakeholders" (Dickinger et al., 2004). This definition includes an important concept of adding value not just for the marketing party, but also for the consumer. The literature shows a variety of technological platforms such as wireless application protocol (WAP), SMS, and multimedia message service (MMS) that are available to support mobile marketing applications (Barnes & Scornavacca, 2004; Dickinger et al., 2004).

SMS is the most popular mobile data application to date, showing phenomenal usage with 580 million mobile messaging users sending over 430 billion messages worldwide in 2002 (TTI, 2003). Text message services have been hugely popular for interpersonal communication, allowing users of all ages to exchange messages with both social and business contacts (Dickinger et al., 2004; Xu, Teo, & Wang, 2003). Xu, Teo, and Wang (2003) identified three consistent success indicators for SMS messaging. The first factor is the cost effectiveness and interoperability of the wireless infrastructure, the second is the high penetration of mobile phones (ubiquitous penetration levels of over 80% in some countries), and the third is the relatively low cost of the SMS messaging service.

Countries such as Japan, New Zealand, Germany, and the UK have cost-effective and interoperable wireless structures, a high penetration of mobile phones, and a relatively low cost for the SMS messaging service have experienced remarkable success with the SMS application (Barnes & Scornavacca, 2004). The success that SMS has had as a messaging service provides a potentially huge SMS messaging customer base which could lend itself as a SMS mobile marketing customer base, making it an attractive opportunity for marketers (Kellet & Linde, 2001).

One of the main challenges and opportunities for mobile advertising companies is to understand and respect the personal nature of the usage of mobile phones (Barnes & Scornavacca, 2004; Barwise & Strong, 2002; Jelassi & Enders, 2004; Heinonen & Strandvik, 2003).

Consumer Acceptance of Mobile Marketing

The acceptance of a mobile marketing message is likely to be influenced by the consumer's acceptance of the mobile medium, the relevance of the content, and the context of the marketing message (Barnes & Scornavacca, 2004; Dickinger et al., 2004; Enpocket, 2003; Heinonen & Strandvik, 2003). Messages that are concise, funny, interactive, entertaining, and relevant to the target group usually achieve higher levels of success (Dickinger et al., 2004; Jelassi & Enders, 2004). The recent m-business literature offers a couple of frameworks that investigate user acceptance of SMS based mobile marketing (Barnes & Scornavacca, 2003; Dickinger et al., 2004).

The guiding model used for this research is the conceptual model of permission and acceptance developed by Barnes and Scornavacca (2004). This model was selected as it looks at a small subset of factors identified in the literature, which are believed to be the *most* important variables influencing consumer acceptance.

Barnes and Scornavacca (2004) believed that *user permission, wireless service provider control (WSP),* and *brand recognition* are the three most important variables that could influence consumers' acceptance of mobile marketing.

Among those, user permission was believed to be the most important variable, the main reason for this being that most consumers are fearful of SMS mobile marketing becoming like e-mail marketing, that is, with high levels of spam. WSP control is found to increase the probability of user acceptance to mobile marketing. This was supported by the fact that users are likely to have high levels of trust with their WSP (Enpocket, 2002b; Ericsson, 2000).

The model also puts forward eight propositions of varying levels of acceptance according to the different combinations of factors. Table 1 presents

Table 1. Scenarios for m-marketing acceptance (Barnes & Scornavacca, 2004)

User's Permission	WSP Control	Brand Trust	Acceptance
High	High	High	High Acceptance
High	High	Low	Acceptable
High	Low	High	Acceptable
High	Low	Low	Acceptable
Low	High	High	Low Acceptance
Low	High	Low	Low Acceptance
Low	Low	High	Low Acceptance
Low	Low	Low	Not Acceptable

Barnes and Scornavacca's (2004) hypothesized acceptability of SMS marketing messages based on high and low levels of permission, WSP control, and brand trust. This model is yet to be empirically tested with primary data.

These propositions provide a starting point in further exploring the factors that could contribute to consumer acceptance of mobile marketing.

METHODOLOGY

The chosen strategy of inquiry for this research is sequential exploratory mixed methods. Sequential procedures are ones in which the researcher uses the findings of one method to elaborate on or expand with another method (Creswell, 2003; Green, Caracelli, & Graham, 1989). The objectives of the sequential exploratory approach for the purpose of this study is to use two qualitative focus groups to explore the perceptions of mobile marketing, focusing on the main variables believed to influence mobile marketing acceptance, and then elaborate on this through experimental research in which the findings of the initial phase will be used. The empirical data will hopefully confirm what has been identified from the literature and the findings from the focus groups.

Focus Groups

The samples for the focus groups were purposely selected based on convenience sampling, availability, and profiling. Participants for both groups were in the age range 20–28 reflective of one of the major target groups for SMS mobile marketing. Four participants were selected for focus group A and five participants for focus group B. The participants in focus group A had a greater knowledge of mobile commerce technologies and applications than the participants in focus group B, which was purposely achieved in order to canvas a range of experiences and provide differing viewpoints. The participants in this study were students of a university in New Zealand as well as professionals working in the local central business district.

Interviews were based on open-ended questions and triggers. Video recording was used to tape the focus group discussions, with additional notes being taken by the facilitator. The advantages of using a focus group was that a range of ideas and perceptions were derived and the dynamics of the group provided a rich understanding of the research problem. These focus groups generated new propositions that were tested in the survey questionnaire phase.

Data analysis for the focus groups involved initially transcribing interviews and sorting the data into groups of information based on various topics. The transcriptions were then read over to look for ideas, depth, and credibility of the information from participants; thoughts were noted down in the margins of the transcript (Creswell, 2003). A coding process was then carried out where the data was organized into clusters before any meaning was derived from it (Rossman & Rallis, 1998). The themes and categories identified from the analysis are the major findings of the qualitative phase, and have been shaped into a general description of the phenomenon of mobile marketing acceptance (see the results section for details). Reliability measures were used to check for consistency of themes and patterns, while validity measures (triangulation, member checking, bias discussion, and peer debriefing) were used to determine the accuracy of the findings (Creswell, 2003).

Survey Questionnaire

This phase involved the use of a cross-sectional survey questionnaire to test the acceptance of mobile marketing messages against 16 various propositions that were formulated from the results of the focus groups. The advantage of using a survey in this study was the economy and rapid turnaround of data collection that a survey provides. Surveys are also advantageous in their ability to make inferences about consumer behaviour for given populations based on a sample (Babbie, 1990).

A survey questionnaire was chosen due to its cost effectiveness, data availability, and convenience. Seventy-eight participants for the quantitative phase of the research were selected using random convenience sampling with eight members of the sample being nonrespondents.

The instrument used in the survey was a modified version of the permission and acceptance model of mobile marketing developed by Barnes and Scornavacca (2004) with four variables: permission, WSP control, content, and delivery of the message. Sixteen propositions were formulated around these variables that were tested with a 4-point Likert scale ranging from "unacceptable" to "accept enthusiastically."

The data that was collected from the surveys was entered into an Excel spreadsheet, and statistical calculations were carried out. The 16 propositions were then placed in a table with the expected and actual levels of acceptance that were found for each proposition (see Tables 2 and 3). Tabular analysis was conducted in order to analyze the change in SMS mobile marketing acceptance through the various combinations of the set of variables (permission, WSP control, content, and delivery). The results from the quantitative phase were then compared against previous literature in order to provide further insight of the findings.

To avoid possible threats to validity, caution was taken when the results of this experiment were generalized to other populations and environments, when conducting statistical analysis on the data, and when the definitions and boundaries of the terms were defined.

RESULTS FROM THE FOCUS GROUPS

While focus group A was more knowledgeable in the area of mobile commerce, mobile technologies, and the potential of mobile marketing; both focus groups had only ever experienced mobile marketing through their wireless service providers. To some extent the participants' experience of receiving marketing messages from their service provider influenced their individual perceptions and perceived importance of varying factors contributing to consumer acceptance. The results of both focus groups were consistent with little disparity between the two.

Factors identified in the focus groups as having a significant impact on consumer acceptance of mobile marketing were permission to receive mobile marketing messages, control of the wireless service provider, relevance of the content, timeliness and frequency of the messages, simplicity and convenience of the messages, the brand or company sending the message, the control of the marketing from the consumer, and the privacy of the consumer. Consistent with Barnes and Scornavacca's (2004) model, permission and WSP control were perceived to have a heavy bearing on the acceptance of a mobile marketing message; however, brand was found to have little or no impact on acceptance than the likes of content, and time and frequency of the messages. The emerging there are classified as follows:

- **Permission:** Permission raised the most discussion in each focus group, and it was concluded by the participants as the most important success factor. Participants stated that consumers should have to "opt in" before they receive mobile marketing messages of any kind, and have the option to "opt out" at any stage.

- **Wireless service provider (WSP) control:** Although there was great emphasis on permission, it was also strongly felt that there needed to be a degree of filtering from the service provider. As participant A stated, "there has to be some sort of protection; they can't just open it up to anyone—if companies want to market to customers they should have to go through Vodafone." The idea was raised that if participants had just one company to go to which was linked to their service provider, then there would be just one point of contact allowing consumers to easily "opt in" and "opt out" rather than tracking down several different companies. Participants agreed that it should be evident in the message that it is being filtered by the service provider and legitimate.
- **Personalization and content:** It was agreed that permission regarding time of day, frequency, and content would also be critical to the acceptance of mobile marketing. Both focus groups agreed that content and its relevance would play a key role in the acceptance of a mobile marketing message, with some participants arguing this as the most important factor. It was believed that marketers should make use of the technology and the advantages it provides over traditional forms of marketing and the Internet, looking to add value other than just advertising. Other ideas discussed in the focus groups were to tie content with location, timing, and ensure that the format of the message works with the limitations of the phone.
- **Frequency:** Participants agreed that there would be a limit to the number of mobile marketing messages they wished to receive, and there should be some control over the number of messages they are receiving depending on what good or service was being marketed or the industry (e.g., food/flowers). Both focus groups agreed that if consumers were to be hounded by marketing messages, it may result in switching providers, or deleting messages without reading them.
- **Time:** Participants raised the issue of time playing an important role in the acceptance of mobile marketing messages. It was believed that it is important for consumers to receive marketing messages at times suitable for them, and consumers are able to not only give permission to receive messages but also choose the times they wish to receive them.
- **Brand:** As far as the brand or company that was marketing was concerned, the general feeling among both focus groups was that as long as the marketing messages were being filtered by the service provider it would not matter too much who it was from; however, if it was third party, they would be annoyed right away. The majority of participants argued that it would be the more well-known brands or brands that the individual consumer recognizes. However, some consumers may prefer to receive messages from a little boutique shop down the road and there should be a way smaller companies can afford mobile marketing. Again if the brand or company doing the marketing was to go through the wireless service provider, this would result in an even higher level of trust. Focus group B believed that consumers should be able to select which companies and brands they receive messages from to a very specific point.
- **Technology/Ease of use:** A number of important issues were raised with regard to the mobile technology and convenience of the marketing message, some of which have already been pointed out in the previous sections. The main point raised that falls under this section is that marketing

messages should not be a hassle for consumers to receive, they should work with the limitations of the phone, and there should be a manageable way to deal with them.

REVISED MODEL AND SURVEY RESULTS

Four conceptual factors emerged as having the most influence on consumer acceptance based on the tabular analysis and findings of the focus groups. Similar topics were merged as conceptualized themes and then these themes were analyzed according to the number of times they were mentioned in the focus groups, whether these comments were implying that they were important factors and whether the participants explicitly stated them as being one of the *most* important factors.

Table 2 presents 16 new propositions based on varying combinations of the identified factors, ranked according to the importance of factors: (1) permission, (2) WSP control, (3) content, (4) delivery, and also the number of factors which are low (0, 1, 2, 3, or 4).

The results obtained in the survey demonstrated that propositions 6, 7, 8, 11, 12, 13, 14, 15, and 16 were supported, while propositions 1, 2, 3, 4, 5, 9, and 10 were not found to be supported by the data collected. Tables 3 and 4 show the revised propositions with the expected and actual levels of acceptance for mobile marketing. Notice that the second table actually shows the propositions reshuffled in order to demonstrate their rank of acceptance according to the results.

Overall, consumer acceptance of mobile marketing messages was much lower than expected. Over 50% of respondents answered unacceptable to more than 10 out of the 16 scenarios put forward to them, with the average number of scenarios answered as unacceptable being 9. On the other hand, nearly 70% of the respondents did not answer "accept enthusiastically" to anything, and of the 30% who did give this response for at least one scenario, more than 80% only gave this response for one or two of the questions (Tables 3 and 4).

Of all the propositions the highest level of acceptance for mobile marketing was as expected for proposition 1. However, it can be seen that even where consumers have given permission, the

Table 2. Revised model with the 16 scenarios for marketing acceptance

Proposition	Permission	WSP Control	Content	Delivery	Expected Acceptance Level
1	High	High	High	High	Accept Enthusiastically (4)
2	High	High	High	Low	Acceptable (3)
3	High	High	Low	High	Acceptable (3)
4	High	Low	High	High	Acceptable (3)
5	Low	High	High	High	Acceptable (3)
6	High	High	Low	Low	Accept reluctantly (2)
7	High	Low	High	Low	Accept reluctantly (2)
8	High	Low	Low	High	Accept reluctantly (2)
9	Low	High	High	Low	Accept reluctantly (2)
10	Low	High	Low	High	Accept reluctantly (2)
11	Low	Low	High	High	Accept reluctantly (2)
12	High	Low	Low	Low	Unacceptable (1)
13	Low	High	Low	Low	Unacceptable (1)
14	Low	Low	High	Low	Unacceptable (1)
15	Low	Low	Low	High	Unacceptable (1)
16	Low	Low	Low	Low	Unacceptable (1)

Table 3. Revised model ranked according to expected results

Proposition	Permission	WSP Control	Content	Delivery	Expected Acceptance Level	Average	Rank	Actual level
1	High	High	High	High	Accept Enthusiastically (4)	3.16	1	Acceptable
2	High	High	High	Low	Acceptable (3)	1.60	8	Accept reluctantly
3	High	High	Low	High	Acceptable (3)	1.99	3	Accept reluctantly
4	High	Low	High	High	Acceptable (3)	2.29	2	Accept reluctantly
5	Low	High	High	High	Acceptable (3)	1.91	4	Accept reluctantly
6	High	High	Low	Low	Accept reluctantly (2)	1.50	9	Accept reluctantly
7	High	Low	High	Low	Accept reluctantly (2)	1.70	5	Accept reluctantly
8	High	Low	Low	High	Accept reluctantly (2)	1.63	7	Accept reluctantly
9	Low	High	High	Low	Accept reluctantly (2)	1.43	10	Unacceptable
10	Low	High	Low	High	Accept reluctantly (2)	1.41	11	Unacceptable
11	Low	Low	High	High	Accept reluctantly (2)	1.66	6	Accept reluctantly
12	High	Low	Low	Low	Unacceptable (1)	1.41	12	Unacceptable
13	Low	High	Low	Low	Unacceptable (1)	1.30	14	Unacceptable
14	Low	Low	High	Low	Unacceptable (1)	1.30	15	Unacceptable
15	Low	Low	Low	High	Unacceptable (1)	1.39	13	Unacceptable
16	Low	Low	Low	Low	Unacceptable (1)	1.19	16	Unacceptable

content of the message was relevant, the delivery appropriate, and the message had come through the WSP, it was found on average to be only acceptable, with just 31% of respondents accepting this message enthusiastically. Thus disproving proposition 1. Alternatively on average the lowest level of acceptance (unacceptable) was found where there was a low level of all these factors. Only 9 out of the 70 participants answered anything other than unacceptable for this question. This result was expected and consistent in proving proposition 16.

Permission and delivery of the message were the two variables that were found to equally have the most influence on the participant's level of acceptance, while content was found to be the next most important factor with control of the WSP having the least amount of impact on the level of acceptance. Participants were more likely to accept messages that had a lower level of WSP control or irrelevant content than messages that they had not given permission for or that came at an inappropriate time or frequency. This was shown again in Table 4, rows 12–15, where participants found scenarios 13 and 14 more unacceptable, despite having high levels of WSP control and content, respectively, than scenarios 12 and 15 where there were higher levels of permission and appropriate delivery, respectively.

It is interesting to note that consistent with the propositions, the level of acceptance declined with the number of factors that had low levels, except in the case of proposition 2, which was expected to generate the second highest level of acceptance and in actual fact dropped down to position 8. Where all other factors were high, yet the delivery of the message was inappropriate, more than 50% of respondents found this message unacceptable, compared to just 26% of respondents who considered a message with low levels of WSP control unacceptable.

Looking at the other rankings of propositions from their expected to actual perceived influence on acceptance, just three propositions stayed in the same ranked position. However, of the propositions that did get shuffled in rank, nine of these moved only within one or two ranks, with just three propositions moving three places or more. Participants found all messages that had three or more factors with low levels to be completely unacceptable. This was consistent with the expected results, and supported the propositions 12, 13, 14, 15, and 16. Messages that had only high levels of WSP control or relevant content were found to be

Table 4. Revised model ranked according to actual results

Proposition	Permission	WSP Control	Content	Delivery	Expected Acceptance Level	Average	Rank	Actual level
1	High	High	High	High	Accept Enthusiastically (4)	3.16	1	Acceptable
4	High	Low	High	High	Acceptable (3)	2.29	2	Accept reluctantly
3	High	High	Low	High	Acceptable (3)	1.99	3	Accept reluctantly
5	Low	High	High	High	Acceptable (3)	1.91	4	Accept reluctantly
7	High	Low	High	Low	Accept reluctantly (2)	1.70	5	Accept reluctantly
11	Low	Low	High	High	Accept reluctantly (2)	1.66	6	Accept reluctantly
8	High	Low	Low	High	Accept reluctantly (2)	1.63	7	Accept reluctantly
2	High	High	High	Low	Acceptable (3)	1.60	8	Accept reluctantly
6	High	High	Low	Low	Accept reluctantly (2)	1.50	9	Accept reluctantly
9	Low	High	High	Low	Accept reluctantly (2)	1.43	10	Unacceptable
10	Low	High	Low	High	Accept reluctantly (2)	1.41	11	Unacceptable
12	High	Low	Low	Low	Unacceptable (1)	1.41	12	Unacceptable
15	Low	Low	Low	High	Unacceptable (1)	1.39	13	Unacceptable
13	Low	High	Low	Low	Unacceptable (1)	1.30	14	Unacceptable
14	Low	Low	High	Low	Unacceptable (1)	1.30	15	Unacceptable
16	Low	Low	Low	Low	Unacceptable (1)	1.19	16	Unacceptable

10% less unacceptable than messages with only high levels of permission or appropriate delivery—thus supporting the theory that permission and delivery of the messages are perceived to be the most important factors.

DISCUSSION

The findings indicated a number of factors that are critical to the acceptance of mobile marketing by consumers. While the empirical testing showed that some factors are more important than others in influencing the overall level of acceptance, it was found that all factors played a significant role.

Consistent with the literature explicit permission was found to be essential (Barnes & Scornavacca, 2004; Enpocket, 2003; Godin et al., 1999). The wireless channel is relatively protected and spam free with consumers having little experience with mobile marketing. Due to the personal nature of the phone, and experiences with unsolicited spam via e-mail users were weary of receiving marketing to their cell phones, and a number of privacy issues were raised in the focus groups. Another finding that emerged from the study was the importance of delivery with the marketing message. Literature has suggested that frequency and time are linked to targeting, where users are happy to receive messages at a higher frequency so long as the relevance to them is maintained (Enpocket, 2002b). This was supported by the empirical testing where it shows messages with a low level of relevant content yet appropriate delivery were found to be much more acceptable than messages with a low level of relevant content and inappropriate delivery (a higher frequency). While participants in the focus groups made a point of saying that it is useless receiving any messages containing content that is irrelevant, there are a number of possible reasons why the respondents may have found delivery to be more important. If a consumer receives a message that is irrelevant to them once in a blue moon, and it does not come at a disturbing time, they may not be that bothered by it. On the other hand, if they were messages on something that was relevant to them but were receiving these messages continuously and at interruptive times, it is likely to be more unacceptable.

It was interesting to see that the control of the WSP had the least impact on consumer acceptance in the survey results, conflicting with the results of the focus groups where participants expressed their strong opinions towards the importance of WSP control. The results may in fact indicate that where consumers receive messages they find disturbing or intrusive, they would rather it had not come from the service provider they trust. The focus groups indicated this, stating that they trust their service provider's judgment and would expect them to behave responsibly. Consumer attention seems more likely to divert to the filter when they are receiving unsolicited messages that they find disturbing.

Despite literature showing mobile marketing to be a successful tool in building brand awareness, and an important factor in consumer acceptance (Dickinger et al., 2004; Enpocket, 2002a), the study revealed that the brand being marketed may have very little impact. Consumers are more likely to care whether a brand has been accepted by their service provider and has come through a filter, than about their level of trust between two different brands. Despite having a high trust in a brand, consumers are still doubtful of the bona fide of these messages when they have come direct. They are also less likely to care about the brand that is being marketed to them than whether the content is relevant. The importance that is placed on brand is likely to increase when all other factors are high, and there is more choice in the market. Currently there are a limited number of brands being marketed through the mobile phone in New Zealand and more attention to brand is likely to arise in the future where consumers receive similar messages, with all other factors being equal, from competing brands.

CONCLUSION

This research highlights the importance of consumer perceptions and acceptance levels of mobile marketing. The literature showed the powerful marketing potential that mobile marketing can offer companies through its anytime and anywhere nature, yet limited research looking at consumers' perceptions and acceptance of mobile marketing has been carried out. This study set out to overcome the apparent gap in the literature, and through the use of both qualitative and quantitative methodology, a model has been adopted, explored, developed, and empirically tested and validated.

This study suggests that marketers should be optimistic about choosing to deploy mobile marketing; however, exercise caution around the factors that will determine consumer acceptance. While consumers can see the potential in the mobile medium, they are weary of receiving unsolicited messages they do not want. Obtaining user trust and permission will be the main challenge faced by marketers, and future research should focus on ways to overcome these challenges. Consumers are more likely to trust messages coming from their service providers than anywhere else, so it is important that service providers provide a high level of filtering and protection as reassurance for their users. Trust and permission are necessary factors of consumer acceptance; however, they should not be seen as the only objectives. Attention needs to be focused around the relevance of the content and the timeliness and frequency of the delivery of marketing messages.

The research showed that simply focusing on contextual, content, or permission/control factors in isolation is unlikely to result in a high or even moderate level of acceptance. Instead, marketers need to take into account all these factors and how varying combinations of these factors will impact consumer acceptance.

The permission and acceptance model, which has been developed and tested in this research, provides a foundation for further SMS mobile marketing research to be built upon. Academics can refer to this model as a guide for further understanding of consumer acceptance to mobile marketing, while practitioners may find this model useful in providing direction for mobile marketing strategies. The device media aspects discussed in the focus groups may also provide an indication as to what new technologies and mobile devices will be of significance in meeting consumers' needs for the future.

The generalizability of this study is limited by it being conducted only in New Zealand as well as the lack of further qualitative interviews to further elaborate on the initial quantitative analysis. This cross-sectional study only looked at consumer acceptance at one point in time, and little is known about the sample frame that was used for the survey questionnaire. Furthermore the sample of the participants for the quantitative phases was only a small number which leaves possibility for self-selection bias. Longitudinal research testing consumer perceptions and acceptance over a set amount of time, and taking into account demographics when testing consumer acceptance levels would provide some deeper insight into these areas.

REFERENCES

Babbie, E. (1990). *Survey research methods* (2nd ed.). Belmont, CA: Wadsworth.

Barnes, S.J., & Scornavacca, E. (2004). Mobile marketing: The role of permission and acceptance. *International Journal of Mobile Communications, 2*(2), 128–139.

Barwise, P., & Strong, C. (2002). Permission-based mobile advertising. *Journal of Interactive Marketing, 16*(1), 14–24.

Creswell, J. (2003). *Research design qualitative, quantitative, and mixed methods approaches* (2nd ed.). Thousand Oaks, CA: Sage Publications.

Dickinger, A., Haghirian, P., Murphy, J., & Scharl, A. (2004). *An investigation and conceptual model of SMS marketing.* Paper presented at the 37th Hawaii International Conference on System Sciences, HI.

Enpocket. (2002a). The branding performance of SMS advertising. Retrieved March 13, 2003, from *www.enpocket.co.uk*

Enpocket. (2002b). Consumer preferences for SMS marketing in the UK. Retrieved March 13, 2003, from *www.enpocket.co.uk*

Enpocket. (2003). The response performance of SMS advertising. Retrieved March 12, 2003, from *www.mda-mobiledata.org*

Ericsson. (2000). *Wireless advertising.* Stockholm: Ericsson Ltd.

Godin, S., Hardcover, p., 1 edition (May 1, & 0684856360., S. S. I. (1999). *Permission Marketing: Turning strangers into friends, and friends into customers.*

Green, J.C., Caracelli, V.J., & Graham, W.F. (1989). Toward a conceptual framework for mixed method evaluation designs. *Educational Evaluation and Policy Analysis, 11*(3), 255–274.

Heinonen, K., & Strandvik, T. (2003, May 22–23). *Consumer responsiveness to mobile marketing.* Paper presented at the Stockholm Mobility Roundtable, Stockholm, Sweden.

Jelassi, T., & Enders, A. (2004, June 14–16). *Leveraging wireless technology for mobile advertising.* Paper presented at the 12th European Conference on Information Systems, Turku, Finland.

Kellet, K., & Linde, A. (2001). EMS, MMS, & the future of mobile messaging, white paper. Retrieved , from, *www.magic4.com.*

Rossman, G.B., & Rallis, S.F. (1998). *Learning in the field: An introduction to qualitative research.* Thousand Oaks, CA: Sage.

Scornavacca, E., & Barnes, S.J. (2004, March). *Raising the bar: Barcode-enabled m-commerce solutions.* Paper presented at the Austin Mobility Roundtable, Austin, TX.

TTI. (2003). *Mobile messaging: Which technologies and applications will succeed?* Retrieved July 5, 2004, from *www.telecomtrends.net*

Ververidis, C., & Polyzos, G. (2002). *Mobile marketing using location based services.* Paper presented at the First International Conference on Mobile Business, Athens, Greece.

Xu, H., Teo, H.H., & Wang, H. (2003, January 7–10). *Foundations of SMS commerce success: Lessons from SMS messaging and co-opetition.* Paper presented at the 36th Hawaii International Conference on System Sciences, Big Island, HI.

This work was previously published in Unwired Business: Cases in Mobile Business, edited by S. J. Barnes and E. Scornavacca, pp. 109-123, copyright 2006 by IRM Press (an imprint of IGI Global).

Chapter XXVI
Trust Models for Ubiquitous Mobile Systems

Mike Burmester
Florida State University, USA

ABSTRACT

This chapter introduces the notion of trust as a means to establish security in ubiquitous mobile network systems. It argues that trust is an essential requirement to enable security in any open network environments, and in particular, in wireless ad hoc environments where there is no network topology. In such environments, communication can only be achieved via routes that have to be trusted. In general it may be hard, or even impossible, to establish, recall, and maintain trust relationships. It is therefore important to understand the limitations of such environments and to find mechanisms that may support trust either explicitly or implicitly. We consider several models that can be used to enable trust in such environments, based on economic, insurance, information flow, and evolutionary paradigms.

INTRODUCTION

Wireless mobile networks are a paradigm for mobile communication in which wireless nodes do not rely on any underlying static network infrastructure for services such as packet routing, name resolution, node authentication, or distribution of computational resources. The communication medium is broadcast. Nodes in range communicate in a direct peer-to-peer manner, while nodes out of range establish routing paths dynamically through other nodes where possible. The recent rise in popularity of mobile wireless devices and technological developments have made possible the deployment of wireless mobile networks for several applications. Examples include emergency deployments, disaster recovery, search-and-rescue missions, sensor networks, military (battlefield) operations, and more recently e-commerce. Since the network nodes are mobile, the network topology frequently changes: Communication links are established or broken as nodes move in and out of range, and the network may get partitioned with the connectivity restricted to the partitions. As a result it may be much harder (or even impossible) to establish trust associations.

The trend in trust management is to view trust implicitly through delegation of privilege via certificates. Certificates can be chain-linked (linking à priori trust relationships) and used to propagate and distribute trust over insecure media, without the danger of being manipulated.

In this chapter, we give an overview of several models that can be used to support trust in mobile networks, based on economic, insurance, information flow, and evolutionary paradigms.

TRUST IN WIRELESS MOBILE NETWORKS

We consider environments in which there may be no fixed underlying network infrastructure, such as static base stations, for services such as packet routing, name resolution, node authentication, or the distribution of computational resources. In such environments, recalling and maintaining trust relationships is particularly challenging. Mobile systems share many of the complexities of fixed infrastructure systems. For example, nodes may have (Burmester & Yasinsac, 2004):

1. No prior relationship or common peers
2. No shared proprietary software
3. Different transmission, memory and processing capabilities
4. Different mobility characteristics
5. Different lifetime properties

Defining Trust

Trust is a highly abstract concept and it is unlikely that any simple definition can comprehensively capture all the subtleties of its essence. Informally we may define trust as a behavioral expectation of one party toward another. There are two perspectives in this definition, one in which a party *awards* trust to another (Alice trusts that Bob's public key is PK(Bob)), the other in which a party *gains* trust from another (Alice has convinced Bob that her public key is PK(Alice)).

Representing Trust: Certificates vs. Tokens

In any stateful trust model, trust must be represented by some type of persistent structure. Certificates are the de facto standard for representing trust relationships that are protected by cryptography. Certificates are portable and bind a cryptographic key (a digital string) to an entity, thus guaranteeing the authenticity of actions performed by that entity. Trust tokens are another structure that can be used to represent trust in a more direct way, analogous to the relation between checks and cash. Checks guarantee payment by tying the purchaser to some identifying information (like a certificate), while the value of cash is self-contained.

Trusted Third Parties

A trusted third party (TTP) can facilitate significantly the establishment of trust in mobile environments. For example, if two parties A and B who do not know each other have a trust relationship with a third party T, then T can be an effective intermediary for transactions between A and B.

However in general, wireless mobile networks may not have any infrastructure components that are typically used as TTPs. In such cases, TTPs have to be elected or assigned by using an appropriate election or assignment protocol.

MODELS FOR TRUST IN WIRELESS MOBILE ENVIRONMENTS

Trust is *context* driven (e.g., A may trust *B* for event *x*, but not for event *y*). Trust may also be qualitative rather than Boolean (e.g., A may trust *B* more than *C*). Finally, trust relationships may be fixed or dynamic. Dynamic trust relationships are most appropriate for the requirements of mobile environments.

Models for dynamic trust must support establishing, changing, and permanently revoking trust between parties, and must also consider network environment issues. In the following sections we shall consider several models that can be used to support trust in wireless mobile networks (Burmester & Yasinsac, 2004).

A Mathematical Model for Trust: The Trust Graph

We may represent the trust in a network by a directed graph, the *trust graph*, whose links (A, B) correspond to the explicit trust that node *A* has in node *B*. Such links are indicated by A $\Rightarrow$ B. The implicit trust that a node *X* has in another node *Y* is then represented by a trust path from *X* to *Y*:

$$X = X_0 \Rightarrow X_1 \Rightarrow X_2 \ldots \Rightarrow X_{n-1} \Rightarrow X_n = Y,$$

in which node *X* awards trust to node *Y* via a chain of intermediary nodes X_i, where X_i awards trust explicitly to the next node X_{i+1} in the chain. Such trust may be supported by certificates. For example, node X_i may certify (digitally sign) that key $PK(X_{i+1})$ is the public key of node X_{i+1}. A chain of certificates can then be used for implicit certification. This is essentially the trust model for the X509 PKI authentication infrastructure (ISO/IEC 9594-8, 1995). This particular trust infrastructure is hierarchical, with trust centrally managed (by a Root Certifying Authority, which is also a single-point-of-failure). PGP (Zimmermann, 1995) uses a web of trust in which trust is distributed "horizontally." See Burmester and Desmedt (2004) for a discussion on security issues of hierarchical vs. horizontal infrastructures.

In the basic trust graph model, trust is transitive but not necessarily reflexive. That is, even though *A* may award trust to *B*, *B* may not award trust to *A*. However, trust is binary: A $\Rightarrow$ B is either true or false. Therefore, there is a natural trust metric which is one unit for explicit trust. This is also the trust of a trust path that links A to *B*. In this model the trust that *A* awards to *B* is represented by the trust flow of *A*, *B*, which is also the connectivity of *A*, *B*. This model is appropriate for Byzantine faults environments in which the adversary can corrupt a bounded number of nodes, and trust has to be based on à priori beliefs, and not statistical profiles.

A Model Based on a Weighted Trust Graph

There are several other ways to define trust. For a stochastic model based on statistical profiling, we can define the explicit trust that *A* awards to (or has in) *B* as the probability with which *A* trusts *B*, based on, say, a history of good behavior by *B*. See the next section for a discussion on trust based on observed behavior. In this model we have a weighted trust graph in which each link A $\Rightarrow$ B is assigned a weight $t \in [0,1]$, which corresponds to the (explicit) trust that *A* has in *B*. If $\pi 1, \pi 2, \ldots, \pi n$ are (all) the trust paths that link *X* to *Y*, then the implicit trust that *X* has in *Y* can be computed as follows (Burmester, Douligeris, & Kotzanikolaou, 2006):

$$\sum_{\Pi_i} \prod_{t_k \in \pi_i} t_k - \sum_{\pi_i \neq \pi_j} \prod_{t_k \in \pi_i \cup \pi_j} t_k + \\ \ldots + (-1)^{n+1} \prod_{t_k \in \pi_1 \cup \ldots \cup \pi_n} t_k$$

For example, if there are three disjoint paths from X to Y with trust weights (t_1,t_2), (t_3,t_4), (t_5,t_6) respectively, then the implicit trust that X has in Y is:

$$t_1t_2 + t_3t_4 + t_5t_6 - t_1t_2t_3t_4 - t_3t_4t_5t_6 + t_1t_2t_3t_4t_5t_6 .$$

One can extend this model to allow for a dynamic model in which trust is regularly updated, by using a trust-ranking algorithm similar to that used by Web search engines (e.g., PageRank of Google [PageRank, 1997]).

A Model Based on Observed Behavior

A natural way to acquire trust is through direct observation. At its most fundamental level, trust is a decision, subject to emotions and intuition. In this scenario, personal observation is preferred to second-hand methods because of hints, nuances, and feelings that can be garnered. Though feelings are not considered in computer trust systems, there are advantages in doing so. Not all actions give insight into trustworthiness. The challenge is to translate such observations into trust decisions.

A challenge to trust management systems is that trust relationships need to be constructed *before* they are exercised. There are four basic categories of activity that affect trust (Burmester & Yasinsac, 2004):

1. Trust earning actions over time
2. Trust earning actions by count
3. Trust earning actions by magnitude
4. Trust defeating actions

Combinations of the first three allow cautious parties to grant trust frugally. Untrustworthy parties will be challenged to conduct a sufficient quality and quantity of trustworthy actions to gain trust. On the other hand, observation of malicious, reckless, or otherwise unpredictable actions allows reduction or revocation of awarded trust.

A Model Based on the Internet Paradigm

The economic opportunity provided by the Internet has driven rapid establishment of many new trust models. Companies like eBay, Amazon, and Priceline conduct all of their business with customers with whom they have no personal relationship or interaction with. Early work on supporting trust models was from a business perspective (Pardue, 2000). Some work has been done more recently to identify models that support cryptographic protection of trust relationships. In Zhong, Chen, and Yang (2003), a token-based trust model is proposed in which parties accumulate trust, transaction-by-transaction. For trust-earning actions, parties are awarded tokens that can be retained and later presented to reflect the earned trust. If no additional trust information is gathered, tokens may be revoked or restricted. This novel approach to trust acquisition has many properties that are well-suited to mobile networks. Tokens can be created, awarded, and verified via distributed algorithms, allowing a global aspect to trust decisions. Conversely, if the trust algorithm is well understood, parties that desire to perform malicious acts can become sleepers, behaving perfectly until they acquire sufficient trust to allow successful mischief.

Transitive Trust

Transitivity is in many respects a natural attribute of trust and is encountered in some of the most used security systems (Steiner, Neuman, & Schiller, 1988; Zhong et al., 2003). With transitive trust models, trust must be explicit (i.e., parties must know that if they place their trust in one party, then they are automatically placing their trust in other potentially unknown parties as well). For example, if Alice trusts Bob and Bob trusts Carol, then Alice must trust Carol. Such models make strong trust requirements on intermediaries or third parties. Unfortunately, there are inherent dangers in models with transitive trust (Christianson & Harbison, 1997).

A Model Based on Trust Classes

Trust may be considered as a two party relationship or there may be environments where nodes take on *class* trust properties, as in the Bell-LaPadula model (Bell & LaPadula, 1973). One way to form trust management functionality is to establish a trust promotion system. For example, consider a simple trust environment in which nodes can be categorized into the following five trust classes (from most to least trusted): *Highly trusted, Trusted, Unknown, Untrusted, Highly untrusted.* We can then establish a set of rules for promoting and demoting members between groups. These rules will be identified by the desired promotion rule. If promotion is not allowed for highly untrusted parties, then no rule is established for this class. The model may be further extended by designating a subset of the class of most trusted nodes as *promoters*. Promoters are responsible for determining if requestors meet the promotion requirements as designated in the promotion rules and in taking action to effect the justified group movement. While promotion is requested directly, demotion must be requested second hand.

A Financial Model

Trust can also be *contractually* secured. In this case, a Trusted Third Party guarantees the trust. As with secured loans, if the guaranteed trust is violated, the guarantor will deliver the promised security to the offended party. Secured trust is a pure form of transitive trust. It is unique in that its trust graph tree has height one and trust is secured by a contractually agreed value. As with secured financial interactions, the secured value may take many forms, including the following: a *co-signed trust certificate, a trust insurance policy, a trust bond and a trust collateral.*

These correspond to security mechanisms of the financial world. For a co-signed certificate, the co-signing party would have credentials that exceed those of the target and would assume liability for any adverse events that occur as a result of a trust breech. The insurance policy model is similar, except that the security is provided by a well recognized organization that promises benefits to the executor of the policy. The last two models are similar in that the trust target provides the value that secures the trust. The value can be monetary, property, or other items or issues of suitable value to the source.

CONCLUSION

We have considered several models that can be used to manage the trust in mobile wireless environments. These models are highly distributed and address many of the trust management properties that are needed to secure mobile environments.

ACKNOWLEDGMENTS

This material is based on work supported in part by the National Science Foundation under grant number NSF 0209092 and in part by the U.S. Army Research Laboratory and the Army Research Office under grant DAAD19-02-1-0235.

REFERENCES

Bell, D. E., & LaPadula, L. (1973). Secure computer systems: Mathematical foundations and model, *MITRE Corp.* M74-244, Bedford, MA.

Burmester, M., & Desmedt, Y. (2004). Is hierarchical public-key certification the next target for hackers? *Communications of the ACM, 47*(8), 68-74.

Burmester, M., & Yasinsac, A. (2004). Trust infrastructures for wireless mobile networks. *WSAES Transactions on Telecommunications* (pp. 377-381).

Burmester, M., Douligeris, C., & Kotzanikolaou, P. (2006). Security in mobile ad hoc networks. In C. Douligeris & D. Serpanos (Eds.), *Network security: Current status and future directions.* Piscataway, NJ: IEEE Press.

Christianson, B., & Harbison, W. S. (1997). Why isn't trust transitive? In *Proceedings of the 4th International Workshop on Security Protocols* (LNCS 1189, pp. 171-176).

ISO/IEC 9594-8. (1995). Information technology, Open Systems Interconnection. *The Directory: Overview of concepts, models, and services. International Organization for Standardization.* Geneva, Switzerland.

PageRank. (1997). Google. Retrieved from http://www.google.com/technology/

Pardue, H. (2000). A trust-based model of consumer-to-consumer online auctions. *The Arrowhead Journal of Business, 1*(1), 69-77.

Steiner, J., Neuman, C., & Schiller, J. I. (1988). Kerberos and authentication service for open network systems. In *Proceedings of USENIX,* Dallas, TX.

Zhong, S., Chen, J., & Yang, R. (2003). Sprite: A simple, cheat-proof, credit-based system for mobile ad hoc networks. In *Proceedings of INFOCOM 2003.*

Zimmermann, P. (1995). *The official PGP user's guide*. Cambridge, MA: MIT Press.

This work was previously published in Secure E-Government Web Services, edited by A. Mitrakas, P. Hengeveld, D. Polemi, and J. Gamper, pp. 63-70, copyright 2007 by IGI Publishing, formerly known as Idea Group Publishing (an imprint of IGI Global).

Section VI
Emerging Trends

Chapter XXVII
Optical Network Survivability

N. S. C. Correia
University of Algarve, Portugal

M. C. R. Medeiros
University of Algarve, Portugal

INTRODUCTION

The telecommunications world is evolving dramatically toward challenging scenarios where the fast and efficient transportation of information is becoming a key element in today's society. Wavelength division multiplexing (WDM) technology has the potential to satisfy the ever-increasing bandwidth needs of the network users on a sustained basis (Mukherjee, 2000).

Network operators must provide uninterrupted service to their customers, that is, network survivability must be guaranteed. This means that networks must be able to handle link or fiber cuts as well as equipment failures, fact that influences the design and operation of networks (Gerstel & Ramaswami, 2000). When using WDM, survivability becomes even more important because of the huge amount of traffic carried by a single fiber. A single fiber failure, even for few seconds, can be catastrophic (Maier, Pattavina, Patre, & Martinelli, 2002). This issue is actually very important since the optical WDM technology is now being deployed in the field. Network survivability is not just an academic subject. In real networks, failures happen quite frequently (fiber cuts, for example, are very common in terrestrial networks since they share other utility transport conduits such as gas or water pipes and electrical cables, and are considered the least reliable component (Gerstel et al., 2000; Maier et al., 2002). The prevention of service interruption, or the reduction of the service loss when failures occur, must now be an integral part of the network design and operations strategy or otherwise severe service losses can happen.

BACKGROUND

For an easier implementation of optical transport network (OTN) functions, the optical layer has been divided into three sublayers according to the recommendation G.872 of the International Telecommunication Union-Telecommunication Standardization Sector (ITU-T) component (ITU-

T Recommendation G. 872, 1999; Gerstel et al., 2000; Maier et al., 2002):

- **Optical channel (OCh):** The managed entity is the lightpath. It takes care of all the end-to-end networking functions such as routing and wavelength assignment, connectivity check, and failure management. Its functions are done at the end-to-end lightpath terminations.
- **Optical multiplex section (OMS):** The managed entity is the multiplex of all the wavelength channels, that is, provides functionality for networking of an aggregate optical signal with multiple wavelengths. Basically, it performs WDM multiplex monitoring. Its functions are done at the link terminations.
- **Optical transmission section (OTS):** The managed entity is the multiplex of all the wavelength channels, as in the OMS, but it manages and supervises optical transmissions devices, such as amplifiers and repeaters, inserted in links. Therefore, it provides functionality for transmitting aggregate optical signals.

Network protection and restoration can be performed by the OCh or OMS sublayers. Schemes at the OCh sublayer protect individual lightpaths while schemes at the OMS sublayer protect all wavelength channels in a link as a group (Gerstel et al., 2000; Mohan & Murthy, 2000).

Besides the link failures, node and channel failures can also occur in a WDM network. A node failure is due to equipment failure at network nodes while a channel failure is due to the failure of transmitting and/or receiving equipment operating at some wavelength. The probability of such failures is much smaller, when compared to link failures, due to the built-in redundancy of most equipment. Besides being more common, link failures have a very high impact on service loss due to the simultaneous failure of several wavelength channels. Therefore, focus will be given to link failures. For more details on node failures, see Wang, Cheng, and Mukherjee (2003).

In non-WDM systems, the protected entity is the link. In WDM systems, due to the availability of multiple wavelength channels in a fiber, the survivability schemes can be more flexible. Either the link (fiber) or the lightpath (wavelength) can be the protected entity at the optical layer. This basically has to do with the sublayer of the WDM layer in which a given survivability mechanism operates. Since more multiplexing/demultiplexing and per-channel switching equipment is necessary for OCh protection, one could think that OCh protection is more expensive than OMS protection. However, this is not true if not all wavelength channels need protection. In this case, OCh protection utilizes capacity more efficiently than OMS protection and a benefit exists on the number of fibers necessary to provide protection. Since future networks tend to be flexible, providing lightpaths automatically as necessary with a variety of protection levels, and as equipment cost decreases, OCh protection seems to be the choice. For these reasons the focus will be on OCh protection. For more references on OMS protection see Maier et al. (2002).

LIGHTPATH SURVIVABILITY TECHNIQUES

The lightpath survivability techniques used in WDM networks can be broadly classified into protection and restoration (the terms proactive and reactive have also been used) (Mohan, Murthy & Somani, 2001; Sridharan, Salapaka, & Somani, 2002). Protection refers to the fact that recovery from network failures is based on preplanned schemes and uses dedicated resources. These resources are reserved for recovery from failures at either connection setup or network design time, and are kept idle when there is no failure. Thus, the use of capacity is not very efficient but the recovery process is fast.

Restoration implies the dynamic discovery of spare capacity in the network to restore the services affected by the failure. Resources are not reserved at the time of connection establishment but are chosen from available resources when failure occurs. When compared to predesigned protection, dynamic restoration makes a more efficient use of the capacity and provides resilience against different kinds of failures, on the other hand, they are more complex and the restoration time is longer than predesigned protection schemes due to the intense activity of the network management system to set up new connections. Additionally, full recovery is not guaranteed because sufficient spare capacity may not be available at the time of failure. Since restoration time is a key issue in most optical networks, protection is usually used rather than dynamic restoration.

Protection and restoration techniques can be further classified into link-based and path-based, illustrated in Figure 1, according to the kind of rerouting done. In link-based schemes a new path is selected between the end nodes of the failed link. This path along with the working segments of the working lightpath will be used to build the backup lightpath. Therefore, the traffic is rerouted around the failed link. Backup lightpaths used to protect different working lightpaths affected by a link failure may use different paths and/or different wavelengths. In Figure 1a) the backup lightpath b_{11} will replace the primary lightpath l_1 if link 1-3 fails while the backup lightpath b_{12} will replace the primary lightpath l_1 if link 3-5 fails.

In path-based schemes there is a backup lightpath between the end nodes of the failed working lightpath. Thus, the whole path between the source and destination nodes is replaced and there is no need to retain the working segments of the primary lightpath. When compared to link-based, these schemes show better resource utilization but require excessive signalling and have longer restoration times. In link-based schemes the choice of backup lightpaths is limited and thus a lot of capacity may be required since backup lightpaths are usually longer. Also, link-based schemes can not handle node failures while path-based schemes can. In Figure 1(b) the primary lightpath l_1 is replaced by the backup lightpath b_1 for any link failure affecting l_1.

LINK-BASED LIGHTPATH PROTECTION

Link-based protection can be further classified into dedicated and shared. As illustrated in Figure 2(a), dedicated protection means that all wavelength

Figure 1. Protection and restoration techniques: (a) link-based; (b) path-based

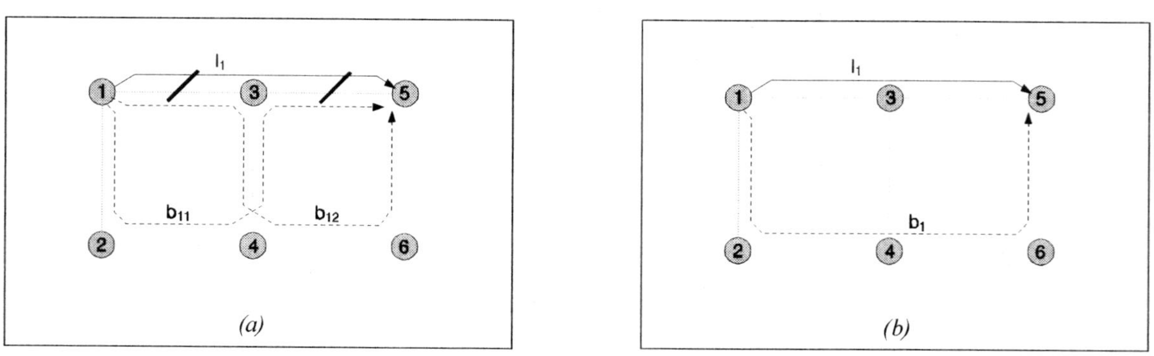

Figure 2. Link-based protection: (a) dedicated; (b) shared

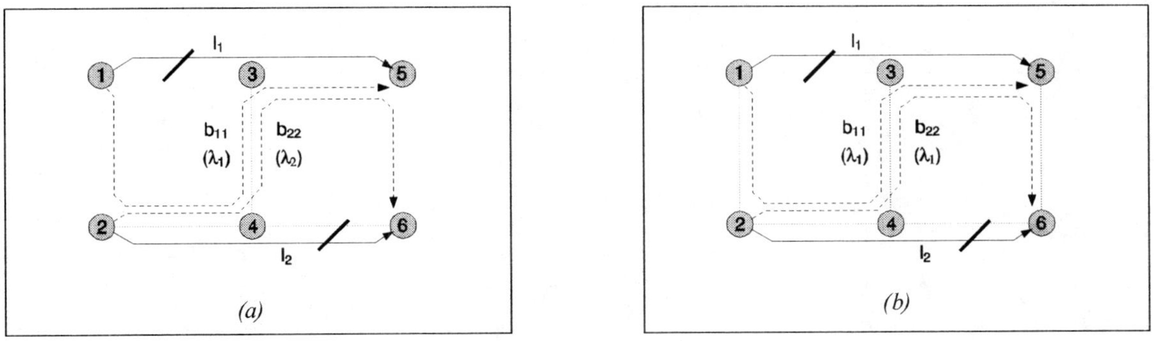

channels of a backup wavelength path will be dedicated to protect a working wavelength channel of a particular link. That is, if the backup wavelength paths of two working wavelength channels overlap then different wavelengths are required for the backup wavelength paths. This happens even if the working wavelength channels being protected are in different links. In Figure 2(a) the backup lightpaths b_{11} and b_{22} must use different wavelengths. The backup lightpath b_{11} replaces the primary lightpath l_1 if link 1-3 fails while the backup lightpath b_{22} replaces the primary lightpath l_2 if link 4-6 fails. Thus, at least 50% WDM channels cannot be used by working traffic.

Shared protection, illustrated in Figure 2(b), explores the idea that the simultaneous failure of two or more links is a very unlikely event. Backup wavelength paths can share the same wavelength on some overlapping portion if the working wavelength channels that they are protecting are in different links, meaning that in a single-link failure scenario these will not be activated at the same time. In Figure 2(b) the backup lightpaths b_{11} and b_{22} can use the same wavelength because they are protecting the primary lightpaths l_1 and l_2 against different link failures, 1-3 and 4-6 respectively, not being activated at the same time for a single-link failure scenario. In a single-link failure scenario, the shared protection can provide 100% protection while using capacity more efficiently than dedicated protection. However, multilink failures are less protected than in dedicated protection, which is able to recover from more multilink failure scenarios. Note that dedicated protection can not recover a working lightpath if the multilink failure affects the working and the backup lightpaths simultaneously.

PATH-BASED LIGHTPATH PROTECTION

Upon a link failure, a mechanism must exist to notify the end nodes of the affected working lightpaths so that they can switch to the corresponding backup lightpaths. This requires more network cooperation than in the link-based schemes. Similarly to link-based, path-based protection can be classified into dedicated and shared. In dedicated path-based protection, illustrated in Figure 3a), every wavelength channel assigned to the backup lightpath is reserved to protect a specific working lightpath and cannot be used by other backup lightpaths even if their corresponding working lightpaths are link-disjoint. In Figure 3(a) the backup lightpaths b_1 and b_2 must use different wavelengths. Thus, at least 50% WDM channels cannot be used by working traffic.

In shared path-based protection, illustrated in Figure 3(b), a wavelength channel in a link can

Figure 3. Path-based protection: (a) dedicated; (b) shared

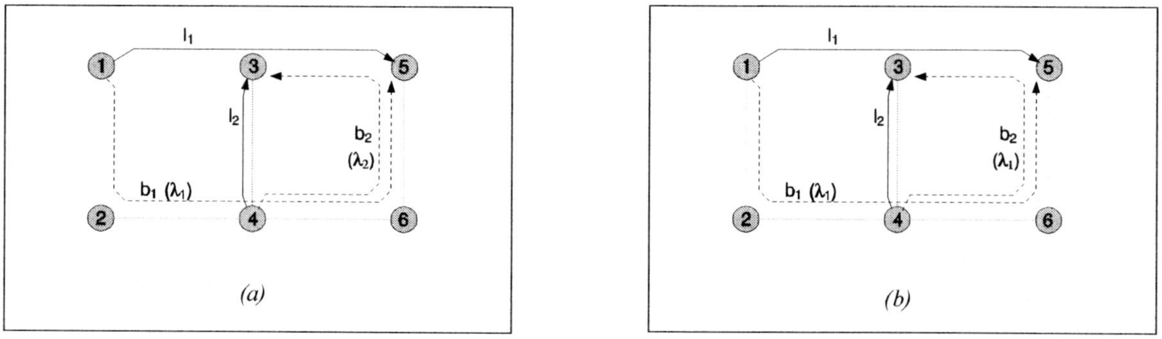

be used for two backup lightpaths if their corresponding working lightpaths are link-disjoint. In Figure 3(b) the backup lightpaths b_1 and b_2 can use the same wavelength because the primary lightpaths l_1 and b_2 do not share links meaning that they will never be activated at the same time in a single-link failure scenario. As in link-based protection, shared path-based schemes can achieve 100% protection from single-link failures whereas dedicated path-based schemes can recover from more multilink failures. In dedicated protection, the resources are kept idle even if no failure occurs, the recovery time is shorter because backup lightpaths are completely set up in advance and only the end nodes will be involved in the recovery process, whereas in shared protection backup resources can be used to protect many working lightpaths. In shared path-based schemes, although the backup lightpaths are preplanned, the configuration of optical crossconnects (OXCs) necessary for backup lightpath activation can only happen after failure occurs, meaning that recovery procedures are more complex and therefore recovery time is longer. In the other hand, dedicated path-based schemes reserve excessive resources, not being recently emphasized in papers, while shared schemes use wavelength channels more efficiently.

Shared path-based protection schemes can be further classified into failure dependent and failure independent. In failure dependent schemes there is one backup lightpath for each link used by the working lightpath. When a link fails the corresponding backup lightpath is activated. In failure independent schemes there is only one backup lightpath that is be activated for any link failure.

DYNAMIC LIGHTPATH PROTECTION

In dynamic traffic scenarios, slightly different schemes can be used so that blocking probability is reduced. Mohan et al. (2001) proposed a scheme called primary-backup multiplexing, that allows wavelength channels to be shared by a working

Figure 4. Primary-backup multiplexing scheme

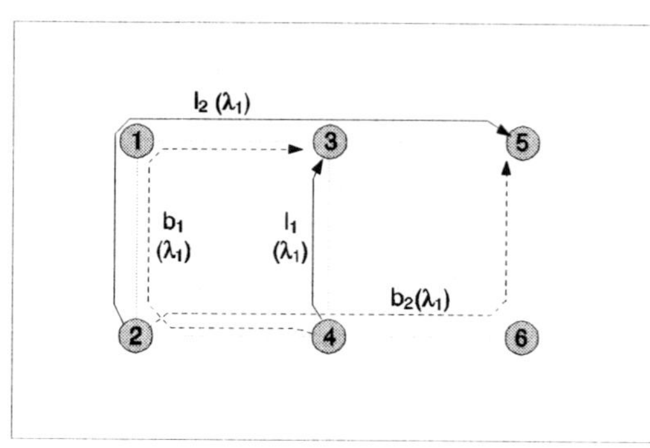

lightpath and one or more backup lightpaths. The authors sustain this approach arguing that for short-lived lightpaths full protection is a waste. In this scheme working lightpaths are allowed to lose recoverability when a channel used by the corresponding backup lightpath is used by another working lightpath. It regains recoverability when the other working lightpath terminates. This is illustrated in Figure 4. The backup lightpath b_1 shares wavelength channels with the working lightpath l_2 meaning that the working lightpath l_1 lost its recoverability when working lightpath l_2 was created. If lightpath l_2 terminates then l_1 regains its recoverability.

SURVIVABILITY IN MULTILAYER NETWORKS

Current networks have several layers interoperating with each other. At the optical layer, the WDM is the emerging technology that must satisfy connectivity and bandwidth requests from client higher layers. As networks become more Internet protocol (IP) data centric and both the IP and WDM evolve, currently used asynchronous transfer mode (ATM) and synchronous digital hierarchy (SDH) layers will become unnecessary (Colle, Maesschalck, Develder, & et al., 2002; Conte, Listanti, Settembre, & Sabella, 2002). Although higher layers such as ATM and IP have recovery mechanisms, their recovery times are very large when compared to the few milliseconds that the optical layer protection can provide. Besides minimizing data losses, due to faster recovery, optical layer protection can provide protection to higher layers not having this capability.

In IP-over-WDM, fault-tolerance can be provided either by the WDM or the IP layer. At the WDM layer protection is given to lightpaths carrying traffic with protection needs while at the IP client layer fault-tolerance can be provided through IP label-switched path (LSP) protection if multi-protocol label switching (MPLS) is used (Sahasrabuddhe, Ramamurthy, & Mukherjee, 2002; Zheng & Mohan, 2003). The advantages of using WDM lightpath protection are that: protection can be applied to all traffic streams in a lightpath; and the restoration is contained within the network where the failure occurs, improving latency and stability. However, lightpath protection is inefficient in what concerns to the use of resources. This is because the bandwidth reserved by backup lightpaths is equal to the total bandwidth of the lightpath being protected even if just a small percentage of traffic is being carried, meaning that unused bandwidth is also protected. This tends to aggravate in the future since the bandwidth of channels is expected to grow to OC-768. An alternative to the traditional WDM lightpath protection is the IP LSP protection approach where protection is provided to individual IP LSPs by the IP layer. When using this scheme resources are used efficiently but many signalling messages can be generated for the recovery of every IP LSP (Ye, Assi, Dixit, & Ali, 2001). Survivable traffic grooming (STG) schemes are emerging to solve this problem.

SURVIVABLE TRAFFIC GROOMING

Protection combined with traffic grooming has recently being considered in Correia and Medeiros (2006) and Yao & Ramamurthy (2005), and STG protection schemes have been proposed and analysed. Survivable traffic grooming addresses the survivability of connections together so that resources are better utilized. The STG protection scheme proposed in Correia et al. (2006), called primary-shared lightpath protection, uses resources efficiently while providing optical layer protection. The idea is to use the available bandwidth of working lightpaths, announced as virtual links, for backup purposes thus improving bandwidth utilization when compared with the traditional lightpath protection. This is illustrated in Figure 5(a).

In this example, the low speed streams c_1 and c_2 are delivered using lightpath l_1, connecting nodes 1 and 6, and another traffic stream c_3 is delivered using lightpath l_2. Assuming that only one wavelength channel between nodes 1 and 4 is available on the network, no backup lightpath would be found to protect l_1 if traditional lightpath protection was used. But if lightpath l_2 has available bandwidth then announcing it as a virtual link to the network would mean that backup computation could use it to build backups. Thus, working lightpath l_2 together with the backup lightpath b_1 are able to protect the working lightpath l_1. In case of failure of lightpath l_1, node 4 must groom local traffic of l_2 with traffic from b_1, activated after failure, thus requiring a G-fabric. Figures 5b and 5c illustrate node 4 before and after a link failure affecting the working lightpath l_1, that is, before and after backup activation. This STG protection scheme uses resources efficiently while fast recovery and good scaling are obtained due to optical layer protection. The implementation of this scheme is possible when using generalized multi-protocol label switching (GMPLS) in a peer model.

Figure 5. Primary-shared lightpath protection: (a) network example; (b) Node 4 before backup activation; (c) Node 4 after backup activation

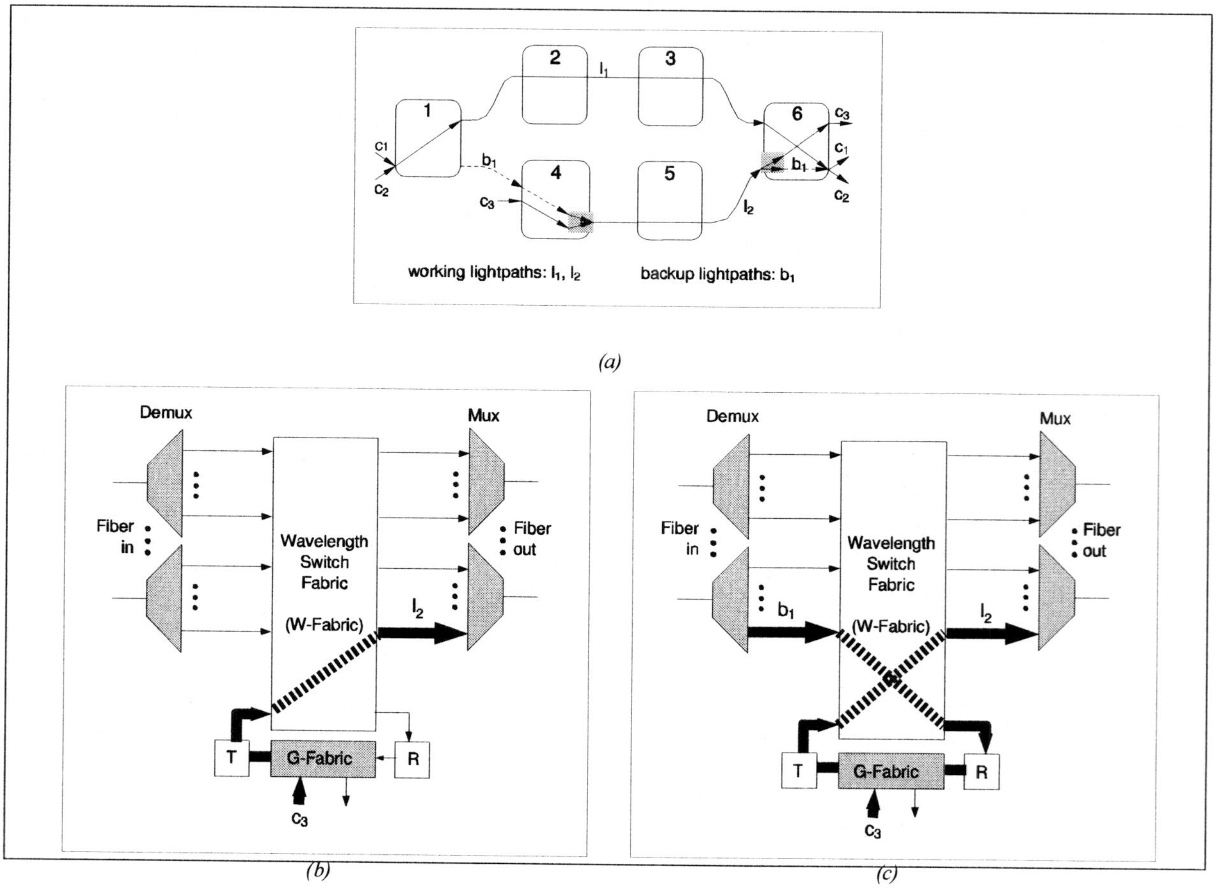

FUTURE TRENDS

As stated by Yao et al. (2005), network providers are facing the pressure of generating revenues by providing reliable multigranularity connection services while reducing network costs. Therefore, recently proposed schemes employing STG protection, such as those proposed by Correia et al. (2006) and Yao et al. (2005) are expected to play an important role in future optical networks. STG protection remains relatively unexplored and is now gaining attention from researchers.

CONCLUSION

In optical networks, survivability becomes very important because of the huge amount of traffic carried by fibers. Therefore, the prevention of service interruption, or the reduction of the service loss, must now be an integral part of the network design and operations strategy. In this article, optical network survivability has been discussed and special emphasis has been given to WDM lightpath protection.

REFERENCES

Colle, D., Maesschalck, S., Develder, C., Van Heuven, P., Groebbens, A., Cheyns, J., Lievens, I., Pickavet, M., Lagasse, P., & Demeester, P. (2002). Data-centric optical networks and their survivability. *IEEE Journal on Selected Areas in Communications*, *20*(1), 6-20.

Conte, G., Listanti, M., Settembre, M., & Sabella, R. (2002). Strategy for protection and restoration of optical paths in WDM backbone networks for next-generation internet infrastructures. *IEEE Journal on Lightwave Technology*, *20*(8), 1264-1276.

Correia, N. S. C., & Medeiros, M. C. R. (2006). Protection schemes for IP-over-WDM networks: Throughput and recovery time comparison. *Photonic Network Communications*, *11*(2), 127-149.

Gerstel, O., & Ramaswami, R. (2000). Optical layer survivability: a services perspective. *IEEE Communications Magazine*, *38*(3), 104-113.

ITU-T Recommendation G.872. (1999). Architecture of Optical Transport Network (OTN).

Maier, G., Pattavina, A., Patre, S., & Martinelli, M. (2002). Optical network survivability: Protection techniques in the WDM layer. *Photonic Network Communications*, *4*(2/4), 251-269.

Mohan, G., & Murthy, C. (2000). Lightpath restoration in WDM optical networks. *IEEE Network*, *14*(6), 24-32.

Mohan, G., Murthy, C., & Somani, A. (2001). Efficient algorithms for routing dependable connections in WDM optical networks. *IEEE/ACM Transactions on Networking*, *9*(5), 553-566.

Mukherjee, B. (2000). WDM optical communication networks: Progress and challenges. *IEEE Journal on Selected Areas in Communications*, *18*(10), 1810-1824.

Sahasrabuddhe, L., Ramamurthy, S., & Mukherjee, B. (2002). Fault management in IP-over-WDM networks: WDM protection versus IP restoration. *IEEE Journal on Selected Areas in Communications*, *20*(1), 21-33.

Sridharan, M., Salapaka, M., & Somani, A. (2002). A practical approach to operating survivable WDM networks. *IEEE Journal on Selected Areas in Communications*, *20*(1), 34-46.

Wang, Y., Cheng, T., & Mukherjee, B. (2003). Dynamic routing and wavelength assignment scheme for protection against node failure. *IEEE GLOBECOM* (pp. 2585-2589).

Yao, W., & Ramamurthy, B. (2005). Survivable traffic grooming with path protection at the connection level in WDM mesh networks. *Journal of Lightwave Technology*, *23*(10), 2846-2853.

Ye, Y., Assi, C., Dixit, S., & Ali, M. (2001). A simple dynamic integrated provisioning/protection scheme in IP over WDM networks. *IEEE Communications Magazine*, *39*(11), 174-182.

Zheng, Q., & Mohan, G. (2003). Protection approaches for dynamic traffic in IP/MPLS-over-WDM networks. *IEEE Optical Communications*, *41*(5), S24-S29.

KEY TERMS

Generalized Multi-Protocol Label Switching (GMPLS): An extension of the MPLS concept to the circuit switching network and the optical fiber network. GMPLS enables unified control management of the network layers (packet / TDM / wavelength / fiber).

Multi-Protocol Label Switching (MPLS): Circuit-switching based mechanism to carry data over a packet-switched network using the concept of label switching.

Network Survivability: Capacity of the network to provide continuous service in the presence of failures.

Optical Crossconnect (OXC): Optical device used mainly in long-distance networks to switch high-speed optical signals in a fiber optic network.

Survivable Traffic Grooming (STG): Approach that provides multigranularity connections that are reliable and resource-efficient.

Traffic Grooming: Aggregation of different low speed connections into high capacity connections, allowing an efficient utilization of resources, higher throughputs, and minimization of network costs.

Wavelength Division Multiplexing (WDM): Technology which multiplexes multiple optical carrier signals on a single optical fibre by using different wavelengths of laser light to carry different signals. This allows for a multiplication in capacity.

This work was previously published in Encyclopedia of Internet Technologies and Applications, edited by M. Freire and M. Pereira, pp. 383-390, copyright 2008 by Information Science Reference, formerly known as Idea Group Reference (an imprint of IGI Global).

Chapter XXVIII
Fractal Geometry and Computer Science

Nicoletta Sala
Università della Svizzera Italiana, Switzerland
Università dell'Insubria, Italy

ABSTRACT

Fractal geometry can help us to describe the shapes in nature (e.g., ferns, trees, seashells, rivers, mountains) exceeding the limits imposed by Euclidean geometry. Fractal geometry is quite young: The first studies are the works by the French mathematicians Pierre Fatou (1878-1929) and Gaston Julia (1893-1978) at the beginning of the 20th century. However, only with the mathematical power of computers has it become possible to realize connections between fractal geometry and other disciplines. It is applied in various fields now, from biology to economy. Important applications also appear in computer science because fractal geometry permits us to compress images, and to reproduce, in virtual reality environments, the complex patterns and irregular forms present in nature using simple iterative algorithms executed by computers. Recent studies apply this geometry to controlling traffic in computer networks (LANs, MANs, WANs, and the Internet). The aim of this chapter is to present fractal geometry, its properties (e.g., self-similarity), and their applications in computer science.

INTRODUCTION

Fractal geometry is a recent discovery. It is also known as Mandelbrot's geometry in honor of its father, the Polish-born Franco-American mathematician Benoit Mandelbrot, who showed how fractals can occur in many different places in both mathematics and elsewhere in nature.

Fractal geometry is now recognized as the true geometry of nature. Before Mandelbrot, mathematicians believed that most of the patterns of nature were far too irregular, complex, and fragmented to be described mathematically. Mandelbrot's geometry replaces Euclidian geometry, which had dominated our mathematical thinking for thousands of years.

The *Britannica Concise Encyclopedia* ("Fractal Geometry," 2007) introduces fractal geometry as follows:

In mathematics, the study of complex shapes with the property of self-similarity, known as fractals. Rather like holograms that store the entire image in each part of the image, any part of a fractal can be repeatedly magnified, with each magnification resembling all or part of the original fractal. This phenomenon can be seen in objects like snowflakes and tree bark.... This new system of geometry has had a significant impact on such diverse fields as physical chemistry, physiology, and fluid mechanics; fractals can describe irregularly shaped objects or spatially nonuniform phenomena that cannot be described by Euclidean geometry."

The multiplicity of the application fields had a central role in the diffusion of fractal geometry (Barnsley, Saupe, & Vrscay, 2002; Eglash, 1999; Mandelbrot, 1982; Nonnenmacher, Losa, Merlini, & Weibel, 1994; Sala, 2004, 2006; Vyzantiadou, Avdelas, & Zafiropoulos, 2007).

BACKGROUND: WHAT IS A FRACTAL?

A fractal could be defined as a rough or fragmented geometric shape that can be subdivided in parts, each of which is approximately a reduced-size copy of the whole (Mandelbrot, 1988). *Fractal* is a term coined by Benoit Mandelbrot (born 1924) to denote the geometry of nature, which traces inherent order in chaotic shapes and processes. The term derived from the Latin verb *frangere, to break*, and from the related adjective *fractus, fragmented and irregular*. This term was created to differentiate pure geometric figures from other types of figures that defy such simple classification. The acceptance of the word fractal was dated in 1975. When Mandelbrot presented the list of publications between 1951 and 1975, the date when the French version of his book was published, people were surprised by the variety of the studied fields: linguistics, cosmology, economy, games theory, turbulence, and noise on telephone lines (Mandelbrot, 1975). Fractals are generally self-similar on multiple scales. So, all fractals have a built-in form of iteration or recursion. Sometimes the recursion is visible in how the fractal is constructed. For example, Koch's snowflake, Cantor's set, and Sierpinski's triangle are generated using simple recursive rules. Self-similarity, iterated function systems, and the Lindenmayer System are applied in different fields of computer science (e.g., in computer graphics, virtual reality, and traffic control for computer networks).

Self-Similarity

Self-similarity, or invariance against changes in scale or size, is a property by which an object contains smaller copies of itself at arbitrary scales. Mandelbrot (1982, p. 34) defined self-similarity as follows: "When each piece of a shape is geometrically similar to the whole, both the shape and the cascade that generate it are called self-similar."

A fractal object is self-similar if it has undergone a transformation whereby the dimensions of the structure were all modified by the same scaling factor. The new shape may be smaller, larger, translated, and/or rotated. Similar means that the relative proportions of the shapes' sides and internal angles remain the same. As described by Mandelbrot (1982), this property is ubiquitous in the natural world. Oppenheimer (1986) used the term fractal, exchanging it with self-similarity, and he affirmed that the geometric notion of self-similarity is evolving in a paradigm for modeling the natural world, in particular in the world of botany.

Self-similarity appears in objects as diverse as leaves, mountain ranges, clouds, and galaxies. Figure 1a shows a snowflake that is an example of self-similarity in nature. Figure 1b illustrates Koch's snowflake; it is built starting from an equilateral triangle, removing the inner third of each side, building another equilateral triangle at the location where the side was removed, and

Figure 1. (a) A snowflake is a natural fractal object, and (b) Koch's snowflake is a fractal generated using simple geometric rules.

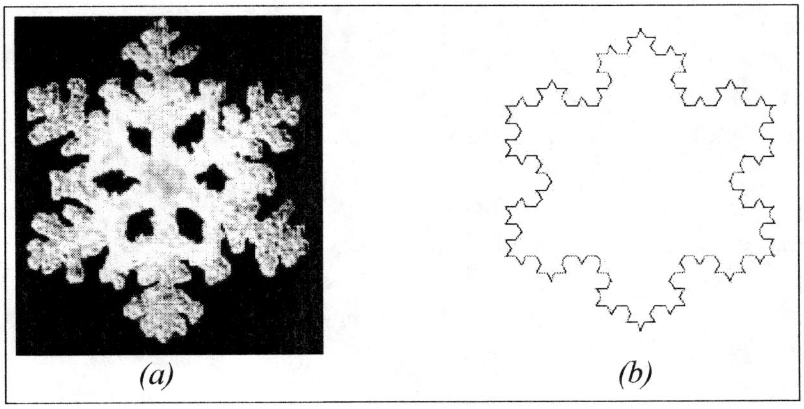

then repeating the process indefinitely. This fractal object represents an attempt to reproduce complex shapes present in nature using few simple geometric rules.

The Koch snowflake is an example of a shape with a finite area enclosed within an infinite boundary. This seems contrary to geometric intuition, but this is characteristic of many shapes in nature. For example, in the human body all the arteries, veins, capillaries, and bronchial structures occupy a relative small fraction of the body. Thus, inside the human body there is the presence of fractal geometry using two different points of view: spatial fractals and temporal fractals. Spatial fractals refer to the presence of self-similarity; for instance, the small intestine repeats its form, observed as various enlargements. Spatial fractals also refer to the branched patterns that are present inside the human body for enlarging the available surface for the absorption of substances (in the intestine) and the distribution and collection of solutes (in the blood vessels, and in the bronchial tree).

Temporal fractals are present in some dynamic processes, for example, in cardiac rhythm. The long-term variability of heart rate observed over a wide range of time scales with scale-invariant power-law characteristics has recently been associated with fractal scaling behavior and long-range correlation properties (Meyer, 2002).

The Iterated Function System

The iterated function system (IFS) is another fractal that can be applied in computer science. Barnsley (1993, p. 80) defined the IFS as follows:

A (hyperbolic) iterated function system consists of a complete metric space $(\mathbf{X}, d)$ together with a finite set of contraction mappings $w_n: \mathbf{X} \rightarrow \mathbf{X}$ with respective contractivity factor s_n, for $n = 1, 2,.., N$. The abbreviation "IFS" is used for "iterated function system." The notation for the IFS just announced is $\{ \mathbf{X}, w_n, n = 1, 2,.., N\}$ and its contractivity factor is $s = \max \{s_n : n = 1, 2, ..., N\}$.

Barnsley put the word *hyperbolic* in parentheses because it is sometimes dropped in practice. He also defined the following theorem (Barnsley, 1993, p. 81):

Let $\{\mathbf{X}, w_n, n = 1, 2, ..., N\}$ *be a hyperbolic iterated function system with contractivity factor s. Then the transformation* $W: H(\mathbf{X}) \to H(\mathbf{X})$ *defined by:*

$$W(B) = \cup_{n=1}^{n} w_n(B) \qquad (1)$$

For all $B \in H(\mathbf{X})$*, is a contraction mapping on the complete metric space* $(H(\mathbf{X}), h(d))$ *with contractivity factor s. That is:*

$$H(W(B), W(C)) \leq s \cdot h(B,C) \qquad (2)$$

for all $B, C \in H(\mathbf{X})$*. Its unique fixed point,* $A \in H(\mathbf{X})$*, obeys,*

$$A = W(A) = \cup_{n=1}^{n} w_n(A) \qquad (3)$$

and is given by $A = \lim_{n\to\infty} W^{\circ n}(B)$ *for any* $B \in H(\mathbf{X})$.

The fixed point A ∈ H(**X**), described in the theorem by Barnsley, is called the attractor of the IFS or invariant set.

Figure 2. Fern leaf created using the IFS

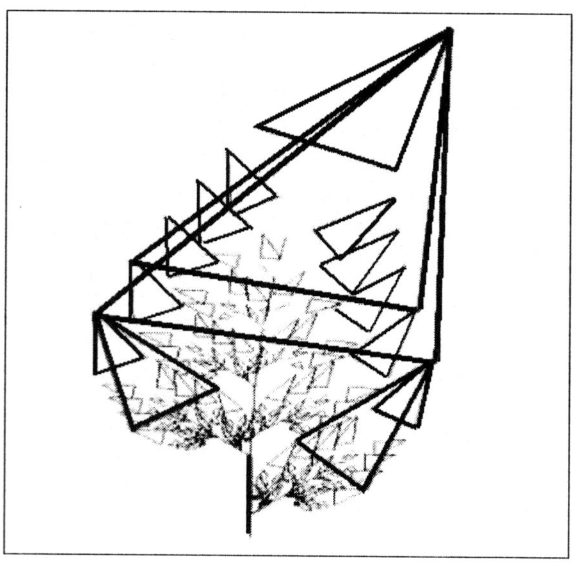

Bogomolny (1998) affirms that two problems arise. One is to determine the fixed point of a given IFS, and it is solved by what is known as the *deterministic* algorithm.

The second problem is the inverse of the first: For a given set A∈H(**X**), find an iterated function system that has A as its fixed point (Bogomolny, 1998). This is solved approximately by the Collage Theorem (Barnsley, 1993).

The Collage Theorem (Barnsley, 1993, p. 94) states:

Let $(\mathbf{X}, d)$*, be a complete metric space. Let* $L \in H(\mathbf{X})$ *be given, and let* $\varepsilon \geq o$ *be given. Choose an IFS (or IFS with condensation)* $\{\mathbf{X}, (w_n), w_1, w_2, ..., w_n\}$ *with contractivity factor* $0 \leq s \leq 1$*, so that,*

$$h(L, \cup_{\substack{n=1\\(n=0)}}^{n} w_n(L)) \leq \varepsilon \qquad (4)$$

Where h(d) is the Hausdorff metric. Then,

$$h(L, A) \leq \frac{\varepsilon}{1-s} \qquad (5)$$

Where A is the attractor of the IFS. Equivalently,

$$h(L, A) \leq (1-s)^{-1} h(L, \cup_{\substack{n=1\\(n=0)}} w_n(L)) \qquad (6)$$

for all $L \in H(\mathbf{X})$.

The Collage Theorem describes how to find an IFS whose attractor is close to a given set; one must endeavor to find a set of transformations such that the union, or collage, of the images of the given set under transformations is near to the given set.

The Collage Theorem states that an iterated function system must represent an image.

Next Figure 2 shows a fern leaf created using the IFS. The IFS is produced by polygons: in this case, triangles that are put in one another. The final step of this iterative process shows a fern that has a high degree of similarity to a real one.

L-Systems

An L-system or Lindenmayer system is an algorithmic method for generating branched forms and structures such as plants. L-systems were invented in 1968 by Hungarian biologist Aristid Lindenmayer (1925-1989) for modeling biological growth. He worked with filamentous fungi and studied the growth patterns of various types of algae, for example, the blue-green bacteria Anabaena catenula. Originally, the L-systems were devised to provide a formal description of the development of such simple multicellular organisms, and to illustrate the neighborhood relationships between plant cells. Later on, this system was extended to describe higher plants and complex branching structures.

L-systems can also be used to generate self-similar fractals that are a particular type of symbolic dynamical system with the added feature of a geometrical interpretation of the evolution of the system. The components of an L-system are the following.

- An alphabet that is a finite set V of formal symbols containing elements that can be replaced (variables)
- The constants that are a set S of symbols containing elements that remain fixed
- The *axiom* (also called the *initiator*), which is a string $\boldsymbol{\omega}$ of symbols from **V** defining the initial state of the system
- A *production* (or *rewriting rule*) P that is a set of rules or productions defining the way variables can be replaced with combinations of constants and other variables. A production consists of two strings: the predecessor and the successor.

The rules of the L-system grammar are applied iteratively starting from the initial state.

L-systems are also commonly known as parametric L-systems, and they are defined as a tuple **G** = {V, S, ω, P}.

Lindenmayer's original L-system for modeling the growth of algae and the blue-green bacteria (Anabaena catenula) is the following (Prusinkiewicz & Lindenmayer, 1990).

- variables: a, b
- constants: none
- start (axiom): b
- rules: (a → ab), (b → a)

The rule (a → ab) means that the letter a is to be replaced by the string ab, and the rule (b → a) means that the letter b is to be replaced by a. The symbols a and b represent cytological states of the cells (their size and readiness to divide). It produces the following.

n = 0: b
n = 1: a
n = 2: ab
n = 3: aba
n = 4: abaab
n = 5: abaababa

This is the simplest class of L-systems, those which are deterministic and context free, called DOL-systems. Using geometric interpretation of strings, it is possible to generate schematic images of Anabaena catenula.

An L-system can be also defined as a formal grammar (a set of rules and symbols) most famously used for modeling the growth processes of plant development, and it has been thought to be able to model the morphology of a variety of organisms.

The differences between L-systems and Chomsky grammars are well described by Prusinkiewicz and Lindenmayer (1990, pp. 2-3), who stated,

The essential difference between Chomsky grammars and L-systems lies in the method of applying productions. In Chomsky grammars productions are applied sequentially, whereas in L-systems

they are applied in parallel and simultaneously replace all letters in a given word. This difference highlights the biological motivation of L-systems. Productions are intended to capture cell divisions in multicellular organisms, where many divisions may occur at the same time. Parallel production application has an essential impact on the formal properties of rewriting systems.

Strings generated by L-systems may be interpreted geometrically in different ways. For example, L-system strings serve drawing commands for a LOGO-style turtle. The turtle interpretation of parametric L-systems was introduced by Szilard and Quinton (1979), and extended by Prusinkiewicz (1986, 1987) and Hanan (1988, 1992).

Prusinkiewicz and Lindenmayer defined a state of the turtle as a triplet (x, y, α), where the Cartesian coordinates (x, y) represent the turtle's position, and the angle α, called the heading, is interpreted as the direction in which the turtle is facing. Given the step size s and the angle increment δ, the turtle can respond to commands represented by the symbols in Table 1.

The Koch snowflake can be simply encoded as a Lindenmayer system with initial string F--F, string rewriting rule F → F+F--F+F, and angle 60°. Figure 3 shows an example of plant-like structures generated after four iterations by bracketed L-systems with the initial string F (angle δ= 22.5°), and the replacement rule F → FF+[+F-F-F] -[-F+F+F].

Figure 3. Plant-like structures generated after four iterations by L-system

Table 1. Commands for LOGO-style turtle derived by L-systems

Symbols	Meaning
F	Move forward a step of length s. The state of the turtle changes; now it is (x', y', α), where x' = x +s ·cos α and y'= y + s ·sin α. A segment between (x, y), the starting point, and the point (x', y') is drawn.
f	Move forward a step of length s without drawing a line.
+	Turn left by angle δ. The positive orientation of angles is counterclockwise, and the next state of the turtle is (x, y, α+δ).
-	Turn right by angle δ. The next state of the turtle is (x, y, α - δ).
[	Push the current state of the turtle onto a pushdown operations stack. The information saved on the stack contains the turtle's position and orientation, and possibly other attributes such as the color and width of lines being drawn.
]	Pop a state from the stack and make it the current state of the turtle. No line is drawn, although in general the position of the turtle changes.

APPLICATIONS OF FRACTAL GEOMETRY IN COMPUTER SCIENCE

Fractal geometry is one of the most exciting frontiers in the fusion between mathematics and computer science. Mandelbrot's geometry permits us to compress the images, and to reproduce in computer graphics and in virtual reality environments the textures and the irregular forms present in nature (e.g., mountains, clouds, and trees) using simple iterative or recursive algorithms. Recent studies also apply this geometry to the control of traffic in computer networks.

Fractal Geometry in Computer Graphics

Fractal geometry has been generalized by the computer graphics community and it includes objects outside Mandelbrot's original definition (Foley, van Dam, Feiner, & Hughes, 1997). It means anything that has a substantial measure of exact or statistical self-similarity. In the case of statistical fractals, it is the probability density that repeats itself on every scale. An application field of fractal geometry is in the compression of images (fractal compression). A fractal compressed image can be defined as follows: It is an encoding that describes (a) the grid partitioning (the range blocks), and (b) the affine transformations (one per range block) (Shulman, 2000). Research on fractal image compression derived from the mathematical ferment on chaos and fractals in the years 1978 to 1985. Barnsley was the principal researcher who worked on fractal compression. The basic idea was to represent by an IFS a fixed point that is close to the image (Barnsley, 1988; Barnsley, Jacquin, Malassenet, Reuter, & Sloane, 1988). This fixed point is also known as a fractal (Fisher, 1995). Each IFS is coded as a contractive transformation with coefficients. The Banach fixed-point theorem (1922), also known as the contraction mapping theorem or contraction mapping principle, guarantees the existence and uniqueness of fixed points of certain self maps of metric spaces, and provides a constructive method to find those fixed points.

An image can be represented using a set of IFS codes rather than pixels. In this way, a good compression ratio can be achieved. This method was good for generating almost real images based on simple iterative algorithms. The inverse problem, going from a given image to an IFS that can generate the original (or at least closely resemble it), was solved by Jacquin according to Barnsley in March 1988. He introduced a modified scheme for representing images called partitioned iterated function systems (PIFSs). The main characteristics of this approach were that (a) it relied on the assumption that image redundancy can be efficiently exploited through self-transformability on a block-wise basis, and (b) it approximated an original image by a fractal image (Jacquin, 1992). In a PIFS, the transformations do not map from the whole image to the parts, but from larger parts to smaller parts. In Jacquin's method, the small areas are called range blocks, the big areas are called domain blocks, and the pattern of range blocks was called the partitioning of an image. Every pixel of the original image has to belong to one range block. This system of mappings is contractive; thus, when iterated, it quickly converges to its fixed-point image. Therefore, the key point for this algorithm is to find fractals that can best describe the original image and then represent them as affine transformations.

All methods are based on the fractal transform using iterated function systems that generate a close approximation of a scene using only a few transformations (Peitgen & Saupe, 1988; Wohlberg & de Jager, 1999; Zhao & Liu, 2005).

Fractal compression is a lossy compression method (compressing data and then decompressing it retrieves data that may well be different from the original, but is close enough to be useful in some way), and most lossy compression formats suffer from generation loss: Repeatedly

Figure 4. Fractal compression: Repeatedly compressing and decompressing the file will cause it to progressively lose quality

compressing and decompressing the file will cause it to progressively lose quality (as shown in Figure 4).

Mandelbrot's geometry and chaos theory can create beautiful images in 2-D and 3-D, as well as realistic, natural-looking structures and fractal textures used to add visual interest to relatively simple and boring geometric models (Ebert, Musgrave, Peachy, Perlin, & Worley, 2003). Fractal algorithms can also be used in computer graphics to generate complex objects using IFS.

Smith (1984) was the first to prove that L-systems were useful in computer graphics for describing the structure of certain plants in his paper "Plants, Fractals, and Formal Languages." He claimed that these objects should not be labeled as fractals for their similarity to fractals, introducing a new class of objects he called "graftals." This class captured great interest in the computer imagery field (Foley et al., 1997; Smith).

Fractal Geometry for Modeling Landscapes

Another interesting application of fractal geometry in computer science is for modeling landscapes. Fournier, Fussel, and Carpenter (1982) developed a mechanism for generating a kind of fractal mountain based on a recursive subdivision algorithm for a triangle. Here, the midpoints of each side of the triangle are connected, creating four new subtriangles. Figure 5a shows the subdivision of the triangle into four smaller triangles, and Figure 5b illustrates how the midpoints of the original triangle are perturbed in the y direction (Foley et al., 1997). To perturb these points, the properties of self-similarity can be used, and so can the conditional expectation properties of fractional Brownian motion (fBm). The fractional Brownian motion was originally introduced by Mandelbrot and Van Ness in 1968 as a generalization of the Brownian motion (Bm). Figure 6 shows a recursive subdivision of an initial polygon using triangles. Other polygons can be used to generate the grid (e.g., squares and hexagons).

This method evidences two problems that are classified as internal and external consistency problems (Fournier et al., *1982)*. Internal consistency is the reproducibility of the primitive at any position in an appropriate coordinate space and at any level of detail so that the final shape is independent of the orientation of the subdivided triangle. This is satisfied by a Gaussian random-number generator that depends on the point's position; thus, it generates the same numbers in the same order at a given subdivision level. External

Figure 5. (a) The subdivision of a triangle into four smaller triangles, and (b) perturbation in the y direction of the midpoints of the original triangle

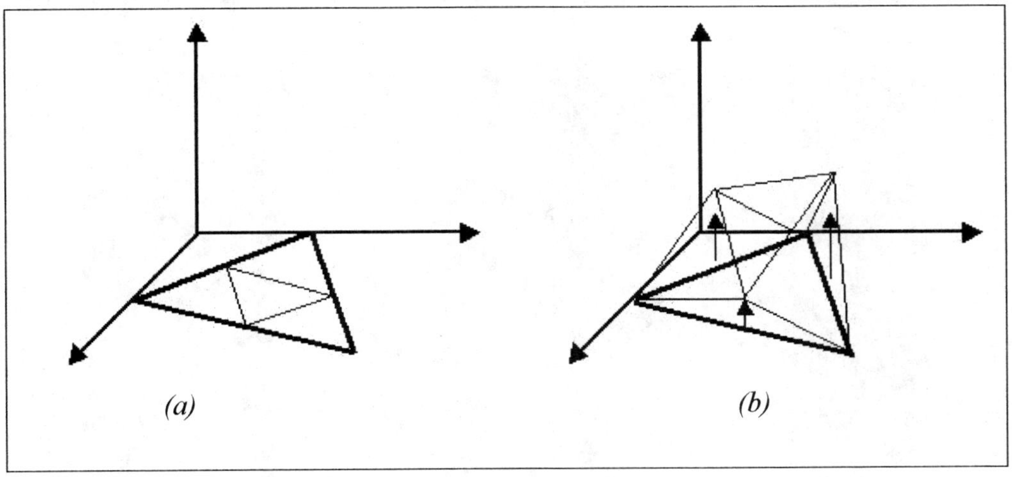

Figure 6. Grid of triangles generated by a recursive subdivision and applying the fractional Brownian motion

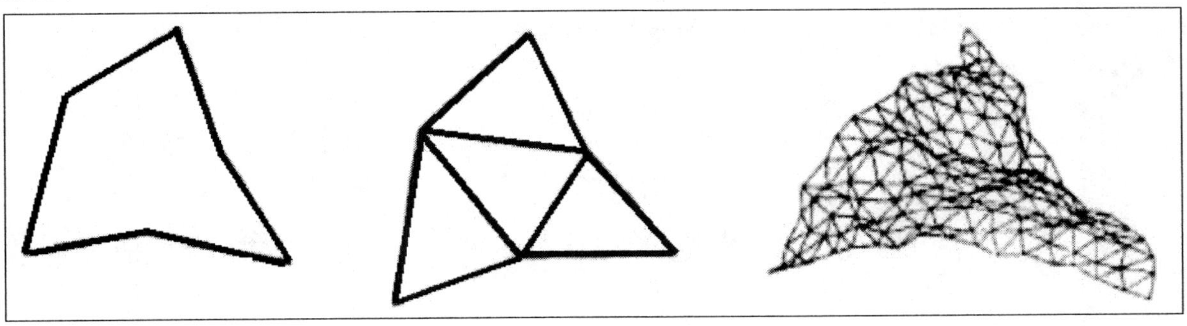

consistency concerns the midpoint displacement at shared edges and their direction of displacement. This process, when iterated, produces a deformed grid that represents a surface; after the rendering phase (that includes a hidden line, colored and shaded) there can appear a realistic fractal mountain.

Using fractal algorithms, it is possible to create virtual mountains described in the virtual reality modeling language (VRML), as shown in Figure 7. VRML is a 3-D graphics language used on the World Wide Web for producing virtual worlds that appear on the display screens using an appropriate VRML browser. This example shows that the connections between fractal geometry, virtual worlds, and the World Wide Web exist.

The last examples describe how to create fractal mountains, but not their erosion. Musgrave, Kolb, and Mace (1989) introduced techniques that are independent of the terrain creation. The algorithm can be applied to already generated data represented as regular height fields, which require

Figure 7. Virtual mountains realized in VRML using fractal algorithms

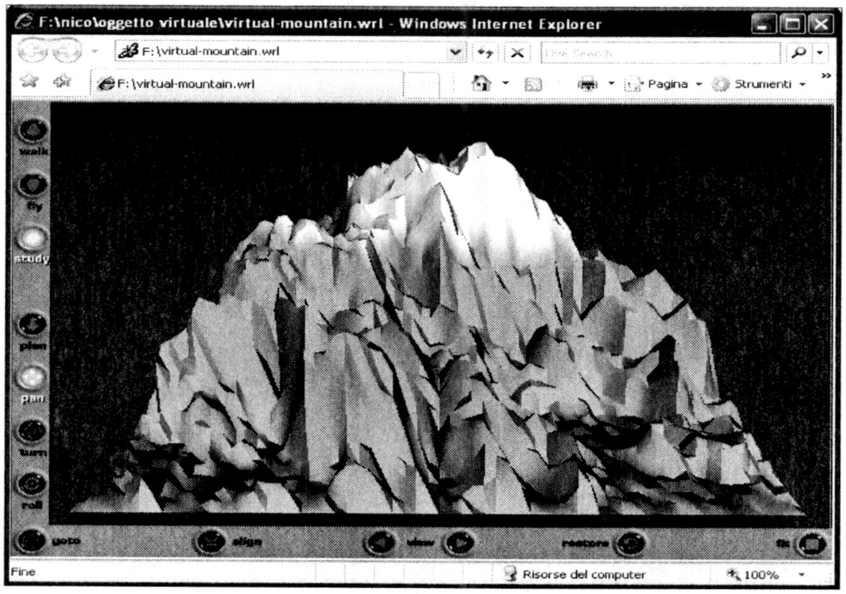

Figure 8. Squig-curve construction (recursion Levels 0-7)

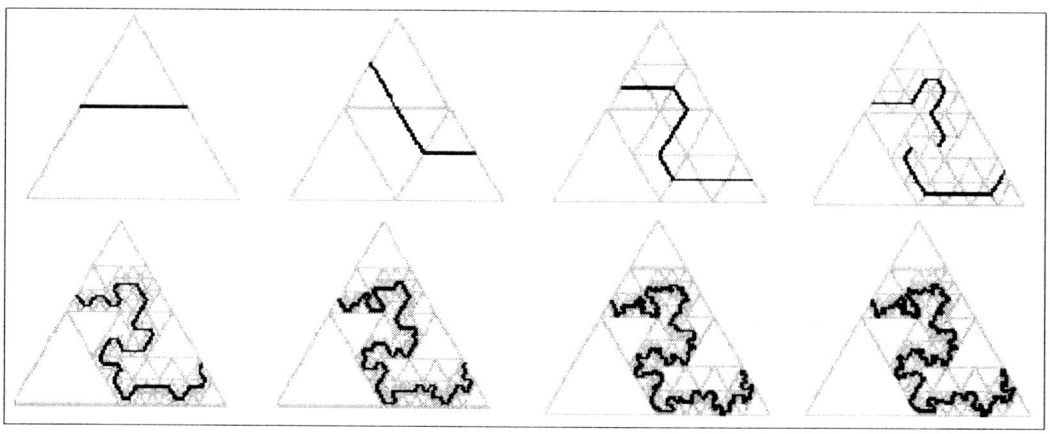

separate processes to define the mountain and the river system. Prusinkiewicz and Hammel (1993) combined the midpoint-displacement method for mountain generation with the squig-curve model of a nonbranching river originated by Mandelbrot (1978, 1979). Their method created one nonbranching river as a result of a context-sensitive L-system operating on geometric objects (a set of triangles). Three key problems remained open: (a) The river flowed at a constant altitude, (b) the river flowed in an asymmetric valley, and (c) the river had no tributaries. Figure 8 shows an example of a squig-curve construction (recursion Levels 0–7; Prusinkiewicz & Hammel, 1993).

Marák, Benes, and Slavík (1997) reported a method for synthetic terrain erosion that is based on rewriting the matrices representing terrain parts. They found a rewriting process that was context sensitive. It was defined as a set of productions A → B, where A and B were matrices of numbers of type N ×N, where N>0.

The terrain parts were rewritten using certain user-defined sets of rules that represented an erosion process. The method consisted of three kinds of rewriting processes (Marák, 1997; Marák et al., 1997). The absolute rewriting permitted us to erode the objects in a predefined altitude. The rewriting with the reference point could erode an arbitrary object in any altitude. The third kind of rewriting process permitted us to erode some shapes in any scale. Its advantage was that the algorithm could be controlled by some external rules and could simulate different kinds of erosion (Marák et al.).

Guérin, Tosan, and Baskurt (2002) proposed an interesting model for fractal curves and surfaces. This method, called projected IFS, combined two classical models: a fractal model (IFS attractors) and a computer-aided geometric design model (CAGD). The model, based on IFS, generates a geometrical shape or an image with an iterative process. This method permitted us to reconstruct smooth surfaces and not only rough surfaces using real data. The original surface generated has been extracted from a geological database (found at the United States Geological Survey home page, http://www.usgs.org).

Fractal Geometry for Controlling Network Traffic

Different models on the nature of network traffic have been proposed in the literature, but in contrast to traditional models, network traffic presents a kind of fractality (J. Liu, 2006; Norros, 1994, 1995; Park & Willinger, 2000).

In the last decades, many fractal traffic processes were described, but the fractality in the traffic was a complex concept, and different approaches were developed. Early models included fractional Brownian motion (Lévy Véhel & Riedi, 1997; Norros, 1994, 1995), the zero-rate renewal process model (Erramilli, Gosby, & Willinger, 1993; Veitch, 1992, 1993), and deterministic chaotic maps (Erramilli & Singh, 1992; Erramilli, Pruthi, & Willinger, 1994; Mondragon, Arrowsmith, & Pitts, 1999).

Many studies have shown that the classical Markovian models were not able to describe the real behavior of the traffic in networks. These models made unacceptable mistakes in the quantitative design in the allocation of the resources, in the connection and in admission control, in the scheduling, and in the traffic regulation. Markovian models supposed an autocorrelation that was decaying in an exponential way, but real examples showed a different behavior (Norros, 1995).

Other studies based on fractal geometry, in particular referring to the property of self-similarity, were realized. Self-similarity is an important notion for understanding the problems connected to the network traffic, including the modeling and analysis of network performances (Fowler & Leland, 1994; Lévy Véhel & Riedi, 1997; Norros, 1994, 1995; Park & Willinger, 2000).

Leland, Taqqu, Willinger, and Wilson (1993, 1994) reported that the Ethernet local area network (LAN) traffic was statistically self-similar. In fact, the Ethernet LAN data were not only consistent with self-similarity at the level of aggregate packet traffic, but they were also in agreement with self-similarity in terms of the basic characteristics of the individual source-destination pair traffics that make up self-similar aggregate trace. None of the other traffic models were able to describe this fractal behavior, which has serious implications for the design, control, and analysis of high-speed networks. Aggregating streams of such traffic typically intensifies self-similarity ("burstiness") instead of smoothing it.

Figure 9 shows the Bellcore pOct traffic trace, a famous fractal trace, with a Hurst parameter of 0.78 (Norros, 1995).

Figure 9. Bellcore pOct traffic trace with Hurst parameter of 0.78 (Norros, 1995)

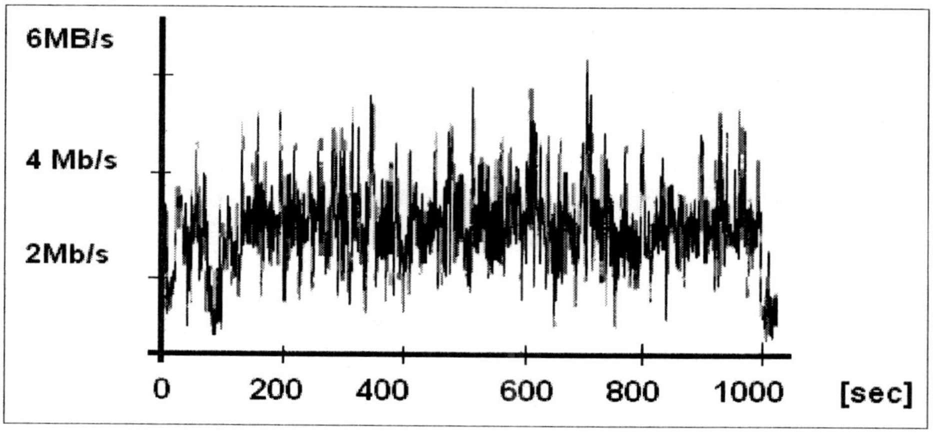

Klivansky, Mukherjee, and Song (1994) presented an examination of packet traffic from geographically dispersed locations on the National Science Foundation Network (NSFNET) backbone. The analyses indicated that packet-level traffic over NSFNET core switches exhibited long-range dependence (LRD), and a subset of these showed the property of self-similarity (an asymptotic second-order self-similarity).

Willinger, Taqqu, Sherman, and Wilson (1995), using an approach suggested by Mandelbrot (1969), showed that the superposition of many on-off sources, each of which exhibited a phenomenon called the Noah effect, resulted in self-similar aggregate network traffic approaching fractional Brownian motion (the so-called Joseph effect).

Giordano, Pierazzini, and Russo (1995) reported some realistic traffic scenarios in the analysis of broadband telecommunication networks. They described the performance evaluation of a broadband network that provided a best-effort, asynchronous interconnection of several remote LANs, observing some relevant effects of the arrival processes on network performances considering a model of the distributed queue dual bus (DQDB) IEEE 802.6 network. The analysis of real traffic in LANs and in metropolitan area networks (MANs) and of its long-range dependence confirmed a self-similar nature of the traffic offered to a broadband network.

Paxson and Floyd (1994, 1995) reported the results of measurement and analysis on wide area network (WAN) traffic (transmission-control protocol [TCP] traffic) for applications like TELNET (teletype network) and FTP (file transfer protocol). Millions of connections and TCP packets from different sites, with traces ranging from 1 hour to 30 days, were collected. These studies showed the self-similar nature of WAN traffic.

Crovella and Bestavros (1995, 1996) reported that the World Wide Web's traffic is self-similar. They observed that (a) the traffic patterns generated by browsers have a self-similar nature, (b) every Web browser is modeled as an on-off source model and data fit well the Pareto distribution, and (c) the files available via the Web over the Internet seem to have a heavy-tailed size distribution (bimodal distribution).

Lévy Véhel and Riedi (1997) reported that the fractional Brownian motion, which has been used to model the long-range dependence of traffic traces, showed the property of self-similarity. They noticed that the multifractal approach to traffic was natural because it was a process of positive increments.

All these studies highlight that fractal traffic in LANs, MANs, WANs, and in the WWW has a correlation existing at multiple time scales, a kind of self-correlation that decays in a low-power way (as shown in Figure 10). This self-correlation has an important impact on network performance.

The Hurst parameter H is able to show a degree of self-similarity (for example, a degree of persistence of the statistical phenomenon under test). H has a value range of $0.5 \leq H \leq 1.0$. A value of $H = 0.5$ indicates the lack of self-similarity, a value for H close to 1.0 indicates a large degree of self-similarity or long-range dependence in the process.

Zhang, Shu, and Yang (1997) applied the Hurst parameter to capture the fractal behavior of the network traffic. They also made some interesting considerations of the multifractal behavior and the value of the Hurst parameter when the traffic is merged.

The fractal characterization of the Internet traffic has an exhaustive description by Park and Willinger in their book *Self-Similar Network Traffic and Performance Evaluation* (2000).

More recently, Salvador, Nogueira, Valadas, and Pacheco (2004) addressed the modeling of network traffic using a multi-time-scale framework. They evaluated the performance of two classes of traffic models: Markovian and Lindenmayer-systems-based traffic models. These traffic models included the notion of time scale using different approaches: indirectly in the model structure through a fitting of the second-order statistics in the case of the Markovian models, or directly in the model structure in the case of the Lindenmayer-systems-based models.

Figure 10. Self-similarity in the traffic trace

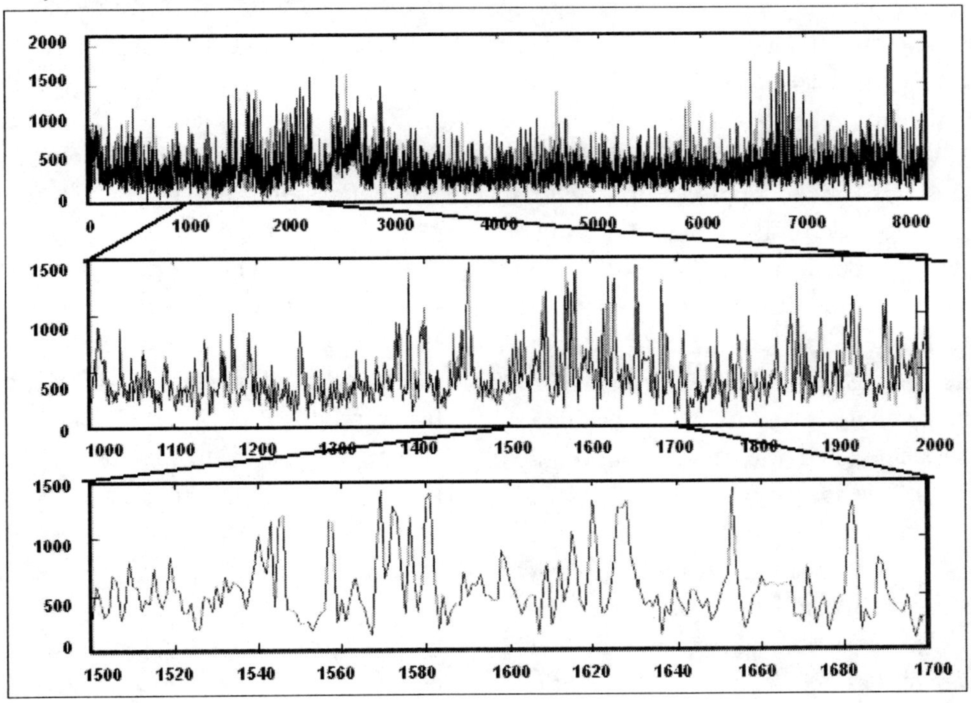

FUTURE TRENDS

In the field of the compression of images, novel methods are studied. For example, van Wijk and Saupe (2004) present a fast method to generate fractal imagery based on IFS. The high performance derives from three factors: (a) All steps are expressed in graphics operations that can be performed quickly by hardware, (b) frame-to-frame coherence is exploited, (c) only a few commands per image have to be passed from the CPU (central processing unit) to the graphics hardware, avoiding the classical CPU bottleneck.

Wang, Wu, He, and Hintz (2006) present an interactive progressive fractal decoding method, where the compressed file can be transmitted incrementally and reconstructed progressively at the users' side. The method requires no modification to encode or decode any fractal image compression algorithm, and it provides the user-controlled decoding procedure with the inherited fractal fast decoding feature. Their experimental results have shown that the method can be applied to various fractal compression techniques and in particular where the transmission bandwidth is relevant.

Other studies are based on the idea of integrating spiral architecture (SA) and fractal compression (He, Wang, Wu, Hintz, & Hur, 2006; D. Liu, 2006). SA is a recent approach to machine vision systems (Sheridan, 1996; Sheridan, Hintz, & Moore, 1991). It is inspired from anatomical consideration of the primate's vision system (Schwartz, 1980). The natural data structure that emerges from geometric consideration of the distribution of photo receptors on the primate's retina has been called the spiral honeycomb mosaic (SHM; Sheridan & Hintz, 1999). Spiral architecture, inspired by the natural SHM, is an image structure on which images are displayed as a collection of hexagonal pixels placed in a hexagonal grid (as shown in Figure 11a). Each hexagonal pixel has six neighboring pixels that have the same distance to the center hexagon of unit of vision. In this way, SA has the possibility to save time for local and global processing. In the human eye, these hexagons would represent the relative position of the rods and cones on the retina. This arrangement is different from the traditional rectangular image architecture, a set of 3×3 rectangles used as a unit of vision, where each pixel has eight neighbor pixels (in Figure 11b).

This hexagonal representation has special computational features that are pertinent to the vision process, and it has features of higher degree of circular symmetry, uniform connectivity, greater angular resolution, and a reduced need of storage and computation in image processing operations.

Sheridan, Hintz, and Alexander (2000) introduced a one-dimensional addressing scheme for a hexagonal structure, together with the definitions of two operations, spiral addition and spiral multiplication, that correspond to image translation and rotation respectively. This hexagonal structure is called the spiral architecture. Each pixel on the spiral architecture is identified by a designated positive number, called a spiral address. Figure 12a shows a spiral architecture and the spiral addresses. The numbered hexagons form the cluster of size 7^n. In Figure 12a there is a collection of $7^2 = 49$ hexagons with labeled addresses. Every collection of seven hexagons is labeled starting from the center address as it was done for the first seven hexagons. The collection of hexagons in Figure 12a grows in powers of seven with uniquely assigned addresses. It is this pattern of growth of addresses that generates the spiral; in fact, the hexagons tile the plane in a recursive modular manner along the spiral direction (Figure 12b).

Wu, He, and Hintz (2004) construct a virtual hexagonal structure that represents an important innovation in this field. Using a virtual spiral architecture, a rectangular structure can be smoothly converted to the spiral architecture. A virtual spiral architecture exists only during the procedure of image processing, which creates virtual hexagonal pixels in the computer memory. The data can be reorganized in a rectangular architecture for display. The term virtual refers to the characteristic that the hexagonal pixels do not physically exist.

Figure 13 shows the phases of the image processing on a virtual spiral architecture (He, Hintz, Wu, Wang, & Jia, 2006). The accuracy and efficiency of image processing on SA have been demonstrated in many recently published papers (Bobe & Schaefer, 2006; He, Hintz, et al., 2006; He, Wang, et al., 2006; Wu et al., 2004). Spiral architecture applied into fractal image compression improves the compression performance in the compression ratio with little suffering in image quality. The methods present the following advantages: Little, if any, distortion is introduced and regular pixels are divided up into subpixels.

In the field of modeling landscapes, the future trends are oriented to use fractal geometry, in particular IFS, generating terrain from real data and extracting from geological databases. This is useful in the reconstruction of real terrain and landscapes (Guérin & Tosan, 2005; Guérin et al., 2002).

Fractal geometry also offers an alternative approach to conventional planetary terrain analysis. For example, Stepinski, Collier, McGovern, and Clifford (2004) describe Martian terrains, represented by topography based on the Mars Orbiter Laser Altimetry (MOLA) data, as a series of drainage basins. Fractal analysis of each drainage network computationally extracts some network descriptors that are used for a quantitative characterization and classification of Martian surfaces.

In the field of network traffic control, recent studies intend to check the fractal behavior of network traffic supporting new applications and services (Chakraborty, Ashir, Suganuma, Mansfield, Roy, & Shiratori, 2004). Internet traffic being extremely variable and bursty in a wide range of time scales is usually characterized by self-similarity, which describes its statistics depending on the scales of resolution.

Marie, Bez, Blackledge, and Datta (2006) report a novel method to capture the fractal behavior of Internet traffic, adopting the random scaling fractal model (RSF) to simulate its self-affine characteristics. They realize the transmission of a digital file by splitting the file into a number of binary blocks (files) whose size and submission times are compatible with the bursty lengths of Internet traffic.

The fractal traffic analyses and their results will be useful to enhance the performance of real-time traffic, in particular in multimedia video applications (J. Liu, 2006). This is possible because video traffic is (a) self-similar and fractal, (b) time sensitive, and (c) bandwidth consuming.

CONCLUSION

This chapter has described some applications of fractal geometry in computer science. The fascination that surrounds fractal geometry seems to exist for two reasons. First, it is based on simple rules. The other reason is that fractal geometry is very suitable for simulating many phenomena. For example, fractal behavior and long-range dependence have been observed in the field of fluctuations in different systems, from black holes to quasars, rivers to fault lines, financial networks to computer networks, and brains to hearts; these and many other kinds of complex systems have all exhibited power law scaling relations in their behaviors.

Self-similarity, which characterizes fractal objects, is a unifying concept. In fact, it is an attribute of many laws of nature and innumerable phenomena in the world around us.

In computer science, fractals can be applied in different fields: to compress images using simple algorithms based on IFS, to model complex objects in computer graphics (e.g., mountains and rivers) using L-systems and the fractional Brownian motion, and to control network traffic. In particular,

Internet traffic time series exhibit fractal characteristics with long-range dependence. This is due to the existence of several statistical properties that are invariant across a range of time scales, such as self-similarity and multifractality, which have an important impact on network performance. The traffic self-similarity existent in network and application traffic seems to be a ubiquitous phenomenon that is independent of technology, protocol, and environment. Therefore, traffic models must be able to include these properties in their mathematical structure and parameter inference procedures. This is one reason why fractal traffic analysis and modeling have been a popular research topic in network engineering for two decades.

Understanding the nature of traffic in computer networks is essential for engineering operation and their performance evaluation.

ACKNOWLEDGMENT

The author wishes to thank Professor Giovanni Degli Antoni (University of Milan) for his precious suggestions during the draft of the chapter.

REFERENCES

Barnsley, M. F. (1988). *Fractals everywhere*. Boston: Academic Press.

Barnsley, M. F. (1993). *Fractals everywhere* (2nd ed.). Boston: Academic Press.

Barnsley, M. F., Jacquin, A. E., Malassenet, F., Reuter, L., & Sloane, A. D. (1988). Harnessing chaos for image synthesis. *SIGGRAPH 1988* (pp. 131-140).

Barnsley, M. F., Saupe, D., & Vrscay, E. R. (Eds.). (2002). *Fractals in multimedia*. Berlin, Germany: Springer.

Bobe, S., & Schaefer, G. (2006). Image processing and image registration on spiral architecture with saLib. *International Journal of Simulation Systems, Science & Technology, 7*(3), 37-43.

Bogomolny, A. (1998). *The collage theorem*. Retrieved September 15, 2005, from http://www.cut-the-knot.org/ctk/ifs.shtml

Chakraborty, D., Ashir, A., Suganuma, G., Mansfield, K., Roy, T., & Shiratori, N. (2004). Self-similar and fractal nature of Internet traffic. *International Journal of Network Management, 14*(2), 119-129.

Crovella, M. E., & Bestavros, A. (1995). *Explaining World Wide Web traffic self-similarity* (Tech. Rep. No. TR-95-015). Boston: Boston University, Computer Science Department.

Crovella, M. E., & Bestavros, A. (1996). Self-similarity in World Wide Web traffic: Evidence and possible causes. *Proceedings of ACM SIGMETRICS'96*.

Ebert, D. S., Musgrave, F. K., Peachy, D., Perlin, K., & Worley, S. (2003). *Texturing and modeling: A procedural approach* (3rd ed.). Morgan Kaufmann Publishers Inc.

Eglash, R. (1999). *African fractals: Modern computing and indigenous design*. Piscataway, NJ: Rutgers University Press.

Erramilli, A., Gosby, D., & Willinger, W. (1993). *Engineering for realistic traffic: A fractal analysis of burstiness*. Proceedings of ITC Special Congress, Bangalore, India.

Erramilli, A., Pruthi, P., & Willinger, W. (1994). *Modelling packet traffic with chaotic maps*. ISRN KTH/IT/R-94/18-SE, Stockholm-Kista, Sweden.

Erramilli, A., & Singh, R. P. (1992). *An application of deterministic chaotic maps to model packet traffic* (Bellcore Technical Memorandum). Bellcore.

Fisher, Y. (1995). *Fractal image compression: Theory and application.* New York: Springer-Verlag.

Foley, J. D., van Dam, A., Feiner, S. K., & Hughes, J. F. (1997). *Computer graphics: Principles and practice* (2nd ed.). New York: Addison Wesley.

Fournier, A., Fussel, D., & Carpenter, L. (1982). Computer rendering of stochastic models. *Communications of the ACM, 25*, 371-384.

Fowler, H., & Leland, W. (1994). Local area network traffic characteristics, with implications for broadband network congestion management. *IEEE Journal of Selected Areas in Communications, 9*(7), 1139-1149.

Fractal geometry. (2007). *Britannica concise encyclopedia.* Retrieved February 26, 2007, from http://concise.britannica.com/ebc/article-9364797/fractal-geometry

Giordano, S., Pierazzini, G., & Russo, F. (1995). *Multimedia experiments at the University of Pisa: From videoconference to random fractals.* Retrieved October 10, 2006, from http://www.isoc.org/HMP/PAPER/109/html/paper.html

Guérin, E., & Tosan, E. (2005). Fractal inverse problem: Approximation formulation and differential methods. In J. Lévy-Véhel & E. Lutton (Eds.), *Fractal in engineering: New trends in theory and applications* (pp. 271-285). London: Springer.

Guérin, E., Tosan, E., & Baskurt, A. (2002). Modeling and approximation of fractal surfaces with projected IFS attractors. In M. M. Novak (Ed.), *Emergent nature: Patterns, growth and scaling in the science* (pp. 293-303). NJ: World Scientific.

Hanan, J. S. (1988). *PLANTWORKS: A software system for realistic plant modelling.* Unpublished master's thesis, University of Regina.

Hanan, J. S. (1992). *Parametric L-systems and their application to the modelling and visualization of plants.* Unpublished doctoral dissertation, University of Regina.

He, X., Hintz, T., Wu, Q., Wang, H., & Jia, W. (2006). A new simulation of spiral architecture. *Proceedings of 2006 International Conference on Image Processing, Computer Vision, & Pattern Recognition (IPCV'06)* (pp. 570-575).

He, X., Wang, H., Wu, Q., Hintz, T., & Hur, N. (2006). Fractal image compression on spiral architecture. *International Conference on Computer Graphics, Imaging and Visualisation (CGIV'06)* (pp. 76-83).

Jacquin, A. E. (1992). Image coding based on a fractal theory of iterated contractive image transformations image processing. *IEEE Transactions, 1*(1), 18-30.

Klivansky, S. M., Mukherjee, A., & Song, C. (1994). On long-range dependence in NSFNET traffic (Tech. Rep. No. GIT-CC-94-61). Georgia Institute of Technology.

Leland, W. E., Taqqu, M. S., Willinger, W., & Wilson, D. V. (1993). On the self-similar nature of Ethernet traffic. *Proceedings of the ACM/SIGCOMM'93* (pp. 183-193).

Leland, W. E., Taqqu, M. S., Willinger, W., & Wilson, D. V. (1994). On the self-similar nature of Ethernet traffic (extended version). *IEEE/ACM Transactions on Networking, 2*(1), 1-15.

Lévy Véhel, J., & Riedi, R. (1997). Fractional Brownian motion and data traffic modeling: The other end of the spectrum. In J. Lévy Véhel, E. Lutton, & C. Tricot (Eds.), *Fractals in engineering* (pp. 185-202). London: Springer.

Liu, D. (2006). A parallel computing algorithm for improving the speed of fractal image compression based on spiral architecture. *Proceedings of 2006*

International Conference on Image Processing, Computer Vision, & Pattern Recognition (IPCV'06) (pp. 563-569).

Liu, J. (2006). *Fractal network traffic analysis with applications.* Retrieved January 10, 2007, from http://hdl.handle.net/1853/11477

Mandelbrot, B. B. (1969). Long-run linearity, locally Gaussian processes, H-spectra and infinite variances. *International Economic Review, 10,* 82-113.

Mandelbrot, B. (1975). *Les objects fractals: Forme, hasard et dimension.* Paris: Nouvelle Bibliothèque Scientifique Flammaron.

Mandelbrot, B. B. (1978). Les objets fractals. *La Recherche, 9,* 1-13.

Mandelbrot, B. B. (1979). Colliers all′eatoires et une alternative aux promenades aux hasard sans boucle: Les cordonnets discrets et fractals. *Comptes Rendus, 286A,* 933-936.

Mandelbrot, B. (1982). *The fractal geometry of nature.* W. H. Freeman & Company.

Marák, I. (1997). *On synthetic terrain erosion modeling: A survey.* Retrieved April 14, 2005, from http://www.cescg.org/CESCG97/marak/

Marák, I., Benes, B., & Slavík, P. (1997). Terrain erosion model based on rewriting of matrices. *Proceedings of WSCG-97, 2,* 341-351.

Marie, R. R., Bez, H. E., Blackledge, J. M., & Datta, S. (2006). *On the fractal characteristics of network traffic and its utilization in covert communications.* Retrieved January 2, 2007, from http://ima.org.uk/Conferences/mathssignalprocessing2006/Marie.pdf

Meyer, M. (2002). Fractal scaling of heartrate dynamics in health and disease. In G. A. Losa, D. Merlini, T. F. Nonnenmacher, & E. R. Weibel (Eds.), *Fractal in biology and medicine* (Vol. 3, pp. 181-193). Basel, Switzerland: Birkhauser.

Mondragon, R. J., Arrowsmith, D. K., & Pitts, J. M. (1999). Chaotic maps for traffic modelling and queueing performance analysis. *Performance Evaluation, 43*(4), 223-240.

Musgrave, F. K., Kolb, C. E., & Mace, R. S. (1989). The synthesis and rendering of eroded fractal terrain. *Computer Graphics, 23*(3), 41-50.

Nonnenmacher, T. F., Losa, G. A., Merlini, D., & Weibel, E. R. (Eds.). (1994). *Fractal in biology and medicine.* Basel, Switzerland: Birkhauser.

Norros, I. (1994). A storage model with self-similar input. *Queueing Systems Theory and Applications, 16*(3-4), 387-396.

Norros, I. (1995). On the use of fractional Brownian motion in the theory of connectionless networks. *IEEE Journal on Selected Areas in Communications, 13*(6), 953-962.

Oppenheimer, P. (1986). Real time design and animation of fractal plants and trees. *Computer Graphics, 20*(4), 55-64.

Park, K., & Willinger, W. (2000). *Self-similar network traffic and performance evaluation* (1st ed.). London: John Wiley & Sons.

Paxson, V., & Floyd, S. (1994). Wide area traffic: The failure of Poisson modeling. *Proceedings of ACM SIGCOMM'94* (pp. 257-268).

Paxson, V., & Floyd, S. (1995). Wide area traffic: The failure of Poisson modeling. *IEEE/ACM Transactions on Networking, 3*(3), 226-244.

Peitgen, H., & Saupe, D. (1988). *The science of fractal images.* New York: Springer-Verlag.

Prusinkiewicz, P. (1986). Graphical applications of L-systems. *Proceedings of Graphics Interface '86: Vision Interface* (pp. 247-253).

Prusinkiewicz, P. (1987). Applications of L-systems to computer imagery. In H. Ehrig, M. Nagl, G. Rozenberg, & A. Rosenfeld (Eds.), *Lecture notes in computer science: Vol. 291. Graph-Grammars and Their Application to Computer Science (3rd International Workshop)* (pp.534-548). Heidelberg, Germany: Springer-Verlag.

Prusinkiewicz, P., & Hammel, M. (1993). A fractal model of mountains with rivers. *Proceeding of Graphics Interface '93* (pp. 174-180).

Prusinkiewicz, P., & Lindenmayer, A. (1990). *The algorithmic beauty of plants.* New York: Springer-Verlag.

Sala, N. (2004). Fractal geometry in the arts: An overview across the different cultures. In M. M. Novak (Ed.), *Thinking in patterns: Fractals and related phenomena in nature* (pp. 177-188). Singapore: World Scientific.

Sala, N. (2006). Complexity, fractals, nature and industrial design: Some connections. In M. M. Novak (Ed.), *Complexus mundi: Emergent pattern in nature* (pp. 171-180). Singapore: World Scientific.

Salvador, P., Nogueira, A., Valadas, R., & Pacheco, A. (2004). Multi-time-scale traffic modeling using Markovian and L-systems models. In *Lecture notes in computer science: Vol. 3262. Proceedings of 3rd European Conference on Universal Multiservice Networks* (pp. 297-306). Heidelberg, Germany: Springer-Verlag.

Schwartz, E. (1980). Computation anatomy and functional architecture of striate cortex: A spatial mapping approach to perceptual coding. *Vision Research, 20*, 645-669.

Sheridan, P. (1996). *Spiral architecture for machine vision.* Unpublished doctoral dissertation, University of Technology, Sydney, Australia.

Sheridan, P., & Hintz, T. (1999). Primitive image transformations on hexagonal lattice (Tech. Rep.). Bathurst, Australia: Charles Sturt University.

Sheridan, P., Hintz, T., & Alexander, D. (2000). Pseudo-invariant image transformations on a hexagonal lattice. *Image and Vision Computing, 18*, 907-917.

Sheridan, P., Hintz, T., & Moore, W. (1991). Spiral architecture in machine vision. Proceedings of the Australian Occam and Transputer Conference, Australia.

Shulman, J. A. (2000). *Fractals and Benoit Mandelbrot: A computer science discovery often considered kind to prime numbers.* American Computer Science Association (ACSA). Retrieved December 28, 2006, from http://www.acsa2000.net/frac/

Smith, A. R. (1984). Plants, fractals, and formal languages. *International Conference on Computer Graphics and Interactive Techniques: Proceedings of the 11th Annual Conference on Computer Graphics and Interactive Techniques* (pp. 1–10).

Stepinski, T. F., Collier, M. L., McGovern, P. J., & Clifford, S. M. (2004). Martian geomorphology from fractal analysis of drainage networks. *Journal of Geophysical Research, 109*(n°E2), E02005.1-E02005.12.

Szilard, A. L., & Quinton, R. E. (1979). An interpretation for DOL systems by computer graphics. *The Science Terrapin, 4*, 8-13.

van Wijk, J. J., & Saupe, D. (2004). Image based rendering of iterated function systems. *Computers & Graphics, 28*(6), 937-943.

Veitch, D. (1992). *Novel models of broadband traffic.* Proceedings of the Seventh Australian Teletraffic Research Seminar, Murray River, Australia.

Veitch, D. (1993). Novel models of broadband traffic. *IEEE Global Telecommunications Conference, 1993, including a Communications Theory Mini-Conference: Technical Program Conference Record, IEEE in Houston. GLOBECOM '93, 2*, 1057-1061.

Vyzantiadou, M. A., Avdelas, A. V., & Zafiropoulos, S. (2007). The application of fractal geometry to the design of grid or reticulated shell structures. *Computer-Aided Design, 39*(1), 51-59.

Wang, H., Wu, Q., He, X., & Hintz, T. (2006). A novel interactive progressive decoding method for fractal image compression. *First International Conference on Innovative Computing, Information and Control (ICICIC'06), 3*, 613-617.

Willinger, W., Taqqu, M. S., Sherman, R., & Wilson, D. V. (1995). Self-similarity through high-variability: Statistical analysis of Ethernet LAN traffic at the source level. *ACM Sigcomm '95* (pp. 100-113).

Wohlberg, B., & de Jager, G. (1999). A review of the fractal image coding literature. *IEEE Transactions on Image Processing, 8*(12), 1716-1729.

Wu, Q., He, X., & Hintz, T. (2004). Virtual spiral architecture. *Proceedings of the International Conference on Parallel and Distributed Processing Techniques and Applications, 1*, 399-405.

Zhang, H. F., Shu, Y. T., & Yang, O. W. W. (1997). Estimation of Hurst parameter by variance time plots. *Proceedings IEEE Pacrim 97, 2*, 883-886.

Zhao, E., & Liu, D. (2005). Fractal image compression methods: A review. *ICITA 2005: Third International Conference on Information Technology and Applications, 1*, 756-759.

This work was previously published in Reflexing Interfaces: The Complex Coevolution of Information Technology Ecosystems , edited by F.F. Orsucci and N. Sala, pp. 308-328, copyright 2008 by Information Science Reference, formerly known as Idea Group Reference (an imprint of IGI Global).

Chapter XXIX
Transitioning from Face-to-Face to Online Instruction:
How to Increase Presence and Cognitive/Social Interaction in an Online Information Security Risk Assessment Class

Cindy S. York
Purdue University, USA

Dazhi Yang
Purdue University, USA

Melissa Dark
Purdue University, USA

ABSTRACT

This article briefly reviews two important goals in online education: interaction and presence. These are important goals in online education because they are linked to learning and motivation to learn. The article provides guidelines and an extended example of how to design an online course in information security in a manner that will enhance interaction and presence. This article's contribution is to provide guidelines with a corresponding extended and concrete example for those who are tasked with designing and delivering online courses. Although the guidelines and example were targeted to the field of information security, they can be readily adopted by other disciplines.

INTRODUCTION

Although online education can offer convenience and flexibility for learners, it is not without challenges. Frequently, online education is no more than instructor notes and lecture materials posted on a Web site, perhaps with some required discussion. Much online instruction is designed, developed, and delivered without careful consideration of foundational instructional design principles. Research has shown that online courses that lack substantive and meaningful interaction, coupled with a sense of presence (feeling as though belonging in a virtual environment), contribute to a sense of isolation, unsatisfying learning experiences, and high dropout rates (Aragon, 2003; Bennett, Priest, & Macpherson, 1999; Glickman, 2003; Moore & Kearsley, 1996). The goal of this article is to provide a set of online course design guidelines based on research findings and best practices to enhance interaction and sense of presence, which are two critical factors that impact learning and motivation to learn in online courses (Moore, 1992; 1993; Muirhead, 1999; Richardson & Swan, 2003). Finally, an example is provided for applying the guidelines to transition a face-to-face class to an online class, using an information security risk assessment class. In order for these guidelines to make sense, we start with a brief discussion of interaction and presence.

Interaction

Moore (1989) identified three major types of interaction: a) learner-content, b) learner-instructor, and c) learner-learner. Learner-content interaction refers to the amount of substantive interaction occurring between the learner(s) and the content. Content could be in the form of text, radio, television, and/or audiotape. Participant interaction (learner-learner and learner-instructor) refers to the engagement of the learners and instructor in the learning and teaching process. It also refers to dialogue between and/or among different participants in online learning environments. Thus, interaction is more than a communication exchange; interaction occurs when objects, actions, and events mutually influence one another (Wagner, 1994). Instructional interaction is meaningful communication that challenges learners' thinking, shapes the acquisition of knowledge in meaningful ways, and changes learners, moving them toward achieving their goals. Effective interaction is not necessarily more interaction, rather it is interaction resulting in learners thinking in new and more profound ways. While the literature and research confirmed the importance of interaction in the learning process (Muirhead, 2001), online learners frequently do not interact at sufficient levels and/or in substantive ways with the instructor or other learners in online courses. The lack of appropriate and deep interactions is a common inadequacy of current online courses (Bennett et al., 1999).

Presence

Closely related to interaction is the concept of presence. From the learner's perspective, presence is the "sense of being in and belonging in a course and the ability to interact with other students and an instructor although physical contact is not available" (Shin, 2002, p. 22). Presence also refers to the "involvement, warmth, and immediacy" (Danchak, Walther, & Swan, 2001, p. 1) learners experience during communication and interaction with others in the online learning environment. According to Picard (1997), an online course that conveys affective or emotional information to learners will lead to a higher sense of social presence and interaction. Leh (2001) found lack of interaction, originally due to lack of physical and face-to-face contact, in online learning environments leads to a sense of isolation (or lack of social presence). On the other side, an appropriate level of interaction promotes a better sense of social presence (Rovai, 2001). Research also has shown social presence is positively related to learner satisfaction, perceived learning (Richardson & Swan, 2003), and learning success (Rifkind, 1992; Tu, 2000). In other words, a good sense of social presence influences interaction and interaction influences students' sense of social presence. Together, appropriate interaction and presence lead to increased cognitive activity and also cognitive activity at higher levels, resulting in more meaningful learning in online learning environments. While these relationships are known, many designers, developers, and instructors of online courses do not consciously implement instructional methods and techniques that will effectively increase interaction and social presence. We asked ourselves why. The answer, we believe, is that they have not had access to pedagogically content-based guidelines grounded in research. With this information, we turn to the guidelines.

GUIDELINES FOR PROMOTING INTERACTION AND PRESENCE IN AN ONLINE COURSE

There are four main components to consider when transitioning a traditional face-to-face course to an online version: (a) introductions, (b) organization, (c) instruction, and (d) feedback. There are techniques to use for all four of these components that will allow students to be more socially and cognitively interactive and present in an online course.

Introductions

Much of the current literature on online courses emphasizes the value of creating a learning community among the online participants. According to Hanna, Glowacki-Dudka, and Conceiçào-Runlee (2000), "a learning community is a group of people who have come together to form a culture of learning in which everyone is involved in a collective effort of understanding" (p. 14). This sounds great, but as an instructor you are probably asking what techniques can be used to accomplish this task. "You need to build a climate that will foster professional learning or collaboration by crafting communications that support a sense of safety in the discussion areas" (Collison, Elbaum, Haavind, & Tinker, 2000, p. 30). You want students to share their experiences with each other, but this is difficult unless they feel comfortable with each other. There are a number of strategies that can be used to foster this feeling of community.

In the course content discussion area, start with a social icebreaker for students to introduce themselves. This should be a non threatening type of interaction that "breaks the ice of using technology to communicate," (Conrad & Donaldson, 2004, p. 47) is participant focused versus academic content focused, and requires reading and responding to other postings (Conrad & Donaldson, 2004). Conrad and Donaldson (2004) list and describe a number of different types of ice breakers. For example, BINGO requires everyone to post a short biography on the discussion board. The instructor then e-mails everyone a bingo card with something from everyone's posting in a box. Students must then determine which box belongs to which student and fill in the correct name. Another possible icebreaker is TWO TRUTHS AND A LIE. Students post two truths about themselves and one lie. Other students then try to determine which is the lie. This is most fun when the truths are so outrageous it is hard to distinguish them from a lie.

Another method to help foster the feeling of community is to have a page dedicated to the biographical sketches and pictures of the students. This could also be a social space with a title, such as lounge, hallway, or water cooler, where students can discuss any off-content topics. Students need a space provided just for "*social dialogue* or simple chitchat" (Collison et al., 2000, p. 20). This helps prevent clutter in the content-discussion area and encourages students to contact each other via e-mail, instant messaging, or chat.

As the facilitator of this community, you will want to send an opening message to each student or post one on the content-discussion board. It should be a warm, welcoming message, perhaps with a friendly photo. The opening message should include a question requiring a response from students. This first message will set the tone for the class; it also can serve as a model for online discussions.

Organization

There are a number of organizational strategies to use to help increase interaction and presence in an online class. As the instructor, you will want to hold "online" office hours. This can either be a specified time when you will be answering e-mail or using instant messenger to "chat" with students either synchronously and privately. Another strategy is to provide job aids on how to use the technology employed in the course, which allows the technology to become invisible as students become more familiar with using it.

The course syllabus should consist of more than taking the face-to-face paper copy and making it digital. In an online course, the syllabus needs to include things such as guidelines for discussions, definition of roles, and so forth, and to function as a contract between instructor and students. In addition to content traditionally included in a syllabus, you should include contact information for student technical difficulties. Instead of listing "participation" or "online attendance" and the point value, define what participation entails. For example, we suggest the participation grade be based on the quality of the postings and not just the quantity. Participation could include posting on an asynchronous discussion board, showing up for a synchronous chat, working on a team project, and so forth.

One strategy to help foster quality asynchronous postings is to ask the students to send private e-mails for single comments, such as "I liked your last posting" or "I agree." According to Moore and Kearsley (1996), this helps prevent cluttering the discussion board; they encourage only postings that will contribute to the "community's pool of knowledge (p. 151)." Do not discourage positive comments like this, because they contribute to the social presence of the community.

Another strategy is to group students into teams of three or four and have them write up what they believe are good ground rules for discussions and participation. For example, what to do if someone reads all the postings, but does not post any. These could be posted in a forum that explains discussion board procedures and guidelines. Guidelines that include a posting with "emoticons" for students unfamiliar with how to express text-based emotions are helpful (e.g., ☺ means smiling or happy). In addition, some students might use abbreviations that are now common in Internet-based chat, such as LOL or "laughing out loud." These small additions can add personality to the text-based "voices" of the participants. In a face-to-face classroom, physical presence is displayed through "voice, body language, intonation, expressions, [and] gestures" and helps communication (Ko & Rossen, 2004, p. 12). In the online environment, participants rely solely on text-based communication and need to avoid words that could be misinterpreted, such as sarcasm, inappropriate jokes, and so forth. Thus everyone in the community must demonstrate a culture of respect, so participants "feel what they say matters and is valued by the other members of the community" (Collison et al., 2000, p. 30).

An additional aspect of the syllabus that needs to be addressed is the schedule. The schedule is the lifeline of the online course. Students will look to this to effectively manage time. Therefore, it is critical to present course content in a consistent manner, either all up front or on a regular schedule. This will reduce confusion and promote consistent checking of the site by students. Keep in mind not everyone is logged on when an assignment is posted. Give approximately a week for assignments, to allow for those who log in later in the week. This is one of the advantages to online learning: the ability to log on anytime of the day or night. You could require everyone to log in every two to three days to ensure they have the most up-to-date information. Supplying information, such as due dates, in more than one location on the course Web site is also a good idea as some students might look in different locations for information.

Instruction

There are a number of different instructional techniques to use when attempting to increase presence and interaction in an online class. Collaboration can be fostered in small or large groups of students. If you choose to have large class projects, there are grouping considerations. Before grouping the students into teams they will work with for the large class project, consider pairing them up for a smaller assignment, such as an article critique or peer review. This helps foster feelings of comfort when learning how to work with others over distance. When assembling teams for a large class project, groups of four are typically the optimum number. Encourage collaboration to prevent the group from splitting up the work, then putting it together to turn in; you want them to "construct their learning together" (Palloff & Pratt, 2005, p. 39). Also, explain to the students why it is important that they work collaboratively and that it is a requirement.

Palloff and Pratt (2005) discuss the importance of collaboration, saying it promotes critical thinking skills and helps to foster the feeling of community. There are a number of ways to do this online. Students first can do the work individually and bring it to the group for critique and to certify the correctness of the papers. Then the instructor can pick one student randomly to answer the questions studied by the group or choose one paper from the group to grade with everyone in the group receiving that grade. A second technique is to provide different team members with the charge of finding different information. This is called information interdependence, or the jigsaw strategy, where students have the different pieces needed to complete the puzzle. In order for all the team members to do well on the assignment, they need to rely on the information the rest of the team members have learned. Hence, the students are held accountable for teaching the material to their team members. Another technique is to have the team devise a "charter" or team agreement delineating the different roles individuals will play, how they will interact, and different project deadlines. It is helpful to provide a sample charter, so students know what is expected of them. Some possible roles are secretary, liaison to the instructor, organizer, discussion board poster, and so forth. These roles might change during the project's phases. Have the team create a team name; this helps with team identification on the discussion board and also with a sense of community. To encourage team buy-in, give the team choices in determining project topic. Monitor the team's progress and intervene if there are participation problems. Johnson, Johnson, and Holubec (1991) state that there are three reasons an instructor should intervene:

1. To correct misunderstandings or misconceptions about task instructions and the academic assignments they are completing.
2. To correct the absence, incorrect use, or inappropriate use of collaborative skills.
3. To reinforce the appropriate or competent use of collaborative skills. (pp. 6:29)

In addition, tips for online conflict resolution could be included as a job aid (Palloff & Pratt, 2005). When assessing the collaborative assignment, perhaps include peer evaluations in the grading criteria. This can help prevent noncontributing team members. In addition, ask each team member to write a reflection on what they did to contribute to the project, and how they tackled their role and task throughout the process.

As many online courses are taught using mainly asynchronous discussion boards, there are discussion strategies and activities that encourage interaction and a sense of presence online. A main goal is to ensure there is a high level of interaction and dialogue. This can be facilitated by using different types of questions, activities, and presentations.

Questions

When posting discussion questions, the instructor does not always need to be the initiator. After the instructor has modeled question facilitation, allow students, or pairs of students, to take turns facilitating different discussion topics. This allows students to see that each participant in the community is as valuable as the instructor because every participant shares personal experiences to help the community learn. It also allows participants to see multiple perspectives. During online discussion, it is important to provide the discussions with a distinct beginning and end to prevent information overload and frustration among students (Conrad & Donaldson, 2004). Different types of questions can help encourage critical thinking, such as questions asking for more evidence, questions asking for clarification, open-ended questions, hypothetical questions, cause and effect questions, and summary and synthesis questions (Palloff & Pratt, 2005). In addition, you or the discussion topic facilitator should write a wrap-up paragraph summarizing the main points of the discussion, including students' names and the different points they made, which contributes to the feeling of presence.

Activities

Different types of activities can take place on a discussion board. The typical threaded discussion can get boring to students who read numerous posts. Some activities to increase interaction are the following: a) role playing, b) debates, c) simulations, d) case studies, e) outside experts, f) sharing related personal/professional experiences, and g) electronic virtual field trips. In addition to these asynchronous activities, consider having a few required synchronous discussions. It should be noted, however, synchronous discussions tend to be more social; therefore, they are usually more effective at fostering social interaction than cognitive interaction. Guest lecturers via audio or video conferencing and synchronous large group sessions, where the instructor uses a whiteboard to demonstrate a problem, also can be used. If the instructor must present some sort of lecture to provide information to the class, include meaningful interactive links, discussion threads, and other activities to make the lecture interactive.

Presentations

Online course technologies often allow for different types of presentations. Individuals can post papers, PowerPoint presentations, and other documents in discussion threads. However, how do you have a group do a presentation to the entire class? If the students have access to software, such as Breeze, Camtasia, or Articulate, they can create multimedia presentations the class can watch. If this type of software is not available, students can prepare a discussion thread led by the team to present their project. Teams also can create simple Web sites to showcase their projects.

Feedback

The final component we are going to discuss is the use of feedback, which is essential to fostering interactivity. Online feedback should consist of both instructor-to-student and student-to-student (or peer) feedback. Responding to individual e-mails asking the same questions can get redundant; therefore, encourage students to post questions on the discussion board, so everyone will benefit from the response. In addition, provide a discussion forum that allows students to provide feedback about the course; perhaps these are recommendations for improvement or lessons learned that can be shared with future classes.

No matter the activity students are involved in, provide opportunities for individual as well as group practice and feedback; this may be the first online class they have taken. In addition, the instructor should respond to all student queries. Make sure responses are prompt if it is a technical question. If there is a delay in responding, explain the reason. Instructor feedback should offer detailed analysis of student work and suggestions for improvement, along with additional hints and information to supplement learning. These can be private, via e-mail, or public to a team via the discussion board. If a student is not accessing the site enough, the instructor can send informal e-mails to see if the student is having problems in terms of the technology. The instructor should send encouraging supportive e-mails to individuals on an ongoing basis. Include questions that require the student to respond, thus drawing them into active participation.

Students should complete peer reviews for student-to-student feedback. This provides the reviewer the opportunity to focus on others' interpretations and the original writer to receive multiple perspectives. Provide guidelines and the rubric to be used for grading. Both the instructor and the students should use "track changes" in Microsoft Word documents to provide feedback, so everyone can see changes made, comments, or notes that include questions. Also try to get feedback from participants about their progress. This can be done through direct questions, assignments, quizzes, polls, and questionnaires.

THE ONLINE INFORMATION SECURITY CLASS EXAMPLE

This section of the article begins with a brief overview of how to introduce and organize the online information security course. Next, there is an in-depth focus on three weeks of instruction, explaining how the course was transitioned from face-to-face instruction to an online format. While this specific example focuses on an information security course, the purpose of the example is simply to enact the guidelines. The guidelines can be generalized to other topics and fields in technology education.

Introductions for Building a Learning Community

When building a learning community in a face-to-face security assurance class, the instructor tends to have class introductions and perhaps an ice breaker activity. For an *online* security assurance class, the instructor needs to facilitate a learning community in a similar, yet different way. The instructor needs to provide a Web-based orientation to both the online environment and the course materials. An opening message should be sent by instructors, including a question requiring a response from students. In order to allow students to get to know each other early in the course, online ice breaker activities that are via discussion boards as well as having students work in virtual teams to produce a visual presentation about the team are useful. In addition, a space is provided on the class Web site for students to post their digitized images. The instructor also should encourage students to contact each other via e-mail and chat.

Course Organization

When organizing a face-to-face security assurance class, the instructor has a syllabus that includes a schedule, required textbook, and office hours. In the online version of the security assurance class, the instructor needs to go further. In addition to the online syllabus, the instructor needs to post a schedule with hyperlinks to that day's information as well as discussion and participation guidelines and requirements. Links to online readings, in addition to the listed textbook, are included. Furthermore, the instructor needs to hold online office hours when the students can be sure to reach him or her immediately.

Instruction—Week One

Perform Asset Identification and Classification

In the first week of the face-to-face security assurance class, the instructor provides the students with readings on the risk assessment process and various models. She also presents a lecture to provide them with additional information. A discussion ensues about asset identification. They look at the different authors and different information provided in order to compare and contrast what each author said. They also discuss the purpose in the risk assessment process. In addition, the class brainstorms assets in the k-12 setting. The instructor assembles small groups and has the students apply asset identification to the k-12 setting. The groups then compare their new list to the other groups' lists. As a class, they then group information assets (types of data, part of classification). FIPS 199 is discussed as a classification scheme for sensitivity of assets. For homework, students are asked to apply FIPS 199 to their list and write a critique of the usability of FIPS 199. They can revisit the first readings to discuss their classifications.

In the online security assurance class, the instructor provides the students with links to readings and Web sites about the risk assessment process and various models. She also posts a short lecture (approximately 10 minutes) with a PowerPoint presentation via Breeze to provide them with additional information. A discussion forum is started in which the instructor poses an initial question about asset identification. The students have two to three days to respond. The instructor assigns different students to read different authors to gain different information about risk assessment models. Concurrently, the instructor creates a new discussion forum for students to a) post a summary of their article, b) then compare their article to other postings, and c) discuss the purpose of models in the risk assessment process. The students again have two to three days to respond. The instructor creates a new discussion The instructor creates a new discussion thread to brainstorm information assets in the k-12 setting and posts an initial questions. She has students individually apply asset identification to the k-12 setting then post their responses. She organizes students into groups of three or four and provides them group discussion forums. She has each group create one new list and has groups compare/contrast lists with each other. She also has groups apply a classification scheme to their list as well as write a group critique of FIPS 199. Groups also discuss their classification, according to the first readings they did. The students have four to five days to respond.

Instruction—Week Two

Perform Threat Identification

In the second week of the face-to-face security assurance class, the instructor provides the students with readings on information security threat analysis and classifications of threat types. As a class, they discuss how different threats might correlate to different assets. This is done first in small groups, then together in one large group discussion. The instructor also presents them with information on methods for identifying types of threats. She provides them with existing reports (FBI CIC Survey to Crime Data) and tells them where they can get this type of information for typical threats in other organizations. She asks the students if they can try to generalize to their organization, and how they would monitor their own employees/network/system. She poses the question, "How are you going to get clients to think about modeling their threats before we go out to the client? Where we do actual threat identification?"

In the online security assurance class, the instructor provides the students with links to readings on information security threat analysis as well as Web sites about classifications of threat types. The students have two to three days to read this information. The instructor creates a new discussion forum about how different threats might correlate to different assets. Students are first assigned to small group discussion areas to answer a posted question. Then students discuss their findings in a large group discussion area. Students have two to three days to respond. Another discussion thread is created about methods for identifying types of threats. The instructor posts open-ended questions about the following: existing reports (FBI CIC Survey to Crime Data); where to get information; typical threats in other organizations; how the students could generalize to their organization; and how to monitor their own employees/network/system. In addition the instructor posts a fourth question: "How are yhou going to get clients to think about modeling their threats before we go out the the client, where we do actual threat identification?" Students have the same two to three days to respond.

Instruction—Week Three

Perform Vulnerability Identification

In the third week of the face-to-face security assurance class, the instructor provides a lecture and PowerPoint presentation on the three types of vulnerabilities— people, policy, and technology—and about establishing criteria for assessing vulnerability. She asks the students to individually develop an evaluation checklist (for policy) to take into a company. The instructor presents information about technical vulnerability. For example, she discusses the reporting tools companies and schools are likely to have as well as passive scanning tools. The class goes to a computer lab as a group and experiments with a variety of these tools. The students are provided criteria to evaluate different types of scanning tools: purpose, when to use, cost, and advantages/disadvantages; this is done in small groups. For homework, students are put into small groups and asked to select a tool. They are then provided with a flawed system with known vulnerabilities to run their tool against. They must then take, analyze, and report their findings.

In the online security assurance class, the instructor provides the students with links to readings and Web sites on the three types of vulnerabilities—people, policy, and technology—and about establishing criteria for assessing vulnerability. She also posts a short lecture (approximately 10 minutes) with PowerPoint presentation via Breeze to provide them with additional information. She asks the students to individually develop an evaluation checklist (for policy) to take into a company and submit this to the online assignment drop box. Students have two to three days for this. Students are then put into small groups, where each student presents his/her checklist to the other group members. The small group is then tasked with coming up with a "Best of Breed" checklist, using their individual checklists. The instructor presents information about technical vulnerability via an audio presentation. For example, she discusses the reporting tools and passive scanning tools that companies and schools are likely to have. The instructor provides links to demonstration software for students to experiment with different types of these tools. She posts a list of criteria along with an example for students to evaluate different types of scanning tools: purpose, when to use, cost, and advantages/disadvantages; she assigns this to be done in small groups and posted within three days. The students also are requested to select one tool per small group. They are then provided a flawed system with known vulnerabilities to run their tool against. This system is accessed via a virtual private network (VPN). The groups of students must then take, analyze, and report their findings on the discussion board within three days.

SUMMARY

The goal of this article was to provide guidance to faculty who are tasked with transitioning face-to-face instruction into distance learning. More specifically, these guidelines for an online course and the example of one are meant to provide readers with action steps that can be taken to improve the level and nature of interaction as well as students' sense of presence. The ultimate goal, of course, is to produce equally, if not more effective, results from online learning. Our hope is that faculty who attempt to use these guidelines will see increased learning and motivation to learn among their distance learning students.

REFERENCES

Aragon, S. R. (2003). Creating social presence in online environments. In S. R. Aragon (Ed.), *Facilitating learning in online environments* (pp. 57-68). San Francisco: Jossey-Bass.

Bennett, S., Priest, A., & Macpherson, C. (1999). Learning about online learning: An approach to staff development for university teachers. *Australian Journal of Educational Technology, 15*(3), 207-221.

Collison, G., Elbaum, B., Haavind, S., & Tinker, R. (2000). *Facilitating online learning: Effective strategies for moderators.* Madison, WI: Atwood.

Conrad, R-M., & Donaldson, J. A. (2004). *Engaging the online learner: Activities and resources for creative instruction.* San Francisco: Jossey-Bass.

Danchak, M. M., Walther, J. B., & Swan, K. P. (November 2001). *Presence in mediated instruction: Bandwidth, behavior, and expectancy violations.* Presented at the annual meeting on Asynchronous Learning Networks Conference, Orlando, FL.

Glickman, N. W. (2003). *The veterinary distance learning program at Purdue University: Factors associated with persistence and dropout. Digital Dissertations, 64(11), 4018B.* (UMI no. 3113802): Purdue University, West Lafayette, IN.

Hanna, D. E., Glowacki-Dudka, M., & Conceiçào-Runlee, S. (2000). *147 practical tips for teaching online groups: Essentials of web-based education.* Madison, WI: Atwood.

Johnson, D. W., Johnson, R. T., & Holubec, E. J. (1991). *Cooperation in the classroom.* Edna, MN: Interaction Book Co.

Ko, S., & Rossen, S. (2004). *Teaching online: A practical guide* (2nd ed.). New York: Houghton Mifflin.

Leh, A. S. (2001). Computer-mediated communication and social presence in a distance learning environment. *International Journal of Educational Telecommunications, 7*(2), 109-128.

Moore, M. G. (1989). Editorial: Three types of interaction. *The American Journal of Distance Education, 3*(2). Retrieved January 24, 2005, from http://www.ajde.com/Contents/vol3_2.htm#editorial

Moore, M. G. (1992). Distance education theory. *The American Journal of Distance Education, 5*(3), 1-6.

Moore, M. G. (1993). Theory of transactional distance. In D. Keegan (Ed.), *Theoretical principles of distance education* (pp. 22-38). New York: Routledge.

Moore, M., & Kearsley, G. (1996). *Distance education: A systems view.* Belmont, CA: Wadsworth Publishing Company.

Muirhead, B. (1999). *Attitudes toward interactivity in a graduate distance education program: A qualitative analysis.* Parkland, FL: Dissertation.com.

Muirhead, B. (2001). Practical strategies for teaching computer-mediated classes. *Educational Technology & Society, 4*(2). Retrieved September 12, 2005, from http://ifets.ieee.org/periodical/vol_2_2001/discuss_summary_jan2001.html

Palloff, R. M., & Pratt, K. (2005). *Collaborating online: Learning together in community.* San Francisco: Jossey-Bass.

Picard, R. W. (1997). *Affective computing.* Cambridge, MA: MIT Press.

Richardson, J. C., & Swan, K. (2003). Examining social presence in online courses in relation to students' perceived learning and satisfaction [Online]. *Journal of Asynchronous Learning Networks, 7*(1), 68-87.

Rifkind, L. J. (1992). Immediacy as a predictor of teacher effectiveness in the instructional television classroom. *Journal of Interactive Television, 1*(1), 31-38.

Rovai, A. P. (2001). Building and sustaining community in asynchronous learning network. *Internet and Higher Education, 3*(2000), 285-297.

Shin, N. (2002). Beyond interaction: The relational construct of 'Transactional Presence.' *Open Learning, 17*, 121-137.

Tu, C. H. (2000). Strategies to increase interaction in online social learning environments. In *SITE 2000.* Proceedings from the Society for Information Technology and Teacher Education International conference. Norfolk, Va.: AACE, 2000. (ED 444 550).

Wagner, E. D. (1994). In support of a functional definition of interaction. *The American Journal of Distance Education, 8*(2), 6-29.

This work was previously published in International Journal of Information and Communication Technology Education, Vol. 3, Issue 2, edited by L. A. Tomei, pp. 41-50, copyright 2007 by IGI Publishing, formerly known as Idea Group Publishing (an imprint of IGI Global).

Chapter XXX
A Mobile Intelligent Agent-Based Architecture for E-Business

Zhiyong Weng
University of Ottawa, Canada

Thomas Tran
University of Ottawa, Canada

ABSTRACT

This article proposes a mobile intelligent agent-based e-business architecture that allows buyers and sellers to perform business at remote locations. An e-business participant can generate a mobile, intelligent agent via some mobile devices (such as a personal digital assistant or mobile phone) and dispatch the agent to the Internet to do business on his/her behalf. This proposed architecture promises a number of benefits: First, it provides great convenience for traders as business can be conducted anytime and anywhere. Second, since the task of finding and negotiating with appropriate traders is handled by a mobile, intelligent agent, the user is freed from this time-consuming task. Third, this architecture addresses the problem of limited and expensive connection time for mobile devices: A trader can disconnect a mobile device from its server after generating and launching a mobile intelligent agent. Later on, the trader can reconnect and call back the agent for results, therefore minimizing the connection time. Finally, by complying with the standardization body FIPA, this flexible architecture increases the interoperability between agent systems and provides high scalability design for swiftly moving across the network.

INTRODUCTION

Many people nowadays use mobile devices such as personal digital assistants (PDA) or mobile phones to access information through the Internet. In addition, they desire to have the ability to participate in e-business anywhere and anytime via their mobile devices. Current e-business applications, such as business-to-consumer (B2C) or Internet-based shopping, are typically developed over the Web for human-computer interaction. These applications require that users must login the intended Web sites from their personal computers or public terminals. Also, users often need to visit lots of sites and are always involved in a time-consuming process. To address these challenges, several wired agent-based e-business systems have been proposed. Kasbah (Chavez & Maes, 1996), for example, is an electronic marketplace where buying and selling agents can carry out business on behalf of their owners. Nevertheless, these systems do not satisfy the users' mobile demand due to their lack of wireless channels.

This article proposes a feasible architecture that combines agent mobility and intelligence for consumer-oriented e-business applications. It allows a user to create a mobile, intelligent agent via a mobile device, and then launch the agent to the Internet to perform business on the user's behalf. The aspect of mobility enables our architecture to support the agent's migration and the user's mobility (the ability to conduct e-business via mobile devices anyplace and anytime). The mobile agent will migrate from market to market, communicating with different trading agents to find the most appropriate one. Once an appropriate agent is found, it will inform the user of the results. This architecture complements the current Web-based, Internet systems by adding the wireless channel of mobile agents. Our current work focuses on lightweight mobile agents which act on behalf of consumers and participate in consumer-to-consumer (C2C) e-business applications. However, the architecture can be extended to business-to-consumer (B2C) or business-to-business (B2B) applications, as discussed later in the article.

Since personal software agents essentially need to communicate with other agents (to accomplish their designated tasks), they have to comply with a set of standards concerning the agent communication language and the protocols to be used. Although there is currently no universally accepted set of standards for developing multi-agent systems, the Foundation for Intelligent Physical Agents (FIPA), which aims at providing one language commonly understood by most agent-based systems (FIPA, 2006), is obtaining a growing acceptance. With FIPA becoming a de facto standard in this field, the architectures such as JADE (Java Agent Development Environment) have become available to allow for the implementation of a FIPA-compliant multi-agent system such as our proposed architecture (Chiang & Liao, 2004).

It should be noted that mobile devices suffer not only from limited battery time, memory, and computing power, but also from small screen, cumbersome input, and limited network bandwidth and network connection (Wang, Sørensen, & Indal, 2003). The proposed architecture, by making use of mobile agent technology, offers a solution to those problems. That is, after creating and initializing a mobile agent to act on the user's behalf, a user can disconnect the mobile device from the server. The user only needs to reconnect later on to recall the agent for results, hence minimizing the use of resources. In addition, mobile agent technology also addresses such challenges as increased need for personalization, high latency, demand for large transfers, and disconnected operation (Kotz & Gray, 1999).

The remainder of this article is organized as follows: the second section introduces background knowledge and related work. The third section illustrates the proposed architecture. The fourth section shows an implementation of the proposed architecture. The fifth section discusses some existing problems and future works. The sixth section concludes the article.

BACKGROUND AND RELATED WORK

Mobile Agent Paradigm

An intelligent agent is a piece of software, which differs from the traditional one by having such features as being autonomous, proactive, social, and so on. One of these characteristics is mobility, that is, the agents' ability to migrate from host to host in a network. Mobile agents are defined as programs that travel autonomously through a computer network in order to fulfill a task specified by its owner, for example, gathering information or getting closer to the required resources to exploit them locally rather than remotely. A mobile agent is not bound to the system on which it begins execution, and hence can be delegated to various destinations. Created in one execution environment, it has the capability of transporting its state and code with it to another host and execute in the same execution environment in which it was originally created. Several mobile agent systems have been designed in recent years. Telescript (White, 1996) is the first commercial mobile agent system developed by General Magic. Telescript provides transparent agent migration and resource usage control. Aglets from IBM (Lang & Oshima, 1998) is also a mobile agent system based on the concept of creating special Java applets (named aglets that are capable of moving across the network). JADE (Bellifemine, Caire, Trucco, & Rimassa, 2006) is one of the agent development tools that can support efficient deployment of both agents' mobility and intelligence in e-business applications. As a middleware implemented in Java and compliant with the FIPA specifications, JADE can work and interoperate both in wired and wireless environments based on the agent paradigm. JADE supports weak mobility; that is, only program code can migrate while no state is carried with programs. NOMADS (Suri et al., 2000) supports strong mobility and secure execution; that is, the ability to preserve the full execution state of mobile agents and the ability to protect the host from attacks.

Recently, mobile agents have found enormous applications including electronic commerce, personal assistance, network management, real-time control, and parallel processing (Lange & Oshima, 1999). Kowalczyk et al. (2002) discuss the use of mobile agents for advanced e-commerce applications after surveying the existing research. There are many advantages of using the mobile agent paradigm rather than conventional paradigms such as client-server technology: reduces network usage, introduces concurrency, and assists operating in heterogeneous systems (Lange & Oshima, 1999).

Related Work

Mobile agents have been recognized as a promising technology for mobile e-business applications. The interest of developing mobile agent systems for mobile devices has increased in recent years. Telescript describes a scenario in which a personal agent is dispatched to search a number of electronic catalogs for specific products and returns best prices to a PDA from where it starts (Gray, 1997). An integrated mobile agent system called Nomad allows mobile agents to travel to the eAuctionHouse site (http://ecommerce.cs.wustl.edu) for automated bidding and auction monitoring on the user's behalf even when the user is disconnected from the network (Sandholm & Huai, 2000). They aim at reducing network traffic and latency.

Impulse (2006) explores a scenario in which e-business meets concrete business through a system of buying and selling agents representing individual buyers and sellers that carry out multiparameter negotiation and running on the wireless mobile devices. Impulse deploys personal agents on mobile devices to help users seek agreement on purchase terms. However, these personal agents are directed to move online to participate in negotiations, and hence resulting in potentially long-time connection with the Internet. We also think that the Impulse system was designed with a single communication protocol for all agents. This presents drawbacks due to the heterogeneity of exchanged information and leads to an inflexible environment, which can only accept those agents especially designed for it. Agora (Fonseca, Griss, & Letsinger, 2001) is a project conducted at HP Labs to develop a test-bed for the application of agent technology to a mobile shopping mall. A typical scenario consists of mobile shoppers with PDAs interacting with store services while in the mall, on the way to the store, or in the store itself. The Zeus agent toolkit, developed by British Telecommunications, was used to implement all agents in the Agora project. Only the infrastructure agents speak the FIPA Agent Communication Language (ACL), causing the architecture to conform partly to FIPA, although more effort in conformance is needed. The purpose of the Agora project is to gain experience in agents' communication protocols and to realize the significance of architectural standards.

It has been shown that modern agent environments such as JADE could be easily scaled to 1,500 agents and 300,000 messages (Chmiel et al., 2004). Thus, it is now possible to build and experiment with large-scaled agent systems. Moreno et al. (2005) use JADE-LEAP (JADE Lightweight Extensible Agent Platform) to implement a personal agent on a PDA. This agent belongs to a multi-agent system that allows the user to request a taxi in a city. The personal agent communicates wirelessly with the rest of the agents in the multi-agent system, in order to find the most appropriate taxi to serve the user. IMSAF (Chiang & Liao, 2004) is an architecture designed to fulfill the requirements of Impulse-introduced mobile shopping and implemented using JADE-LEAP tools. LEAP can also be used to deploy multi-agent systems spread across mobile devices and servers; however, it requires a permanent bidirectional connection between mobile devices and servers. Considering the current expensive connection fees for cell phones, such a required permanent connection is not affordable for consumers in practice.

In contrast to the above works, we are motivated to propose a mediator-based architecture that attempts to enable users' wireless participation in several e-marketplaces through their mobile devices. The mobile agents can move across the network and perform trading tasks on behalf of their users when the users are disconnected from the network. We believe that it is important to consider the limitations of mobile devices, such as low-battery, low bandwidth, high latency, limited computing ability, and expensive connection fees. The fact that consumers in our physical world may need to access the worldwide markets and distributed e-business environments requires the agents to operate in heterogeneous and dynamic environments as well as to talk a common language. By complying with the FIPA specifications, the proposed architecture provides an interoperable solution to allow users dynamically to connect to the network by means of their mobile devices only when needed. Also, the mobile devices will not suffer those limitations mentioned above. In addition, the benefit of using mobile agents to a user becomes more obvious in our architecture if the user has a mobile phone and is interested in minimizing expensive connection costs. The next section explains the architecture in more detail.

SYSTEM ARCHITECTURE

Overview

Figure 1 shows a distributed C2C wireless e-business environment and a traditional wired e-business one. A consumer can connect a mobile device, such as a PDA or mobile phone, to the mediator server through wireless connection and then send a request for creating a mobile buying or selling agent to undertake a specific business task (e.g., auction bidding) on the user's behalf. A

Figure 1. Distributed e-business environment

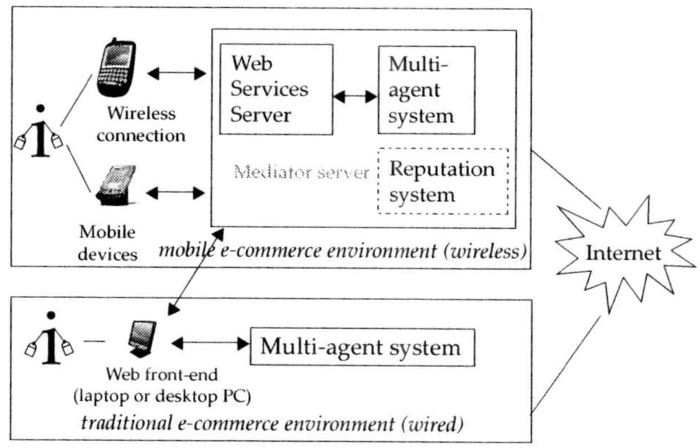

personal agent that resides on the mobile device is needed to interact directly with the consumer and to consider the consumer's personal preferences. Considered as a true representative of the consumer, the personal agent represents the consumer's interests and allows the consumer to have a choice of dispatching either a buying or a selling agent. The mediator server sits in the fixed network and provides services such as generating mobile agents according to consumers' requests. After being created, the mobile agents will autonomously travel to multiple agent-based servers on the Internet. The agent-based servers offer places for selling and buying agents to meet and negotiate with one another. The proposed mediator server contains two main components: the Web services server, which facilitates mobile agents to interface with other agents, and the multi-agent system, which manages the agents and plays the role of a marketplace similar to an agent-based server. An additional component, a reputation system, will be necessary in our architecture. Using this reputation system, agents could sign binding contracts and check user's credit histories and reputations. The trust problem will be further studied in future research (e.g., Jøsang & Ismail, 2002 present a Beta Reputation System). Also, the mediator server provides a Web-based interface, and as shown in Figure 1, a consumer can also connect a laptop or a desktop PC to the network and launch an agent to execute on the mediator server.[1] In this article, we focus on the electronic trading of second-hand products for owners of mobile devices.

The main idea is that a consumer will request the mediator server to create a buying or selling agent and then dispatch it to agent-based servers on the Internet. The main operation that occurs in an agent-based server is price negotiation where buying agents negotiate price with selling agents. According to the consumer's preferences, the buying agent may travel to different e-market sites known by the white-page agent[2] to seek goods, when the consumer desires to conduct a global multiple markets comparison. The W3C's XML schema specification (www.w3.org/XML/Schema) provides a standard language for defining the document structure and the XML structures' data types. The consumer's preferences can be represented in an XML format. In a real business situation, we would have to ensure that messages are reliably delivered to the mediator server from the personal agents. Although this communication protocol's reliability is not detailed in our architecture currently, we could use a reliable transport at the very least, such as Reliable HTTP

(HTTPR) (Todd, Parr, & Conner, 2005), for the communication between the personal agents and the mediator server. Another consideration is to encrypt the communication. Encryption technologies can also help ensure that even intercepted transmissions cannot be read easily.

A Scenario of Our Architecture

To understand the environment best, let us consider a typical scenario taken from daily life, where two hypothetical customers, named Mary and Tom, try to participate in an eBay-like auction. Mary wants to sell her used Sony MP3 player. At her office, she initiates a selling agent from a PDA, through a wireless LAN connection with the mediator server in the building. Then this selling agent lives in the server and waits for potential shoppers. Due to some unpredictable event, Mary may have to leave her office and cannot access the selling agent via her PDA (as there may be no available wireless LAN network coverage). However, she will be able to reconnect later on.

Haphazardly, Tom enters his buying preferences into his Java-enabled mobile phone, trying to buy a second-hand Sony MP3 player under a maximum price. The personal agent on his mobile phone establishes a connection with the mediator server and asks the server to launch a mobile buying agent according to his preferences. Then Tom disconnects his cell phone from the server. The mobile agent knows where and how to migrate, as instructed in the migration itinerary. As days pass, while this buying agent is roaming around the Internet, it enters into Mary's mediator server and searches for services provided. After the negotiation between the selling agent and buying agent, they reach an agreement on the item and price. With that, the buying agent will return to its host server and send a SMS (Short Message Service)-based notification to the personal agent running on Tom's mobile phone, about the potential seller gathered from the Internet. Also the selling agent sends an e-mail-based notification to the personal agent running on Mary's PDA. Finally, things left to Mary and Tom seem to be simple and easy since they could have either the cell phone number or the e-mail address from the information reported by their personal agents, respectively. As we have seen, mobile consumers only need a small bandwidth connection twice, once for initiating a migrating mobile agent and once for collecting the results when the task is finished.

Architecture Description

We explain how the whole system works in this section. Figure 2 illustrates the system architecture and the operation process. As shown in Figure 2, mobile devices are supported by personal agents and connected to the mediator server via a wireless connection. A personal agent is a static agent running on a mobile device and offers a graphical user interface (GUI) for its user to communicate with the system. The mediator server is connected to the Internet where other mediator servers or other FIPA-compliant systems exist. In the mediator server, a servlet answers any requests from the personal agent and is linked to the behavior of a proxy agent[3] in charge of handling the requests. The proxy-agent interfaces with the servlet and constructs a bridge between the Web service server and the multi-agent system. Each instance of the behavior[4] connects not only to the AMS agent (Agent Management Service as defined in FIPA, i.e., the white-page agent mentioned above), asking for the creation of a buying or selling mobile agent in the multi-agent system as well as providing a response, but also connects to the agent DF (Directory Facilitator as defined in FIPA, i.e., the yellow-page agent), retrieving the list of agents advertising services with the DF. In this architecture, the multithreaded-servlet server is mirrored by a multibehavior proxy agent to allow for handling multiple requests in parallel.

Figure 2. System architecture and process

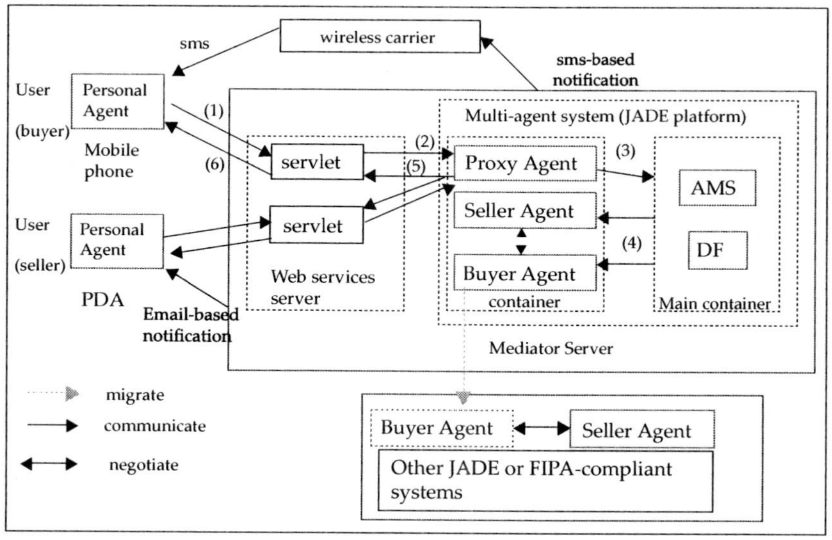

As illustrated in Figure 2, the procedures from (1) to (6) depict how a buying or selling mobile agent is created by a user according to preferences:

1. At the first step, the user configures the preferences via the personal agent (residing in the mobile device). The personal agent then sends an XML-based request to the mediator server.
2. An instance of the servlet accepts the request and communicates with the proxy agent.
3. The proxy agent cooperates with the AMS agent who lives in the main container of the JADE platform to create a buying or selling mobile agent.
4. If the buying or selling agent is created successfully in the container, it might be mobilized to other systems to undertake the user's task.
5. and (6) The personal agent receives a response from the proxy agent via the servlet and informs the user of the relevant mobile agent being created.

The above is an asynchronous process after which the user can disconnect from the network at will. Even if the user decides to disconnect from the network, the user will still receive an SMS-based notification from the mediator server via an interface with the wireless carrier, or an e-mail-based notification from the mediator server via an interface with a mail server, as long as the user reconnects to the network.

The mediator server provides the required support for the creation of mobile agents, messaging among agents, agent migration facility, collaboration, protection, destruction, and control of mobile agents. Mobile agent platforms such as JADE have been proposed to provide the supporting environment. Obviously, any multi-agent system can be used here as long as it provides the required support.

Different Types of Agents in Our Architecture

The following agents co-exist in our architecture: personal agents, proxy agents, buying or selling agents, yellow-page agents, and white-page agents. Among them, only buying or selling agents are mobile agents, while personal agents and proxy agents are stationary agents. Both the yellow-page and white-page agents are fixed on a component of the mediator server. Details of these agents are described as follows:

A personal agent is a stationary agent that runs on a user's mobile device and provides a graphical interface to allow the user to configure a mobile buying or selling agent (from the mobile device). When starting the personal agent on the mobile device, the user can choose either to initiate a new mobile agent or to recall a previous mobile agent. One may argue that such a personal agent is nothing more than an interface. From the agent's viewpoint, however, the personal agents are able to autonomously communicate with the proxy agent which is running in the mediator server.

A proxy agent is also a stationary agent which links the multi-agent system to the Web service server. It is one of the agents that is always up and running in the multi-agent system. The proxy agent cooperates with the AMS (white-page) agent to create a mobile buying or selling agent for each user. There is only one proxy agent per mediator server due to its unique multibehavior ability.

A yellow-page agent (such as the DF agent in the JADE platform) provides the service of yellow pages, by means of which an agent can receive information about available products or find other agents providing necessary services to achieve its goal. A white-page agent (like the AMS agent in the JADE platform) represents the authority and provides naming services. It stores information about addresses and identifiers of all agents in the system. In our architecture, sellers have permission to advertise their products; and buyers are allowed to query the sellers which post the products they are looking for. Selling agents update yellow pages by publishing their services via the yellow-page agent. Buying agents query relevant services from the yellow-page agent. Both buying and selling agents update white pages by registering in or deregistering from the system. They communicate with each other via querying agent's information from the white-page agent.

Both buying and selling agents are mobile agents, which are also called *service agents*. A service agent is the counter part of a personal agent

Table 1. Attributes of a mobile agent

Attribute	Description
Agent type	The agent type that a user can select, that is, either a buying agent or a selling agent
Agent server	The configuration of the mediator server address.
User id	The user identification which can be email address, cell phone number, or IMEI (International Mobile Equipment Identity).
Quantity	Quantity of the predefined product.
Price	For a buying agent, this is the maximum price that the agent can bid: for a selling agent, this is the minimum price that the agent can accept.
Current Price Inquired	For a buying agent, this is the best price offer colledted from the Internet.
Lifetime	The total time an agent can be away before being recalled or terminated.
Mobility	Specification of whether a user desires to enalbe the agent's migration ability (i.e., in the context of a local, single or global, multiple market comparison).
Server Activity Time	The time an agent can spend on each server before migrating ot another.

and is involved in the migration from host to host on the Internet. A service agent first negotiates with other service agents in the same host mediator server before migrating among multiple Web sites to talk to other service agents, provided that they can talk a common language.

To demonstrate a useful mobile agent system, we present a prototype for buying and selling agents, with attributes depicted in Table 1. This means that a user will configure a mobile buying or selling agent on a mobile device, precisely according to the characteristics in Table 1.

Behaviors of Mobile Agents

As illustrated in Figure 3, a mobile (buying or selling) agent starts with its registration in the system and ends with a timeout of its lifetime. There are three time events that indicate the behaviors of a mobile agent: (1) the agent starts its negotiation process at a regular interval (e.g., every minute); (2) the agent starts its migration when activity time per server is reached; and (3) the agent ends its life cycle when its lifetime is exhausted. An argument may arise; how can one be sure that the mobile agent will be terminated according to the parameter and lifetime, as users prefer? This parameter may be changed by a third party (including the mediator server). The assumption we made is that the mobile agent can be protected from the attacks (e.g., from the host or other agents) once a future security mechanism is imposed on our architecture. (The security problem is discussed in the Discussion and Future Work section.)

Negotiation Process

The proposed interaction between agents complies with the FIPA-Contract-Net Protocol (FIPA, 2006). This protocol allows a buying agent (initiator) to send a call for proposals (CFP) to a set of selling agents (responders), evaluate their proposals, and then accept the most preferred one (or even refuse all of them). Both initiators and responders should register in the system before they negotiate with each other.

In this article, we consider a classical situation in which a selling agent offers a single item to the highest bidder (similar to eBay), and the simplest type of bid is an offer to buy or sell one unit at a specified price. As shown in Figure 4, the buying agent sends a CFP to all the available selling agents (obtained from the yellow-pages

Figure 3. Activity diagram of mobile agent

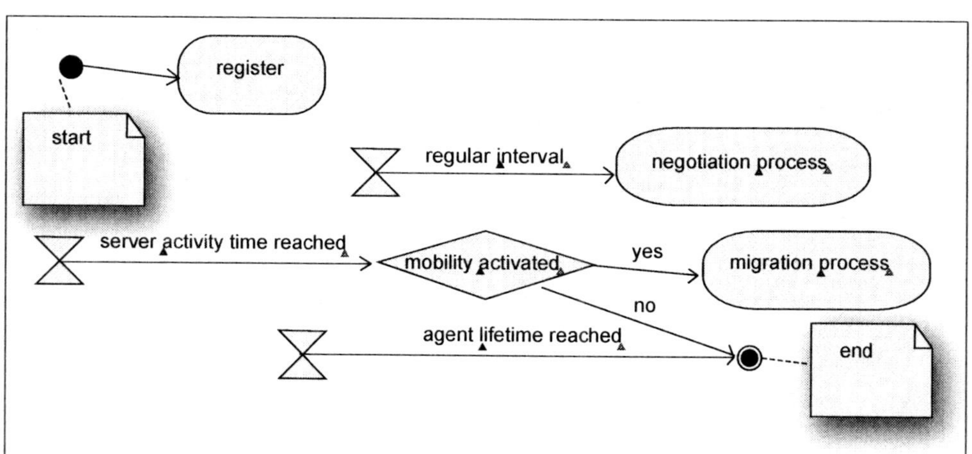

Figure 4. Negotiation process

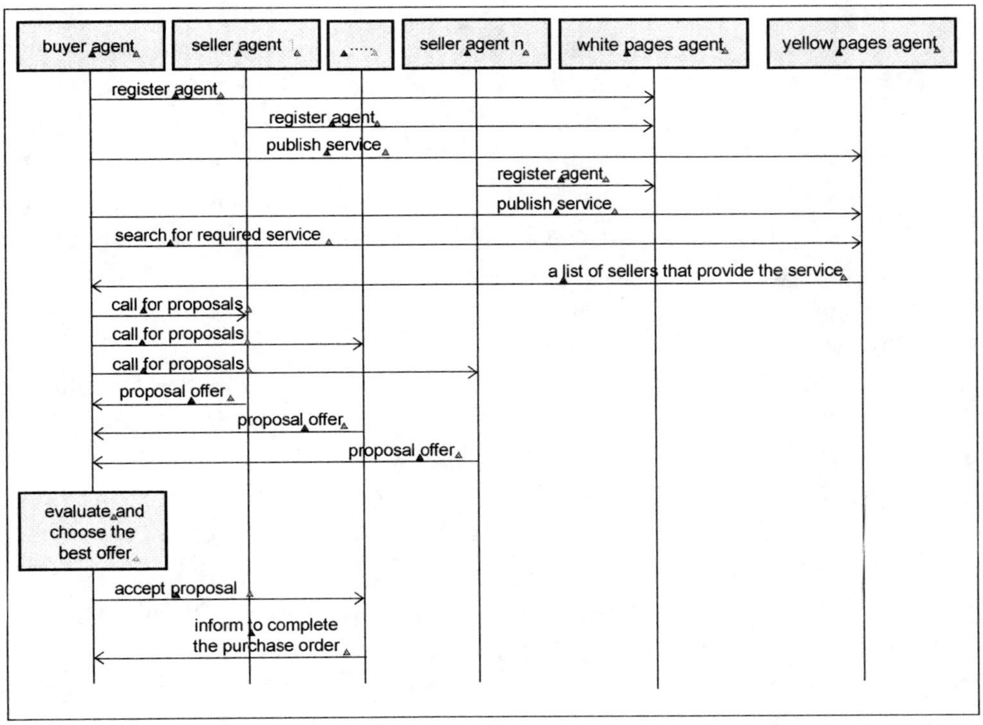

Figure 5. Agent migration process

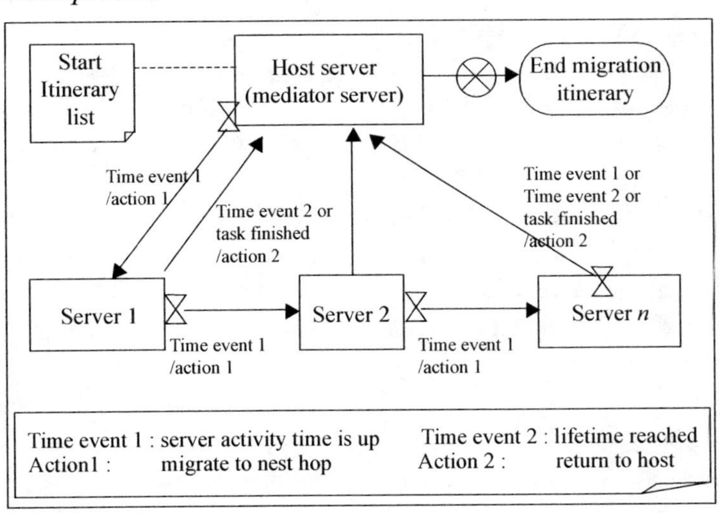

service). After receiving the message, a selling agent can send the buying agent a proposal with the price for the product. If the product is not available or sold, it does not need to send any proposal. The buying agent will place a purchase order if the offer price is within the maximum price that the customer has specified. Results of price negotiations are sent back to the personal agent and showed in a graphical interface to the user. Since the system is fully asynchronous, an intention to make a purchase does not have to lead to a successful transaction. By the time the offer is made, other buying agents may have already purchased the last available item.

Agent Migration

The general process of migration is depicted in Figure 5. An agent starts its migration from its host server (i.e., the mediator server) with the itinerary list acquired from the host. We assume that there are *n* servers, which will be visited by the agent in sequence. In each server, two time events happen resulting in two actions respectively: if the agent reaches its lifetime, it will return to its host where it was created, and then end the migration process; if the agent exhausts its server activity time, it will migrate to the next server. Additionally, before the agent migrates to the next server, it should also make the decision if it has fulfilled the task at the current server. As we know, a task is finished when an agent receives an acceptable offer from another agent. The migration process actually describes a scenario of price comparison (finding a price less than a buyer's reservation price for buying, or searching for a price greater than a seller's reservation price for selling). The agent may access its host server repeatedly during its lifetime and updates its itinerary list every time when visiting its host server. One interesting problem here is how the mediator server maintains the itinerary list that includes a series of service-providing servers to be visited by the agent. Curbera, Duftler, Khalaf, Nagy, Mukhi, and Weerawarana (2002) state that "several individual companies and industry groups are starting to use 'private' UDDI directories to integrate and streamline access to their internal services" (p. 90). UDDI (Universal Description, Discovery and Integration) (UDDI, 2006) enables businesses to publish service listings and to discover each other. We assume that the white-page agent can interact with the UDDI server to obtain other service-providing servers' addresses (the feasibility of this function will be further studied) and therefore mobile agents can update the itinerary list during their migration. Only the mobile agents, which are originally created in this mediator server, are allowed to access this resource (a list of servers).

System Implementation

We have implemented a simple prototype to evaluate the concepts proposed in our architecture, using the Java programming language. Figure 6 shows the screenshots of a personal agent and a JADE-based multi-agent system, respectively. The personal agent was developed as a J2ME MIDlet[5] application that offered a graphical interface for its user to initiate or recall the mobile agent, and to dialogue with the mediator server. The mediator server played an important role in our architecture, running a Tomcat Apache Servlet Engine on a JADE platform. JADE is an open-source with good scalability, one of the best modern agent environments compliant with FIPA. As shown in Figure 6, there are two containers on the JADE system, Main-container and Container-1. Main-container holds the basic management agents defined by FIPA (AMS, DF, and RMA, which manages the GUI of the JADE platform). The proxy agent, buying agents, and selling agents run in Container-1. We can deploy the mediator-based architecture in one or several PCs.

The Web services architecture communications are based on JSR172, J2ME Web services, which include two independent parts: the JAX-

Figure 6. Screenshots of a personal agent and the JADE platform

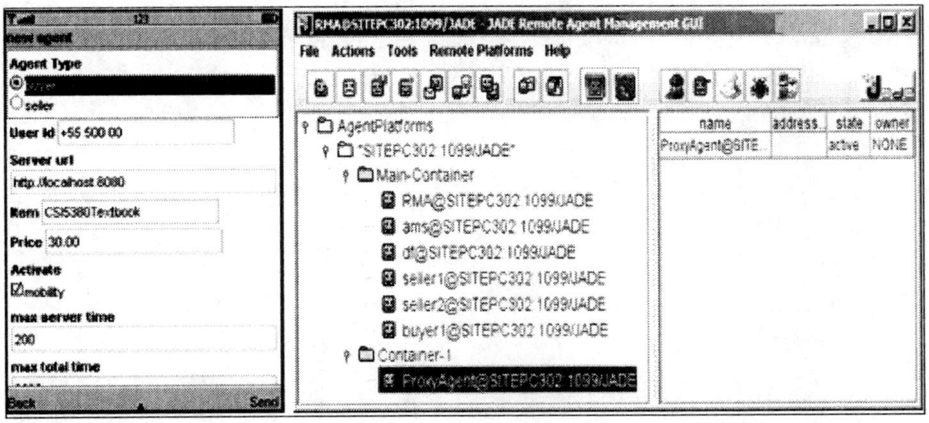

Figure 7. Experiment environment

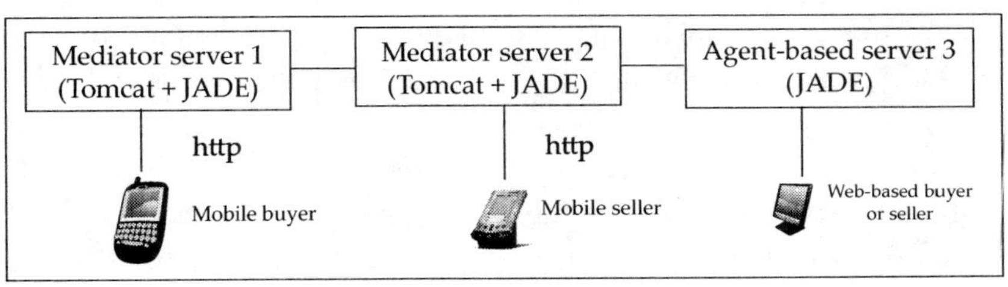

RPC and JAXP. XML is chosen as the standard way for clients to interact with backend servers so as to use the remote services. J2ME JAX-RPC APIs subset solves how to access the SOAP/XML Web services and JAXP APIs subset solves how to process the XML messages. Messages exchanged by agents in the multi-agent system have a format specified by the ACL language defined by FIPA for agent interoperability.

As shown in Figure 7, we deployed three servers in the Local Area Network, installed J2ME MIDlet in two mobile phone simulators, provided one GUI for the Web-based seller, and simulated a simple used-item electronic trading scenario similar to the one we described in A Scenario of Our Architecture section previously. The mobile phone emulator is a tool provided by the Sun J2ME wireless toolkit 2.2. Both mediator servers deployed the Tomcat server and the main container of JADE platform was initialized. The third computer played the role of an agent-based marketplace on the Internet. For the buyer's emulator, the user activated the mobility of the buying agent, but it was not the case for the seller's emulator. We observed the following results:

- Mobile users can connect to mediator servers via HTTP and initiate mobile buying or selling agents in the mediator server.
- Mobile users do not need to instruct their mobile agents of what to do after configuring their preferences.
- Mobile users can add new items anytime and relevant mobile agents will be created

to handle the trading of these new items respectively.

- Mobile users can kill their mobile agents to cancel their tasks by sending instruction to their personal agents.
- Mobile agents are active in their servers within the specified server activity time and then migrate to other servers.
- Buying agents can reach agreements with selling agents when the required item and price are matched. Mobile users then receive text messages from their agents, displayed on the screen of the simulators.
- Mobile agents end their life cycles when finishing their tasks.

As confirmed by the experiments, mobile users connect to their servers only when they need to add new items or to cancel their tasks. This obviously results in such benefits as reduced bandwidth utilization, increased battery life for mobile devices, and no complicated computation conducted in mobile devices. Also, mobile agents can move to various servers to negotiate autonomously, and mediator servers can accept mobile agents from outside their systems. This feature enables users to participate in multiple markets on the Internet.

In particular, we observed the migration process of a mobile agent. Mobile agents should be active in their servers within a specified time and migrate among the servers. Thus, we developed a scenario where we supposed that a buying agent started from Serverl and continued searching for the required product or service in Server2 and Server3. We set up two parameters for this mobile agent: Maximum server active time was set to 100 seconds and total lifetime was set to 850 seconds. As expected, the buying agent contacted the other agents in Serverl and then migrated to Server2 after approximately 100 seconds. Similarly, the buying agent communicated with other agents in Server2 and then traveled to Server3. The same things happened in Server3. Because there were no more sites to be visited, the buying agent migrated back to Serverl, ending its first round of migration. The second round was started since the total lifetime was not reached. We assumed that no sellers offered the required product or service to this buying agent. With the time elapsed, the buying agent was in its third round and roamed into Server2. At this stage, the buying agent used up its lifetime of 850 seconds and predicted an ending of its life cycle. Therefore it migrated back to the host Serverl, even though the third round trip was not finished. In another scenario, we used the same parameters for the buying agent, except that the total lifetime was enlarged to 1,000 seconds. The difference was that we dispatched a selling agent in Server2 at the moment the buying agent was ready to launch its third round trip. This selling agent offered exactly the service that the buying agent needed. As we expected, the two agents met and reached an agreement after negotiating with each other. This experiment confirmed that after completing its task, the buying agent migrated back to Serverl, regardless of its remaining lifetime that had not yet been exhausted.

DISCUSSION AND FUTURE WORK

In this article, we propose a feasible mobile agent architecture that assists users in C2C e-business. It enriches the resources for users to perform comparison shopping activities at the point of purchase. Users' mobile devices connect to the network only when needed, thus making efficient use of limited bandwidth and reducing the network traffic. In addition, it helps cell phone users save money from their expensive bills. At any time, users may add items via their personal agents and specify their preferences such as time limit and preferred price for trading. Through the negotiation process between mobile buying and selling agents, users also gain valuable information for making trading decisions.

Our proposed architecture is extensible: On the one hand, XML-based communication is used to enhance extensibility; on the other hand, the architecture could be easily extended to B2C, or even B2B business models. That is, not only individuals but also business companies can be attached to the architecture. With mobile phones and PDAs already being used as extended enterprise tools, business companies, such as retailers and suppliers, can publish their products and/or services on their servers via mobile devices. As long as these businesses take part in our architecture parties, they could benefit from the automatic discovery of other business partners. Also, it is possible for businesses, especially for retailers, to sell their products to potential buyers in the manner described in the proposed architecture as an extra way to their traditional ones. In this sense, our architecture is an integration model of C2C, B2C, and B2B e-business. Nonetheless, using mobile devices for complex tasks can be quite frustrating (e.g., difficult to enter data), so probably people will not use it. An idea is to incorporate targeted messaging or advertising into our model, where businesses could send a message to users who are physically located in their vicinity. Agents could negotiate a transaction, and the buyer would already be located nearby to complete the purchase and pick up the item.

Currently, we present a conceptual framework that needs to be refined. Using this work as a starting point, we have outlined a number of future research directions:

(1) Negotiation protocols do not have to be hard-coded into the agents. Instead, mobile agents can adapt to any intelligent negotiation strategies when they arrive at a new remote location. Thus, our architecture paves the way for future research in which more general architectures can be explored to allow mobile agents to participate in a variety of negotiation protocols, such as factor negotiation (price, quality, delivery time, etc.), electronic contracting, and so on. Currently, the negotiation strategy module consists of only a purchase determined by price (agents seek a preferable price by a fixed amount). FIPA defines auction protocols (e.g., Dutch and English auctions) as well as simpler strategies such as fixed pricing, fixed pricing with a discount, and so on. We will add them into the negotiation protocols in our future research.

(2) Items are described only by their names. Obviously, other attributes, such as color, age, terms of warranty and delivery should also be considered. We believe that ontologies can help to solve this problem. It should be noted that the small screen of mobile devices will bring inconvenience to users when they specify many attributes of an item. A possible solution is to make use of the persistent memory of mobile devices to store the users' preferences.

(3) Mobile agent technology currently has some limitations, such as identity management, fault tolerance, protection of agents, and resource security. These limitations have brought up some concerns about the practical utilization of mobile agents. For example, in the area of security, e-business applications are often involved with money and thus users may hesitate to use mobile agents, unless mobile agents are secure enough to be trusted.

In the situation presented in this article, the mobile agents representing different buyers or sellers migrate over the Internet and then execute themselves on remote computers. These mobile agents are thus exposed to open environments and may become vulnerable. Since the mobile agents execute on unknown computers and interact with unknown agents, a reliable security infrastructure is vitally needed for the design of the system. The mobile agents must be able to deal with situations where they have been shipped off to the wrong address or to a hostile environment (Neuenhofen & Thompson, 1998). Listed below are some possible security concerns:

- Malicious mobile agents can try to access services and resources without adequate permissions. In addition, a malicious agent may assume the identity of another agent in order to gain access to platform resources and services, or to cause mischief or even serious damage to the platform.
- Mobile agents may suffer eavesdropping attack from other mobile agents. A malicious agent can sniff the conversations between other agents or monitor the behavior of a mobile agent in order to extract sensitive information from it.
- Mobile agents may suffer alteration attack from malicious hosts. To execute the agent and update its state, the host must definitely be capable of reading and writing the agent. A malicious host may steal private information from the agent or modify the agent to compute the wrong result or to misbehave when it jumps to another site.

Current research efforts in the area of mobile agent security adopt two different perspectives (Kotz, 2002): First, from the platform perspective, we need to protect the host from malicious mobile agents (such as viruses and Trojan horses) that are visiting it and consuming its resources. Second, from the mobile agent perspective, we need to protect the agent from malicious hosts. There are many mechanisms to protect a host against malicious agents. Digital signatures and trust management approaches may help identify the agent and evaluate how much it should be trusted. The malicious host problem, in which a malicious host attacks a visiting mobile agent, is the most difficult problem. We found in the literature some works on powerful techniques such as Sandboxing and Proof-Carrying Code (PCC). Sandboxing (Wahbe, Lucco, Anderson, & Graham, 1993) is a software technique used to protect a mobile agent platform from malicious mobile agents. PCC (Lee & Necula, 1997) introduces the technique in which the code producer is required to provide a formal proof that the code complies with the security policy of the code consumer. Therefore, we envisage that the security of mobile agents is an important issue that will encourage techniques and mechanisms for e-business in the future.

CONCLUSION

We propose in this article an e-business architecture that allows traders to do business at remote locations by means of mobile intelligent agents. Our architecture, which adheres to standardization efforts in the multi-agent field such as FIPA paves a possible way towards a near future when mobile buying (and selling) agents can smoothly travel among different agent-based marketplaces to carry out tasks on their users' behalves. Our purpose of presenting this idea is to improve our understanding of the value of mobility and to encourage the conceptual construction of a global community. We do not claim that buyers and sellers around the world would have to buy into this to make it work, and that worldwide C2B e-commerce would be revolutionized thereby. In practice, however, we hope that our work would be useful on a smaller scale and lead to new investigations that may result in new solutions to the problems we addressed. Our proposed architecture, aimed at providing new capabilities for advanced e-business solutions, employs an approach that integrates intelligent and mobile agents. Intelligent agents can provide automation support for decision-making tasks, while mobile agents can extend that support by allowing users to participate in several marketplaces in a networked e-business. We believe that intelligent and mobile agent technology is also a promising solution to the problems of low speed, high latency, and limited computing ability that the current wireless network is facing.

REFERENCES

Bellifemine, F., Caire, G., Trucco, T., & Rimassa, G. (2006). JADE programmer's guide. Retrieved July 7, 2007, from http://jade.cselt.it/docs

Chavez, A., & Maes, P. (1996). Kasbah: An agent marketplace for buying and selling goods. In *Proceedings of the 1st International Conference on the Practical Application of Intelligent Agents and Multi-Agent Technology*, London, United Kingdom.

Chiang, H., & Liao, Y. (2004). An agent-based architecture for impulse-induced mobile shopping, *Computer and Information Technology*.

Chmiel, K., et al. (2004). Testing the efficiency of JADE agent platform. In *Proceedings of the 3rd International Symposium on Parallel and Distributed Computing* (pp. 49-57). IEEE Computer Society Press.

Curbera, F., Duftler, M., Khalaf, R., Nagy, W., Mukhi, N., & Weerawarana, S. (2002, March-April). Unraveling the Web services web: An introduction to SOAP, WSDL, and UDDI. *IEEE Internet Computing, 6*(2), 86-93

FIPA. (2006). Retrieved July 7, 2007, from http://www.fipa.org

Fonseca, S., Griss, M., & Letsinger, R. (2001). *An agent-mediator e-business environment for the mobile shopper* (HP Tech. Rep. No. HPL-20010157).

Gray, R.S. (1997). Agent Tcl. *Dr. Dobb's Journal*, pp. 18-26.

Impulse. (2006). Retrieved July 7, 2007, from http://agents.media.mit.edu/projects/impulse/

Jøsang, A., & Ismail, R. (2002, June). The Beta Reputation System. In *Proceedings of the 15th Bled Electronic Commerce Conference*, Bled, Slovenia.

Kotz, D. (2002). Future directions for mobile agent research. *IEEE Computer Science*.

Kotz, D., & Gray, R. (1999). Mobile code: The future of the Internet. In *Proceedings of Autonomous Agents'99: Workshop on Mobile Agents in the Context of Competition and Cooperation*.

Kowalczyk, R., et al. (2002). Integrating mobile and intelligent agents in advanced e-business: A survey. In *Proceedings of Agent Technologies, Infrastructures, Tools, and Applications for E-Services, NODe'2002 Agent-Related Workshops*, Erfurt, Germany.

Lange, B.D., & Oshima, M. (1998). *Programming and deploying Java mobile agents with aglets*. Addison-Wesley.

Lange, D.B., & Oshima, M. (1999). Seven good reasons for mobile agents. *Communications of the ACM*.

Lee, P., & Necula, G. (1997). Research on proof-carrying code on mobile-code security. In *Proceedings of the Workshop on Foundations of Mobile Code Security*.

Moreno, et al. (2005). Using JADE-LEAP to implement agents in mobile devices. Retrieved July 7, 2007, from http://www.zdnet.de/itmanager/whitepapers

Neuenhofen, K.A., & Thompson, M. (1998). Contemplations on a secure marketplace for mobile Java agents. In K.P. Sycara & M. Wooldridge (Eds.), *Proceedings of Autonomous Agents 98*, Minneapolis, Minnesota. New York: ACM Press.

Sandholm, T., & Huai, Q. (2000). Nomad: Mobile agent system for an Internet-based auction house. *IEEE Internet Computing*, pp. 80-86.

Sun. (2006). Java. Retrieved July 7, 2007, from http://java.sun.com/javame/

Suri, N., et al. (2000). NOMADS: Toward a strong and safe mobile system. In *Proceedings of the 4th International Conference on Autonomous Agents* (pp. 163-164). New York: ACM Press.

Todd, S., Parr, F., & Conner, M. (2005). An overview of the reliable HTTP protocol. Retrieved July 7, 2007, from http://www-128.ibm.com/developerworks/webservices/library/ws-phtt/

UDDI. (2006). Retrieved July 7, 2007, from http://www.uddi.org/

Wahbe, R., Lucco, S., Anderson, T.E., & Graham, S.L. (1993). Efficient software-based fault isolation. In *Proceedings of the 14th ACM Symposium on Operating Systems Principles* (pp. 203-216).

Wang, A.I., Sørensen, C.F., & Indal, E. (2003). A mobile agent architecture for heterogeneous devices. *Wireless and Optical Communications.*

White, J.E. (1999). Telescript technology: Mobile agents. In Mobility: Processes, computers, and agents (pp. 460-493). New York: ACM Press/Addison-Wesley.

ENDNOTES

[1] In the experiment, we developed a GUI in the mediator server for users to launch a buying or selling agent.

[2] The white-page agent maintains different service provider sites. Section 3.4 will describe this agent in more detail.

[3] Detailed description of the proxy agent is provided in Section 3.4.

[4] In an object-oriented context, a behavior is an inner class of the proxy agent.

[5] MIDlet is a Java program generally running on a cell phone, for embedded devices, more specifically the Java ME virtual machine.

This work was previously published in International Journal of Information Technology and Web Engineering, Vol. 2, Issue 4, edited by Ghazi I. Alkhatib, pp. 63-80, copyright 2007 by IGI Publishing, formerly known as Idea Group Publishing (an imprint of IGI Global).

Index

E

F

H

I

J

K